AF788997

PROCEEDINGS
OF THE
FORTY-NINTH
PUGWASH CONFERENCE ON
SCIENCE AND WORLD AFFAIRS

CONFRONTING THE CHALLENGES
OF THE 21ST CENTURY

PROCEEDINGS
OF THE
FORTY-NINTH
PUGWASH CONFERENCE ON
SCIENCE AND WORLD AFFAIRS

RUSTENBURG, SOUTH AFRICA
7 – 13 SEPTEMBER 1999

CONFRONTING THE CHALLENGES
OF THE 21ST CENTURY

EDITOR

JOSEPH ROTBLAT

President Emeritus of the Pugwash Conferences

World Scientific

New Jersey • London • Singapore • Hong Kong

Published by

World Scientific Publishing Co. Pte. Ltd.

P O Box 128, Farrer Road, Singapore 912805

USA office: Suite 1B, 1060 Main Street, River Edge, NJ 07661

UK office: 57 Shelton Street, Covent Garden, London WC2H 9HE

British Library Cataloguing-in-Publication Data
A catalogue record for this book is available from the British Library.

CONFRONTING THE CHALLENGES OF THE 21ST CENTURY

Copyright © 2001 by World Scientific Publishing Co. Pte. Ltd.

All rights reserved. This book, or parts thereof, may not be reproduced in any form or by any means, electronic or mechanical, including photocopying, recording or any information storage and retrieval system now known or to be invented, without written permission from the Publisher.

For photocopying of material in this volume, please pay a copying fee through the Copyright Clearance Center, Inc., 222 Rosewood Drive, Danvers, MA 01923, USA. In this case permission to photocopy is not required from the publisher.

ISBN 981-02-4773-7

Printed in Singapore.

PUGWASH CONFERENCES ON SCIENCE AND WORLD AFFAIRS

President: Sir Michael **Atiyah** *Secretary-General:* Professor George **Rathjens**

PUGWASH COUNCIL

Chairman:	Professor Francesco **Calogero**	(Italy)
Members:	Professor Ulrich **Albrecht**	(Germany)
	Sir Michael **Atiyah**	(UK)
	Professor Gabriel **Baramki**	(Palestine)
	Professor Julio **Carasales**	(Argentina)
	Professor Ana Maria **Cetto**	(Mexico)
	Lt.-Gen. Emmanuel **Erskine**	(Ghana)
	Dr Esmat **Ezz**	(Egypt)
	Academician Vitalii **Goldanskii**	(Russia)
	Maj.-Gen. Leonard **Johnson**	(Canada)
	Dr Venance **Journé**	(France)
	Dr Martin M. **Kaplan**	(Switzerland)
	Professor Catherine **Kelleher**	(USA)
	Professor Michiji **Konuma**	(Japan)
	Mr Sverre **Lodgaard**	(Norway)
	Dr Steven **Miller**	(USA)
	Professor Marie **Muller**	(South Africa)
	Professor Maciej **Nalecz**	(Poland)
	Dr Alexander **Nikitin**	(Russia)
	Professor Amnon **Pazy**	(Israel)
	Dr Sebastian **Pease**	(UK)
	Professor George **Rathjens**	(USA)
	Professor Luiz Pinguelli **Rosa**	(Brazil)
	Sir Joseph **Rotblat**	(UK)
	Air Commodore Jasjit **Singh**	(India)
	Dr Mark Byung-Moon **Suh**	(South Korea)
	Professor Fenggan **Zhuang**	(China)

PUGWASH EXECUTIVE COMMITTEE

Chairman:	Professor Ana Maria **Cetto**
Members:	Sir Michael **Atiyah**
	Professor Francesco **Calogero**
	Lt.-Gen. Emmanuel **Erskine**
	Academician Vitalii **Goldanskii**
	Dr Venance **Journé**
	Professor Maciej **Nalecz**
	Professor George **Rathjens**
	Sir Joseph **Rotblat**
	Air Commodore Jasjit **Singh**

PREFACE

The material in this volume is for the personal use of the recipient of these Proceedings and not for publication, unless permission is obtained from the Secretary-General or the author of the paper in question. The Statement from the Pugwash Council and the Reports on the Working Groups can be published without permission, with appropriate attribution.

Participants in Pugwash Conferences attend as individuals, in their private capacity, and represent only themselves. The privacy of the meetings, except for the special plenary sessions open to the public, ensures a free exchange of views. Only the Pugwash Council or its Executive Committee can represent Pugwash in public statements.

I should like to express my appreciation to the Consiglio Nazionale delle Ricerche for their generous contribution towards the publication of the Proceedings.

The main burden of preparing this volume was carried by Sally Milne and Tom Milne and I want to express my profound gratitude to them.

Joseph Rotblat

CONTENTS

Members of the Conference 1
Programme of the Conference 11
Working Group Topics and Sub-Topics 13
Members of Working Groups 14

Part I. Statements and Reports from the Conference

Statement of the Pugwash Council 19

Report on Working Group 1 27
A Nuclear-Weapon-Free World

Report on Working Group 2 30
Emerging Security Threats: Africa and Globally

Report on Working Group 3 34
Development

Report on Working Group 4 39
The Environment

Report on Working Group 5 45
International Governance

Part II. Proceedings of the Plenary Sessions

Proceedings of the Opening Plenary Session 51

Welcoming Address 53

Messages of Greeting 55

Response by President of Pugwash 58

Tributes to Deceased Pugwashites by Dr. Martin Kaplan 59

Address by the Secretary-General 61

Introduction of Dr. Waldo Stumpf by Sir Joseph Rotblat 66

Keynote Address: 67
South Africa's Nuclear Weapons Programme:
From Deterrence to Dismantlement
Dr. Waldo Stumpf

Proceedings of the Second Plenary Session 75
Security in Africa

The African Renaissance 77
Ambassador Moushira Mahmoud Khattab

The Security Context in Southern Africa and Beyond 82
Dr. Jakkie Cilliers

Proceedings of the Special Plenary Session 87

The Situation in East Timor

Proceedings of the Third Plenary Session 89

Science and Technology in South Africa

Science and Technology in South Africa 91
Mr Jacob Zuma

Development through Science 94
Dr. B.S. Ngubane

Proceedings of the Fourth Plenary Session 99

Light Weapons and Conflict in Africa

Introductory Address by Dr. Jeffrey Boutwell 101

Light Weapons and Conflict in Africa 102
Mr. Jackie Selebi

Light Weapons and Conflict in Africa 106
Dr. Martinho Chachiua

Light Weapons and Conflict in Africa 109
Lt. Col. Jacob Kamenju

Light Weapons and Conflict in Africa 112
Dr. Peter Lock

Proceedings of the Fifth Plenary Session 117

Presentation and Discussion of Working Group Reports

Proceedings of the Closing Plenary Session 119

Sovereignty and Intervention: International Perspectives

Sovereignty and Intervention: International Perspectives 121
Dr. Alexander Nikitin

Sovereignty and Intervention: International Perspectives 125
Air Commodore Jasjit Singh

Sovereignty and Intervention: International Perspectives 128
Professor George Rathjens

Cooperative Governance, Partnership and Sovereignty in a Future Democratic Society 131
Mr. Derek Powell

Presidential Address by Sir Michael Atiyah 137

Part III. Papers Submitted to the Conference

Working Group 1

A Nuclear-Weapon-Free World

Bart van der Sijde	*Nuclear Disarmament: A Lost Cause?*	145
Bart van der Sijde and Karel Koster	*A Road to Nuclear Disarmament*	151
Liu Gongliang	*Movements Towards Comprehensive Nuclear Disarmament*	159
Tom Milne	*Verifying the Transition from Low Levels of Nuclear Weapons to a Nuclear-Weapon-Free World*	163
Jasjit Singh	*Indian Nuclear Doctrine*	171
Shu Yuan Hsieh	*A Nuclear-Weapon-Free Zone in the Taiwan Strait*	180
Young Hoon Kang	*Rationale for North Korea's Development of a Nuclear Programme*	183
Liu Min	*Development of Theatre Missile Defence and its Effect on International Stability and Arms Control*	185
Francesco Calogero and Giancarlo Tenaglia	*The Risk of the Misuse of Highly-Enriched Uranium (HEU) by Terrorists*	187
Jozef Goldblat	*Yet Another Lesson to be Learned from the NATO-Yugoslav War*	197

Working Group 2

Emerging Security Threats: Africa and Globally

Peter Lock	*Comprehensive Measures to Reduce Illicit Small Arms Availability*	201
Jean-Paul Hébert	*Economic Analysis of the Arms Trade and Arms Production*	215
Tara Kartha	*Trends in Terrorism*	218
Ulrich Albrecht	*Organized Crime as a New Threat to Societies in Transition*	224
Jasjit Singh	*Strategic Impact of Transnational Crime*	232
Jorma Miettinen	*Weapons of Mass Destruction, Doctrines for Their Use, and the Threat from Terrorism*	238
Diana Cammack	*Approaching the Threshold: Conflict, Ingenuity and Environmental Degradation in Malawi*	246
Annemarie Dupré	*Globalization – Migration – International Partnership*	253
Jacob Kamenju	*Illicit Arms Trade in the Greater Horn of Africa and its Impact on Development*	262
Lucy Mulli	*Security as an Issue when Addressing the Refugee Crisis in Africa*	266
Undule Mwakasungura	*Firearms Proliferation and Violent Crime in Malawi*	269
Hussein Solomon	*Endless Night: Conflict and Conflict Resolution in Angola*	272
Lynne Hopkins and Martin Kaplan	*Elimination of Anti-Personnel Landmines: The Agenda for the 21st Century*	276
Vladimir Knapp	*Demining: A Challenge to Science and Technology and to International Solidarity*	285

Working Group 3

Development

Morris Miller	*Beyond Aid: Cooperative Approaches to Achieve Effective International Programmes for Poverty Alleviation*	293
Hugo Estrella Tampieri	*Nationality and Citizenship in a Globalized World*	313
Stefan Luby	*Science in the Process of Globalization*	315
Paul Guinnessy	*Information Technology in Development Situations*	317

D.Lucius O. Mendis	*South Africa as Leader of the Non-Aligned Movement - Some Issues from Sri Lanka*	319
Tannia Falconer	*The Integral Plan of Economic Development for the Tehuantepec Isthmus, Mexico: Some Facts Behind the Mask of Progress*	322
Maria Elena Montero-Cabrera	*The Microtron MT-25 Regional Laboratory Project: An Effort in Cooperation for Development*	325

Working Group 4

The Environment

Peter Gleick	*Conflict and Cooperation Over Fresh Water*	329
André Mechelynck	*Water Supply and Demand: A Tentative Inventory*	339
Peter Otim	*Water Resources and Survival: A Look at Pastoral Livelihood*	346
Carsten Rohr	*War Over Water - The Case of the Ilisu Dam Project in Turkey*	350
Anthony Turton	*Water and Social Stability: The Southern African Dilemma*	353
Alexander Ginzburg and Veronica Ginzburg	*City Environmental Profile as a Basis for Programmes of Urban Sustainable Development*	363
Eva-Jane Haden	*The View of the Environment from a Student Pugwashite*	371
Luiz Pinguelli Rosa	*Sustainability and Development Patterns in the South Under Global Change and Market*	373
Catterina Ferreccio	*Avoiding an Epidemic in Santiago*	378
Alan Isaacman	*Large Dams, Sustainable Livelihoods and the Use of Oral Histories: The Case of Cahora Bassa*	382
Aharon Zohar	*Water and Hunger in Africa*	397
Brendon Gooneratne	*To Whale Or Not To Whale: Is that the Question?*	400

Working Group 5

International Governance

John Avery	*A World Free From War*	405
Bengt Broms	*The Present Plans to Restructure the Security Council of the United Nations*	407

Joseph Rotblat	*Reforming the United Nations*	412
Goran von Bonsdorff	*International Governance*	416
Micheline Egge Grung	*The Resort to Military Force, International Law, and the Just War Tradition*	418
Jasjit Singh	*United Nations: Manager of International Peace and Security or a Silent Spectator?*	421
Jasjit Singh	*United Nations Peacekeeping Operations: The Challenge of Change*	430
Ivo Slaus	*UN Reform and Civil Disobedience*	436
Mariano Egana and Walter Scheuer	*Some Aspects of the Governance Question in South America*	441
Anna Moden	*Information Technologies and International Governance: Possibilities and Problems*	447
Bas de Gaay Fortman	*Good Governance in an African Perspective: An Inquiry into Actors and Actions*	450
Nceba Gomomo	*Community Participation A Sine Qua Non for An Effective Anti-Corruption Strategy: Youth Partnership Focus*	456

Abbreviations and Acronyms 469

Author Index 471

MEMBERS OF THE CONFERENCE

A. Participants

Afzal, Mr. Muhammad
Minister (Technical), Permanent Mission of **Pakistan**, Geneva, Switzerland

Albrecht, Prof. Ulrich
Acting Director, Department of Political Science, Free University, Berlin, **Germany**; Member, Pugwash Council

Atiyah, Sir Michael
President of Pugwash; Honorary Professor, University of Edinburgh, **UK** [formerly: President, Royal Society of London; Master, Trinity College, Cambridge]

Avery, Prof. John
Associate Professor of Theoretical Chemistry, H.C. Ørsted Institute, University of Copenhagen, **Denmark**

Baramki, Prof. Gabriel
Consultant to the Ministry of Higher Education, Ramallah, **Palestine**; President, Palestinian Council for Justice and Peace; Treasurer, Palestinian European American Cooperation in Education (peace) Program; Member, Pugwash Council

Baregu, Prof. Mwesiga
Senior Research Fellow, Southern Africa Regional Institute for Policy Research (SARIPS), Harare, **Zimbabwe**; Coordinator, International Relations and Peace |and Security Research Programme

Behar, Prof. Nansen
Member of Parliament, Republic of **Bulgaria**; Director, Institute for Social and Political Studies (ISPS), Sofia

Bonsdorff, Prof. Göran Edvard von
Professor Emeritus, University of Helsinki, **Finland**; Honorary President, Foreign Policy Association; Association for the United Nations [Former Member of Finnish UN Delegation]

Boutwell, Dr. Jeffrey
Associate Executive Officer, American Academy of Arts and Sciences, Cambridge, MA, **USA** [formerly: Staff Aide, National Security Council, Washington, DC]

De Bretagne, Mr. Yves
Head, Science, Technology and Society Program, Charles Leopold Mayer Foundation for the Progress of Humankind, Paris, **France**

Broms, Prof. Bengt (**Finland**)
Judge, Chairman of Chamber One, Iran-United States Claims Tribunal, The Hague, The Netherlands

Calogero, Prof. Francesco
Professor of Theoretical Physics, University of Rome "La Sapienza", Rome, **Italy**; Chairman, Pugwash Council

Cammack, Dr. Diana (**Malawi/USA**)
Research Consultant, Redwood Partnership, Lilongwe, Malawi; Malawi Election Media Monitoring Project Coordinator

Chilundo, Prof. Arlindo
Director of Planning, and Director of Land Studies Unit (NET) in the Faculty of Arts, Eduardo Mondlane University, Maputo, **Mozambique**

Cilliers, Dr. Jakkie
Executive Director, Institute for Security Studies (ISS), Pretoria, **South Africa**

Dippenaar, Prof. Nola
Associate Professor, Department of Physiology, Faculty of Medicine, University of Pretoria, Pretoria, **South Africa**; Vice-Chairperson, South African Pugwash Group

El-Shafei, Amb. Omran
Adviser, National Center for Middle Eastern Studies, Cairo, **Egypt**; Retired Ambassador

Epstein, Dr. William (**Canada**)
Expert Consultant, Department for Disarmament Affairs, United Nations, New York; Pugwash NGO Representative at UN

Eriksson, Prof. Karl-Erik
Professor of Theoretical Physics, Institute of Physical Resource Theory, Chalmers University of Technology and Göteborg University, **Sweden**

Erskine, Lt-General Emmanuel A.
Member, Pugwash Executive Committee, Accra, **Ghana**; [formerly: First Force Commander, Unifil; Chief of Staff, UNTSO]

Ezz, Dr. Esmat
Professor, Military Medical Academy, Cairo, **Egypt;** Member, Pugwash Council

Ferguson, Dr. Eric T.
Consultant on Energy and Development, MacFergus bv, Eindhoven, **The Netherlands**

Ferreccio, Dr. Catterina
Director, Health Research Development Group (GREDIS), Santiago, **Chile**; Professor, Department of Public Health, School of Medicine, Pontificia Universidad Catolica de Chile

Fischer, Dr. David
Consultant on Nuclear Nonproliferation and Safeguards, **UK**

De Gaay Fortman, Prof Bas
Chair, Staff Group "States, Societies, World Development", Institute of Social Studies, The Hague, **The Netherlands**

Ginzburg, Prof. Alexander
Research Professor, Institute of Atmospheric Physics, Russian Academy of Sciences; Professor of Climate Change, International University in Moscow, **Russia**

Gleick, Dr. Peter H.
Director, Pacific Institute for Studies in Development, Environment, and Security, Oakland, California, **USA**

Goldblat, Dr. Jozef (**Sweden**)
Vice President, Geneva International Peace Research Institute (GIPRI), Geneva, Switzerland

Goma, Prof. Lameck K.H.
Emeritus Professor of Zoology, University of **Zambia**, Lusaka

Gomomo, Advocate Nceba
Institute of Security Studies (ISS), Pretoria, **South Africa**

Gooneratne, Dr. Brendon
Medical practitioner and President, Project Jonah, Australia; Chairman, Friends of the Ancient Cities of **Sri Lanka**

Hébert, Dr. Jean-Paul
Engineer, Ecole des Hautes Etudes en Sciences Sociales (EHESS), Paris, **France**; Member of Economic Defense Council

Hennop, Mr. Ettienne Louis
Researcher, Institute for Security Studies (ISS), Nen Muckleneuk, Pretoria, **South Africa**

Hinde, Prof. Robert
Royal Society Research Professor (now retired), **UK** [formerly: Master, St. John's College, Cambridge; Hon. Director, Medical Research Council Unit on the Development and Integration of Behaviour]

Hsieh, Dr. Shu Yuan
Independent Researcher/Adviser on Defense, Legislative Yuan/Congress, Taipei, **Taiwan**

Husbands, Dr. Jo
Director, Committee on International Security and Arms Control (CISAC), National Academy of Sciences (NAS), Washington, DC, **USA**

Ionno, Ms. Sandra
Executive Director, Student Pugwash USA; Washington, DC, **USA**

Isaacman, Prof. Allen
Professor of African History and Distinguished Scholar of the College of Liberal Arts, and Director, Interdisciplinary Center for the Study of Global Change, Minneapolis, Minnesota, **USA**

Islam, Dr. Saiful (**Bangladesh**)
President, International Study Group on Self-Organization, Munich, Germany [formerly: Professor of Physics, University of Dhaka, Bangladesh]

Jacchia, Prof. Enrico
Director, Center for Strategic Studies Libera Università (Luiss) Rome, **Italy**; Senator of the Italian Republic

Johnson, Maj.-Gen.(Ret.) Leonard V.
Member, Pugwash Council, **Canada** [formerly: Commandant, National Defence College of Canada]

Journé, Dr. Venance
Researcher, National Scientific Research Council (CNRS), Paris, **France**; Member, Pugwash Executive Committee

Kang, Mr. Young-Hoon
Chairman, The Sejong Foundation, Sungnam, **South Korea**; [formerly: Prime Minister of the Republic of Korea

Kamenju, Lt. Col. (retired) Jacob Alex Ndiritu
Director, Security Research and Information Centre (SRIC), Nairobi, **Kenya**; Facilitator and Trainer in Reconciliation, Negotiations, Active Non-violence and Small Arms

Kaplan, Dr. Martin (**USA/Switzerland**)
Member, Pugwash Council; Director, Pugwash Conferences Geneva Office [formerly: Director, Research Promotion and Development WHO; Secretary-General of Pugwash (1976-1988)]

Kartha, Dr. Tara
Research Fellow, Institute for Defence Studies and Analyses (IDSA), New Delhi, **India**

Kayanja, Prof. Frederick
Vice Chancellor, Mbarara University of Science and Technology, Mbarara, **Uganda**; Chairman, National Agricultural Research Organization of Uganda

Kelleher, Dr. Catherine M. (**USA**)
Director, Aspen Institute Berlin, Germany [formerly: Deputy Assistant Secretary of Defense, Office of the US Secretary of Defense (1996-1998)]

Khalil, Dr. Ayman
Director, Center for Research on Arms Control and Security (CRACS), Amman, **Jordan**

Khupe, Mr. John
Principal Environmental Engineer, Gaborone City Council, Gaborone, **Botswana**; Principal Researcher African Energy Policy Research Network

Kirsten, Prof. Johann
Head, Department of Agricultural Economics, Extension and Rural Development, University of Pretoria, Pretoria, **South Africa**

Konuma, Prof. Michiji
Dean, Faculty of Environmental and Information Studies, Musashi Institute of Technology, Yokohama; Member, Pugwash Council; **Japan**

Kuźnicki, Prof. Leszek
Chairman of the Committee for Future Studies, Warsaw, **Poland** [formerly: President, Polish Academy of Sciences]

Liu, Prof. Gongliang
Vice Chief Engineer, Institute of Applied Physics and Computational Mathematics (IAPCM), Beijing, **China**

Liu, Prof. Min
Advisor, Program for Science and National Security Studies, China Academy of Engineering Physics, Beijing, **China**

Lock, Dr. Peter
Coordinator, EART e.V., Hamburg, **Germany**

Lodgaard, Mr. Sverre
Director, Norwegian Institute of International Affairs (NUPI); Member, Pugwash Council **Norway**; [formerly: Director, United Nations Institute for Disarmament Research (UNIDIR)

Luby, Acad. Stefan
President, Slovak Academy of Sciences (SAS), Bratislava, **Slovak Republic**; Head, Slovak National Pugwash Group

Maretzki, Dr. Hans
Professor of International Relations, Potsdam, **Germany**; Co-Editor of *WeltTrends Journal on International Relations*

Masera, Prof. Omar
Professor of Ecology, Instituto de Ecología, National Autonomous University of Mexico (UNAM), Campus Morelia, Michoacan, **Mexico**; Advisor to Secretary of Environment and Natural Resources

Masperi, Prof. Luis
Director, Centro Latinoamericano de Fisica, Rio de Janeiro, Brazil; Professor and Researcher in Theoretical Physics, Centro Atomico Bariloche and Instituto Balseiro, S.C. de Bariloche, **Argentina**

Mechelynck, Mr. André L.
Retired Engineer, now Consultant in Energy Matters, Huldenberg, **Belgium**

Mendis, Mr. Lucius (D.L.O.)
Development Consultant, Colombo, **Sri Lanka**; Secretary/Convenor, Sri Lankan Pugwash Group

Miettinen, Acad. Jorma K.
Academician and Emeritus Professor of Radiochemistry, University of Helsinki, **Finland**

Miller, Prof. Morris
Professor and Consultant, University of Ottawa, **Canada** [formerly: Executive Director, World Bank]

Milne, Mr. Tom
Staff Member, Pugwash Conferences on Science and World Affairs; PhD Student, Programme of Policy Research in Engineering, Science and Technology, University of Manchester, **UK**

Molitor, Prof. Michael
Director, Climate Change Program, University of California San Diego, La Jolla, CA, **USA**

Montero Cabrera, Prof. Maria Elena
Titular Professor, Nuclear Physics Department, Instituto Superior de Ciencias y Tecnologia Nucleares, Havana, **Cuba**; Past President, Cuban Physical Society

Mshigeni, Prof. Keto Elitabu (**Tanzania**)
Pro-Vice Chancellor (Academic Affairs and Research), University of Namibia, Windoek, Namibia; Editor-in-chief, *Discovery and Innovation*

Muller, Prof. Marie
Professor and Head, Department of Political Sciences, University of Pretoria, **South Africa**; President, South African Political Studies Association; Chair, Pugwash South Africa; Member, Pugwash Council

Mulli, Ms. Lucy (**Kenya**)
Researcher, Institute for Security Studies (ISS), Pretoria, South Africa

Muniz Gutiérrez, Prof. Onaney
Senior Researcher, Institute of Tropical Geography, Havana, **Cuba**; Adjunct Titular Professor

Mwakasungura, Mr. Undule
Zone Officer, Centre for Human Rights and Rehabilitation (CHRR), Lilongwe, **Malawi**; Coordinator on Small Arms and Light Weapons Control/Landmines Research and Documentation Officer, NGO

Nalecz, Prof. Maciej
Director, International Center of Biocybernetics, Polish Academy of Sciences (PAS), Warsaw, **Poland**; Member, Pugwash Executive Committee

Natvig, Prof. Bent
Department of Mathematics, University of Oslo, Blindern, **Norway**; Chairman and Secretary, Norwegian Pugwash Group

Nikitin, Dr. Alexander
Director, Center for Political and International Studies (CPIS), Moscow, **Russia**; Deputy Chair, Russian Pugwash Committee of Scientists for Disarmament and International Security

Okumu, Prof. Washington A.J.
Founder and Executive Director, African Renaissance Institute, Nairobi, **Kenya;** Chairman and Executive Director, African Centre for Preventive Diplomacy

Otim, Mr. Peter
Research Fellow/Student, Centre for Basic Research, Kampala, **Uganda**

Palankaai, Prof. Tibor
Rector, Budapest University of Economic Sciences (BUES), Budapest, **Hungary** ; Corresponding Member, Hungarian Academy of Sciences

Park, Prof. Keun
Special Professor, Hanyang University, Seoul, **Republic of Korea**; Chairman, Executive Committee, Movement for Bright and Dynamic Nation

Pinguelli Rosa, Prof. Luiz
Professor and Vice-Director, COPPE-Post Graduate School of Engineering, Federal University of Rio de Janeiro (UFRJ), **Brazil**; President, Latin American Association of Energy Planning

Rathjens, Prof. George
Secretary-General, Pugwash Conferences on Science and World Affairs; Professor Emeritus, Massachusetts Institute of Technology (MIT), Cambridge, Massachusetts, **USA**

Rezakovic, Prof. Dzenana (**Croatia/Bosnia & Herzegovinia**)
Research Professor, Associate Professor of Medicine, School of Medicine, University of Zagreb; Head and Editor-in-Chief, CARDIOet Program

Rotblat, Sir Joseph **(UK)**
Emeritus Professor of Physics, University of London; Emeritus President of Pugwash; 1995 Nobel Peace Prize Laureate; Member, Pugwash Executive Committee

Rupiya, Mr. Martin
Director, Centre for Defence Studies, and Lecturer, Department of History, University of **Zimbabwe**, Harare; ret. Lt. Col., Zimbabwe Defence Forces

Salomon, Prof. Jean-Jacques
Honorary Professor, Centre Science, Technologie et Societe, Conservatoire National des Arts et Metiers (CNAM), Paris, **France**

Scheuer, Dr. Walter
Senior Project Manager (now retired), Invap S.E. (a government-owned, advanced technology company) **Argentina**

Singh, Air Commodore Jasjit
Director, Institute for Defence Studies and Analyses (IDSA), New Delhi, **India**; Member, Pugwash Executive Committee

Slaus, Prof. Ivo
Professor of Physics, Rudjer Boskovic Institute, Zagreb, **Croatia**; Secretary General, Croatian Movement for Democracy and Social Justice

Solomon, Mr. Hussein
Research Manager, African Centre for the Constructive Resolution of Disputes, Umhlanga, **South Africa**; Council Member, South African Political Science Association

Suh, Dr. Mark Byung-Moon (**South Korea/Germany)**
Member, Pugwash Council; Senior Researcher, Free University of Berlin, Berlin, Germany; Member, Advisory Council on Democratic & Peaceful Unification of Korea

Sukovic, Dr. Olga
Consultant, UN Department for Disarmament Affairs, New York; Scientific Adviser, Institute of International Politics and Economics, Belgrade, **Yugoslavia**

Suzuki, Prof. Tatsujiro
Associate Professor, Sociotechnics of Nuclear Energy, University of Tokyo, **Japan**; Senior Research Fellow, Central Research Institute of Electric Power Industries

Swahn, Dr. Johan
Assistant Professor, Dept. of Physical Resource Theory, Chalmers University of Technology and Göteborg University, Göteborg; **Sweden**; Secretary, Swedish Pugwash Group

van der Sÿde, Dr. Bart
Lecturer on Physics and Society, Faculty of Physics, Eindhoven University of Technology, Eindhoven, **The Netherlands**; Secretary and Board Member, Pugwash-Netherlands

Tavkelidze, Acad. Prof. Albert
President, Georgian Academy of Sciences, Tbilisi, **Georgia**

Tenaglia, Dr. Giancarlo
Voluntary Staff Member, Pugwash Conferences on Science and World Affairs, Rome, **Italy**

Turton, Mr. Anthony Richard
Head, African Water Issues Research Unit (AWIRU), Pretoria University, Pretoria, **South Africa**; Development Consultant

Udgaonkar, Prof. Bhalchandra
Retired scientist, Homi Bhabha Centre for Science Education, Mumbai, **India**

Wilson, Ms. Henrietta **(UK)** [formerly: Researcher, Berlin Information-Center for Transatlantic Security (BITS), Berlin; Researcher: British Pugwash Group

Zohar, Dr. Aharon
Consultant, Regional and Environmental Planning, Carmei-Yosef, **Israel**

B. Student/Young Participants

Algotsson Ms. Emma **(Sweden)**
Intern, Centre for Applied Legal Studies, Witwaterstrand University and the South African Human Rights Commission

Estrella Tampieri, Mr. Hugo Daniel
Advisor to Córdoba State Senate on democratization policies and people's participation, **Argentina**; Executive Director, Student/Young Pugwash Argentina; Professor of Theory of Communication and Public Opinion

Etherden, Mr. Nicholas
Fourth-year Student in Engineering Physics, Uppsala University, Uppsala, **Sweden**

Falconer, Ms. Tannia
Undergraduate Student in Geography/Scholar at the Institute of Geography, UNAM (National Autonomous University of Mexico), Mexico D.F., **Mexico**

Forberg, Mr. Ekkehard
Student/Research Assistant, Free University of Berlin, **Germany** [formerly : Research Assistant and Legal Advisor, State Parliament of Saxony, "Buendnis 90/Die Gruenen" (Civil Rights Party)]

Fujita, Ms. Pauline
Undergraduate student in Microbiology and Immunology at McGill University, **Canada**

Grung, Ms. Micheline Egge
Senior Student in Political Science, University of Oslo, and Research Assistant, International Peace Research Institute (PRIO), Oslo, **Norway**

Guinnessy, Mr. Paul
Editor, *PhysicsWeb*, Institute of Physics Publishing, Bristol, **UK**

Haden, Ms Eva-Jane
President, Swiss Student Pugwash, Geneva, **Switzerland**

Hechter, Mr. Willem H.
Researcher, Rabies Laboratory, Onderslepoort Veterinary Institute, Agricultural Research Council and Laboratory of Molecular Virology, University of Pretoria, **South Africa**

Hörnebrant, Ms. Helena
Financial Consultant in Development Economics, Stockholm, **Sweden**

Jao, Mr. Richard
Undergraduate Student in Electrical Engineering & Computer Sciences (EECS), with Public Policy minor, University of California, Berkeley, CA, **USA**

Kang, Dr. Daeseok
Researcher, Environmental Research Institute, Ewha Women's University, Seoul, **South Korea**

Kurosaki, Mr. Akira
Research Associate, Tohoku University, Sendai-shi, **Japan**

Moden, Ms. Anna-Karin (**Sweden**)
Associate Director, Student Pugwash USA, Washington DC

Mutschler, Ms. Claudia
Latin America Researcher, South African Institute of International Affairs (SAIIA), Johannesburg, **South Africa**;

Oosterlaken, Ms. Elisabeth Theodora (Ilse)
Student, Eindhoven University of Technology, Eindhoven, **The Netherlands**

Østby, Ms. Lise
Student in Sociology/Social Anthropology, University of Oslo, Oslo, **Norway**

Pomatta, Ms. Maite R.
Psychologist, Graduate Student, Department of Psychobiology (Aggression Research Unit), Univ. Complutense, Madrid, **Spain**

Pretorius, Ms. Joelien
Student/Assistant Lecturer, Department of Political Sciences, University of Pretoria, **South Africa**

Rajain, Mr. Arpit
PhD student, Centre for International Politics, Organisation and Disarmament, School of International Studies, Jawaharlal Nehru University, New Delhi, **India**

Rohr, Dr. Carsten **(Germany)**
PhD Student in Physics (Photovoltaics), Imperial College of Science, Technology and Medicine, London, UK; Chair, Student Pugwash, Imperial College

Susiluoto, Ms. Taina Maria
Researcher and Reporter, Channel 4 News, Helsinki, **Finland**; Director, Student Pugwash Finland;

C. Guest of the Conference

Tawana, Mr. Motomisi
Foreign Service Officer, Department of Foreign Affairs, Pretoria, **South Africa**

D. Staff

Cambridge (MA) office:	Ms.Tracy Sanderson
Rome office:	Ms. Claudia Vaughn, Ms.Mimma de Santis
London office:	Ms. Jean Egerton

PROGRAMME OF THE CONFERENCE

Wednesday 8 September

08.30-11.30	**Opening Plenary Session** Chair: Professor Marie Muller and Sir Joseph Rotblat Welcoming Address by Professor Marie Muller Messages of Greeting Response by President of Pugwash Tributes to Deceased Pugwashites by Dr.Martin Kaplan Address by the Secretary-General Keynote Address: *South Africa's Nuclear Weapons Programme: from Deterrence to Dismantlement* by Dr. Waldo Stumpf
12.00-13.30	**Second Plenary Session** ***Security in Africa*** Chair: Sir Michael Atiyah *The African Renaissance* Ambassador Moushira Mahmoud Khattab *The Security Context in Southern Africa and Beyond* Dr. Jakkie Cilliers
15.00-18.30	Meeting of Working Groups
21.00-22.00	**Special Plenary Session** ***The Situation in East Timor***

Thursday, 9 September 1999

08.30-11.30	Meeting of Working Groups
15.00-18.30	Meeting of Working Groups
20.00-20.30	Meeting of Student-Young Pugwash Groups
20.30-21.30	Meeting of Secretary-General and Pugwash Council with Student-Young Pugwash Groups
21.30-22.30	Meeting of Secretary-General with National Group Contact Persons

Friday, 10 September 1999

09.30-11.30	**Third Plenary Session** ***Science and Technology in South Africa*** Chair: Professor Francesco Calogero
	Address by Mr. Jacob Zuma, Deputy President of South Africa
	Address by Dr. B.S. Ngubane, Minister of Arts, Culture, Science and Technology of South Africa
11.45-13.50	**Fourth Plenary Session** ***Light Weapons and Conflict in Africa*** Chair: Dr. Jeffrey Boutwell
	Discussion Panellists: Mr. Jackie Selebi, Dr. Martinho Chachiua, Lt.Col. Jacob Kamenju, Dr. Peter Lock,
14.45-15.10	**Special Session**
	The Archeological Finds Made at Mapungubwe (Hill of the Jackals) Professor Andrie Mayer

Saturday, 11 September 1999

09.30-12.30	Meeting of Working Groups
Afternoon:	Excursion

Sunday, 12 September 1999

10.30-15.30	**Fifth Plenary Session** ***Presentation and Discussion of Working Group Reports***
10.30-12.45	Chair: Professor Michiji Konuma
14.00-15.30	Chair: Professor Ulrich Albrecht
20.30	Banquet

Monday, 13 September 1999

08.45-12.30	**Closing Plenary Session** ***Sovereignty and Intervention: International Perspectives*** Chair: Dr. Catherine Kelleher
	Discussion Panellists: Dr. Alexander Nikitin, Air Commodore Jasjit Singh, Professor George Rathjens, Mr Derek Powell
11.30	**Presidential Address by Sir Michael Atiyah**

WORKING GROUP TOPICS AND SUBTOPICS

1. A Nuclear-Weapon-Free World

1.1 Dismantlement of nuclear weapons, disposal of weapon-grade fissile materials and cut-off of their production
1.2 From CTBT and NPT to a Nuclear-Weapon-Free World
1.3 Nuclear-weapon-free zones

2. Emerging Security Threats: Africa and Globally

2.1 Non-nuclear threats: light weapons, CW/BW, and landmines
2.2 Light weapons and conflict in Africa
2.3 Refugees and conflict resolution in Africa
2.4 International terrorism and organized crime

3. Development

3.1 The role of global financial institutions and multinationals in development
3.2 Beyond aid: Co-operative solutions to problems
3.3 Human resource development, education and health

4. The Environment

4.1 Water resources and energy: availability and control
4.2 Agricultural production and land use
4.3 Conflicting priorities in dealing with environmental problems
4.4 North-South conflict and co-operation in the implementation of the United Nations Framework Convention on Climate Change (UNFCCC)

5. International Governance

5.1 UN Reform and its implications
5.2 Peacekeeping and peacemaking
5.3 Governance in Africa

MEMBERS OF WORKING GROUPS

Working Group 1: A Nuclear-Weapon-Free World

Conveners: Francesco Calogero
Catherine Kelleher

Rapporteur: Jozef Goldblat

Muhammad Afzal
Omran El-Shafei
Bill Epstein
David Fisher
Liu Gongliang
Suzie Hsieh
Young Hoon Kang
Sandy Ionno
Michiji Konuma
Min Liu
Hans Maretzki
Tom Milne
Bent Natvig
Alexander Nikitin
Jean Jacques Salomon
Bart van der Sijde
Jasjit Singh
Olga Sukovic
Taina Susiluoto
Johan Swahn
Bhal Udgaonkar

Working Group 2: Emerging Security Threats: Africa and Globally

Conveners: Esmat Ezz
Martin Kaplan

Rapporteurs: Hussein Solomon
Jo Husbands

Ulrich Albrecht
Emma Algotsson
Mwesiga Baregu
Jeff Boutwell
Diana Cammack
Jakkie Cilliers
Ekkehard Forberg
Lameck Goma
Jean-Paul Hébert
Etienne Hennop
Jacob Kamenju
Tara Kartha
Peter Lock
Jorma Miettinen
Lucy Mulli
Undule Mwakasungura
Joelien Pretorius
Martin Rupiya
Giancarlo Tenaglia
Henrietta Wilson

Working Group 3: Development

Conveners: Gabriel Baramki
Robert Hinde

Reporteur: Morris Miller

Nansen Behar
Karl-Eric Eriksson
Hugo Estrella
Tannia Falconer
Pauline Fujita
Paul Guinnessy
Helena Hornebrant
Saiful Islam
Richard Jao
Stefan Luby
Ayman Khalil
Lucius Mendis
Maria Elena Montero Cabrera
Keto Mshigeni
Lise Øestby
Tibor Palankaai
Maite Pomata
Dzenana Rezakovic
Tatsuo Suzuki
Motumisi Tawana

Working Group 4: The Environment

Conveners: Omar Masera
Venance Journé

Rapporteur: André Mechelynck

Yves de Bretagne
Arlindo Chilumdo
Nola Dippenaar
Eric Ferguson
Catterina Ferreccio
Alexander Ginzburg
Peter Gleick
Brendon Gooneratne
Eva Jane Haden
Allen Isaacman
Daesok Kang
Frederick Kayanha
John Khupe
Johan Kirsten
Leszek Kuznicki
Michael Molitor
Onaney Muniz Gutiérrez
Peter Otim
Luiz Pinguelli Rosa
Carsten Rohr
Albert Tavkhelidze
Tony Turton
Aharon Zohar

Working Group 5: International Governance

Conveners: Sverre Lodgaard
Leonard Johnson

Rapporteur: John Avery

Michael Atiyah
Göran von Bonsdorff
Bengt Broms
Bas de Gaay Fortman
Nceba Gomomo
Micheline Egge Grung
Enrico Jacchia
Akira Kurosaki
Luis Masperi
Anna Karin Moden
Marie Muller
Claudia Mutschler
Maciej Nalecz
Washington Okumu
Keun Park
George Rathjens
Joseph Rotblat
Walter Scheuer
Ivo Slaus
Mark Suh

PART I.

STATEMENT AND REPORTS FROM THE CONFERENCE

STATEMENT OF THE PUGWASH COUNCIL

The 49th annual Pugwash Conference on Science and World Affairs met in Rustenburg, South Africa from September 7 to 13, 1999, under the title "Confronting the Challenges of the 21st Century." Being the final Pugwash conference of the 20th century, it was highly fitting that the 49th Conference was hosted by South Africa, whose people have shown remarkable initiative and courage in both managing the difficult transition from apartheid to democracy and in voluntarily dismantling the country's nuclear weapons capability and forswearing a security policy based on nuclear deterrence. The Conference was attended by 124 participants from 49 countries, including 23 members of Student/Young Pugwash. As is customary with Pugwash, participants took part as individuals, not as representatives of their governments or institutions.

Also taking place during the 49th Pugwash Conference were meetings of the Pugwash Council and a two-day Student/Young Pugwash conference on 5 - 6 September that was attended by more than 40 participants from South African Student Pugwash and the international Student/Young Pugwash groups.

The opening plenary of the 49th Conference began with welcoming messages from UN Secretary General Kofi Annan; Salim Ahmed Salim, Secretary General of the Organization of African Unity; Kader Asmal, South African Minister of Education; and Nelson Mandela, former President of the Republic of South Africa. Michael Atiyah, President of Pugwash, responded to these messages, and Martin Kaplan presented the obituaries of deceased Pugwashites. George Rathjens then gave the Secretary General's address to the conference, focusing in particular on the important and sensitive interactions between sovereignty, intervention, and global security.

The first guest speaker, Waldo Stumpf, Chief Executive Officer of the Atomic Energy Corporation of South Africa, addressed the topic of "South Africa's Nuclear Weapons Program: From Deterrence to Dismantlement." In providing a fascinating overview of the factors that influenced South Africa's decisions to first acquire and then eliminate its nuclear weapons capability, Stumpf's address was a stimulating and encouraging example of how political will can overcome entrenched interests and "old ways of thinking."

Next, the conference was addressed by Moushira Mahmoud Khattab, Ambassador of the Arab Republic of Egypt to South Africa, whose talk on "The African Renaissance" movingly described a number of positive social and cultural trends throughout the continent. This was followed by Jakki Cilliers, Executive Director of the Institute for Security Studies, South Africa, who gave comprehensive analysis of "The Security Context in Southern Africa and Beyond."

At the plenary session on Friday, September 10, participants were addressed by Jacob Zuma, Deputy President of South Africa, and B.S. Ngubane, Minister of Arts, Culture, Science a Technology. Both Deputy President Zuma and Minister Ngubane provided fascinating insights i the achievements of and challenges facing the South African government on the eve of the century.

Other plenary sessions included a panel discussion on Light Weapons and Conflict in Af involving Jackie Selebi, Director General of the Foreign Ministry of South Africa, Jeffrey Bou (US), Martinho Chachiua (Mozambique), Jacob Kamenju (Kenya), and Peter Lock (Germany panel discussion on Sovereignty and Intervention: International Perspectives, was chair Catherine Kelleher and included panellists Alexander Nikitin (Russia), Derek Powell (South A George Rathjens (US), and Jasjit Singh (India). In his presidential address, Michael highlighted the need for the UN to deal with the abuses that take place in many states. He con that a world based on injustice can never be at peace.

The Conference Working Groups that met in parallel closed sessions focused on the following topics: (1) Nuclear Weapon-Free World, (2) Emerging Security Threats: Africa and Globally, (3) Development, (4) Environment, and (5) International Governance. The following statement has been informed by the working group and plenary session discussions, but its formulation is the responsibility of the Pugwash Council alone, and the statement has not been reviewed by other Conference participants. The statement is not intended to be an exhaustive account of the working group discussions; rather, it presents conclusions and recommendations on which the Council agrees and which we believe important to communicate to a wider audience. Summaries of the discussions of the working groups have been prepared by rapporteurs and will be included as part of the published Proceedings of the 49th Conference, along with the texts of the speeches from the plenary sessions and the papers presented in the Working Groups.

Pugwash and the Crisis in East Timor

A special plenary, chaired by Michael Atiyah, was convened on the unfolding crisis in East Timor, where the international community was witnessing massive violence by Indonesian paramilitary groups, and possibly the Indonesian Army (TNI) itself, against the majority population of East Timor.

Recognizing the unfolding human tragedy in East Timor, the Pugwash Council sent an urgent message to Kofi Annan, calling upon the UN Secretary General to "use all the means in your power to accelerate proposed action, or take new initiatives, to prevent the situation from turning into an utter human catastrophe." The Pugwash Council sent similar pleas to the UN Security Council, to the heads of government of the Asia Pacific Economic Cooperation (APEC) then meeting in Auckland, New Zealand, and to President B.J. Habibie of Indonesia. As co-recipients of the 1995 Nobel Peace Prize, Sir Joseph Rotblat and Pugwash joined Jose Ramos Horta of East Timor and other fellow Nobel Laureates in urging the World Bank, the International Monetary Fund, and the European Commission to serve notice on President Habibie that "all non-humanitarian aid to Indonesia will be halted until Jakarta agrees to the deployment of international peace enforcing troops in East Timor."

Several days later, President Habibie announced that Indonesia would accept a UN peacekeeping force, and negotiations began on the logistics of deploying such a force to East Timor, but the outcome of these talks as of the final day of the conference, September 13, was anything but clear.

1. A Nuclear-Weapon-Free World

Despite the example set by South Africa in being the first country in the world to have voluntarily renounced nuclear weapons, there continues to be a serious and dangerous deadlock in the nuclear arms control process. Start II remains unratified by the Russian Duma, the Comprehensive Test Ban Treaty (CTBT) has not yet entered into force, nuclear weapons continue to be a central component of the military strategies of the nuclear weapons powers, both declared and undeclared, and political pressures are building for the development and deployment of ballistic missile defense (BMD) systems that risk undermining the ABM Treaty, the cornerstone of the US-Russian strategic relationship and the foundation of the global nuclear arms control regime. If deployed, BMD systems are likely to lead to offensive countermeasures and nuclear proliferation.

To break this impasse and re-start a process of substantial cuts in nuclear forces on the path toward a nuclear-weapon-free world, the Pugwash Council calls upon the nuclear powers to ratify currently signed agreements, to de-alert their strategic nuclear forces (provided this does not stimulate increased force levels), and to withdraw and eliminate all tactical nuclear weapons stationed outside national territory. The Council also calls upon the signatories of the CTBT, most especially the USA,

Russia, and China, to ratify the treaty as soon as possible. In addition, all states required to ratify the CTBT should do so to ensure the treaty's entry into force at an early date.

Recent developments in South Asia are particularly disturbing. The Pugwash Council is concerned over what may be a spiraling nuclear competition between India and Pakistan, and calls upon both countries to formalize their unilateral moratoria on testing by concluding a formal bilateral agreement; to not operationally deploy their nuclear weapons; to keep fissile material stocks to a minimum; and to negotiate a "restraint regime" in South Asia covering both nuclear and conventional forces that could be enlarged to include neighbouring areas.

More broadly, all countries of the world should work toward the elaboration and implementation of nuclear-weapon-free zones, including the Treaty of Pelindaba encompassing Africa and the proposed nuclear-weapon-free zone for Central Asia. Establishing such zones covering the Korean peninsula, the Taiwan Straits, and Central and Eastern Europe, could provide substantial benefits in reducing tensions and limiting the possibility of a conventional conflict escalating to the use of nuclear weapons. In the Middle East, the objective is a zone free of all weapons of mass destruction.

We urge that the forthcoming Review Conference of the Non-Proliferation Treaty (NPT) agree on concrete measures, such as sharply reducing world stockpiles of Highly-Enriched Uranium (HEU), that could reinforce the non-proliferation regime.

Failure to implement these various measures can only mean that the world's nuclear powers and those countries seeking nuclear weapons will continue a highly dangerous process of subjecting the international community to the risk of the catastrophe of a nuclear war, or the terroristic use of nuclear explosive devices by subnational groups.

2. Emerging Security Threats: Africa and Globally

Human security at the end of the 20th century is threatened by traditional military weapons, including anti-personnel landmines (APM) and small arms and light weapons, and by chemical and biological weapons, as well as by emerging threats that range from international terrorism and organized crime to the ravages of the HIV/AIDS virus and to the instability of civil society caused by massive refugee flows.

Relationships among these various threats are especially evident in Africa. Endemic poverty and the lack of sufficient development resources are fundamental causes of widespread social instability and civil conflict in many African countries as in other parts of the world. Societies characterized by tragically high unemployment, especially among the young, and weak government structures are especially prone to devastating civil wars and high rates of crime and social violence. When combined with the easy availability of small arms and light weapons, it is little wonder that such societies can be penetrated by international organized crime, the drug trade, and terrorist groups. These trends are not confined to Africa, however. As seen in many other regions, the global expansion and de-regulation of markets made possible by new communications technologies has facilitated the growth of international criminal and narcotics organizations and their entry into the illegal light weapons trade.

To combat the lethal trade in small arms and light weapons, a multi-pronged approach will be needed that imposes greater controls on supplier states while also addressing the social and economic factors that make light weapons such a valuable commodity in poorer societies. The international community can do its part by providing adequate resources and training for more effective border control, policing, and weapons destruction programmes in the countries affected. These policies can only work, however, if they are paired with community-based reconciliation and development programmes that reduce the cultural, social, and economic significance of possessing a weapon in the first place.

The pairing of development and security policies is also crucially important for dealing with the international refugee crisis spawned by widespread intrastate conflict. Millions of refugees in Africa, and many more millions worldwide, represent both a humanitarian crisis of the first order as well as a continuing source of instability and conflict. Two initiatives that could address both problems would be the relocation of refugees away from the borders of countries in conflict and the use of international peacekeeping forces to separate legitimate refugees from armed militias operating in refugee camps.

Anti-personnel landmines, the continued worldwide trade in major conventional weapons, and the threat of chemical and biological terrorism represent different challenges for the international community. Despite the entry into force, in March 1999, of the Ottawa Convention, landmines continue to be used in conflict (as in Kosovo and Angola) and continue to be deployed. Major powers such as the United States, Russia, and China, have not yet signed the convention. In addition to pressing these states to join the ban on APMs, the Pugwash Council urges the scientific community to contribute to research to improve landmine detection and removal technologies. We also urge the international community to devote more resources to currently effective demining efforts.

With regard to the danger of chemical and biological weapons, Pugwash urges the early completion of negotiations, long underway, for an effective verification protocol for the 1972 Biological Weapons Convention.

3. Development

Attempts to eradicate pervasive poverty while promoting human rights must be predicated on providing basic education and health services, an adequate diet, suitable housing and the other basic amenities of life. Such goals can only be achieved, however, within a framework of development policies that recognize the importance of both increased per capita economic growth and a reasonably equitable distribution of aggregate income, as well as the building of a societal framework characterized by democratic rights and "civility" in the broadest sense of the term.

These aims would be strengthened by the establishment of an international convention to enforce human rights, which could help reduce the likelihood of civil conflicts arising from unacceptable actions by political authorities in denying such rights to large segments of their population. The creation of the International Criminal Court is one positive step of many that will be needed to implement international legal norms that protect basic liberties and the integrity of the individual.

A related issue is that of the "borderless world," especially with respect to the movement of scientific and technological talent and knowledge across borders and the implication of this for developing countries. The scientific and academic community must take the lead in formulating policies that address some negative effects of the "brain drain" phenomenon, where developing countries risk perpetual second-class status in the knowledge economy of the 21st century. Such policies should not, however, infringe on the right of individuals to freedom of movement.

Access to capital and related technical assistance is another top priority for developing countries. Current flows of official development assistance and foreign direct investment have largely bypassed Africa, which receives only about five per cent of all such aid to developing countries. New sources of capital investment, including proposals for a carbon tax and other forms of international taxation (for example, the Tobin tax on international currency transactions and the use of global biological and mineral rights in the oceans, international airline flights and shipping), need to be seriously evaluated if development policies are to succeed in those poorest countries that are also heavily indebted and are experiencing net outflows of capital.

In our increasingly integrated global marketplace, foreign direct investment must also be far more sensitive to equitable standards regarding minimal wages, safety and other working conditions, and the environmental consequences of siting commercial enterprises in developing countries. Such investment also needs to address the creation of environmentally-benign, low cost energy, transportation, and communications infrastructures that can support a society's health and education as well as economic needs.

Recognizing the need for a transformative change in the global economic/financial system that can help meet these goals, the Pugwash Council supports the concept of conducting feasibility studies on the launching of an international initiative that could have a short-term payoff in providing education, health, and energy services for the desperately poor living in rural areas. The utilization of rapidly advancing communications and other technologies toward this end would be only the first step toward achieving a more just and equitable world, a world where many of the fundamental sources of today's conflicts have been eradicated.

4. Environment

Environmental insecurity and conflict stem from causes that are both historically ancient and as new as the 21st century. While the potentially dire ramifications of global climate change may not be known in the near future, throughout history water supply has been a major source of conflict between nations sharing the same hydrological basin, and sometimes between regions of the same nation. The unequal geographical distribution of water resources between states and regions, as well as varying precipitation levels at different seasons of the year, continues to affect strongly both human development and social conflict.

Water resources and systems have been both targets and tools of conflict. Although some global agreements on shared water use have been reached (the 1997 Convention on the Non-Navigational Use of Shared International Watercourses), enforcement is difficult. Bilateral and multilateral agreements over water supply and use offer alternatives for equitable sharing of resources, if enforcement mechanisms are included and such agreements are precise and comprehensive enough to take into account, for example, extreme conditions such as drought.

The direct and indirect influence of human activity on water availability and flow, whether through the construction of dams and reservoirs or through agricultural use, deforestation, soil erosion, residential and recreational uses and the draining of swamp lands, must be better understood. Africa faces special regional problems and challenges in this regard, including irregular and unreliable rainfall, high temperatures, and weakness of support systems, as well as political instability.

We recognize that, in promoting sound development policies for Africa, reliance on dams and reservoirs as a renewable energy source is often preferred over coal-fired and other CO_2 emitting energy stations. Yet large scale dam projects raise their own set of problems, including damage to local ecosystems, displacement of populations, loss of water, disruption of aquatic life, mosquito and pest proliferation, downstream problems from changes in water flow, regional or international disputes, and the vulnerability of dams during wartime or in case of earthquakes and geological instability.

Accordingly, it is imperative that the scientific community analyses all relevant factors inherent in the construction of major dams, including the proposed second dam 100 km below the Cabora Bassa on the Zambesi River in Mozambique, and the proposed Ilisu dam on the Tigris River in Turkey. More research is needed to evaluate the social and ecological consequences of these projects in the overall context of the benefits promised for energy generation and irrigation. In particular, the Pugwash Council calls on the parties concerned to ensure the completion of thorough analyses of water use patterns, the need for adequate staffing and financing of the water operations, the strengthening of existing bilateral and multilateral water conventions, as well as greater sensitivity to the environmental, social, cultural and political implications of large water projects.

Looking ahead, the scientific community is continually seeking to evaluate the political, economic and security implications of global climate change. Under the Kyoto Protocol of the United Nations Framework Convention on Climate Change, the industrialized countries alone – the so-called 'Annex I' countries – have the obligation to reduce their greenhouse gas emissions. The developing countries have, on average, low emissions per capita because the poorer part of their population consumes energy at a very low level, far below that of the industrialized countries (while recognizing that developed country lifestyles are imitated by the richest segments of the population in the developing world).

A flexibility mechanism, the Clean Development Mechanism (CDM), has been put forward to allow the industrialized countries to meet part of their reduction commitments by investing in projects in the developing countries, where reductions can be achieved at a lower cost. This issue raises some important equity concerns; for instance, the CDM negotiation has reflected a bias towards the interests of the industrial countries for low cost-abatement opportunities, to the detriment of the development priorities of the developing countries. Therefore, we recommend that this mechanism be controlled so as to ensure and promote the improvement of the living conditions of the poorer part of the population in these countries. Industrial initiatives, in addition to national and international legislation, could prove beneficial as well. Finally, as the required substantial greenhouse gas emission reductions will entail some modifications of the consumption patterns, it will not be possible to meet them without the cooperation of the general public. As with the issue of water management, more information, education, and media efforts will be needed to mobilize the necessary worldwide support.

5. International Governance

Increasing global interdependence, characterized by a rapidly changing world economy and advances in modern telecommunications, have made the world a very different place from when the United Nations was founded in 1945. Today, the UN is called upon to respond to a host of new challenges, not the least of which is the increase in civil wars and the persecution of minorities.

There are numerous proposals for transforming UN operations, infrastructure, and financing. These include changing the composition of the Security Council to eliminating or altering the veto power of its permanent members. There are similar recommendations for changing the voting system of the General Assembly (perhaps by regions instead of countries) or for establishing a popularly elected world parliament. Whatever changes are finally adopted, we applaud the efforts of the UN Secretary General to increase the efficiency of the organization and to increase its transparency and accountability.

One key element of this reform effort will be strengthening the UN's financial resources. Several proposals have been advanced in this regard, including various forms of international taxation (for instance, the Tobin tax) and levies on states in proportion to their arms expenditures. In the meantime, the Pugwash Council calls upon all member states to meet their budgetary obligations to the UN.

As evidenced by the crises in Kosovo and East Timor, there is a demonstrated need for the United Nations to mount timely and successful peacekeeping operations. The Pugwash Council believes that comprehensive analyses should be made of the feasibility of creating an effective, rapid-response UN military force, and elaborating the political and legal mechanisms for its use. In crisis situations where a larger force is needed, the UN could call upon earmarked national forces under the provisions of Article 43.

In terms of global stability, however, far more effort needs to be invested in preventive action, of resolving conflicts before they erupt into war. With adequate financing, a reformed and strengthened UN, as well as regional organizations recognized by the UN, could become more active in mediating disputes and in stationing peacekeepers so as to allow successful conflict resolution negotiations to take place.

As shown by the African experience in the 1990s, the legacy of colonialism, superpower rivalry during the Cold War, and continued economic dependence and poverty, have all contributed to spawning terribly destructive internal conflicts. These conflicts also reveal, however, the extent to which nepotism and corruption have undermined effective governance across the continent. As demonstrated by the evolution of South Africa from apartheid to democracy, these are some of the many issues that need to be addressed in order to realize the goal of an African Renaissance in the 21st century.

All over the world, the international community confronts the tensions between its obligations to protect human rights and threatened minorities with the historical principles of state sovereignty. The Pugwash Council believes strongly that the United Nations and regional organizations can play an even more effective role in this area in the next century. Ultimately, the Council considers that an important element of state sovereignty – the capability to wage war – should be eliminated. An important step in this direction would be to make the first use of nuclear weapons a crime against humanity, which in turn would support the process of achieving a nuclear-weapon-free world.

The Pugwash Council would like to extend its deep appreciation to the South African Pugwash Group in general, and to Marie Muller and Nola Dippenaar in particular, for their efforts in organizing and hosting the 49th Conference. A great many individuals associated with both South African Pugwash and the Student/Young Pugwash Group of South Africa gave generously of their time and talents in helping to ensure a successful conference. Pugwash would also like to acknowledge the efforts of the Swedish Pugwash Group in helping to secure substantial funding for the conference from the government of Sweden.

REPORT ON WORKING GROUP ONE

A NUCLEAR-WEAPON-FREE WORLD

At its meetings here, in South Africa, the first country which had unilaterally given up a nuclear weapon programme and destroyed the weapons it had manufactured, it was opportune for Working Group One to examine the state of nuclear arms control in the world at large.

Having produced several important international treaties after the termination of the Cold War, the arms control process has come to a halt. The US-Russian Start II Treaty remains unratified by the Russian Federation, while its disarmament provisions are being implemented. Dismantlement of US and Russian nuclear warheads does take place, but negotiations for further reductions of nuclear arsenals have not begun. NATO has restated its strategic concept that the nuclear forces of the Alliance guarantee the security of its members. The Russian Federation, too, considers nuclear forces to be a key element in ensuring its security. It even shows a renewed interest in the production and deployment of tactical, short-range nuclear weapons for a "limited" war. The United States continues the development of a national missile defense (NMD) system, as if the Cold War has not ended. If deployed, the NMD would undermine the Anti-Ballistic Missile (ABM) Treaty, the cornerstone of the US-Russian strategic relationship and the foundation of the global nuclear arms control regime and the preservation of which is in the interest of many nations. The international inspection regime to guard against the production of weapons of mass destruction in Iraq has broken down. The events in former Yugoslavia, especially the bombing of Kosovo and Serbia by NATO forces, brought about a deterioration of US-Russia and US-China relations and diminished the prospects for a rapid resumption of meaningful nuclear arms control. Moreover, because of the approaching parliamentary and presidential elections in the Russian Federation and the United States, arms control processes in both countries have become captives of domestic politics.

The Comprehensive Test Ban Treaty (CTBT), signed three years ago, is not yet effective. It would, therefore, be appropriate for the Pugwash Council to urge its signatories, especially the United States, the Russian Federation and China, to ratify the Treaty as soon as possible. The Council should also call upon the October 1999 Conference of States which have ratified the CTBT to take measures facilitating its entry into force.

The Conference on Disarmament (CD), which a long time ago agreed to negotiate a ban on the production of fissile materials for weapon purposes, is, instead of negotiating, bogged down in controversies over its programme of work. The inactivity of the CD, since 1997, prompted a group of eminent disarmament experts to propose a suspension of its work. It may prove unavoidable to radically improve the functioning of the multilateral negotiating machinery.

The publication of the draft nuclear doctrine of India gave rise to the following questions: Will the Indian doctrine, if officially adopted, destabilize the situation in South Asia? What is the planned size of the Indian nuclear arsenal and what is the meaning of "minimum deterrence"? Will India use nuclear weapons in response to any threat of use of such weapons? What is meant by "failure" of deterrence? Will test explosions be needed to build up the Indian nuclear arsenal to the envisaged level of a triad of aircraft, mobile land-based missiles and sea-based assets? What kind of early warning system does India envisage? Can India afford the cost of nuclear forces, given its present economic situation? After a lively debate, many of the above questions were addressed in a paper specially prepared and submitted to the Group. No common understanding of the consequences of the Indian nuclear doctrine was achieved.

Considering the gravity of the situation, it was proposed that the Pugwash Council should state the following: India's assurance of no first use of nuclear weapons notwithstanding, the implementation of the Indian nuclear doctrine might start a dangerous arms race in South Asia. Neither India nor Pakistan should conduct further nuclear test explosions. Pending the coming into force of the CTBT, India and Pakistan should formalize their unilateral moratoria on testing by concluding a binding bilateral agreement. Neither India nor Pakistan should operationally deploy their nuclear weapons. The fissile material stocks of both India and Pakistan should be kept at a minimum level. India and Pakistan should negotiate a "restraint regime" in South Asia, which would cover both nuclear and conventional forces. Serious efforts should be made to resolve the contentious issues between India and Pakistan, in the first place, the issue of Kashmir.

There are expectations that the current session of the UN General Assembly will reiterate its call of last year upon the nuclear powers to demonstrate a commitment to the speedy and total elimination of nuclear weapons - as Pugwash has long advocated - and that it will specify the steps leading to that goal. There is also some hope that the forthcoming Review Conference of the Non-Proliferation Treaty (NPT) will agree on concrete measures to reinforce the nuclear non-proliferation regime and to bring about nuclear disarmament. One such measure (in addition to the elimination of weapon-grade Plutonium) could be to drastically reduce the world stockpiles of high-enriched Uranium (HEU) by rendering it unusable for explosive purposes. This could be achieved by mixing weapon-grade Uranium with natural Uranium. The resulting low-enriched Uranium (LEU) - a valuable commercial commodity - could be stored for use as a non-explosive source of energy. Whereas enrichment of Uranium is a difficult and costly job requiring sophisticated technologies, de-enrichment is easy and relatively inexpensive. In addition to serving the cause of nuclear non-proliferation among states, it would greatly reduce the risks of nuclear terrorism. For no great competence is required to manufacture a primitive nuclear device having an explosive yield in the kiloton range, provided an adequate quantity - a few tens or several tens of kilograms of HEU (depending on the sophistication of the device) - is available. As a measure of comparison - the Russian stock of HEU (allegedly poorly guarded) exceeds one million kilograms. The 1993 US-Russian agreement for the elimination of 500 tons of HEU possessed by Russia constitutes a step in the right direction. Regrettably, however, it is now treated as a commercial deal rather than a security measure, it is implemented very slowly, and it may even collapse if purely economic interests prevail. Although physical protection of fissile materials is the responsibility of individual governments, possible thefts and misuse of such materials are of concern to the entire international community.

Another measure could be the de-alerting of strategic forces, followed by an observable separation of warheads from their delivery vehicles. Yet another measure could be the "delegitimization" of nuclear weapons by banning their use (or first use) against any state or, at least, against any non-nuclear-weapon state. Co-operative development of means of verification by the nuclear powers and, possibly, by other states, as well as greater openness could help build international confidence in nuclear disarmament.

The Group is of the opinion that the Pugwash Council should issue a statement expressing deep concern about the present deadlock in the nuclear arms control negotiations. It should call upon the nuclear powers to ratify the signed agreements and to carry out significant cuts in their nuclear arsenals. Particularly important and urgent appears to be the withdrawal of all tactical (substrategic) nuclear weapons stationed abroad and their subsequent elimination.

The contribution of nuclear weapon-free zones to the prevention of the wider spread of nuclear weapons is generally recognized and widely supported. In addition to prohibiting the acquisition of nuclear weapons by zonal states, the treaties that establish such zones proscribe the deployment and storage of nuclear weapons. Zonal procedures to monitor compliance with the non-proliferation obligations are in certain respects stricter than the procedures prescribed by the NPT. Moreover, zonal states benefit from some legally binding security assurances of the nuclear powers. Four agreements setting up nuclear weapon-free zones have so far entered into force: in Latin America and the Caribbean, in the South Pacific, on the Korean Peninsula (not yet operational), and in South-East Asia. The required number of states has not yet ratified the Treaty of Pelindaba setting up a nuclear-weapon-free zone in Africa.

Denuclearization of Korea - a measure indispensable for the security of the Northeast Asian region - will require profound political and economic transformations in the northern part of the Peninsula. Denuclearization of the area around the Taiwan Strait - as proposed by some - seems unlikely for obvious political reasons, but its proponents hope to bring it about some time in the future. A nuclear-weapon-free zone in the Middle East would be particularly desirable, because it would involve the denuclearization of a non-avowed nuclear weapon state; its establishment will depend on the development of the ongoing peace process in the region. There was also a brief discussion of the idea to create a nuclear-weapon-free zone in Central and Eastern Europe. Hope was expressed that all countries in the region, as well as NATO as an organization, would soon recognize that the establishment of such a zone was in everybody's interest.

There is, however, a possibility to set up a nuclear-weapon-free zone in Central Asia that would cover the territories of Kazakhstan, Kyrgyzstan, Tajikistan, Turkmenistan and Uzbekistan. The negotiations conducted to reach the above goal have considerably progressed since 1998. A few problems that have arisen are related to the characteristics of the region. They concern, among others, the geographic area of application of the denuclearization obligations and the possibility of enlarging the zone. The intention of the negotiators is to finalize the treaty text before the 2000 NPT Review Conference.

Finally, a proposal coming from the Japanese scientific community was brought to the attention of the Group. It calls on scientists to pledge not to participate in research, development, manufacture, acquisition and utilization of nuclear weapons as well as of other weapons of mass destruction.

REPORT ON WORKING GROUP TWO

EMERGING SECURITY THREATS: AFRICA AND GLOBALLY

For the Working Group, security threats related to both the more traditional forms of security as well as wider human security; all the threats, however, had developed new dimensions and hence could be regarded as "emerging". These include non-nuclear threats (light weapons, Chemical and Biological Weapons, and Landmines); refugees and conflict resolution in Africa; and international terrorism and organized crime. Despite the disparate nature of the sources of insecurity, the group discovered relationships among these threats. Taking full advantage of meeting in Africa and the inclusion of African perspectives on security, the Group benefited from the comparative experiences shared and came to a new appreciation of the intricate relationships between security threats at global and regional levels. In addition, there was general agreement that an urgent need exists to explore the relationship between security and development. This is especially true in the context of the developing world where poverty often is the basis of more traditional security threats.

Organized Crime

The Working Group defined organized crime as "the planned execution of criminal acts directed by the search for profit or power." The discussions on organized crime made clear the tremendous amount of similarity among criminal syndicates in Eastern Europe, Africa, and other regions of the world. Three reasons account for this. First, it was noted that unregulated globalization with the rapid opening up of markets allows organized crime syndicates greater opportunities in terms of new markets for their products. It also facilitates such activities as money laundering. A second, related point: the current global neo-liberal economic order results in social exclusion of large sectors of society, as shown in the rising number of people unemployed. These easily form a recruitment pool from which organized crime bosses can draw their foot soldiers. Third, whilst organized crime is not limited to societies in transition, it is also true that their penetration of the state apparatus is easier in such societies where the state has little or no capacity to enforce its will. Thus in Africa, it was noted that 50 per cent of the economy is informal and this could easily be used by criminals for the smuggling of contraband, and so on. As mentioned above, polities in such transitional societies are generally fragile and this allows criminal organizations the space within which to consolidate their power. Thus it was noted that the Russian *mafiya* may control four-fifths of the current Russian economy. Their economic affluence is transformed into political influence as politicians are corrupted and good governance is undermined. Within Southern Africa, it was argued that organized crime syndicates also specialize in small arms proliferation. This weakens the coercive apparatus of the state; under circumstance of economic deprivation, ethnic tensions, and the like, this can contribute to armed violence that can threaten the existence of the state itself.

Following extensive discussions on this issue, several proposals were presented to curb the spread of such organized crime syndicates. Two of these were:

- given that organized crime syndicates exist and thrive with elaborate, sophisticated links to the formal economy, combating organized crime syndicates should start in the formal/legal economy. In other words, policies should concentrate on severing the economic linkages between the shadow (illegal) and the regular (legal) economies;

- given that organized crime syndicates are a transnational phenomenon, international and regional efforts to curb this menace must be strengthened. For instance, the security forces of South Africa, Mozambique and Swaziland have engaged in joint operations (Operations Rachel I – IV) to identify and destroy weapon caches and thereby undermine organized crime syndicates' smuggling these in the region.

Light Weapons

Light weapons, however, are far more than an issue linked to organized crime. Often in societies which lack a recognized system of egalitarian social participation, weapons mean power: the ability to force one's will upon others. In this sense, arms could be viewed as an instrument of violent social exclusion. Given these circumstances, it is hardly surprising that up to 90 per cent of the world's deaths in conflict since 1992 have been civilians killed by small arms. In addition to the tragic loss of human lives, the proliferation of arms undermines economic development.

It was also argued that light weapons are not an independent variable but a dependent variable of certain political and socio-economic circumstances that lead to violence. In this regard, it was noted that the presence of large number of arms in countries where social cohesiveness is high, such as Finland in the late 1940s, did not result in serious consequences for the Finnish state or people.

Various proposals were put forward to curb the spread of light weapons, especially on the supply side. These include a transaction tax, obligatory third party liability insurance, an ammunition tax, the payment of a large recycling deposit, the creation of a trust fund for the destruction of surplus stocks of small arms, restrictions on advertisements for small arms and the need for the dual standardization of small arms calibres between those for civilian and military/police use. It was also noted that in countries such as Malawi, many gun collection schemes (amnesties, gun buy-backs or exchanges for services or equipment) have been ineffective because they failed to gain the active support of local communities, or because they were not associated with other measures to improve confidence in the police or security services. These problems will be reduced if there are effective peace-building and development programmes and security sector reforms to improve confidence and reduce the incentive to hold arms for personal security or to make a living.

Landmines

The indiscriminate use of mines in the conflicts of the post-Cold War period represents a global humanitarian tragedy. The Ottawa Convention banning the stockpiling, transfer, and use of antipersonnel landmines entered into force in March 1999. Sadly, this achievement is offset by the widespread use of mines during the Kosovo conflict and their renewed use in Angola and elsewhere. Moreover, major nations such as the United States, Russia and China have not yet signed the ban.

Given the Pugwash Conferences' commitment to the contributions of science to peace, the Group spent much of its time on the importance of increased R&D to improve technologies to detect and clear mines. Several group members made the case that current methods are slow, dangerous, and not cost effective. The latter assertion, however, was strongly challenged as based on estimates using costs of developed country labour and infrastructure. They also pointed to the need to assess the costs of demining relative to the costs to development of leaving mines uncleared. Until better technologies emerge, current methods represent the only means to address this urgent problem.

There was general agreement nonetheless that significant improvements in demining technologies are desirable – and that scientists should be urged to contribute to these efforts. Priority should be given to developing methods for aerial detection and mapping and for remote destruction. It was noted that experts generally agree that these would contribute the most to enabling rapid, economical, and effective demining. Unfortunately, R&D efforts to date have been disappointing, but this only serves to highlight the need for enhanced efforts. The Centre created with UN support in Switzerland in 1997 to coordinate such R&D was noted as a promising development. Several participants also stressed the need for greater assistance to landmine victims, whose injuries are beyond the capabilities of local health systems, and who, if they survive, have great difficulty in leading productive lives.

The Group agreed that Pugwash must continue to monitor and support the efforts to implement the ban and improve demining technologies. given the strong NGO networks already in place and working, however, in which national Pugwash groups take part, the Group did not agree on whether additional efforts were needed by the organization as a whole.

Chemical and Biological Weapons (CBW).

The group heard a brief report on the urgent need to complete negotiations for an effective verification protocol for the Biological Weapons Convention, but its main focus was whether terrorists will use Chemical or Biological Weapons (CW/BW). This thesis assumes that: (1) there is a "new terrorism" in which political groups or cults and millennial groups such as Aum Shinrikyo are now willing to inflict mass casualties; and (2) such groups would and could use CW/BW to do so.

Both assumptions provoked sharp disagreement within the Group, highlighting the need for objective research and analysis. Several cases of alleged CW/BW use were also debated, emphasizing the importance of reliable information and independent scientific investigations of such accusations.

Terrorism.

The Group also discussed more conventional terrorism and noted that, although the number of terrorist incidents worldwide has declined, the number of casualties from them has risen. The Group discussed a number of other possible trends, including the blending of terrorism and other forms of warfare, continuing state sponsorship, increasing tactical alliances with organized crime, and the apparent increase in radical religious ideologies among terrorists.

In addition to light weapons, the Group discussed but did not agree on two proposals for Pugwash activities regarding major conventional weapons. These were to organize:

- a workshop to evaluate the implications for global security of three post-Cold War trends among the military industries of the developed world – privatization, diversification from a reliance on producing military goods, and consolidation into fewer, far larger firms;

- an international analytical effort to sort out the many discrepancies among data about arms exports. In addition to hampering the general goal of greater transparency, these discrepancies can have political effects. For example, one standard US source recently revised its earlier estimate that the United States has been the world's leading arms supplier, which could affect NGO efforts to win stricter controls on US arms exports.

Conflict Resolution in Africa.

From Algeria to Angola, from the Atlantic to the Indian ocean, the African continent continues to be plagued by a variety of conflicts. These range from resource-based conflicts (as in Sierra Leone and Angola), to the politics of identity (Algeria, Sudan), to the politics of incapacity (South Africa) to the politics of exclusion (DRC) and the politics of collapsed states (Somalia). . Using Angola, as a case-study, it was argued that much more needs to be done to strengthen conflict resolution mechanisms at sub-regional (Southern African Development Community) and regional (Organization of African Unity) levels. It was also noted that there needs to be greater interaction between the various levels of preventive diplomacy and conflict resolution; for example, actors involved in Track 1 (state, formal level) and Track 2 (NGOs, informal level) need to better coordinate their interventions in a given conflict situation. In addition, several conflicts in Africa have regional dimensions and form a conflict system, as the cases of the Democratic Republic of the Congo and Angola vividly demonstrate. This suggests a more holistic, integrated, and regional approach to conflict resolution as opposed to resolving a particular conflict in a particular national setting in a piece-meal approach. Such a holistic approach to conflict resolution would also seek to address the root causes: politics, economics, socio-cultural variables, military dimensions, and environmental factors within the context of globalization.

One of the sad results of such conflict is the emergence of large numbers of refugees, sometimes through expulsion by force. The African continent faces a refugee crisis, for example with an estimated seven million refugees. It was argued that this could not only be seen as a humanitarian crisis but also as a security threat to the host state, and in some cases to the country of origin. For example, 1.2 million Hutu refugees fled from Rwanda in 1994 and sought refuge in the former Zaire. Amongst these refugees, however, were armed Hutu militias who used the refugee camps as recruitment centres and also to stage attacks on the Kigali regime. In addition, they terrorized Banyamulenge Tutsis in the former Zaire. Ultimately, the neighbouring states responded by attacking these camps, causing further suffering for *bona fide* refugees. Two of the proposals suggested to deal with this situation were:

- ensuring that refugees are settled away from the borders in order to make it more difficult for those militants amongst them to launch attacks into a neighbouring state; but also to protect them from any possible military reprisal from a hostile neighbouring state;

- providing regional or international peacekeeping forces to clearly separate armed militants from *bona fide* refugees (as defined by the 1951 UN Convention Relating to the Status of Refugees, the 1967 New York Protocol, and the 1969 OAU Convention Relating to Refugees), to adequately police these camps, and to protect humanitarian aid workers as well as refugees.

REPORT ON WORKING GROUP THREE

DEVELOPMENT

The Workshop discussion focused on a range of issues that pertained to the eradication or, at least, the serious alleviation of global-wide and pervasive poverty and, at the same time, the promotion of "human rights" as that concept relates to expanding economic, social and cultural opportunities. This objective was deemed, at the very least, to involve access to education and health services, adequate diet and housing and the other basic amenities of life that are essential for the physical well-being and the dignity of the individual, and the cohesiveness of the community at various levels of governance. Attaining this multi-faceted goal could be considered as one of the key challenges of the 21st century and, accordingly, constituted the agenda for the discussion of the workshop.

1. At the outset the Group considered it important to clarify the meaning of the "development" concept as understood in the context of this Conference. This concept is assumed to mean the expansion of opportunities for individuals to realize their potentials that, in turn, is a function of a) adequate income for all citizens and, therefore, a sufficiently rapid rise in the rate of *per capita* growth of the economies within which they live and a reasonably equitable distribution of that aggregate income and regard for environmental constraints, b) a societal framework characterized by democratic rights and by "civility" in the broadest sense of that term.

2. The first issue discussed pertained to the possibility of establishing an international convention to enforce human rights with respect to the granting of citizenship and protection against denial of fundamental rights. Their absence and/or the exercise of discriminatory oppressive practices by governments often leads to societal stress and thus becomes a source of civil or ethnic conflict within nations when, as often happens, there is a denial of such a fundamental right to large segments of a country's population. The desirability and feasibility of international agreement by the community of nations on standards of "acceptable" behaviour by political leaders with respect to respect for human rights and other attributes of a "civil society" is not a theoretical issue of our times but a practical and urgent one: *vide*, the International Court of Justice and indictments for crimes against humanity, bringing General Pinochet to trial by a non-Chilean court.

This raised consideration of the possibility of establishing a category of "global citizenship" that would enable people to move freely without regard to borders. To do so would require a very much greater degree of equality of incomes and opportunities as between nations. Without that condition of greater equality in living conditions as between countries, people would undoubtedly gravitate in great numbers to those countries that had higher incomes and job opportunities. Such a movement on a historically unprecedented scale occurred at the end of the last century when between 1890 and 1910 over 40 million persons entered the USA from Europe. This would not be possible today when capital flows between nations on a historically unprecedented scale, as people are severely constrained from large-scale movements on a voluntary basis despite the urge to do so because of the extreme and rapidly increasing polarization of incomes: the ratio of average *per capita* income of the developing countries to that of the industrialized countries was 1:30 in 1960 and is now about 1:60+ (under $400 per year as compared to over $24,000).

The following questions were raised:

- *if the polarizing trends were to be continued, what would be the implications for violent conflict both within and between nations?*

- *how would the "breaking points" likely manifest themselves and within what likely time frame? and what, if anything, could be done beforehand?*

3. There is a related issue pertaining to the concept of a borderless world, namely, with respect to the movement of scientific and technological talent across borders and its implications for the developing countries who thereby suffer from the "brain drain" phenomenon. There already is a great and growing imbalance between nations with regard to the availability of qualified scientists in both the fundamental and applied sectors. This imbalance is being exacerbated when, for example, the USA, at the request of the multinational corporations, raised its annual immigration quota for qualified computer personnel from 65,000 per year to almost double that number.

This situation is surely a matter of great concern since serious weakening of the capacity of most nations to undertake research at the cutting edge would doom them to a permanent status of second-class countries in the era of the "knowledge economy". When this diminished capability is viewed with regard to research and policy-making capabilities in the fields of the social sciences pertaining to education, health, social organization, and such, it could be recognized as a matter of such great importance as to merit being placed as one of the top priorities on the international agenda. It is a situation that could be characterized as "technological imperialism."

The question:

- *what could be done at the level of international governance to achieve a more equitable geographic balance with regard to scientific and technological research capacity and capability?*

4. Access to capital and related technical assistance by the developing countries is another of the issues that need to be placed among the top priorities on the international agenda. The flows of official development assistance (ODA) and foreign direct investment (FDI) are far from adequate in meeting the needs of the developing countries if they are to meet the *per capita* target rate of growth of three per cent that was recently set at a UN conference. FDI exceeds ODA by a ratio of 5:1 (about $250-$300 billion to $50-$60 billion per year) and this inadequacy is compounded by the fact that both types of capital flows are overwhelmingly directed to about a dozen developing countries, leaving over 160 countries with a trickle. For example, both types of capital flow have largely bypassed Africa (which receives about five per cent of the flow to developing countries) and, as well, the 47 poorest countries that have been receiving about one per cent of the flow. In net terms (taking account of the outflow to meet very onerous debt servicing obligations) the available capital from abroad is even less adequate.

One approach with regard to increasing the total capital flow to the developing countries is to tap new sources beyond that contributed for ODA by individual countries through bilateral and multilateral aid agencies and FDI that is contributed by private investors, principally multinational corporations. Thus, in addition to proposals of a general nature such as "a Marshall Plan type of capital transfer", there are proposals extant such as a global carbon tax (that would, at the same time, discourage emissions). There is also the proposed global tax on financial flows between foreign exchanges that is now estimated to be about $1.8 trillion per day (a tax proposed by Nobel Laureate economist, Professor James Tobin known as "the Tobin tax" that, he suggests, might be levied at a miniscule rate of about 0.01 per cent to both dampen volatility that is due to the speculative component not related to trade and investment of the capital flow and which, at the same time, could raise very substantial funds that could be devoted to developmental and environmental purposes). The Tobin proposal has the characteristic of deriving funds from private users of the "global commons", outer space and the deep oceans, who in the absence or under-pricing for such use as parking satellites or laying cables thus capture enormous economic rent. These sources have a great financial potential and merit serious consideration that could take the form of feasibility studies under international auspices. There are technical, political and social barriers to be overcome with respect to each such proposal of global scope. For example, the US Congress has passed a resolution forbidding the discussion of such sources so as to prohibit tapping a source of funds that lessens dependence on their contribution and that would thereby weaken their classic withholding tactics. (It may be noted that the Canadian Parliament has passed a resolution to give serious consideration to the Tobin tax and similar proposals.)

The questions:

- *how could ODA and FDI flows be substantially increased and, at the same time, shifted in their allocation towards the poorest countries that are most in need of capital, especially those that are also heavily indebted and have to contend with a net outflow of capital?*
- *how could the international community organize to undertake feasibility studies to examine the desirability and feasibility of the proposed initiatives to increase the availability of funds for developmental and environmental purposes?*

5. There are troubling side-effects of some foreign direct investment (external diseconomies) as in the cases of large-scale projects that confer some benefit to the local population often in the form of income from low-paying jobs of an unskilled nature but, at the same time, impose some environmental costs for the community-at-large in the region. Many governments are evidently anxious for foreign investment but insensitive to injuries thereby being inflicted on their own people. Non-governmental organizations had often been effective in avoiding or mitigating these negative impacts through political action, a phenomenon characterized as "grassroots power".

The question:

- *in the case of developing countries where such countervailing actions are not tolerated, what could be done at the level of international governance to ensure that the foreign investors adhere to standards related to minimal wages, safety and other working conditions and to regional environmental standards?*

6. The aspect of health and educational deprivations of the poor merit special attention as critical attributes of "absolute poverty" defined roughly as incomes *per capita* of $2 per day. Those who suffer from such deprivations are a very large segment of humanity, about 40 per cent of humanity, the greater part living in rural areas that are marginalized by virtue of distance and lack of access to modern technologies and related amenities. Alleviating these key deprivations is fundamentally an issue of "human rights". In the case of health care, the treatment of cardiovascular diseases is illustrative of the acute dilemma facing developing countries and those in transition. This class of diseases is the Number One killer in both industrialized, developing and transitional countries but while it merits a priority on that account, especially since it affects the most active part of the working population, the diagnostic and curative treatment - and training of doctors and other medical practitioners as a continuous process - is very expensive and places a huge financial burden on the health care services. This poses a special problem for the poorer countries where budgets and talents are severely constrained.

The questions:

- *how could these countries attain greater access to needed medical diagnostic and other essential equipment, medications and talent?*
- *could modern communication technologies play a role in this process of increasing access and reducing costs?*
- *should/could assistance be provided in the form of cost/benefit analysis of medical programmes as a basis for designing national health care policy in developing and transitional economies?*

7. If scientific and technological progress is to play an increasing role for the direct benefit of the poor, especially in the developing countries, some formidable barriers need to be overcome, such as a lack of supporting infrastructure with respect to reliable environmentally benign low-cost energy sources and a low-cost wireless telecommunication network that would be appropriate for meeting some priority needs as primary education. Though the new technologies have the potential to play an important role in closing the income gap between the richer industrialized and the developing countries, the difference in terms of access to these synergistic technologies (the communication technologies cannot work without the energy sources) will undoubtedly lead to exacerbating that gap that is already figuratively as wide as the Grand Canyon. Modest beginnings have been made in introducing these technologies into developing countries but as the world enters into the era of the knowledge or informational global economy of the 21st century, the pace must be accelerated before the trends in education for the computer age gets to a point of no return.

The question:

- *how could the industrialized countries design aid programmes so as to ensure that the new technologies are adapted in price and design and concomitant supporting measures (that is, as a system) to directly serve the needs of the poor on a scale and speed that would make a significant advance in reducing the critical deprivations of the poor in the developing countries with regard to access to energy and informational/educational technologies?*

The answer to the last question merits elaboration and suggested approaches:

- Given the fact that, with a few special exceptions, despite the transfer of capital and technical assistance of official development assistance (ODA) on a country-by-country basis of over a trillion dollars over the span of fifty years, the group of countries designated as "developing" remain on the whole still very poor in terms of average *per capita* income;
- given the fact that the global economic/financial environment is not congenial (in terms of levels and volatility of interest rates, the terms of trade and other features) for almost all of the developing countries to enable them to grow *sufficiently rapidly* (defined as at least a three per cent rate of *per capita* growth to enable their citizens to escape from deep poverty *within a generation*) and that sufficient changes in this regard will take a long time;
- given that the deprivation of the poor in the developing countries is most acute with respect to primary education and that the highest rate of return (in the broadest sense) for any investment in the social sector is in programmes to educate the young - and especially young women - and that in a short time span aspects of modern technology could be harnessed to bridge distance so that teachers and doctors could "reach" rural villages while living in urban centres;
- given that there is a suggestive precedent of a successful ODA initiative that goes beyond aid on a country-by-country basis to address the food deficiency aspect in terms of supply, namely, the Consultative Group on International Agricultural Research (CGIAR) that helped triple global food output while global population doubled- and that it would appear to be logical to consider a strategy of replicating CGIAR (of course, with appropriate modifications) with internationally-sponsored research programmes focused on both energy and informational technologies as synergistic systems that would address the educational and health deprivations of the absolute poor.

The following questions are posed:

- *pending the achievement of a sufficiently transformative change in the global economic/financial system such as would enable developing countries to grow at the target rate set in UN conference meetings –a long-term proposition - why not launch an international initiative with the prospect of a short-term payoff in providing energy and education and health services for the desperately poor living in rural areas?*

- *if this proposal is considered as worth exploring as a prima facie promising means to alleviate poverty, why not launch feasibility studies that would endeavor to answer questions as to the organizational arrangements, funding, and other questions that would constitute the terms of reference for the feasibility studies?*
- *should/could Pugwash play a catalytic role with respect to raising money for the feasibility phase and participating in its design, staffing, etc.?*

REPORT ON WORKING GROUP FOUR

ENVIRONMENT

Recommendations to the Pugwash Council

Water: Considering that water-sharing problems are essentially *regional*, rather than worldwide problems, and that water supply and conflict are strongly related, it is respectfully suggested that Pugwash initiate a study of these particular cases, based on a) a collection of the relevant technical, social and ecological facts relative to each project; b) a confrontation of the viewpoints of the parties; c) a scientific evaluation of the facts and positions; d) a search for the means of resolving the differences and avoiding armed or economic or other conflicts resulting from water issues; e) a workshop or symposium dedicated to the awakening of the general public to the need for adequate water management and a more efficient use of resources at all levels, including in the home and in agriculture, a reunion to which the media could be asked to participate; this reunion could be held in Africa.

Climate change: Considering that climate change problems (in opposition to nuclear, chemical and other weapons problems, which were almost impossible to discuss more or less openly during the Cold War) need a strong and direct improvement of the public's behaviour in general towards greenhouse gases (including CO_2) emissions, it is respectfully suggested that Pugwash adopt a more open attitude towards communications, by using, for instance, the following means: a) contact the multiple environmental agencies (UN, regional, governmental, NGO's) to offer them the expertise of Pugwash members; b) organize a conference on the scientific technical aspects of the UNFCCC/Kyoto protocol for the benefit of the media; c) organize a series of workshops (as it did in the case of the Chemical Weapons Convention talks) where scientific negotiators could meet informally to confront their viewpoints outside of the official meetings.

These suggestions are based on the discussions held in the Working Group, as follows.

I. Water Resources: Availability and Control

Water is the principle of all things
Thales of Miletus, 635-548 B.C.

Resources

A problem lies in the very unequal distribution of resources between regions and between seasons of the year, some regions receiving a fairly regular flow over the year (temperate and equatorial zones), others a sometimes heavy flow concentrated over a short period (near-tropical zones), and some practically none at all (tropical deserts). The influence, direct or indirect, of human activities on water availability and flow must be recognized. Regulation of water flow through the use of canals, dams and reservoirs raises several problems, the latter two of which are considered below. But other human factors must be considered: agriculture (including irrigation, the largest single consumer), deforestation, soil erosion, draining of swamp lands[1] or shallow seas.

The special case of Africa (poverty, high temperatures, irregular and unreliable rainfall, illiteracy, political instability, corruption, weakness of support systems) was given particular consideration. It was pointed out that there was no food insecurity in Africa until the 19th century colonialist period, when new non-food crops (mainly cotton) were introduced. A further problem, which will be considered below, is that most river basins are shared between several countries or regions, and that water supply can become an important source of conflict.

The problems raised by dams and reservoirs were discussed fairly extensively. Dams have multiple purposes, which are often joined: power production, water supply, recreation, flood control and more. These purposes are sometimes conflicting, the need for electrical output occurring at times different from the call for water release or retention. The two aims of power and water supply will be considered separately, while the problems they raise will be considered jointly, being the same for the two applications. The major cases discussed were the Cahora Bassa (Zambezi) and Ilisu (Tigris) dams; the Aswan High Dam (Nile), Inga (Congo) and some others were also mentioned.

The *Cahora Bassa* dam on the Zambezi (Mozambique) has been the source of considerable controversy, being described by its protagonists as a potentially indispensable source of non-polluting energy and a stimulant for the economic development of the Southern African region. Critics of the dam stress that it was designed as a colonial project to export cheap energy to apartheid South Africa and that it has been a hydrological, environmental, economic and human disaster. Preliminary data support the latter interpretation. Nevertheless, the government of Mozambique is planning to build a second dam approximately 100 km below Cahora Bassa to meet anticipated energy needs within the country and in the region over the next decade.

The problem seems to lie in the absence of proper research and study of the dam's social and ecological consequences, followed by a public discussion.

The *Ilisu* dam on the high reaches of the Tigris (Turkey), currently under construction, is part of a 22-dam project, spanning the Tigris and Euphrates rivers, and meant to provide irrigation water and electric power to the south-eastern region of Turkey. The project leads to strong opposition on several grounds: politically, the project seems to be at least partly motivated by a strategic interest of the Turkish government to strengthen its power position in the Middle East and to control the unruly Kurdish area, and economically, the downstream states (Syria and Iraq) complain about a decrease in the water flow through the irrigation and the evaporation from the reservoirs. Negotiations about fair sharing of the water appear urgently required.

Dams (and their corresponding reservoirs) are often recommended by the promoters of "*soft*" (that is to say, renewable) *energies*, as an economic and environmentally friendly substitute to the use of carbon-containing fuels; once the dam is built, the direct cost (as distinct from amortization and financial costs) is very low, the flexibility of the production is high, enabling the installation to follow easily the peaks and troughs of power demand. There are only a limited number of suitable sites, and the tendency is to use a maximum of those available to alleviate the forthcoming conflict between an ever-increasing power demand and the ceiling imposed on CO_2 emissions (see Ch. III, below).

Dams - as most coal or oil-fired and nuclear power plants - mean centralized production, as opposed to some renewable energy sources (photovoltaics, wind energy). This implies, particularly in regions with low populations and little industry, transportation of electricity over long distances and the installation of costly, unsightly and vulnerable power lines. Dams and reservoirs are required when the instantaneous *flow of a stream* varies considerably in the course of the year; the reservoir fills up in periods of high flow and empties at low flow, ensuring a fairly constant downstream supply of water. Dams have also been used to divert water to formerly arid areas in order to promote agriculture.

Use *versus* Consumption

Consumption of water is essentially tied to agriculture (the largest user by far). Other users deliver mainly polluted water, in varying degrees (heat, organic or inorganic materials, micro-organisms). Some pollutants are real dangers to the environment and the population (chemicals such as As or Hg salts, micro-organisms); the particular case of cholera, typhoid and hepatitis infection in the Santiago metropolitan region of Chile, attributable to the use of domestic sewage for food crops irrigation, was described in detail.

It was pointed out that water scarcity was often a symptom of the real problems of population growth and desire for improvement; water use is strongly r*elated to population importance and degree of development*; both are on the increase, and measures will have to be taken in the near future to avoid catastrophes, some of which are imminent.

Water Problems and Conflicts

Water has been a source of conflicts; water resources and systems have been both targets and tools of conflicts; disparities in water use can lead to conflict. It was pointed out that water supply was essentially a *regional,* rather than a worldwide, problem. It could be - in fact it is, and has been for centuries - a major source of conflicts between nations sharing the same hydrological basin, and sometimes between regions of a same nation; many cases were cited, going back far into history, of disputes over the sharing of water between adjacent regions, or between pastoral and agricultural communities. Several specific cases were discussed: the Zambezi, Euphrates and Tigris, Nile and Jordan rivers, for instance, and many others (the Rhine, Danube, Rhône, Amazon) were mentioned in passing. Although some worldwide agreements have been made over the use of water (for instance the Convention on the Non-Navigational Uses of Shared International Watercourses, 1997, UN General Assembly A/51/869, 11 April 1997), enforcement appears difficult. More can be expected from bi- or multilateral agreements between the concerned nations, if they can be adequately enforced. Precision in wording seems to be sometimes lacking, leading to subsequent difficulties, in cases of drought, for instance.

Transportation of Water. over fairly long distances can be a costly process, which is however sometimes – but not always - justified between two regions where an excess and a dearth of available water respectively exists. (The case of South Africa was considered in detail). Difficulties may arise when the resources or the needs of either region show sizeable variation over the years. Exportation or importation of water may sometimes be replaced by "virtual" exports and imports, in the form of cereals, fruit or other products. This appears to be a key mitigating factor in water conflicts.

Dams. It was pointed out that dams raise quite a number of particular problems, which could become sources of conflicts, as follows:

- *damage to the environment*; the size of the reservoir lake can vary considerably, but always affects natural ecosystems and often prime agricultural land.
- *displacement of populations*; the inhabitants of the submerged area must be lodged elsewhere, often in hardly acceptable conditions, with new ways of life and a disruption of their traditions, with a cultural loss, a decrease in security and a lowering of their living standard.
- *loss of water*; the increase in evaporation from the increased surface can lead to severe water losses, depending on the climactic conditions.
- *disruption of aquatic life*: fish migration is often prevented, unless special precautions are taken; some species fail to adapt from their former running water surroundings to the still (and often warmer) water of lakes.
- *mosquito and pest proliferation;* schistosomiasis and other water-related diseases sometimes appear after completion of projects.
- *downstream problems*: the change from variable to constant flow may create serious downstream problems to the riparian populations.

- *international or regional disputes*; when, as is often the case, the river or lake crosses international or regional boundaries, the sharing of the power and water flows between the participants can lead to serious disputes, particularly when the actual flows differ severely from the forecast assumptions.
- dams and reservoirs could constitute highly vulnerable *targets* and can also be used as *weapons* in case of war.
- dams have been known to collapse and be the source of considerable downstream damage in case of *earthquakes or geological instability.*

The conflict related to the sharing of water between pastoral and agricultural populations in the Sahel region of Uganda was described. The immediate solution appears to be in the development of suitable water supplies, but, on a longer term, alternative survival plans, changes in living habits and improved education are definitely called for.

Reclamation and Conservation

Reclamation. means treating effluent water to restore its purity; the choice of the method depends on the degree of pollution, the degree of purity desired, and the cost of the operation. Desalinization of seawater, for instance, is a fairly costly process, whatever method or energy source is used, and justified only in places where clear water is scarce and energy easily available (Kuwait being a case in point). It is in general much more economical to purify effluent water from industrial or domestic sources, if available.

Conservation. implies using the right quality of water, in the minimum quantity, as a function of the intended use. Unless clear water is really plentiful, there are many uses for which untreated or slightly polluted water is quite adequate, such as domestic use (washing machines, toilet flushing, lawn sprinkling), industrial use (rinsing or cooling water), and agricultural use (irrigation). This latter point is particularly important; a need was felt to improve agricultural methods by several means, tending to an increase in the yield per ton of water and km^2 of land used: planting and harvesting methods, controlled use and improvement of pesticides, herbicides and fertilizers, use of animal or mechanical power, development of new crop types by selection (or even genetic improvement, although this method met with some opposition within the group).

Care must be taken, however, to fit the quality of water to the final aims; cases were mentioned of disease spreading (cholera, typhoid, hepatitis) through the use of contaminated water for the growing of vegetables. Some countries have set up regulations defining which type of water could be used according to the crop being produced (ranging from vegetables meant to be eaten raw to trees, for instance).

Conclusions

It was suggested that a detailed quantitative study be made, by country and by region, of the several uses of water, and of the minimum quality of water corresponding to each use. Also, that more attention be given to the efficiency of water use, mainly in agriculture, but also in industrial and domestic uses; that more attention be given by governments to water governance, notably by adequate staffing and financing of the concerned services; and that existing international conventions be improved, made more precise, and correctly enforced. It must be recognized that water supply and use is not a technical problem, but a social, economic, cultural and political one; it is easier to build a dam or lay a pipeline than to change the habits and living patterns of a population. Accordingly, dam proposals must be subjected to far more detailed study than at present, not only from the hydrological, engineering and economic points of view, but encompassing ecological, environmental and social considerations, preferably (and certainly when involving two or more countries) under supervision of an international authority.

II. Urban Development and Environmental Problems

The *Moscow Sustainable Cities Project* (MSCP, in the frame of the UN Sustainable Cities Program, SCP) was described in some detail. The power of attraction of cities must be recognized, but the development of large cities calls for regulation aside the play of economic factors, if environmental disasters are to be avoided. Moscow is one of the most actively developing European mega-cities. A Moscow Master Plan, extending to the year 2020, was approved by the city government in 1999, based on a future city population of about 9.5 million inhabitants (15 million with the surrounding region). It aims at the increase in life quality, the preservation of the architectural and historical heritage, the decrease in negative environmental effects of industrial and domestic activities, the control of crime, and the balanced development of all the branches of the city's infrastructure. All interested parties - authorities, businesses, NGOs - are co-operating in the project. The first step is the definition of an environmental profile, including a study of the current situation, a definition of the city's economy, a study of the natural resources, and a description of the management setting.

After a choice of the desired activities to be developed and of the indicators to be used, planning and land use policies can be defined, taking into account the various existing constraints (budget, state authorities) and eventually implemented. Moscow should be one of the most environmentally friendly cities in the world. The plan could conceivably be emulated by many other mega-cities.

III. North-South Conflict and Co-Operation in the Implementation of the United Nations Framework Convention on Climate Change (UNFCCC)

The aim of the convention is to limit the greenhouse gases concentration of the atmosphere to a "safe level", by reducing worldwide emissions. Under the Kyoto protocol, the UNFCCC obligations include only *carbon emission* abatement on the so-called "Annex 1" countries, that is to say,. practically all industrial countries. The non-Annex 1 countries (essentially the developing countries) have on average low emissions *per capita*, because the poorer part of their population consumes energy at a very low level, far below that of the industrialized countries (imitated by the richest. part of the developing countries population). The UNFCCC/Kyoto protocol set no constraints on carbon emissions of non-Annex 1 countries.

Since the Annex 1 countries appear unable to achieve the former goals of emission reductions by the year 2000 (recently postponed by the Kyoto protocol), the Clean Development Mechanism (CDM) has been put forward to allow Annex 1 industries to invest into carbon emission reduction in non-Annex 1 countries, where more reductions can be obtained at a lower cost than in their own country.

The aims pursued appear eminently laudable. However, the initial targets for emission reductions under the UNFCCC and the later Kyoto protocol are inadequate to effectively respond to the problem on a meaningful timescale; the current systems of numerous approaches to emission reductions outlined in the Kyoto protocol are overly complex and may lead to an unworkable regime incapable of adequately addressing the problem; and the Clean Development Mechanism (CDM) allows industrialized countries to meet some part of their emission reductions through investments into projects located in developing countries, but raises some important ethical issues. For example, the CDM negotiations have reflected a bias towards the needs of the industrialized countries for low-cost abatement opportunities, to the detriment of the development priorities of developing countries.

There are encouraging examples of *private sector responses* to the climate change challenge, in that they have been initiated in the absence of any national or international regulations. In this regard, the detailed plans of a large multinational oil corporation, which could serve as template for, and indeed are being considered by, a larger group of companies, were discussed. They call for a reduction of direct CO_2 emissions by the year 2010 to a level ten per cent below that of 1990 (corresponding to a level of 25 per cent below what would normally be reached in the absence of action). The plans include improvements in operating efficiency (and therefore company performance), a switch from coal or oil to natural gas, diversification in alternative energy sources and generating methods, emission trading, sinks, and, maybe just as important, monetary incentives or penalties according to success or failure in achieving the budgeted results.

This approach would essentially apply to Annex 1 countries. As a marketing tool, it would provide highly visible effects, appealing to governments and populations alike. It would be up to the large corporations and industries to initiate action, increase their performance and thus incite their competitors in the world's economy to follow suit.

Concerning *CDM*, it is recommended that this mechanism be controlled, so as to ensure and promote the development and the improvement of the poorer part of the population of the non-Annex 1 countries, besides avoiding carbon emissions. New, fresh ideas will have to be developed: avoid an exportation of wasteful consumer patterns from Annex 1 to non-Annex 1 countries, define local needs and solutions, use a maximum of renewable energy resources, improve education at all levels, innovate in science and technology, improve forest management, limit the population growth. Four aims are defined: modernization, education, employment and increase in income of the poorest sectors of the population. There are also four tools: management of natural resources (including agriculture), adequate energy supply, market force dynamics and governmental action[2].

The industrial investors will get the credit, but at least the needs of the non-Annex 1 populations will be satisfied. These responses could form an integral part of the long-term success of any UNFCCC regime.

However, more is needed for the final goals - whatever they will be - to be attained. Both in Annex 1 and non-Annex 1 countries, the attitude of the general public must be changed in many ways, including systematic verification of the operation of used automobiles; smaller cars, more efficient; a switch to other fuels (natural gas, methanol, or even electricity[3]); use of public transportation; more efficient home appliances (lighting fixtures, refrigerators, TV sets); and a decrease in heating and air conditioning expenses (home insulation, changes in design temperatures). This might call for governmental action, but will also need the help of the media.

In all cases, more precise rules, regulations and criteria are needed, concerning not only the obligations of the parties, but also the monitoring and verification of these obligations. It will be the responsibility of the competent authorities to provide these, in consultation with all sectors of the economy: industry, labour, consumers, scientists.

Notes

[1] The particular case of the association of mangrove swamps with coral reefs – a particularly efficient CO_2 sink, apart from their environmental interest- and of their destruction, was discussed.

[2] Either through regulation or taxation, the former approach being favoured.

[3] This electricity would of course have to be produced in a conventional power plant, but with a far better efficiency and much less pollution than an ordinary car engine.

REPORT ON WORKING GROUP FIVE

INTERNATIONAL GOVERNANCE

UN Reform and Its Implications

The Group felt that although the United Nations has made many important contributions to peace, its functioning ought to be improved by a number of changes. The original charter was drafted more than fifty years ago, and it was signed before Hiroshima. The enormous destructiveness of modern weapons gives urgency to the task of eliminating war. Other great changes have taken place: colonialism has ended, information technology has given us the internet; and school children in Latin America can exchange e-mail messages with children in Europe and Asia. A new sense of global citizenship has been made possible by information technology, and it will certainly become much stronger in the future. In addition to instantaneous communications, we now have a borderless world economy. The Cold War is over, and conflicts between nations have to some extent been replaced by problems involving civil wars or persecution of minorities within nations. As we approach the new millennium, the great increase in global interdependence challenges us to develop a new vision of governance at the global level.

Reform of the United Nations is already underway. The Secretary-General, Kofi Annan, hopes that in the coming session of the General Assembly it will be possible to make changes in the Security Council which will make it more democratic and transparent and which will make its performance more reliable. Instead of representing the five nuclear powers, the permanent members could perhaps be given a more equitable geographical distribution. It was felt that the time is right for such changes; and in particular it was hoped that the veto power in the Security Council, if not eliminated, could at least be restricted by agreements among the permanent members, and be made more accountable to the world community. An alternative way of dealing with the veto problem would be a rule that if a proposed resolution is consistently defeated by one member, and thus the Security Council fails to exercise its primary responsibility, the issue be passed to the General Assembly, where it could be enacted by a two-thirds majority.

It was felt that among the proposals which have been made for improving the financial basis for UN operations, one of the most promising is a very small tax on international currency exchanges - the so-called Tobin tax. The amount of money involved in such exchanges is enormous. A very slight tax on international currency transactions would provide an adequate basis for the increasingly demanding tasks which are being given to the UN, and it would have the additional positive effect of damping currency speculation. The International Monetary Fund could administer the tax, with the understanding that the money would go to the United Nations. Other proposals for improving the financial basis of the UN included a tax on international airline flights, a tax on international shipping, ownership by the UN of the biological and mineral rights of the deep oceans, and a levy on states which would be proportional to their arms expenditures.

The Group applauded the signing of the Rome Statute establishing an International Criminal Court with jurisdiction over war crimes and genocide. It was felt that all nations should be urged to ratify the treaty as soon as possible, and that those nations who have already done so should be congratulated. The International Criminal Court represents an important first step towards establishing a system of international law, which acts on individuals rather than on states. The case of General Pinochet was also seen as an important step towards establishing the principle of individual responsibility. The Group heard an account of genocide in Bosnia, Croatia and Kosovo, and a description of the feelings of the people in the region. Reconciliation will require both truth and justice; and in particular it was felt that those responsible for the genocide should be tried for this crime. It would be desirable to be able to arrest the guilty - even heads of state residing in their own countries.

The system of voting in the UN General Assembly was also discussed. The present system seems unjust to some, since representation is given to national governments rather than to people. One proposal for reform was to have final votes cast by regions. The Group also heard a contributed paper, which proposed the establishment of a popularly elected world parliament. In the election of representatives, it would be possible for a citizen of one country to vote for the election of a candidate from another country.

Peacekeeping and Peacemaking

The Group recommended that the UN should have a permanent force, made up of volunteers from many nations, paid by the UN, wearing UN uniforms, and under the direct command of the United Nations. Such a UN Police Force would be constantly prepared for rapid deployment in situations where international peace is threatened or in a humanitarian crisis. There was some discussion of the size of such a force. A low figure of 15,000 was mentioned, with 5,000 stationed in Latin America, 5,000 in Africa, and 5,000 in Asia. Others in the Group felt that while 15,000 might be sufficient for a crisis such as East Timor, the UN Police Force ought to be larger in number in order to be able to address a more general type of crisis. There exists a possibility to raise a larger force under Article 43 of the UN Charter, according to which the Security Council should make bilateral agreements on the forces to be made available to the UN. It was suggested that Article 43 should be brought to life.

An interesting paper was presented reviewing the Augustinian tradition of the "Just War". This tradition has sometimes been used as a normative standard, and it may still have some value; but it is easy to misuse. The majority of the Group felt that the UN Charter and international law have now replaced this concept.

Regarding interventions by the UN, it was felt that preventive strategies are best. Crises should be anticipated and prevented by diplomacy, combined perhaps with preventive stationing of the UN Police Force. However, since it is difficult to obtain financing for preventive actions, the possibility of preventing crises is linked with an improved system for financing the UN.

It was also noted that democracies seldom fight with each other, a fact, which implies that conflicts can to some extent be prevented by encouraging democracy. The present Trusteeship Council of the UN should perhaps be changed to a Council for Human Rights and Democracy.

The Group felt that human rights must take precedence over the rights of nation-states. In an increasingly interdependent world, the concept of absolutely sovereign nation-states has become an anachronism. To some extent, the absolute sovereignty of nations must yield when it is in conflict with the interests of humanity as a whole. Looking towards the next century, one can foresee a time when national armies will be replaced by a global force. Ultimately, the most important element of state sovereignty - the capability to wage war - will have to be eliminated. The only coercive power in the world should be vested in the United Nations in the form of an international police force. Indeed, under the present UN Charter, no nation has the right to initiate a war, and an intervention is illegal except under a specific resolution of the Security Council.

In discussing NATO, it was noted that while the organization has the military strength and logistic capabilities required for deciding conflicts, it can only legally be used to defend the territory of its member states against attack, and even then only after the Security Council has had time to act. NATO also lacks impartiality, and thus it cannot be a substitute for the development of crisis-resolving capabilities within the United Nations.

The Group felt that the United Nations Charter should be extended to make the first use of weapons of mass destruction a crime against humanity. Such an extension of the Charter might be an important first step towards the complete abolition of nuclear weapons.

In connection with achievement of a nuclear-weapon-free world, Gandhi's civil disobedience tradition may be important to encourage societal verification. Civil disobedience may also be important in setting bounds to the degree of tyranny which a state can impose on its citizens. For example, soldiers ordered to commit atrocities have a duty to disobey.

It was felt that in strengthening the United Nations, one should bear in mind the principle of subsidiarity. Only global issues should be decided globally, and as much as possible should be handled locally. A system of checks and balances will also be needed, since, although we should be idealists in our vision of the future, we should be realists concerning human nature. Both realism and idealism were seen in the papers presented and in the discussion. The Group recognized that there are severe obstacles to UN reform, but felt that Pugwash ought to have the vision to see that what is needed to achieve is a world which is free from war and based on the principles of democracy and law. It was proposed that Pugwash should organize a workshop on UN reform.

Governance in Africa

Although Africa is potentially an extremely rich continent, its potentialities have not yet been realized. This is partly due to the legacies of colonialism and the Cold War. Boundaries were drawn in colonial times with little regard for ethnic realities; and the riches of Africa were exploited by international commerce. During the Cold War, the superpowers poured arms into Africa, and these weapons remain today, causing great suffering and impeding development. However, although many of Africa's problems may have been caused by colonialism and by the Cold War, the continent must now take responsibility for some self-made problems, for eliminating nepotism and corruption and for improving governance.

The Group heard three papers on this subject. The first paper pointed out that while good governance has been the criterion for investment and for development aid from the world outside Africa, the purchasing power of the poor is of more relevance to most of the people of the continent. The majority of Africans belong to an informal economic system, without formal employment. For them, it is irrelevant whether conditions are favourable to banking and commerce. The issue is daily survival - access to reasonably nourishing food, clean water, and health services. The needs of the majority of Africans rather than the promotion of commerce ought to be the primary focus of donor agencies. Also the phenomenon of enrichment - enormous and completely unjustifiable accumulations of wealth by a few people - ought to be investigated.

The second paper focused on strategies to combat the phenomenon of corruption. The practice of offering and taking bribes perhaps has its roots, not only in the absence of effective institutions to detect and punish corruption, but also in the fact that bribery may not even be perceived as corruption. Foreign firms in Africa have some complicity, since many allow the listing of money spent on bribes as a business expense. In addition to small-scale corruption, there are instances of very large-scale corruption, where heads of state have taken national treasuries for private use. In order to combat corruption, it is vital to address the youth of Africa. They must be educated to see that corruption blocks development. The youth of Africa were active in the fight for freedom, and their help must now be enlisted to fight corruption and to promote development.

The final paper explored parallels between the governance problems of South America and those of Africa. Among the common features was the existence of a criminal economy, linked to the formal economy through money laundering activities. Corruption and flight of capital were also common problems of the two regions, and parallels could be found in the privatization of security in South America and in Africa. In South America, one of the biggest problems is the great contrast between the lowest and highest income sectors. This has led to growing social violence and civil insecurity. Addressing this problem is vital, both in South America and in Africa. The Group concluded that in both regions, the great challenge is to use state power to address poverty, and to work not only for good governance, but also for economically just governance.

PART II

PROCEEDINGS OF THE PLENARY SESSIONS

PROCEEDINGS OF THE OPENING PLENARY SESSION

Held on Wednesday 8 September 1999 at 08.45
at the Kagiso Convention Centre

Chair: Professor Marie Muller and Sir Joseph Rotblat

1. **Professor Marie Muller** welcomed the participants and formally opened the Conference (text on pp. 53-54).

2. **Messages of greetings**

   a. From the Secretary-General of the United Nations, Mr. Kofi Annan, read by Dr. William Epstein (text on pp. 55-56)

   b. The Secretary-General of the Organization of African Unity, Dr. Salim Ahmed Salim, sent a message of congratulations to the South African Pugwash Group on hosting this Conference, emphasizing the importance of addressing the challenges of the new century.

   c. The South African Minister of Education, Professor Karder Asmal conveyed to the Conference his very best wishes for its success.

   d. A message from the former President of South Africa, Mr. Nelson Mandela, was read by Professor Marie Muller (text on p. 57)

3. **The President of Pugwash**, Sir Michael Atiyah, responded to the messages of greeting (text on p. 58).

4. **Tributes to deceased Pugwashites**. Dr. Martin Kaplan read obituaries of the Pugwashites who had died since the Querétaro Conference (text on pp. 59-60).

5. **Address by the Secretary-General**. Professor George Rathjens addressed the meeting on Pugwash activities during the year (text on pp.61-65).

6. **Discussion.** There was a discussion on the Secretary-General's Address in which the following took part: Prof. Francesco Calogero, Mr. David Fischer, Dr. Peter Gleick, Dr. Jozef Goldblat, Prof. Dzenana Rezakovic, Prof. Balchandra Udgaonkar.

7. **Sir Joseph Rotblat** introduced the main speaker, Dr. Waldo Stumpf, with some general comments (text on p. 66).

8. **Dr. Waldo Stumpf** presented his paper "South Africa's Nuclear Weapons Program: From Deterrence to Dismantlement" (text on pp. 67-74).

9. **Discussion**. In the discussion on Dr. Stumpf's paper the following took part: Sir Michael Atiyah, Dr. Peter Gleick, Dr. Tara Kartha, Air Commadore Jasjit Singh, Mr. Motomisi Tawana.

The session adjourned at 11.30.

WELCOMING ADDRESS FROM PROFESSOR MARIE MULLER

Welcome to the last annual Pugwash Conference of the millennium and the first on African soil for more than thirty years.

The theme of our Conference is "Confronting the Challenges of the 21st Century". This epitomizes the commitment of Pugwash to go forward and meet new challenges as they arise. As we count down the months, weeks, days and eventually the hours to the new century and the new millennium, Pugwashites are already identifying, debating and looking for possible solutions to these changes – some would say problems, problems of the coming era. This is clear from the agenda for this Conference as well as the themes of the smaller Pugwash meetings symposia and workshops being planned for the coming year.

However, Pugwash does not look to the future and its challenges only. Pugwash attempts not to forget the past and its lessons. This is clearly illustrated by the exhibition of photographic panels you see on the walls of this plenary hall. We are very honoured to have this exhibition here and to show it here in the plenary hall of this Conference. The man responsible for acquiring and sending this collection to us, Professor Shoji Sawada, is a survivor of the Hiroshima bomb. He could unfortunately not be here himself for the Conference, however, he is here in spirit conveying to us through this set of panels entitled "Atomic Bombs and Human Being" his urgent message. I urge all of you, but in particular the new Pugwashites, and those who did not have the honour and the privilege of attending the Pugwash Conference in Hiroshima in 1995, to study these panels very carefully. The panels show the actual events as well as the after effects of the dropping of the atomic bombs on Hiroshima and Nagasaki at the end of the Second World War. These panels will stay in South Africa after this Conference and will continue to remind us, in Africa, of the horror that befell the two Japanese cities.

The exhibition of these panels also emphasizes that Pugwash will not abandon its traditional concern, that of nuclear non-proliferation and disarmament. Though we have many problems and face many challenges we are also fortunate in many things in Africa, and one of those is that we are today free of the scourge of nuclear weapons on our continent. However, we remain part of the world where the dangers are still very real and even growing. We will therefore not rest on our laurels and congratulate ourselves on our good fortune and/or our wisdom and pretend that the nuclear weapons problem no longer concerns us. This is why our first speaker today will address the issue of nuclear deterrence and dismantlement from a South African perspective.

The focus of this Conference will be both global and African. We will be deliberating on universal problems as well as the special problems of the developing world. You will have had a chance to look at the complete agenda, at the list of keynote speakers and plenary panellists, as well as the papers to be delivered in the course of the next few days. You know that Pugwash Conferences bring together from around the world scientists, scholars and others concerned with reducing the danger of armed conflict and seeking cooperative solutions for global problems. We meet as individuals rather than as representatives of governments or institutions to exchange views, to explore alternative approaches to arms control and tension reduction, and today also an increasing number of other security related problems in the hope that we may in this way contribute to a better world for all. Though the Pugwash Movement had its origin during the Cold War and played an important role in bridging divides during that era, and was indeed awarded the 1995 Nobel Peace Prize in recognition of this role, Pugwash has also adopted a broader agenda of human security, and it is this that we will be addressing in many of the sessions to follow. As the 1999 Conference will be the first to take place on African soil since the 1965 Conference in Addis Ababa in Ethiopia, it is also intended as a celebration of the important role which we believe the continent will play in the coming millennium. Hosting this Conference is a very special honour that has been awarded to South Africa, but we gladly share this honour with the other African Pugwashites. We are therefore very happy that Her Excellency the Ambassador of Egypt will be joining us a little later this morning to address the topic of the African Renaissance – a dream often elaborated on by our President Mr. Thabo Mbeki but shared by many other Africans.

Though we will today and on Friday - when we expect the Deputy President of South Africa, Mr. Jacob Zuma, and the Minister of Arts, Science, Culture, and Technology, Dr. Ben Ngubane, as well as Mr. Jackie Selebi, the Director General of the South African Department of Foreign Affairs – focus on local African issues, we will also always do so in a broader context. This will be particularly evident at our Final Plenary Session when both African and other panellists will address the difficult issues of Sovereignty and Intervention.

We in Africa need the expertise you from the rest of the world can offer us. However, we believe that we also have a real contribution to make in the common search for a better and more peaceful world. It is precisely because we believe so strongly in the two-way movement of knowledge insight and wisdom, that we value the Pugwash Movement so highly and that the South African group, a rather young and still fairly small group, was so determined to bring Pugwash back to our continent before the turn of this century.

We know that much of the value that will come from this Conference will come from both the formal deliberations and the informal discussions, which already commenced last night. We also hope that those who came from outside our country and our continent will find this a pleasant experience and will return many times hereafter. We extend to you all a hearty welcome to this conference, to our country, and to our continent, Africa.

I would also like to take this opportunity to thank the Swedish Government represented here today by Ulrika Soneson of the Swedish Embassy in Pretoria, for the financial support given to this conference. We are extremely grateful for this and I would request that this be conveyed to the Ambassador Bo Heineback and to the Swedish Government. I would also like to take this opportunity to thank the Swedish Pugwash Group for the role they have played, in particular I would like to thank Carin Atterling Wedar for her tireless work on our behalf. We are very sorry that Carin was, in the end, unable to come to this Conference – a Conference which she has supported and worked for so energetically. We are of course in direct contact with her but would also like to ask the Swedish Pugwashites present to formally thank her from us.

Ladies and Gentlemen, many people contribute to making a conference such as this happen and we have many people to thank for their efforts and support. We will take the opportunities to do so as they arise in the course of the Conference but at this moment I want to thank you all for coming here and wish you a happy time in South Africa.

MESSAGE OF GREETING
From Mr. Kofi Annan
Secretary-General of the United Nations

As you gather for this last Pugwash Conference of the 20th century, your agenda is timely and crucial. Solutions to the challenges facing the world today need the kind of expertise that you can offer. The issues you will be addressing concern us all: freedom from nuclear danger and non-nuclear threats; security; development; the environment; and international governance. They are issues that are on the minds of thoughtful men and women everywhere.

As we prepare to enter the 21st century, the nuclear threat still hangs over the world. Although the major powers have reduced their arsenals, huge numbers of nuclear weapons remain on hair-trigger alert, and no negotiations are under way on a START III. On the closely related matter of delivery vehicles for weapons of mass destruction, I would like to underscore once more the need for multilaterally negotiated norms regarding missile development. As depositary to the Comprehensive-Nuclear-Test Ban Treaty, I will convene next month a meeting of parties to consider what measures should be taken to expedite the treaty's entry into force.

It is fitting that this conference should take place in Africa, a continent which is striving to forge solutions to today's challenges according to its own context and needs. African countries have taken steps to protect themselves from nuclear danger by concluding the Treaty of Pelindaba; they must not ratify it quickly to consolidate the continent's status as a nuclear-weapon-free zone.

As Africa knows all too well, most victims of today's conflicts are the victims of conventional weapons and small arms rather than sophisticated weapons of mass destruction. They are mostly civilians, women and children. Even after emerging from conflict, society faces the risks posed by the flow of small arms and light weapons as a source of continuing violence. As we have seen in countries such as Nicaragua, Mali, Cambodia and most recently Liberia, public displays of destruction of weapons that have been surrendered voluntarily can be of both symbolic and practical value. But destruction of weapons does not happen in a vacuum. It requires an environment conducive to peace, created by the careful implementation of a comprehensive policy of disarmament, demobilization and reintegration of ex-combatants, including child soldiers. And it requires the support of the international community through assistance for development.

Many aspects of human endeavour must converge in the building of durable peace: economic and social development; education; health care; respect for human rights and the environment; and good governance rooted in a profound sense of responsibility for our communal life on this planet. This has never been more important than today. We are about to leave a century engulfed by the whirlwind of globalization. While this has opened new opportunities for many, it has also pushed many others to the margins of society. In confronting poverty and other problems, governments and the international community as a whole are faced with competing priorities and limited resources. Addressing this challenge in the next century will require all our ingenuity and courage.

And it will require an effective and responsive United Nations. Let me say in this context how pleased I am that you will be discussing, for the first time, the reform of the Organization. The "quiet revolution" we have embarked upon was intended to revitalize the Organization and to better position it for the highly complex context of the new era. Today, we can say with some satisfaction that the United Nations family acts with greater unity of purpose and coherence of effort than before.

And we can feel sure that "We, the peoples of the United Nations" will begin to speak with louder and more informed voices in the international arena, and that their views will be heard. Starting with the next year's Millennium Forum at the United Nations, I know I count on the Pugwash conferences on Science and World Affairs to be among them.

And so, on behalf of the United Nations, it gives me great pleasure to send my best wishes and encouragement to the 49th Pugwash Conference and all its participants. I add my gratitude for the productive partnership we have enjoyed over the decades, and the hope that it will prosper further in the next millennium.

MESSAGE OF GREETING
From Mr. Nelson Mandela
Former President of South Africa

It is an honour to extend a very warm welcome to you all on this special occasion. We take great pride in South Africa being chosen as the venue for this important meeting. This will be the first time that this prestigious group will come to our country. It will be a unique opportunity for eminent scientists and scholars from all over the world to meet their South African counterparts, to focus attention on vital challenges facing the continent of Africa on the eve of the new Millennium. We hope our visitors will enjoy their stay and take the opportunity to explore our beautiful country and return home with wonderful memories. I am confident that this Conference will make an important contribution in reaffirming a joint commitment to world peace. I wish you every success in your deliberations and hope that your stay will be a pleasant one.

RESPONSE TO THE MESSAGES OF GREETING

Sir Michael Atiyah

It is a great pleasure for me on your behalf to thank all those who have sent messages of support and encouragement for our meeting here today. First, let me thank Former President Nelson Mandela, without whose remarkable leadership South Africa would not have been such an obvious choice for Pugwash to meet. His emphasis on conciliation and peaceful resolution of conflict is very much in the spirit of Pugwash. It is, as you know, a tradition of Pugwash that anyone attending one of our meetings becomes automatically a Pugwashite. Although Mr. Mandela has not been able to come, we invited him to speak here, he is here in spirit, so I think he should be deemed at least an Honorary Pugwashite.

Let me next thank the Minister of Education for his message of support. Education is one of the most important factors for democratic development and for rational solutions of problems, and we hope that South Africa will be able to offer full educational opportunities to all sections of its population.

We are meeting, as Madam Chairman has said, for the first time in South Africa, and for the first time in many years in the Continent of Africa, and the particular problems of Africa will be one of the areas on which our discussions will focus. We appreciate therefore the messages of support from Dr. Salim, the Secretary-General of the Organization of African Unity. We realize that Africa, as a whole, has to do what it can to deal with the internal problems of the continent. Fortunately, the days when external powers were constantly interfering in Africa are past, and we are now in an area of mutual cooperation and respect.

Finally, I must thank the Secretary-General of the United Nations for his long and detailed message of support. Pugwash has for many years enjoyed a good relationship with the UN since we both share the common objective of a peaceful world. The difference is that in Pugwash we are essentially an academic body of thinkers, and the UN has to take action. This affects the timescale and mode of operation of the two bodies. The UN has to react positively to crises, for example, right now in East Timor, whilst Pugwash has the luxury of thinking at leisure; although of course this does not mean that we are not concerned with crises when they arise, and you might like to know that just now we at the Council of Pugwash sent a message of support to the Secretary-General in his efforts to try to solve the problems in East Timor. Although we are a body of thinkers we are not purely academics engaged in an intellectual exercise. We hope by our thinking and deliberations to influence affairs, particularly by influencing those at the UN and elsewhere who have to take decisions. Conversely, I believe the United Nations, precisely because it is always under constant pressure for immediate actions, appreciates the contributions that bodies like Pugwash can make by addressing problems at a more fundamental level. So we regard ourselves as partners and supporters of the UN and we are encouraged by the interest shown by the Secretary-General. I will be replying fully to the Secretary-General and I will at the same time take the opportunity of extending to him an invitation to come to our next annual meeting in Cambridge, England.

TRIBUTES TO DECEASED PUGWASHITES

Martin Kaplan

Alexander V. Fokin (1913-1998)

The Praesidium of the Russian Academy of Sciences and the Institute of Organoelement Compounds are deeply grieved to announce that Academician Alexander V. Fokin died in Moscow on 4 July 1998 at the age of 85. He was one of the most distinguished Russian organic chemists with an outstanding national and international reputation. He was an active member of Pugwash participating in six annual meetings and symposia. Ill health in recent years prevented him from attending later meetings, but his interest in Pugwash lasted until the end.

Leonard Rieser (1922-1998)

Leonard M. Rieser, a long time friend of Pugwash, who worked on the Manhattan Project and witnessed the first atomic explosion at Alamogordo, died 15 December 1998 in Hanover, New Hampshire. As Chairman of the Board of *The Bulletin of Atomic Scientists* from 1985-1988, Rieser was photographed annually as he reset the hands of the Domesday Clock in the Bulletin's graphic indicator of threat posed by nuclear weapons and a need for new ways of thinking to avert a nuclear catastrophe. Rieser came to several Pugwash meetings including the Annual Conference in Cambridge in 1989, Hiroshima 1995, and Lillehamer 1997, as well as the Pugwash workshop in Budapest held in conjunction with the centenary of Leo Szilard's birth in February 1998. Rieser's interests included not only physics, nuclear arms control, and international securities, but Latin American society and culture, the environment and science education. A member of the faculty at Dartmouth College for 45 years, Rieser served as Dean of the Faculty, Provost, and the first Director of the Dickey Centre for International Understanding. He was also President of the American Association for the Advancement of Science between 1972 and 1975, and a founding member and President of the Interciencia Association, a federation of fifteen science organizations in the Americas. With a delightful combination of dry wit and probing intellect, Leonard Reiser personified the integrity and moral conscience of those nuclear physicists who knew firsthand the dangers posed to humanity by nuclear weapons. He will be sorely missed.

Franco Dupré (1933-1998)

Franco Dupré, a physicist, with a physics department at the University of Rome "La Sapienza", died suddenly on Tuesday evening, 15 December 1998. He had been serving for many years as Secretary of the Italian Pugwash Group. He had attended several Pugwash Conferences and had played important roles in the development in Italy of The International School on Disarmament and Research on Conflicts (ISODARCO) and the USPID the Unione Scienziati per Il Disarmo (the Union of Scientists for Disarmament). He was still serving on the governing bodies of both organizations as well as on that of the association created in 1989 to provide a legal status for International Pugwash in Italy. Born in 1933 he was an active member of the Evangelical Valdensian Church in Italy and a man of conscience. Strongly altruistic, he realized this commitment in many ways including his involvement in Pugwash and his scholarly and proactive interest in scientific education. Although his passing away while conducting an active and happy life without any suffering or protracted illness is the kind of exit many might wish for themselves, it came quite unexpectedly and when he was still relatively young. It is hard to come to terms with it.

Franklin A. Long (1911-1999)

Frank Long died at the age of 88 on 8 February 1999 in California, where he had lived since 1988. He had a long and distinguished career in the field of chemistry, and was a member of the Department of Chemistry at Cornell University for 42 years, before being named Professor Emeritus in 1979. His most notable contributions, however, were to public policy, particularly on matters concerning arms control. He was an active Pugwashite from his first meeting in 1963 in Dubrovnik through to the 1991 Beijing Conference, and he was also a member of the Steering Committee of the US Pugwash Group for over 20 years. In 1962 he was named Assistant Director for Science and Technology of the US Arms Control and Disarmament Agency, and he participated as a scientific consultant in the negotiations leading up to the 1963 Limited Test Ban Treaty. He was member of the President's Science Advisory Committee in 1961, and again from 1964 to 1967, and he also served on several government commissions concerned with science and technology. He was Co-Chairman of the Joint US/ Korea Advisory Committee to the Korean Minister of Science and Technology and a Member of the Indo/US Sub-Commission on Education and Culture, serving as US Co-Chairman from 1977-82. In 1969 he was nominated for the post of Director of the National Science Foundation but his nomination was rejected by President Nixon because of his public position opposing the United States ABM system, then under development. The public controversy erupted over the apparent politicization of what had been considered a non-political post and Nixon eventually offered him the job again. But Long declined. He remained instead at Cornell University where he served as the first Director on the Program on Science Technology and Society, and played a crucial role in establishing the Peace Studies Program. Frank tackled all these assignments with his characteristic energy and common sense. He never flagged in his conviction that political engagement was a duty of all citizens and that individual efforts could make a difference. He was also a person of great good humour whose generosity and loyalty made him a valued friend to many. His wise advice and steady presence will be greatly missed.

Wilhelm Frank (1916-1999)

Professor Wilhelm Frank, born in 1916, died on 14 May 1999. His death is a great loss for the Austrian Pugwash group. He was an enthusiastic Pugwashite with a special interest in energy problems and research policy since the day he joined the organizers of the first World Energy Conference. Energy problems were also at the centre of his duties during his long professional career in different Austrian ministries. Among other leading positions he served as Section Head of the Energy Section in, what was then, the Austrian Ministry of Trade and Industry. After his retirement, he returned to his old love, mathematics, and took up lecturing at the University of Salzburg. He died from a sudden heart attack while waiting for the bus to bring him to the lecture hall.

ADDRESS BY THE SECRETARY-GENERAL

George Rathjens

I can report that we had a very active year in Pugwash, with nine workshops and symposiums. Two resulted in books, which I commend to you.

Putting the other workshops and symposia aside, I would like to comment briefly on the one we had in Castellón Spain in early July on NATO and European Security. I do this because it turns out to be relevant to the main theme of my remarks today, which might perhaps be titled "Reflections on Kosovo". I would note that the Castellón meeting was perhaps the best workshop in my experience in Pugwash in at least the last 20 years, and the report that will appear in the next issue of the Newsletter, and which is available on the web, is probably the best I've ever seen of a Pugwash workshop: more insightful, thorough and longer than we usually get.

Let me refresh your memories of the events relating to Kosovo; then, I want to discuss at some length how Pugwash reacted to them; and then turn to some observations, again going back to the Castellón meeting. Finally, a few remarks about lessons and quandaries that are left over from the Kosovo experience, and about future related Pugwash activities.

Although the international community was concerned about developments in Kosovo much earlier, I go back only to 1998. I do no not think that many of us in Pugwash, myself included, really focused very heavily on what was happening there until that year. Then, especially after the bombing started, it became an issue of great concern. The critical events happened then and in the first half of 1999. Noteworthy were the Security Council resolutions in March 1998, calling for an embargo on arms shipments to Yugoslavia, and one of September calling for a cessation of action by the Serb security forces against the civil population of Kosovo. Then, there were the negotiations of mid-October by Richard Holbrook, which resulted in a partial withdrawal of those forces, to be monitored by an OSCE observer force. Although monitors were deployed, continuing Serb activity led to their quick withdrawal. Then, after the failure of the Rambouillet negotiations, and repeated, but unheeded, threats to Milosevic of bombing in the event of continuation of Serbian ethnic cleansing operations against the Kosovar Albanians, the bombing campaign began on March 24, 1999. It continued for 11 weeks with, as you no doubt recall, repeated statements, particularly from US spokesmen, that NATO would bomb and bomb but would deploy ground forces in Kosovo only for peacekeeping and only after the termination of hostilities.

Now, why such a commitment to the use of airpower alone and, indeed, only from high altitudes? Experience in Somalia and elsewhere had made it pretty clear that on an issue such as Kosovo, any casualties at all would be politically intolerable in the United States. Moreover, there had to be questions about whether all of the other eighteen NATO members would go along with the early introduction of ground forces considering sensitivities of their publics, too, to casualty possibilities and, in the case of at least Greece, about NATO ground forces being sent into Kosovo through Salonika, the only reasonable route for a large scale deployment other than through Hungary (which would have entailed attacking Serbia directly, an option that would almost certainly have had little support in any NATO government).

It is remarkable, in my view, that the attempt was made to coerce Milosevic into surrender by bombing alone and that the alliance held together when the bombing campaign took so long: much longer than the Clinton Administration apparently anticipated, and longer than it no doubt led the other NATO states to believe would be required.

That some in the Administration could profess such optimism as they did about the efficacy of bombing was very likely a consequence, in part, of an erroneous lesson having been drawn from Bosnia. It was widely understood by many, including possibly, but inexcusably, by some in the upper echelons of the Administration that airpower was entirely decisive in bringing conflict there to an end. The reality is that Croatian ground forces, armed and trained largely by the United States to the point where they had superiority over the Serbs, were at least as important as airpower in the realization of the Dayton accords and in the cessation of open hostilities in Bosnia. But the misguided belief, fed by some airpower advocates, coupled with White House perceptions about the political unacceptability of casualties that might result from military operations not widely seen to be essential to American interests, led apparently to the airpower-only approach to Kosovo. I cannot conceive of any general officers - or admirals, for that matter - in whom I have confidence, believing that bombing alone could be decisive in a situation such as obtained there. One of the astonishing things is that none of the people in the higher levels of the American military and foreign policy establishments - none of the generals, admirals or senior civilians in the Defense or State Departments - chose to resign rather than go along with a campaign that seemed so nonsensical from a military point of view. Other American officials have resigned in instances when they disagreed with policy. I think of General Maxwell Taylor's resigning as Chief of Staff of the Army when he disagreed with President Eisenhower's commitment to airpower and nuclear weapons and of Cy Vance's resigning as Secretary of State when he decided he could not go along with the Carter Administration's Iran rescue effort. I'm just surprised that this did not happen in the Kosovo case, because the feelings were very strong that though the objectives might be sound, even laudable, the means chosen made little sense.

You all know that finally after the 11 weeks of bombing, Milosevic did cave in. Did he do so because of the bombing? Certainly, it had some effect, but surely it was not the only reason, perhaps, not the decisive one. There will be speculation about this, and someday we will perhaps have a better understanding than we now do. But it is my impression that NATO use of airpower alone was not the only important factor. By the end of the 11 week period, ground operations by KLA forces in at least one significant instance forced the Serb forces into the open where bombing against them could be effective. Until that happened, as we later learned from post-bombing assessment, the amount of damage done to the Yugoslav military was really minimal. There was a lot of destruction of fixed structures, civil and military, but one of the lessons of all this is that, although precision-guided munitions can destroy fixed targets, against military forces that have some mobility their effectiveness is not likely to be great. There were also, toward the end of the operations, indications that ground forces were perhaps going to be introduced. There were increases in levels of forces that were specified for occupation when hostilities might be terminated, and there was improvement in the transport system from the Adriatic coast that would have been necessary if a large scale intervention by ground forces were to be attempted. Perhaps these developments and evident pressure on the United States, notably from the United Kingdom, might have been viewed by Mr. Milosevic as signals that, notwithstanding earlier remarks, ground forces might be introduced to bring the hostilities to an end. Finally, and I think probably most decisive, but again this remains to be determined, there was the withdrawal of Russian support. Recall Martti Ahtisaari's meetings with Viktor Chernomyrdin and Strobe Talbott, after which Russian policy changed. It is clear from some sources that the Serbs felt betrayed and perhaps that their situation was hopeless.

Well, so much for a brief history except to warn of the possibility of wrong conclusions being drawn about the efficacy of bombing, even as was the case in Bosnia.

Now let me say a bit about how Pugwash reacted. The bombing started on the 24th of March and, although I have not checked my records, I believe I began to be bombarded by e-mail on the 25th from some people asking what was Pugwash going to do about it, and from others urging various courses of action. Should we release a statement condemning the bombing or one trying to get negotiations going again to bring hostilities to an end?

There ensued a rather lengthy and difficult effort on the part of the Executive Committee of Pugwash to try to come up with a consensus statement relating to these questions, but after extended exchanges of email and phone calls, I concluded, as Secretary General, that I could not see consensus in sight. Yet, we felt some obligation to inform the Pugwash constituency of our efforts, including about points of difference among us. So, you will find in the current issue of the Newsletter, the one that came out last April, seven pages at the beginning containing five statements by our four officers and Jo Rotblat, reflecting our different views of what Pugwash should or might say. If you read them, and I recommend you do, you will, I hope, understand why we differed, how we differed, and why it was not possible to get a consensus. You will note too that some of us also differed, as we generally do, on the issue of how important it was to issue a statement, but I do not want to get into that today.

As to what we ought to say, I believe there is a single most significant issue of cleavage, although, as I say, you will find the five statements in the Newsletter and on the web. There was sentiment, obviously very strongly felt, that we ought to call for a return to the negotiating table, which had been vacated after the Rambouillet failure, to try to reach some kind of a settlement, and that we should call for immediate termination of hostilities: in particular, that we ought to ask NATO leaders to stop the bombing, conditional on ethnic cleansing being discontinued. Others, including me - and it is no secret who was on which side of these issues because it is all in the Newsletter - felt very strongly that they could not live with these calls. Why not? In my case, because it implied a total cave-in to Milosevic and the abandonment of the Kosovar Albanians. If my memory is correct, when we were trying to find consensus on a statement, about a half a million Albanians had already been forced out of Kosovo, and I don't know how many more from their homes into the hinterlands within Kosovo. My feeling was that if there were an armistice within, say, the first weeks after the bombing started - and recollect that with its initiation, the expulsion of the Kosovar Albanians accelerated greatly - we, i.e., the world, would be left with of the order of a million people having been forced from their homes and with winter approaching: pretty grim prospects. I couldn't see accepting this. The alternatives of course were pretty grim too: continuing bombing and further ethnic cleansing. I felt, in some ways, that much more vigorous military action could be desirable: that ground forces ought to be employed to conclude the operation in a hurry before more people got expelled from their homes or killed. But this seemed to me to be totally unrealistic, so I had no good recommendation as to what should be done; rather, only the bad one of just continuing with the bombing in the hope, but with grave doubts and no confidence, that the Serbs might decide that continuing the conflict would not be worthwhile. Abandonment of the Kosovars seemed to me to be intolerable, but the conquest of Serbia and the removal of Milosevic by force seemed to me to be politically so unlikely as not something to be advanced seriously.

When I concluded that the Executive Committee was irreconcilably split on the substance of a possible statement, the only alternatives seemed to be to say nothing or to release, as we did, statements reflecting our different views. Any post-mortem on the behaviour of NATO, the behaviour of the Russians, the behaviour of the Serbs and the KLA forces, and the behaviour of Pugwash, will, I suspect, reveal us still in disagreement, even after it is all over, as to what should have been done. These are the realities. It would have been nice to have had consensus but we didn't reach it in this case. We still talk about it. At the meetings of Council in the last two days, the issue has come up again and again. It came up incidentally in connection with whether we should say something about current events in East Timor. We did agree on a simple, and I think non-controversial, statement to be sent to the U.N. Secretary General, an open letter, if you like. But, if events develop - God help us - so that East Timor begins to look like a replay of what we saw in Kosovo, where increasing numbers of people will become refugees, and large numbers will be killed, I expect we could again find ourselves in disagreement about what kinds of military action should be employed, about under whose auspices, and about under what conditions there should be an armistice or negotiations to try to bring hostilities there to an end.

I would like now to return briefly to the report of the Castellón workshop. The decision to hold it was made before the maturation of the crisis in Kosovo, at a time when a second round of NATO expansion was a prospect. The possible adverse affects on relations between Russia and the West of further enlargement seemed worrisome, at least to me. And, I was concerned about questions regarding the future for NATO. Was it an anachronism? Perhaps, a menace? Could, its major role - even its raison d'etre - be in dealing with problems of the kind we have seen in the former Yugoslavia? Shouldn't the big review of NATO's future that was much touted a year or so ago, have been more open-ended?

I got support for having a meeting that would address these kinds of questions but as it turned out, we met one month after the Kosovo bombing started so, as I noted earlier, about half our discussion was Kosovo-related. The last parts of Jeffrey Boutwell's report deal largely with the kinds of problems we faced there and that we might face in future Kosovos. I had expected that within the group we had assembled there would be different views on many questions, participants having been invited partly with the objective of bringing out different perspectives, but there was one surprise. I found less support, though there was some, for the view that NATO had out-lived its utility and should be scrapped.

Besides Jeffrey's report I want to recommend to you one other piece of reading, if I may. It is a very thoughtful piece by Adam Roberts of Oxford University in the current issue of Survival, which I actually got the day I left for this conference. I found it very useful in refreshing my own memory and clarifying my thinking about what had happened in Kosovo.

Now, some remarks about what has come out of all of this. Our failure to reach consensus within the Council led to a suggestion by one of the members that we ought to look at military intervention in defence of human rights from a longer term perspective, and that we ought to take the question pretty seriously because we are not out of the woods yet in Kosovo and because we are likely to see other somewhat similar problems all over the world. We have seen a number of them in the last decade, and we see one now in East Timor. These observations and the discussion in the meeting in Castellón, reinforced my view that we ought now to undertake as a major Pugwash activity at least one workshop, indeed, probably a series of efforts going over at least a couple of years, to look into many of the kinds of questions that were raised by Kosovo and by our inability to deal very effectively with them. Nobody can be happy with the way Kosovo worked out. There is much to criticize in detail, but there are more general questions also raised.

I should say before I finish that I spent time during the last several years looking at a number of other cases of intervention that have involved the use of military force: events in the Congo in the 60s, those incident to the former Yugoslavia's splitting apart, and at Somalia, Cambodia, Rwanda and Haiti. I hate to say this, but when all was said and done, I'm not sure if any of them should be counted real successes if the question at issue is "are the regions better off now than they would be had there been no intervention". Questions have to arise, then, that have to be faced very seriously. Are we just going to tolerate affronts to life and human dignity that now increasingly occur in intra-state conflict? Or, should the world community intervene militarily, recognizing that doing so will often be an affront to sovereignty and that such intervention may often be, at best, only partially successful. And, if we are to intervene, under what conditions?

A last observation about Kosovo. In a gross sense, I'm not sure things were done wrongly. What happened was tragic, but I also think it would have been tragic if there had been no intervention and if ethnic cleansing of the Kosovar Albanians had continued, with Milosevic now being in a stronger position than he is to pursue his objective of a "greater Serbia" beyond Kosovo.

I certainly do not want to defend the bombing, but I do believe that the objectives of stopping the ethnic cleansing of the Kosovar Albanians and of reining in Milosevic were laudable, though the means of military intervention (bombing, mostly of targets in Serbia, with inadequate concern about damage to innocent Serbs and to civil infrastructure and with disproportionate concern about possible NATO military casualties) and the failure to plan realistically for necessary post-conflict involvement in Kosovo was execrable. But, it is not clear that effective intervention could have been carried out in any other way for the reasons I mentioned earlier: little support in the United States for the employment of ground forces and very deep - and understandable - aversion to casualties, not to mention there being no prospect of getting enthusiastic support from all of the other countries within NATO for a quick, decisive defeat of the Serbs, much less of getting support, or even acquiescence, of the Chinese and Russians that would have made it possible to carry out intervention under UN auspices. So, while Kosovo was very badly handled in a tactical sense, maybe that was the only way it could have been handled at all: a very tough choice.

I would hope that Pugwash's getting involved systematically in the questions of intervention, military and otherwise, in defence of human rights - and in the problems of tension between such intervention and respect for sovereignty - might permit us to contribute to the world's being confronted with less stark and undesirable choices. It would be unconscionable were we not to try.

Now, I should say we are going to have a panel where we'll have some opportunity later on to discuss questions of intervention further and we are also going to have an opportunity tonight, though it is not in the program, for some discussion of East Timor, in particular. We decided on this yesterday in the belief that, it being a current issue of very great concern - it ought to be of great concern to everybody - we ought to have an opportunity for people to sound off. So, we are changing the schedule for this evening somewhat to accommodate this.

Introduction of Dr. Waldo Stumpf by Sir Joseph Rotblat

Fellow Pugwashites, Friends, I hope I am not abusing my privilege as Chairman if I start with a few general comments.

I am thrilled to be having my second visit to this country. The first visit was a long time ago, 34 years ago, when I attended the Fifteenth Pugwash Conference which was held in Addis Ababa. That was still the time of the Haile Selassie regime, the last decade of his reign, and the Emperor himself opened the Conference and invited us to his palace for the reception. He was an absolute ruler and I was very disconcerted when I saw senior military people, old people, prostrating themselves flat on the floor and kissing his feet. From Addis Ababa I flew to Johannesburg and from there I went to Pretoria to try to establish some Pugwash contacts with South African scientists. There I was not only disconcerted, I was literally nauseated when I saw the evidence of apartheid in practice, the severe limitations on people, people of this country not being allowed to go into parks or to travel on certain trains. But the worst thing was a conversation I had with a senior person, a scientist in medical disciplines, who tried to justify apartheid on scientific authority - I should say scientific arrogance – by stating that black people were inherently inferior to white people, and that they were incapable of being educated beyond elementary school level. I felt despair, utter despair, that a time would come when these people would be free, if such views were held by the authorities in this country.

And yet, the unbelievable did happen. The people here are free. They are masters of their own destiny, if not yet affluent. They have shown that the colour of the skin is only skin deep. They exhibit the same mixtures of wisdom and foolishness, of making progress and making mistakes, as we do in the North, but showing perhaps more zest for life than we do in the North. This is why I say that I am thrilled to be here. But there is something more to this, in particular for me. I am often accused of lacking in realism, of being too optimistic about achieving the objectives that I set for myself and which I preach about all the time: elimination of nuclear weapons, elimination of all war. And when I look at what is happening in the world I sometimes begin to doubt. Maybe I am wrong? Maybe I live in cuckoo-land? Maybe my objectives are inherently unachievable? But when I see what is happening here, it gives me renewed faith, renewed courage to carry on with my efforts. I hope, Friends, that the same will apply to all of you; that the visit to this country will give you that extra impulse to continue to persevere, because I believe that if we persevere we shall achieve our objectives.

One example which this country has shown the rest of the world, is in fact in the field about which we are very much concerned, namely, nuclear weapons. There were a number of countries which at one time or another were seriously considering the acquisition of nuclear weapons but decided against it. But there is only one country which actually acquired nuclear arsenals and then made the decision to destroy these weapons. This is a very important event, and we are greatly privileged indeed that the keynote speaker today is the man who has been very much involved in the whole operation.

Dr. Waldo Stumpf is a Professor of Metallurgical Engineering at the University of Pretoria – he's one of us – but he is also, in his main job, the Chief Executive Officer of the Atomic Energy Corporation of South Africa. He is going to talk to us on this transition: from deterrence to dismantlement.

SOUTH AFRICA'S NUCLEAR WEAPONS PROGRAMME: FROM DETERRENCE TO DISMANTLEMENT

Waldo Stumpf

The purpose of my talk today is not just to repeat chronologically the events of the South African situation, but rather that people like you and I can learn the lessons. What are the forces that shape leaders of our world to take decisions, even if they are wrong decisions? If you and I understand those forces, maybe in a small way, we can contribute to shaping those forces in a better way.

As always one has to give one's own background, the rules of the game. I have to tell you this story from the viewpoint of the former South African Government. It would make no sense if I adopted another viewpoint. However, this does not mean that I endorse the decisions of the past. I have been critical of them. I share your views in terms of the role of nuclear weapons, but I must give you the story from their point of view. And I have found in the past that sometimes this creates some difficulties with some people when I mention certain countries and so on. I am not here this morning to judge the decisions of the past or to defend them. I am just a technical person who, through a bit of a twist of history, was put in a position to play a meaningful role here. I will leave the judgment to historians and to political analysts. So, please bear with me from that point of view.

On 24 March 1993 the former State President of South Africa, Mr. F.W. De Klerk, informed the South African Parliament and obviously the world that, yes, South Africa had had a nuclear deterrent, a limited one, and a small one, but this had been dismantled before signing the NPT and signing the Safeguards Agreement. I think with this admission South Africa put to rest a lot of world speculation, that was partly right, that had firmly believed that South Africa had built up a small deterrent. The public invitation that Mr. De Klerk gave to the IAEA on that rather historic occasion for full access to all details of the past programme, I think, was an indication of South Africa's point of view, even today with our new democratic elected government, of being a meaningful and responsible member of the international community that plays a meaningful role in turning the world away from nuclear weapons. The policy of full transparency has been maintained - even today the IAEA has in South Africa full access: they can go anywhere, anytime. That was a political decision taken by the former government and maintained by the present government, and is the only way one can really build up a credible position of having dismantled such a full programme. We can see what is happening in countries that have not adopted this political point of view, like Iraq; the difficulty of ascertaining whether there has been full dismantlement of the deterrent.

The story starts in 1948 when the Atomic Energy Board was established by Act of Parliament. There were no real activities in South Africa at that point but, very importantly, South Africa started supplying uranium to the nuclear superpowers of the world, so it is probably correct to say that some of the early nuclear weapons of the superpowers have contained, and maybe still contain, some South African origin material. Activities started in the early 1960s when the Atomic Research Establishment was built, about 1½ hours drive from here, very much with a peaceful focus. However, with South Africa being a very significant uranium producer, it is almost natural that the focus started turning towards enrichment, although for peaceful purposes. I will come back to the point when the programme started swinging over to a military purpose. But this was very much a peaceful programme. It did make sense to work in the area of uranium enrichment to add value to South Africa's exports.

First laboratory results were in the 1960s. Then a separate corporation was established in 1970, the so-called Uranium Enrichment Corporation, to undertake the enrichment task and develop this on an industrial scale. And this was no secret. On 20 July 1970, the then Prime Minister B.J. Voerster, announced in parliament that South African scientists had developed enrichment technology, and he declared at that point South Africa's willingness to accept international safeguards subject to certain conditions, namely, protection of the technology that was used here. Of course, looking back, and I will come to this point a little bit later, international safeguards on South Africa at that point in time, without a fundamental political transformation in the country, would also have been not really worthwhile. And the South African government understood that very well. The two really go hand in hand here.

Key to this whole programme was the Pelindaba site. This housed the research reactor that was supplied by the USA in the early 1960s and was under IAEA safeguards from the start, and obviously never played any role in the nuclear weapons programme. Today, it is used on a very large scale for production of medical isotopes for the world and we have about 15 per cent of the world market.

The so-called Y Plant was the pilot plant in which the highly-enriched uranium (HEU) was made. It is obviously a key component in the whole history and therefore I will briefly point out some of the key dates. Construction commenced in 1971 and the first portion of the first cascade was commissioned in 1974; the full cascade was commissioned in 1977, and in January 1978 the first highly-enriched UF_6 (uranium hexaflouride) was withdrawn. It was not of very high enrichment, about 80 per cent, although that is fairly high but certainly not of the 90 per cent plus that is required for nuclear weapons. The reason for this rather lengthy timescale is that in such a plant it takes a long time to set up what is called the enrichment gradient across the cascade. You do not have it from Day 1 onwards: you have to build it up over a long drawn out process.

Then in August 1979 a catastrophe happened in the plant. The technique used then made use of a mixture of hydrogen gas and UF_6 for certain technical reasons, and in certain circumstances these two can react chemically; this did happen and all the UF_6 in the plant plated out, as we call it, went back into a solid product and the plant had to be abruptly stopped. It took the engineers more than a year to clean out the plant, replace some items and get it started up again. This is fairly key and we are coming back to that date a little bit later, but it was only July 1981 when the plant really commenced making HEU on a meaningful scale. In 1986 the plant was even switched back to make low enriched uranium for the four lead test assemblies for Koeberg, and then for about a year it was switched back to making HEU, and in February 1990 the plant was shut down after a decision by the former government. It is being dismantled at the moment.

Looking back at the history of the Y plant, which operated very smoothly apart from the one catastrophe we had there, when it was constructed, its purpose was not for nuclear deterrent work – that only came later – its purpose was to test the technology; to make fuel for the SAFARI research reactor which does require HEU; and to act as a pilot facility for maybe building a bigger plant. The plant certainly provided enough 45 per cent enriched material for the SAFARI Reactor from about 1981 onwards. It also served as a technology test bed for the later construction of a much larger plant.

Where did the nuclear deterrent arise from? As happened probably also in some other countries, it really arose out of the former US Programme on Peaceful Nuclear Explosives (PNEs), which, as you know, in the early 1960s even the USA promoted. With hindsight, of course, one can argue that maybe that was not a good thing to do. But nevertheless, South Africa was a big mining country and nuclear explosives were then promoted as possibly of great use in mining programmes. So it even made sense for South Africa in 1971 to embark on a very modest investigation. Basically, it was only a literature study at that point in time, it was openly admitted that South Africa was looking into this. And in 1974 a report was sent to government where they approved a very limited PNE programme at that point in time. Obviously, if you want to do that you may have to test the explosives, so a site was developed for underground tests where two shafts were dug.

By August 1977 the focus of the world had shifted very much against peaceful nuclear explosives, after India exploded its first so-called peaceful explosive device in 1974, and the test site was abandoned. There is a bit of history around that. At the time, in August, the South African government was subject to extremely great pressure from the two big superpowers, the former Soviet Union and the USA. Supposedly the sites had been detected by satellite. It was announced that Russian satellites had picked up the site. That may be so but let me give you some further details. I spoke to some of the people who were in charge of the site developing those two shafts. They said, yes, maybe the satellite picked something up, but that was not the real point. He said one afternoon, just as they had finished their work there, they were actually camping in a caravan and sitting and having a beer, when low over the horizon an unmarked plane appeared – a Cessna plane. It flew low-down over the site, turned to the right and disappeared into Botswana. It has never been established from where that plane came and I think a lot of suggestions have been made. No such flight had been lodged with the civil aviation authorities. And we believe that the real purpose of the flight was to find out, at low level, what was going on at this site. Nevertheless, the South African government at that point did respond to international pressure and the head of government instructed the team to abandon the site forthwith and in about two days they had to abandon the site totally. The site was never revisited until late in the 1980s, about 1987, but I will come back to that point again.

Obviously, the South African deterrent must be viewed against the political forces within South Africa; the domestic policies one can quite rightly argue were totally to blame for this whole situation. That would certainly be a correct conclusion. But there were also some international politics around South Africa which made the government at that time somewhat nervous. Portugal's withdrawal from Mozambique and Angola certainly was a factor which influenced the government's decision at that point. The Warsaw Pact countries had a declared expansionist policy in Southern Africa and that certainly produced some fear, but of overriding importance was the 50,000 Cuban forces which were building up in Angola from about 1975 onwards. People have very often asked me, when did the focus go over from a peaceful programme to a military deterrent programme? There is not a fixed date because keep in mind that the first HEU only came available in 1979, so the question is a bit academic, but I would say that it is about 1974-75, because with the IAEA we had to dig into all the cabinet meetings; we had insight into cabinet minutes, and speaking with and interviewing all the people it appeared to us that, around about that time, it was when these Cuban forces started appearing in Angola, when the thinking started to switch within government circles that maybe there should be a deterrent programme. But as I said earlier, it really became a *de facto* deterrent programme only when the HEU became available in 1979. Obviously, the total conventional arms embargo, because of domestic policy, also coloured the decisions of the government.

But I think more important is the so-called nuclear isolation of South Africa. In 1976, keep in mind that SAFARI, which is the isotope producing reactor, was under IAEA safeguards all the time and is obviously today even. South Africa had a long-standing contract with the US government on supplying fuel to the SAFARI and there was a consignment of fuel which already had been manufactured in the USA and had been paid for by the authorities and was thereafter held back by the US government. Now the view of the South African government was: look, if the USA does not want to supply us with their fuel, that is their decision. What really irked them at that point was that the US government even kept the money – didn't want to repay the money – and this was really seen as bullying tactics by a big superpower. But there was not much the South African government could do except tell the engineers and scientists to start preparing themselves for making their own fuel for SAFARI. All right, South Africa lost its designated seat on the board of governors at the IAEA, and Egypt assumed that seat, and I must say that Egypt has played a very prominent and positive role on behalf of Africa all these years.

In 1978 another key event happened. The US Congress adopted the so-called Nuclear Non-Proliferation Act (NNPA) and applied it retroactively and that was to basically avoid any contractual damage claims here on keeping SAFARI fuel. But another interesting thing happened then. ESKOM, our electricity utility had already contracted with FRAMATOME of France to build a civilian nuclear power station, also under IAEA safeguards. ESKOM had bought their uranium on the world market, probably South African uranium, I would think. They had had it transformed to UF_6 and had it shipped to the USDOE in the USA for enrichment. The USDOE started enrichment material and when ESKOM applied for an export permit to export this material to France for fabrication the US government put an embargo on that also. Interestingly enough what happened then was that ESKOM was held liable for the enrichment costs used by the DOE. This was a very similar situation to that of SAFARI. This was ESKOM's material sitting in the USA and it was therefore held basically hostage by the US government and ESKOM had to pay enormous costs for its enrichment and were not allowed to sell it on the world market and not allowed to take it out of the USA. These two events were viewed politically very negatively by the government at that time. I think we can understand where the decision by the US government came from. If we want to learn a lesson from this, we should also understand how a government here would react to those sort of pressures which were perceived to be somewhat bullyish – bullying by a big superpower. In the end, when the Carter administration changed to the Reagan administration, President Reagan relented somewhat and he paid the South African government back for that money and a few years later he allowed ESKOM to sell that material, not for use in South Africa, but they were allowed to sell it to an NPT country, and ESKOM had to absorb a considerable loss from this material.

Now, all these events convinced the South African government at the time that accession to the NPT would bring no worthwhile benefit unless the domestic political policies would also change. This was really understood well by that government, and therefore, although there were many discussions with particularly the big powers on partial accession to the NPT, it was never seriously considered because it was seen that it would not bring benefits unless the politics also transformed fundamentally. I think that was a correct conclusion.

Let us talk about the nuclear deterrent. Where did it come from? As I said before, there was no real magic date, no cabinet decision, but looking back from 1975 onwards, when the Cubans started building up in Angola, that was where the thoughts started going over to: should we not have a nuclear deterrent? As I said it was a bit academic, of course, because the enrichment material only became available around about then. But there was a first full-scale dummy device completed, but obviously without enriched uranium. Then the South African government adopted a so-called three-phase strategy. There have been many arguments about this strategy, about whether it was a workable strategy. I am not a political analyst and I do not want to comment on that. The strategy was: Phase One: it would maintain uncertainty. It would not acknowledge or deny the existence of a programme in South Africa. Phase Two: if South Africa's territory were being threatened by the Cubans in Angola, they would acknowledge the capability covertly to the big superpowers in order to convince them that something should be done to make the situation in South Africa more manageable. Phase Three: if that didn't help they would have to then demonstrate this by an underground test, in other words then announce to the world that, yes, we have got this capability. Of course, it never proceeded beyond Phase One, and then I think we are all very thankful for that, that they didn't have to proceed much further. People often ask me also wasn't there an unofficial Phase Four? I cannot vouch for what was in people's minds, but there was no official Phase Four. There was no means to deliver it. Obviously, yes, one could have pushed this out of the door of a big aircraft, like it happened in Hiroshima, but there was no official Phase Four. It was stated very very clearly, and I think that was understood by the government at the time, if there were ever a Phase Four the world would have flattened South Africa. There would have been such massive retaliation. So that at all the meetings it was underlined that there was not a Phase Four; that these weapons would never be used tactically in real life, and they were only to provide a political pressure point to somehow coerce the international community to do something within the Southern African region.

In 1979 the programme was assigned to ARMSCOR. My own organization was only a support then, and later in 1979 the first device was fitted with HEU. This is a key date because very often when I give these talks people ask me about that famous flash over the South Atlantic on 22 September 1979. South Africa was certainly not involved there; ours were very crude devices. The first HEU was not even very high enrichment - about 85 per cent. Secondly, South Africa had a measuring station near Cape Town, and actually my own organization managed to measure fall-out from the atmospheric tests in the 1960s, and that station is even today still in operation and during the atmospheric testing times that station could pick up fall-out in Cape Town within two weeks after a test in the Northern hemisphere. After the so-called flash in the South Atlantic Ocean no fall-out was measured at any time; so either it was no nuclear weapon, or it was a so-called clean device. To attribute to South Africa, with its severely limited capability, credit for designing a "clean" device, using HEU of relatively low enrichment, is, I think, beyond the bounds of reasonableness. South Africa was not involved there – one may have one's own ideas as to who was involved.

In 1982 the first device was completed in the new ARMSCOR facility and in 1985 the programme limits were reconfirmed. This was a key event, the first event in turning this programme around. A lot has been written and said about the former State President Mr. P.W. Botha, and I do not want to go into that debate here, but I think we should give him one point of credit. He saw that South Africa must cross the Rubicon, although he could not make that crossing himself. He realized from the early 1980s onwards that this Rubicon will have to be crossed on the political front. And, therefore, in September 1985 he called together a meeting of a few cabinet ministers in which they reconfirmed the limits of this programme, because there were signs that the engineers had thought of going into further advanced designs, maybe even neutron initiators, going into implosion devices and going away from the old so-called gun-type devices. He put a stop to all that because he also realized, well in advance, that this programme at some point would become a liability and could even impede a political transformation.

The question then arises, at that point, as to what happens to all the nuclear scientists and engineers who were working on this programme. He did allow them, therefore, to carry on with paper studies on advanced concepts. But the writing was on the wall and we could see scientists leaving that programme from about that date onwards because they realized that that programme was not going anywhere and probably in a few years' time would be terminated. However, I think we should understand that the three-phase strategy was still once again confirmed at that point. The reason for that is because in 1987 the defence force revisited the Kalahari test site and they put a shed over one of the test shafts. This was picked up by the cameras. The reason for that is that the three-phase strategy was still very much alive at that point in time. If things were really to deteriorate very rapidly a demonstration of this capability was still certainly on the cards. In the end only 6 devices were produced, instead of the 7 which were planned. Now the question was being asked where does the figure 7 come from? I am not a weapons expert, not even conventionally, but they tell me that if you design a weapon of any design without being able to test it you have to build in redundancy. Normally, what they do is to build double the figure just to allow for uncertainties if you cannot test something. So the argument went as follows: if you have to test this capability you actually have to use one of the devices, you need at least one. Now if you have demonstrated you at least need a second one, otherwise you have no back-up credibility, so at least two, but it may be better to have at least 3 in case the first test failed. And that's where the 3 came from and then you have to double it up because of the uncertainty. That is where the 6 to 7 came from. A bit of a tortuous argument but that is the way the thinking went at that point.

The events leading up to its turnaround are also very relevant. The tension in South Africa started relaxing towards the end of the 1980s. The ceasefire on the 1 August 1988 on the northern border of Namibia certainly was one of the main contributing factors. There was a tripartite agreement between the South African government, Angola and Cuba on 22 December of that year leading to the planned withdrawal of these Cubans from Angola. This was the key event that really also showed us that the deterrent would not be necessary; in fact, it would become a liability. Moreever, on 1 April 1989, the UN resolution 435 of 1978 was implemented leading to Namibia's independence, and the fall of the Berlin Wall and the end of the Cold War, and therefore superpower rivalry in Africa also disappeared. Also, South Africa had a change in head of government when President F.W. De Klerk took over from the former State President in September 1989, and he immediately started domestic political reform.

I was just a technical person but I was drawn into this entire thing almost by accident. What had happened is that I had been assigned in September 1989 to take over as Head of the Atomic Energy Corporation. I was assigned to take over on 1 January 1990, but by September I had already been assigned. So I got a phone call that said I must attend a meeting with the new State President Mr. F. W. De Klerk in his office. I did not know why I should be there; when I arrived and there were about three or four other cabinet ministers and that is where he made the announcement; that was only two weeks after he had assumed office. He said: "Gentlemen, in my term of office I am going to turn this country around to be a respected member of the international community once more. So that means two things: we have to turn around the political system totally, and we have to dismantle the nuclear deterrent." That was why I was there. So although he only announced the political reforms somewhat later, in early 1990, immediately from his assumption of office this was his planning, and we were told then to go away and draw up a programme of dismantlement and present it to him. That was done in November of that year.

The decision was taken, as I said, early in September 1989; that was the date when this programme became a liability to the political transformation process and therefore had to be turned around. In September of that year we submitted the first project plan. I was put in charge of the dismantlement programme and they assigned a steering committee of my own organization, ARMSCOR, and the South African defence force to oversee this whole programme. He also appointed an independent auditor to oversee the whole process, which we actually proposed to him because we knew there would be many questions asked later on as to whether everything had been dismantled. The Y plant ceased operating 1 February 1990 and dismantlement started basically immediately on the high enrichment end of the cascade. The written instructions were received from Mr. De Klerk on 26 February 1990. The instructions basically said dismantle all six devices and the seventh uncompleted device, re-cast all the HEU and return it to the AEC for safe storage, decontaminate all the ARMSCOR facilities and convert them to commercial or conventional weapon use, destroy all hardware components and technical designs, and then advise the government on accession to the NPT.

I may say something about that a bit later but let me just add here, that the NPT is not a friendly treaty for somebody who deproliferates and then accedes to the treaty – it is not made for that - and it poses a lot of questions looking at legal interpretations of how do you accede to the NPT. A lot of questions have been asked: should not South Africa have acceded first and then dismantled? Technically, no: you were in violation of the treaty the moment you join it. Second, the question was asked could not the Committee have said this was a special case and let the IAEA then dismantle this programme. Talking to the IAEA people they were not convinced that they wanted such a project because it would expose them to enormous international pressures. So on the whole, we do still believe that the right way to do it was to dismantle this programme in secret and then accede to the NPT after its dismantlement. The question has also been asked why did not the government at that time announce this programme at the point where they acceded to the NPT. That is a valid question and I can see that would have been the opportune moment to do it. There were, however, two factors that were against that. One of them was that they acceded to the NPT at the worst possible moment when the Iraqi situation was at its height and we could see on the TV the people in blue caps going in blowing up bunkers and whatever. The South African government were maybe quite rightly afraid that they would be branded as another Iraq although technically, of course, they did not break the NPT like Iraq did. But the news media and the world could not have made that distinction and for that reason it was decided not to announce the programme. But there were also some internal fears from Mr. De Klerk that the internal transformation process was not far enough progressed to absorb this piece of news. So we had to do it in secret. It took us about six months to draw up all the documents because you obviously have to do it very carefully. In July 1990 the first device was dismantled and a year later the last device was dismantled. Now, we pushed very hard for these dates so that we could sign a Safeguards Agreement at the Annual General Conference of the IAEA. Thus on 10 July 1991 South Africa signed the NPT. I advised strongly that we should go in as a full NPT member and adopt the INFCIRC 153 type of safeguards. This advice was adopted and I think it was the right decision. The last HEU was returned to the AEC on the nights of the 5/6 September 1991; it was stored there and obviously is under safeguards now.

As I said, South Africa acceded to the NPT on 10 July. From the moment you accede you have a number of years before you sign the Safeguards Agreement but you have got 18 months to complete your full initial inventory of nuclear materials, and that was a formidable task, and my safeguards manager at the time came to me and he pleaded with me that we should use the 18 months. And I said to him: "Sorry Nick. We don't have 18 months politically. You are going to have to do it in a few weeks' time." He had long tears to cry, but he went away; we just pushed him a little bit harder and on 16 September 1991, only 7 weeks after the date, we signed a Safeguards Agreement on the morning of the opening of the Annual General Conference in Vienna. Behind the scenes there were frantic efforts then to complete this initial inventory because the safeguards agreement says that by the end of the following month after you have signed the safeguards agreement you have to submit your inventory. On 15 October 1991 we did submit our initial inventory to the IAEA and I must say this was a very difficult task because we were sitting with lots of grades of materials, lots of different enrichments, different purities, different chemical forms, these all had to be inventoried in great detail.

But let me say that in the history of the IAEA even up until today our inventory stands as probably one of the most comprehensive and complete inventories of all the member countries, and that was done in a very short time period.

In November 1991 the first verification team from the IAEA arrived in South Africa and we stated publicly the policy of full transparency of visit anytime, anywhere, anyplace. Now let me say a few things about this phase. This was not an easy phase even for the IAEA to handle. Keep in mind, at that point we had not publicly acknowledged the existence of a past programme, and the IAEA had no real legal right to delve into the past. The NPT looks only forward from the date of accession. But I must say that I had the authority from the State President at that point that if they were to ask me outright I would have said yes there was such a programme and then we could have revealed it to them. They never asked us and, in retrospect, some months later we sort of laughed around this; and they said to me that maybe they should have asked me and I said yes maybe you should have because I had the authority to say yes. They were afraid to ask because they had no legal right. So between us and the IAEA it was very difficult: yes, all the evidence was there, all this material, HEU, was under safeguards now, but why so much material and so on - all these questions were asked. It also made it somewhat necessary but also more difficult by the General Conference calling for a so-called completeness evaluation of the South African initial declaration. And the word "completeness" actually is inherently associated with looking back. I mean how can you ascertain completeness if you do not look back and delve into the past. I am saying all this because after the Iraqi situation where they had the same very unfortunate learning experience, in the South African situation this was fully realized and therefore the IAEA has now adopted the so-called Strengthened Safeguards regime which was based on the experience they gained in South Africa. This will happen in future again, if some of the current countries that are non-nuclear weapon states accede to the NPT; but it will certainly also happen one day, and I am not sure when that will be, when even nuclear weapon states have dismantled their last nuclear weapon.

A revised Nuclear Energy Act was passed in parliament in 1993 to embody obligations brought about by South Africa's accession to the NPT and signature of a safeguards agreement. We have signed the National Convention for the Protection of Nuclear Material. I think you all know we proposed an indefinite extension to the NPT and maybe played a small part in getting this through. Then there is the so-called Pelindaba Treaty, which has not been fully ratified yet, but certainly is on its way, and we played a very constructive role in getting that treaty off the ground - and maybe the recognition of the role we played is that it is now called the Pelindaba Treaty. Pelindaba is the local name for my facility and, very interestingly, in the local language it means "we have finished talking". I think that is very fitting because this treaty took many years of talking and it was concluded on our site and I think that is why it is also called the Pelindaba Treaty. As I said earlier on all this is now very interesting, but the real issue I want to raise is: what are the lessons that we can learn from this, because otherwise it just remains an interesting story. But as you know I cannot learn these lessons so I would like to summarise them very quickly.

It has been very often thought that the technology of uranium enrichment is so difficult that this would be a natural self-limiting element to any would-be proliferator. This is not necessarily so, although you do need some pretty advanced engineering expertise. I think this expertise is available in most middle-ranking countries and I do not believe that technology as such is necessarily a self-limiting criterion here. Secondly, people very often look at the vast Iraqi programme - and there has been talk that they spent vast sums of money – and they think that that would be a self-limiting factor. The South African experience proved otherwise. Over the total ten years of the Y Plant operation plus the ARMSCOR group, there were not many people, about 200 people at any time were working on the Y Plant and at the ARMSCOR designing and building these things. The total cost of the ten years was less than 680 million Rand. This is a lot of money, I must concede that, but it is just a fraction of the supposed cost of the Iraqi programme. If you have a well-focused programme, well-managed, and with fairly limited technical objectives, it certainly does not have to cost as much as the Iraqi programme cost. Thirdly, while international political isolation certainly is an accepted instrument to try to coerce a country, and in South Africa's case obviously the political domestic policies had a very large role to play there, a point is reached in such a campaign where a would-be proliferator is maybe pushed over the edge towards real proliferation. If you were to ask me when was that point reached, I believe it was at the point when the USA cut off the Koeberg fuel elements and the SAFARI fuel elements. This was viewed very negatively by the South African government and convinced them that they would have to establish there own enrichment technology and the sad thing is that the USA lost its leverage at that point. Now I concede that it is easy to be clever after the fact. If the USA had agreed to continue supplying, these were all safeguarded facilities, since the material would be under safeguards, at least they would have some leverage. They had lost their leverage and South Africa went ahead on its own.

Where proliferation, and this is the fourth point, has occurred due to a real or perceived political threat (I am not saying the Cubans were a real threat but they were certainly perceived to be a threat), a reversal towards demilitarization may occur only upon removal of neutralization of this threat. Without the Cubans having returned back to Cuba, I do not believe the decision by the South African government would have been so easy to dismantle this deterrent. This means that international pressure by a superpower certainly has a role to play, but in the end you have to defuse regional tensions. I believe that just as that was the case in South Africa, the same applies to the Middle East, to South Asia and to the Korean Peninsula.

Fifth point. The reversal from a position of nuclear proliferation to a true and permanent state of non-proliferation within the NPT will not be achieved by a technical decision; it will not be achieved by a military decision or a strategic decision; but it requires a fundamental political decision by the head of that particular government at the time. I believe this strongly and this was borne out by the South African case, where it required a political decision to turn everything around.

Sixth point. The rollback option for a so-called threshold state is not an easy process and it is fraught with many difficulties. Of course, as I said earlier, the NPT instruments are not made for handling such a case, so this is something we should keep in mind as the world slowly but hopefully surely progresses towards a nuclear-weapon-free world. South Africa experienced a lot of unnecessary pressure during the so-called completeness investigation and under different circumstances it could have even derailed the process. In South Africa's case certainly not. It was never in danger of derailing the process but certainly it was unnecessary pressure which under other circumstances could derail the process.

Seventh point. For a threshold state that has taken the decision to rollback and then to achieve international credibility and acceptance as an NPT member is not an easy process, as I said. This process can only be eased by the country itself if they take a political decision of full transparency. It serves no purpose to play games like Iraq did with the IAEA. The only way to do this is to open up and be fully transparent.

Finally, I do believe that valuable lessons were learned between South Africa and the IAEA in the so-called completeness investigation and these certainly have already been built into some of the new strengthened safeguards, the additional protocol, but they can also play a role, hopefully some time in the future, when there will be a cut-off convention for weapons-grade material. I think the lessons learned in South Africa certainly can play a big role there.

This concludes my talk. South Africa certainly occupies a somewhat unique position in world history as being so far the only country with nuclear weapons that has unilaterally swung this around and acceded to the NPT. Public debate over the merits or the demerits of this programme will certainly continue for many years. I am not a historian, I do not really want to get into that debate, but I do believe that this is a chapter of South African history that we should close now. We should open it from time to time like this morning just to re-examine it, to see what can we learn from it, but South Africa has embarked on a new future as we heard this morning too. We are very glad that we can play a more meaningful role in the world of non-proliferation and we are looking forward to the Pelindaba Treaty now coming fully into force so that we can, on the African continent at least, have nuclear technology playing its peaceful role as it was meant to be.

PROCEEDINGS OF THE SECOND PLENARY SESSION

held on Wednesday, 8 September 1999 at 12.00

Chair: Sir Michael Atiyah

Security in Africa

1. **Ambassador Moushira Mahmoud Khattab** addressed the meeting on the subject of "The African Renaissance" (text on pp. 77-81).

2. **Dr. Jakkie Cilliers** addressed the meeting on the subject "The Security Context in Southern Africa and Beyond" (text on pp. 82-86).

3. **Discussion**. The following took part in the discussion: Amb. Omran El-Shafei, Ms. Barbara Isaacman, Prof. Michiji Konuma, Lt.Col. Jacob Kamenju, Mr. André Mechelynck, Ms. Claudia Mutschler, Mr. Hussein Solomon, Mr. Motomisi Tawana.

The session closed at 14.00.

THE AFRICAN RENAISSANCE

Moushira Mahmoud Khattab

Everyone is talking about African Renaissance. So what is happening now? What are we doing to achieve this African Renaissance? In the minds of the South Africans the notion of African Renaissance has been a link to the name of President Thabo Mbeki. He is the one who is really carrying this slogan and trying to invite enthusiasm from all Africans across the continent, from north to south, to gather around this African Renaissance. And we believe in Egypt that the emergence of the new South Africa is an act of African renaissance itself. The fact that we have the new South Africa invite us to have confidence in ourselves, and as Africans. Also the leadership of President Nelson Mandela will always remain as a great inspiration for all Africans as he is an inspiration for the world at large. With the new president at the helm, President Mbeki, we expect the African Renaissance notion to take new dimensions and to be an integral part of the foreign policy of the new South Africa.

The first question that I wish to raise is, where do we come from? If we are going to assess the notion of African renaissance, where do we come from? We have to remind ourselves of the happenings that have taken place in this century which is about to elapse. The twentieth century has witnessed an era in which the African initiative was impeded for a long time. Firstly by the yoke of colonialism, followed by the constraints and rivalry of the Cold War. The ramifications of this in Africa's economic, political, and social development lingers with us up to this day. We can see it, or it is manifested, in a number of aspects such as low economic growth, continuing civil strife, degrading environments, weak institutions and imperfect markets, high population growth with poor health conditions, HIV to mention one, and a mountain of debts - but some of the structural obstacles to Africa's development. The end of the Cold War and the changes that swept the world has not given Africa the peace and prosperity it had the right to expect. To make matters worse, we witnessed a wave of civil strife in our continent. To stop here would be to blame others for all our mistakes. As Africans, some of us must share the blame for giving in to manipulation that instigated the civil strife; for not providing good governance, democracy and respect of human rights; for not making the best use of our rich natural resources. So, these are the realities we are living now, and there is a new emerging international economic order, globalization, challenges and opportunities. We all know that globalization provides a lot of opportunities for the world as a global village but it puts a lot of very serious challenges in front of us African countries. How can we compete in this very wide market which has really become one? We see the barriers falling, so you cannot really protect your industry anymore, and you have to compete with the products of the most advanced economies. Also we see as a result of this that the strong are getting stronger and the poor are getting poorer, and we see also that we suffer from an even worse situation with the civil wars that divide and are actually tearing our continent apart. So at the end, we see the gap widening between the developed and the developing. With less than three months before the dawning of the new century, many have expressed fear that Africa will be left behind, away from the global revolution, marking the evolving or the evolution of the new international order. So what we need to do is to put our act together, and become pro-active in shaping our present and future. We should no longer allow ourselves to be marginalized.

My second question is, how far have we gone in our quest for this African renaissance that we are all talking about? As I mentioned before, the present century which is about to finish has witnessed tragic events in many parts of Africa, ranging from outbreaks of epidemics and famine to civil strifes which have dampened the optimism and expectations that many have attached to the twenty-first century ushering in a new and brighter page for the future of our continent. Yet it's not all doom and gloom. We must not lose sight of positive achievements in a number of areas, among which are the emergence of the new South Africa, and before that Namibia; the end of the civil war in Mozambique; and positive developments in the DRC with the signing of the ceasefire by all warring factions. We are very optimistic that we can finally see peace in this very important country.

A growing number of African countries have undertaken economic and social reform aimed at improving governance and creating an enabling environment of mobilizing resources, both domestic and international. In recent years three African countries have exceeded eight per cent growth rate in GDP. Eight countries attained six to eight per cent, and a dozen have achieved three to six per cent, signalling many signs of hope among the turbulence. These signs are an indication of our ability to change and we must capitalize on that. Allow me here to quote the report from the World Bank for the year 1998. The report says that despite cautious optimism about the region's progress, "Africa's growth must be boosted if poverty is to be reduced for large numbers of people. The macro-economic reforms that facilitated positive growth for three years now must be deepened to reshape the role of the state and enhance its effectiveness." The report continues to say that "special attention was given to human development especially to helping African people take advantage of global communication systems for accessing information." It says that, "for the third year in a row, the average Sub-Saharan African country experienced positive *per capita* economic growth in 1997. This, combined with improved economic policies and increased political openness has created greater opportunity for development."

"As Africa's leaders address the different development needs of their countries and strive for, what some are calling, an African renaissance," and again I am still quoting the report of the Bank, "the World Bank has been challenged to find new ways to respond to these opportunities and to meet the changing needs of its African clients. The Bank's fiscal 1997 Annual Report was optimistic about the region's economic recovery, and this is reinforced by the fiscal 1998 results." While the region's gross domestic product (GDP) grew at 3.8 per cent, this was heavily influenced by South Africa's growth of 2 per cent in 1997, and I think now the figures would be 1.7 positive growth rate in the GDP of South Africa. Some 37 countries registered positive *per capita* GDP growth in 1997, and 21 of them grew at 5 per cent and more. Exports expanded roughly twice as fast as the GDP in recent years and lower fiscal differences and inflation also boosted growth. The report continues to say that optimism must remain guarded while economic performance falls short of the level needed to reduce poverty substantially and deliver growth to all levels of society. We know that in South Africa it is believed that to create jobs and achieve the optimum development you need at least a 6 per cent growth rate, but at least the figures of the growth and the GDP are now positive and 1.7 I think is much better than the figures that were issued in the month before which indicated that the growth rate was negative. The report continues to say that serious challenges to recent progress might be recognized, and they list these challenges as the HIV/Aids as a social and economic threat; the dependence on development assistance; the rapid rise of domestic debt in some countries; and the potentially dampening effect of the East Asian crisis. The majority of the African economies have responded positively to growing social stability and sound policies, but the plight of Africans in some eight or so countries still severely affected by social and political instability, must not be forgotten. Some countries, like Angola, show impressive growth figures but they do not translate into either stability or better living standards.

So if I would like to summarize the question I put forward, how far did we go, the short answer would be, not too badly, especially if I add to the above the very important fact that we tend to overlook or really take for granted, and that is the establishment of the Organization of African Unity in 1963. We, in Egypt, believe that this has been the time of the emergence of the African renaissance. Since then, the African continent has gone from one upheaval to another, but the OAU survived. In this, we must commend the African leaders determination to convene annual African summit conferences in the context of the OAU providing a forum for exchange of views and discussion of African issues and concerns, unlike other regions which have not been able to do likewise. Egypt is an African and also an Arab country. The Arab summit has not been as regular as the African summit, and this is a fact we tend to take for granted but really it is very positive fact that we must take pride in and must foster and must make the organization more efficient.

There has been no complaint of lack of transparency in the work of the organization, and again this is a good thing for the OAU because we know now the reform of the Security Council of the United Nations we are not only talking about increasing the number of developing countries in the Council, but also improving the methods of work in the Council. We want more transparency in the work of the Council. This we cannot complain about in the OAU.

For those who might think that I am painting too rosy a picture of the OAU, let me tell you that the OAU is what we Africans make of it. It is incumbent upon all its members to give it the support and the power to be instrumental in dealing with the problems and challenges that face our Africa. In other words, if we want a powerful OAU, we can have it. We in Egypt believe that we should not re-invent the wheel nor should we stretch ourselves too thinly. All of Africa, its North, its South, its East and its West should work hand in glove to achieve what is good for all of us. We must really guard against the division of the continent: we should not work in isolation from each other. We must put our hands together as a continent because there is a lot of complementarity between us. The existing regional groupings should be building stones and pillars for a unified Africa cooperating together under the banner of the OAU.

Another aspect which shows positive developments in the African continent in our quest for the true African renaissance has been the establishment of the OAU African mechanism for the prevention, management and resolution of conflicts which was signed in Cairo. This has also been a very positive element, and we hope that this mechanism will be integrated so that it can play a role in finding solutions to the civil strife and the problems in the countries suffering from internal problems, and in this the OAU must work in cooperation with sub-regional organizations, and it should also work under the umbrella of the UN because the OAU on its own cannot really cope with these problems. There is a financial aspect that should be covered by the United Nations. But we must work together, and we must use the existing machineries rather than creating new structures and mechanisms.

Another positive achievement in our quest for an African renaissance has been the signing of the Pelindaba Agreement in Cairo. Actually this has been a very bright time in the African endeavour, and in particular in the cooperation between Egypt and South Africa. South Africa played a very positive role as well as Egypt in encouraging everyone in the continent to sign this agreement. So all the African continent agreed on a unified position concerning this very vital issue, or process, by which in April 1995 the agreement was signed declaring Africa as a nuclear-free continent. And I think that this is an achievement that we should be proud of.

Another achievement is the launch of the African Economic Community. Another is the African Information Initiative. The still another achievement is the unified African position concerning the reform and enlargement of the UN Security Council. So these are but a few examples of our achievements to demonstrate that in our quest for African renaissance we can do it if we really want to put our hands together. The most important thing, as I said, is to put our hands together and work closely, because if everyone starts working alone we will be fragmented and we will not get anywhere.

If I can quickly give you a very brief overview of the Egyptian position concerning the issues I have just mentioned. I want to say that Egypt's social and economic bonds with Africa are as ancient as the Egyptian civilization, and we are fully committed to the goals that were agreed on when we established the African Economic Community during the summit that was held two years ago in Harare. Egypt has been actively involved in joint projects that provide new horizons for cooperation with several African countries. We signed agreements to protect the investments; prevent dual taxation; and to promote cooperation in the fields of trade, agriculture, industry, construction, mining, maritime transport, civil aviation, and others. We strongly believe that such a pattern of cooperation provides a model for what we hope to achieve with all African countries. Egypt has become a member of COMESA. We are actively consolidating cooperation within COMESA but at the same time we are actively pursuing fostering closer ties with other groupings. Aware of the need for an increased effort towards further enhancing inter-African cooperation, the Egyptian government has established an inter-ministerial committee which comprises representatives from all government ministries to consider and put forward concrete steps towards enhancing cooperation between Egypt and African countries.

I would like now to conclude by presenting a number of recommendations that could further endeavours towards African renaissance. They are composed of two elements: first our responsibilities as Africans; and the second is the responsibility of the international community. As Africans we need to recognize and realize the importance of economic integration, open markets and free trade among African countries as a pre-requisite to achieve economic development. It is incumbent upon all of us, Africans, to formulate and co-ordinate policy priorities with a view to enhancing political stability and conflict prevention on the one hand; and to promote and harness the new opportunities which will empower us to meet the challenges of peace and development in a more collective and structured manner, on the other. African countries should start harmonizing their social policies; remove their inter-African trade barriers; and enhance the Continent's human resources and infrastructure needed to link its various sub-regions. To be more specific, we need to work towards the unification of Africa's power grid. In the North, we started long before and now we are connecting with Europe and some of the Arab countries to the east, and we hope that we can stretch back and connect with the rest of the African Continent. Of course, the Democratic Republic of Congo was really, or is, a cornerstone in this endeavour, because we all know how rich its resources are in electricity, so we hope that reaching, or finding, peace in the DRC will encourage this unified grid all over the African Continent. Also, we must work for the establishment of a continental highway, and a road network providing an efficient transport system, because this has been a great impediment against the growth of inter-African trade. We see that Africa trades more with the outside world than they trade amongst themselves, and I think this is a shame. We must do something about it. Also the inter-African communication, and in this area also the Organization of African Unity is working diligently, and we have the African initiatives in the information technology area. Also a problem area is common disease and pest control.

These are some of the objectives that African states decided to attain through the establishment of the African Economic Community in 1997. The launch of the AEC was a great step in reaching African economic integration. The existing regional economic groupings in Africa constitute the pillars of the AEC, and we believe that these groupings should be consolidated. We should also encourage interaction and cooperation between them. They should not work in isolation from each other. Another challenge is the enhancement of self-confidence and self-reliance and adoption of a rational approach to conflict resolution, making the OAU and other existing mechanisms our tool to achieve the aspired for renaissance.

Now, another aspect which is very important is our culture. And I must confess that when I gave a talk to the University of Pretoria some time ago about the African Renaissance, I touched briefly on culture because, you know, when you have a problem, or at least this is a defect in me when I have a problem, I concentrate more on the problem than on the assets which I really have. So I touched briefly on the issue of culture and one of the audience asked me where was the African culture in the African renaissance; and the irony is that now I am going to be Assistant Minister especially for Cultural Affairs, which made me realize how wrong it was to go briefly over this. I must reiterate here that culture is a very important asset for us in our quest to achieve this African renaissance so we must utilize it. We must be proud of African culture and heritage, and use it at least to sell our culture to the world, to make it a source, not only of pride, but a source of earnings for our national income. We have a lot that we can offer the world, but in order to be able to sell it to the world, we have to be able to sell it to ourselves. We have to have pride in our culture as Africans, and I must say, we must be proud but critical at the same time. We must know that our cultural heritage has some - I don't want to say something that is very harsh - but we have some biases in our culture. If we look, for instance, at how some of the African cultures look at women, this is a negative issue that we must work on, and I will touch on this when I come to my next point. But our cultural heritage must be one of our strongest assets in our quest for African renaissance.

Now, I must speak about the responsibility of the international community: I already spoke about the responsibility in conflict resolution and this touches on the role of the United Nations, especially the Security Council. Now the responsibility of the donor country, which is very important; first of all, debt relief and partnership in development, because we all know that Africa's external debt is a major obstacle, hindering its development. Africa, alone, cannot overcome this problem. The international community has to cooperate in facing it. Another responsibility is the elimination of poverty through the promotion of labour-intensive industries, investing in education and health. The state must cushion the adverse effects of economic transformation on the most vulnerable groups within their societies, and this can be done with the assistance of the donor countries, empowering the private sector as an engine of economic growth, reinforcing civil society as a catalyst of popular participation, and the empowerment of women. I believe that this is a cornerstone of African renaissance. Without women as genuinely respected equal partners, we will not achieve true African renaissance. We must admit, as I said, that our culture holds an undeniable notion of prejudice against women, and their active contribution in their societies, and we must remedy this. I believe that African renaissance will remain an unattainable dream until gender stops being subservient to culture.

The other point is achieving sound macro-economic indicators and creating an environment that is investment friendly, and then enhancing democracy and respect for human rights and this is a very crucial issue, and I think we are already seeing positive developments in the Continent towards that end.

In conclusion, I believe that African renaissance is not a dream: it is a challenge; it is a reality unfolding. Africa started its realization some time ago. Some progress was made; some problems and hiccups, as grave as they may be, should not dissuade or discourage us from pursuing the long walk. The Continent is filled with dynamism, with new pragmatic leadership, and emerging commitment to connect Africa to the global economy and women undertaking a more prominent role in development and growth. These positive trends must be nurtured and consolidated to enable Africa to lay a solid foundation for its future and pave the way for a real African renaissance as it enters the twenty-first century. We must draw on our existing resources and show strong resolve and surely we will get there.

THE SECURITY CONTEXT IN SOUTHERN AFRICA AND BEYOND

Jakkie Cilliers

I am going to deal with four areas. I am firstly going to speak very briefly on Southern Africa as a region to give some feel for what is the region of Southern Africa, which to many of our foreign participants may not be that well known. And then I am going to move on to a very brief discussion on regional security architecture in the region. I then want to say something on the changing nature of the security debate and the security challenges that face us in this part of the world. I also want to make some remarks about changing patterns of donor assistance, assistance towards trying to help Africa deal with its problems, before concluding.

In the first part then, what I've done there is that I have made an extract from the latest report, the 1999 Report of the UNDP on the Human Development Index, which is based, primarily, as you know, on longevity *per capita*, income *per capita*, and levels of education, which indicates where the country's footing is, in terms of the quality of life, within in a ranking of 1 to 174. Canada is Number 1 and Sierra Leone is Number 174. South Africa ranks as Number 101. Our GNP in 1997 was about was about 103 billion, slightly less than that of Belgium. Our position in the region is that of a regional hegemon. We are to an extent to Africa what the USA is to North America. It is a bit of an over-statement but, if you look at the trade patterns, it is not actually incorrect. Zimbabwe, the next largest country within the Southern African Development Community (SADC), has a GNP of 8.2 billion, that is about one quarter, if I am correct, of that of Ford, which is a company, not a country. The Democratic Republic of Congo at 114 is also a member of SADC, which I will say something about, has a GNP *per capita* of $110 *per annum*. That is what you have *per annum* on average per person. Zambia is lower than the Democratic Republic of Congo. Malawi, Angola, Mozambique, and then Sierra Leone is considered by the UNDP to be the country where it is the least pleasant to stay, and where your quality of life is perhaps the poorest. The region is at the bottom of the Human Development Index.

Let us look at the challenges which face us, such as low levels of economic growth. If I am correct, 89 of the countries of the world have got a lower GNP now than they had ten years ago. GNP growth in the majority of sub-Saharan African countries is declining. High levels of debt, and declining levels of foreign direct investment. Foreign direct investment is certainly not flooding into any emerging markets in Africa. There are changes in these patterns but certainly no real advantages to countries in Southern Africa which are more susceptible to globalization. To give an example, in sub-Saharan Africa you will find that more than 30 per cent of GNP is related to trade, mostly primary commodities. In the OECD countries, you will find that it is something like 17 per cent, which means that we, in one sense, despite being extremely marginalized in a global world, are more susceptible to globalization than most people realize and are more at the mercy of the global markets. In other words, what I am saying is that there is an extreme case of economic marginalization in this part of the world.

We have in the region recurring problems with problematic military relations. There are continuous threats of military intervention in a number of the countries that are members of Southern Africa. To a large extent there is very little difference in the Democratic Republic of Congo and Angola between the quality of the armed forces and the level of armaments and training of the general population. Peter Lock referred to this as being "militianization" of African society; the slow arming of African society and the disintegration of African armed forces in a number of select countries.

High and increasing levels of corruption, increasing corruption and penetration of the African state by organized crime on levels that could, if not halted, come, in a matter of years, to rival that perceived to be the case in Latin and Central America, and is the case very much in the former Soviet Union; the penetration of the state by organized crime; in certain cases, the use of the state as a frame for organized crime. Having said all these nasty and negative things, my purpose is to talk about the challenges; therefore, the emphasis is very much on the challenges that confront us. It is true that we have seen tentative liberalization in a number of African countries. Over all the globe we see an increasing number of people are living under some degree of multi-party democracy. Of course, without continued economic growth, sustainability must be questioned.

The final factor that is probably the single largest problem facing Southern Africa with the highest rates of AIDS and HIV infection in the world is exactly that threat. The UNDP has been arguing for some time that the life expectancy in the majority of countries of this region has declined by about ten years in the last five or six years. Average life expectancy in a country such as Zimbabwe and Zambia is now something like 42 years. HIV infection rates in excess of one-third of the population and the majority of those are productive employees.

Let us look at regional security architecture. What is the region doing to try to deal with the problems of security? Without security, no development will occur and no investment will happen and so on. I am going to go over this fairly briefly, and I am going to talk about the Southern African Development Community; the so-called Inter State Defence and Security Committee, which is a co-ordinating committee; the organ on Politics Defence and Security; Southern African and the Southern African Regional Police Chiefs Co-ordinating Organization (SARPCO). Then I will say something on the challenges and problems that face us.

I will start with a brief overview of the SADC. The South African Development Co-ordinating Conference, which was the precursor to SADC was established in 1980 to pursue economic liberalization, integration, and equitable development to try to reduce economic dependence on South Africa as part of the anti-apartheid struggle. It changed to SADC in 1992, and it has a formal legal status, a treaty. It has 14 member countries. It includes a vast number of countries. NATO is going to put 50,000 troops in Kosovo. The UN may put 25,000 troops in the Democratic Republic of Congo, if we are lucky. So let us just place things in perspective. The SADC is an inter-governmental organization aimed at economic development. It is, therefore, not an institution which is aimed at dealing with security issues primarily, although its charter does provide for its involvement in that.

SADC institutions. It, of course, has a Summit of Heads of State and Government; a Council of Ministers - the treaty says it should be a Council of Development Ministers preferably or Ministers dealing with economic affairs but that is not prescribed; a number of Commissions; a Standing Committee of Officials; and a Secretariat, which is located in Gaborone. The Executive Secretary has just been fired, apparently on South Africa's insistence, because of allegations about ineffectualness. The Heads of State is the supreme body of course. It meets once a year.

Just a few words generally about SADC. These purely relate to how SADC decision-making processes should occur in terms of its treaty. Its treaty has been flouted quite vigorously by almost all the member countries of SADC including South Africa. But not only by this country, but by almost all its most important member countries, through their interventions both in Lesotho and in the Democratic Republic of Congo. Decisions should be taken by consensus but the most important for your purposes is that SADC operates in terms of 16 sectors; each sector is given to a certain country. In other words it has got a very small Secretariat, something like 30-40 people, and then various countries within the region are assigned certain sectors for co-ordination of tourism and so on. There has been for some time now extensive pressure for reform of SADC, for a variety of reasons, and there are proposals to rationalize the organization, since there is considerable evidence that the majority of developmental programmes that are run by SADC are in actual fact bilateral or national programmes that are not truly regional.

As examples of some of the sectors that SADC operates with it would give agriculture, natural resources, research and training, for example, to Botswana; industry and trade to Tanzania; finance and investment to South Africa. And that country is then supposed to co-ordinate; for example, in the case of South Africa, issues around finance and investment, in the rest of the region.

The way that the SADC treaty works is that a protocol is signed on a particular sector and that protocol becomes an integral part of the SADC treaty and then has a recognized status in international law. Some of the SADC protocols that have been signed deal with mundane matters such as immunities and privileges, and some deal with matters such as illicit drug trafficking. Despite the rhetoric that surrounds SADC, it is struggling to move forward. You only have to travel from South Africa to Mozambique by road or from South Africa by road or rail to Zimbabwe, to see the extent to which regional economic integration is more myth than reality.

Very briefly, let me talk about the ISDSC – the Inter-State Defence and Security Committee, which is a ministerial committee, chaired by the Departments of Defence of the various SADC member countries, established in 1983 as part of the Front-Line States Organization. The Front-Line States Organization was an organization originally of seven countries established to help in the fight against colonialism and against apartheid in South Africa. Its membership now parallels that of SADC, but the main problem with the ISDSC is that it has got no legal status. It is purely an informal vehicle for cooperation, and its agreements carry no weight in international law. The Chair rotates between countries (presently held by Swaziland). It has defence, state security and public security sub-committees. Defence is self-explanatory; state security is just another word for intelligence, but it does not really succeed in exchanging intelligence information; and the public security committee polices immigration and customs cooperation. The defence sub-committee's purpose is to review and share experiences, to explore areas for further multilateral military cooperation, and to exchange views and propose mechanisms for the prevention, management and resolution of conflicts.

The ISDSC faces a number of challenges other than the absence of legal status. It has no involvement in foreign affairs, despite the fact that it is meant to deal with preventive diplomacy. There is very little transparency in the organization, no permanent secretariat, and it overlaps with a number of other regional organizations. Having said that, it has made a remarkable contribution to building mutual understanding, cooperation and trust in the region.

A few words on the so-called "organ on politics, defence and security." The word "organ," an unfortunate choice of word most of us would argue, is taken from the Organization of African Unity which has a central organ. Its establishment originated from a SADC workshop in 1994. Its progress has been sullied, particularly by competition between South Africa and Zimbabwe around issues around the chairing and status of the organ. The objectives of the organ are: to work towards a common defence; development of a collective security capacity; peacekeeping in the region; cooperation on crime prevention and intelligence; and to work towards a common foreign policy. There are also objectives with regard to human rights, but these have received the least attention, given the lack of a strong civil society in many southern African countries. With respect to challenges, the organ has no legal status or formal mandate. And the organ, SADC and ISDSC have been used by various governments, including the South African government, to legitimize a variety of military interventions and incursions into neighbouring countries, which simply do not stand up to scrutiny in terms of the South African constitution or international law.

Southern African Regional Policy Commissioners Co-ordinating Organization (SARCO) is the primary operational mechanism to fight cross-border crime. It has a legal status as an association and in terms of international law. It sees itself as the implementation mechanism for the ISDSC – although there is quite a bit of overlap, in particular with the public security committee of the ISDSC. It has 12 member states as opposed to the 14 of SADC – the Democratic Republic of Congo and Seychelles are not members of SARCO. SARCO is structured as follows: a council of police chiefs; a secretariat in Harare; and a permanent co-ordinating committee, composed of the heads of crime investigation of the various members. SARCO is probably the most effective organization within Southern Africa in terms of dealing with security issues affecting the region. SARCO is trying to move towards common legislation among members to deal with issues such as extradition, drug control, etc.

There are serious problems within the police forces of the region. A lot of attention is generally given, in Africa, to the military, very little to police reform, reform of the criminal justice system, prisons, and so on. The military also absorb most of the resources, with the result that basic criminal justice system infrastructure (even in South Africa) is very limited. You will find South African police stations without electricity; you will find that a third of South African police officers do not have a vehicle license; about 20 per cent of the South African police are functionally illiterate. The private security agencies are taking over much of the work that should be carried out by the State, both in South Africa and elsewhere – and their turnover in South Africa rivals that of the police budget at about 12 billion Rand.

That has given you a flavour of the challenges and problems that we face in Southern Africa. Issues of nuclear proliferation, after South Africa has decided to join the NPT, are quite far from our minds.

The function of violence in this part of the world is changing – as it is in a lot of areas of the world. The function of violence is no longer ideological, but it is the accumulation of wealth, particularly by state agencies, by leaders who call themselves Presidents. This is the situation in Sierra Leone, for example. I am giving an example from outside South Africa in order to be politically sensitive. The purpose of the state is to accumulate wealth; it is not ideological, not political; it is not necessarily to extend control over the entire country; it may not even be to end violence because if violence ends you have to share the resources that you get from the diamond fields and may even have to consider the use of some of these for development purposes.

You see the privatization of violence in much of Africa, which is purely a reflection of the weakness of the State. Part of the structural development agenda in the last few years has been to outsource particularly development assistance to NGOs and private consultants. Very often these NGOs are led by the same government officials that sit from 8.00-5.00 as Minister of Energy and from 5.00-8.00 as head of a petroleum exporting company. Or he is the Chief of Armed Forces from 8.00-5.00 and from 5.00-8.00 he has vested interests in private security companies that provide competition to his own armed forces in the most lucrative areas of the diamond fields of Angola – to use a factual example.

In this way, the African state is being weakened on a consistent basis. The greatest challenge facing Africa is state building. We all know the argument that globalization is threatening to weaken the state, but this argument does not really hold in much of Africa (I am, of course, over-generalizing) because the State in Africa in many cases is not consolidated.

Having said all these negative things, what is important to realize in these times of change in this part of the world, is that where in the past the problem was always the result of the meddling of foreign powers and multinationals, what is happening in Africa is that we are gaining an increased sense of responsibility for our own destiny. Sometimes in a very bloody way - in the Democratic Republic of Congo there are African countries fighting one another for African security. There may be a parallel with what happened in Europe in the sixteenth and seventeenth centuries, when going from about 500 fiefdoms to, I think, 17. The scaffolding of the Cold War has been removed, and of colonialism, and Africa is coming to terms with its own security identity – and that has strategic significance for the stabilization of this part of the world.

There is, of course, a declining role of the State generally, with NGOs playing a more prominent role in the provision of development assistance. And we are increasingly seeing a challenge to a State-centred debate where security is seen as purely a function of the State. Security is now seen as it should be: primarily, the security of the individual.

And it can be seen that it is impossible to discuss even preventive diplomacy purely around State-centred approaches. Let me give an example: if you want to bring peace to either the Democratic Republic of Congo or to Angola, you cannot do this only using a diplomacy that is constructed for intervention between States. In these conflicts, political power is only one component; others include the interests of multinational companies (Chevron, BP, Shell, and so on) which are probably more important in Angola than the Angolan government; the interests of the diamond mining companies; Interpol and the nexus between organized crime; the flows of illegal weapons. You have to bring those components to bear at the negotiating table if you want to try to leverage peace from organizations, institutions and individuals who, very often, have a vested interest in the continuation of war.

I am not saying that Shell and so on have a vested interest in the continuation of the war, just that it is a statement of fact that $US8bn that the Angolan government will receive over the next few years from the exploitation of deep water oil finds will go towards debt repayment and the purchase of arms.

Patterns of donor assistance have changed. Humanitarian assistance is increasing; emergency assistance, from the developed world, is increasing; while, of course, we see that development assistance is decreasing. And what is happening is that, in very many areas humanitarian assistance is replacing political action. The world has tried to help Africa. It tried to do that particularly in Somalia. And Somalia was not exactly a great success. It previously tried to do it on the cheap in Angola, but this did not really work. The war in Angola, as in Sierra Leone, is essentially a resource war. The mood has moved to "building African capacities"; helping Africans to help themselves. The world has tried; it is not prepared to allow its peacekeepers to die in Africa; therefore, what it is doing is building African capacities to bring security to Africa. Building on foundations of sand, because the institutions that you want to use to provide security in themselves are so weak, corrupt, torn and ill-disciplined that to think that you can create centres of excellence within otherwise decaying institutions seems a very strange concept.

What has happened in recognition of this, is that there is a trend in the donor community working towards so-called "security sector reform." They are recognizing that you have to provide security first; you have to help State capacity building; and you have, therefore, to reform the total security spectrum – criminal justice system included, not only the military.

So, we have a large, diverse and potentially very rich region, which often, one might argue, is cursed by its minerals and its natural resources, because those fuel the very wars that consume us. You have two failed States in Southern Africa – Angola and Democratic Republic of Congo – which resemble more Somalia than they resemble the United States or any other country you may wish to think of. You have a number of very weak and fragile states in Southern Africa. You are talking about countries with economies that are the size of mid-sized companies in the rest of the world. And you can ask yourself whether this is either a viable security community or a viable developmental community? I would argue that it is not – but that is very contentious and politically incorrect.

My talk has been deliberately hard and harsh, because I think it is important for an organization such as Pugwash, which comes to discuss issues such as nuclear proliferation, to look at the real challenges that face us in this part of the world. The challenges that face us here are the challenges of small arms proliferation, drugs, international crime, State-building, the reform of State institutions at their most basic and fundamental level. What is happening, to come to a positive note, is that there is a clear sense of increased ownership in Southern Africa of our own destiny and that we have to accept ownership of problems that we are both the victim of, and in the case of South Africa, part of the cause.

One of the challenges here clearly is the role of South Africa, given its size in relation to rest of the region. Will we lead, will we follow, or will we just go to sleep? That is the challenge.

PROCEEDINGS OF THE SPECIAL PLENARY SESSION

Held on Wednesday 8 September 1999 at 21.00

Chair: Sir Michael Atiyah

The Situation in East Timor

The session was convened to discuss the crisis in East Timor where massive violence was threatened against the people in East Timor – who had overwhelmingly voted for independence – by paramilitary groups, with the tacit approval of the Indonesian army.

Following the discussion, the Pugwash Council decided that an urgent message be sent to the Secretary-General of the United Nations, as well as to the heads of government of the Asia Pacific Economic Cooperation, at that time meeting in New Zealand, to President B.J. Habibe of Indonesia, to the World Bank, the International Monetary Fund, and the European Commission.

The text of the letter to Mr Kofi Annan, signed by the Secretary-General George Rathjens, read as follows:

> Following the referendum, under UN supervision, in East Timor, in which the population in the territory voted overwhelmingly for independence, unruly elements, with the tacit approval of the security forces, began a systematic program of terror aimed at overturning the voice of the people.
>
> The Council of the Pugwash Conferences on Science and World Affairs, meeting in Rustenburg, South Africa, in conjunction with the Annual Pugwash Conference, views with deep concern the tragic events unfolding in East Timor with great rapidity and threatening the lives of the people who voted for independence. It is of the utmost urgency that action be taken to restore peace and security for the people in the territory.
>
> We call on you, Mr. Secretary-General, to use all the means in your power to accelerate proposed action, or take new initiatives, to prevent the situation from turning into an utter human catastrophe.

PROCEEDINGS OF THE THIRD PLENARY SESSION

Held on Friday, 10 September 1999 at 09.30

Chair: Professor Francesco Calogero

Science and Technology in South Africa

1. **Mr. Jacob G. Zuma**, Deputy-President of South Africa, addressed the meeting (text of his address is on pp. 91-93).

2. **Dr. B.S. Ngubane**, Minister of Arts, Culture, Science and Technology, delivered his address (text on pp. 94-97).

3. **Discussion.** The following took part: Sir Michael Atiyah, Prof. Göran Edvard von Bonsdorff, Dr. Martin Kaplan, Prof. Allen Isaacman, Prof. Mwesiga Baregu, Prof. Bent Natvig, Sir Joseph Rotblat.

The session closed at 11.30.

SCIENCE AND TECHNOLOGY IN SOUTH AFRICA

Jacob Zuma

Thank you for inviting me to this important gathering of scientists, scholars, and government officials from all over the world meeting in a unified spirit to forge appropriate ideas for use of science to develop our countries. I am pleased to be able to share some ideas with you.

The theme "Confronting the Challenges of the 21st Century" is very relevant and comes at a time when the whole world is preparing itself for the new millennium. It is a time when we all are concerned about international security, peace, stability, sustainable economic growth, and development. These are global challenges.

We are living today in what has been called the global village. We are all linked to world developments. The events on the other side of the world concern us as never before. We share a common fate: what is a threat to others is a threat to all of us. We must therefore rise up to the challenges of the global village. We must act now and decisively. Our vision must be one that is driven by our firm determination to act together as we strive to raise awareness of the importance of science and technology and the significant role it can play in the 21st century. I believe that science and technology are important assets to economic growth and infrastructural development of all nations. There are splendid achievements of science and technology which we have witnessed in our lifetime. It is through science and technology that we can bridge the gap between the wealthy nations of the north and the poor nations of the south. I believe that science and technology should be the possession of all of us.

However, scientific discovery and inventions also have come to be identified with death and destruction due to the lethal weaponry that has been invented by some scientists. I think of the nuclear weapons which have caused untold human suffering and death to many people. Hiroshima is a classical example of this. We must therefore call upon scientists in all countries to refrain from creating, developing, improving and manufacturing further nuclear, chemical and biological weapons. They are a menace to our society and pose a serious danger to socio-economic development and human life.

I believe that resources acquired through science must be used for socio-economic development. They should not be used to inflict suffering on innocent children, women, elders and society in general. Science and technology must be used in socially responsible ways - ways that will guarantee a world that is safe for all of us.

There is at present a great emphasis on the economic benefits that should be extracted from our scientific base. It seems quite consistent with this shift that we, as responsible nations of the world, should be trying to divert some of our scientific resources from military to civilian purposes. We must be seen to be highly involved in medical advances to save lives and in social upliftment. As we pursue our vision of development through integration with other countries especially in the Southern African region, and the path of sustained growth which will enable us to eradicate poverty, diseases, and unemployment, we must act sooner rather than later, lest the world leaves us behind. In South Africa, which is regarded as a developing nation, it is our aim to assert ourselves with confidence in the global economy. One of the critical conditions for doing so is that we should embrace and adapt for our own use the latest in technological and scientific knowledge.

Information technology is one of the principal forces making our world an integrated and interdependent one. We are determined to ensure our region is able to exploit this technology in such a way that we can draw advantage from the process of globalization, but sustainable, substantial investment in science and technology will bring us benefits only if we also invest in our most precious asset – development and enhancement of the scientific skills of our people.

South Africa, since it became a democratic state, has undertaken to promote peace and stability in the Southern African region, in Africa as a continent, and in the rest of the world. This is illustrated by our involvement in bringing about solutions to political crises in our region. We believe that there is no military solution to conflicts between the peoples of the world. We also believe that trying to find a diplomatic solution to internal problems through talks and in negotiations is the key to peace and stability. We are all aware how destructive war can be. Any war leaves behind many casualties and that includes women and children. No people can truly say they are blessed with happiness, peace and prosperity where others, as human as themselves, continue to be afflicted with misery, armed conflict, terrorism and deprivation. Thus we can say the challenge posed by the next century is to ensure that violations of human rights shall never be tolerated.

A whole range of areas of conflict confronts us in Africa, Europe, and Asia. All of us are familiar with these which range from the Democratic Republic of Congo, Angola, Sierra Leone, and the Sudan here in Africa, to the Balkans in Europe, and Afghanistan, East Timor and Sri Lanka in Asia. Clearly the Pugwash Movement, together with people of goodwill around the world, has a responsibility to act decisively to contribute to the termination of these destructive conflicts. Continuously we have to fight to defeat the prevailing tendency towards the glorification of arms, the adulation of force born of the illusion that disputes are necessarily best resolved by violent means. I must add that at times business interests have tended to fit into this tendency and thinking.

We are determined to find practical ways in which to increase mutual beneficial cooperation among countries of the South in all spheres of human activity and informed by a spirit of solidarity, shared interests and a common destiny. At the same time, we seek closer cooperation with the countries of the North on the basis of respect for the sovereignty of all countries, equality among nations, and mutual beneficial cooperation. To achieve this, we need to focus on the critically important questions of the restructuring of the United Nations Security Council. Representation needs to be balanced so that all countries, rich and poor, have a say in international security matters.

At this stage, developing nations do not have permanent membership of the UN Security Council. Re-organization is necessary to ensure that it addresses the aspirations of all countries. This is important because if we are to take decisions about destructive types of scientific discoveries and development, even if we all agree as poor nations, if the nations with a bit of power do not agree, nothing will happen. Participants in this conference may be aware that South Africa moved on this question of dismantling the nuclear elements in this country, but how much can we achieve as a country, said to be developing, with no veto power? That is a problem that this important international organization needs to address. This is an important forum where the people of the world meet under one roof to take decisions to determine the destiny of the world. We must face the fact that after countless initiatives and resolutions we still do not have concrete and generally accepted proposals supported by a clear commitment by the nuclear weapons states to the speedy final and total elimination of nuclear weapons and nuclear weapons capabilities.

We live in an era where disarmament and dismantlement of nuclear weapons should be seen as a human rights matter. The challenge facing scientists is, therefore, to bring an end to this self-destruction by selectively using their skills in a manner which is true to their work ethic. We all have a moral obligation and responsibility to adhere to international security laws for peace and stability in all countries. It is therefore important for all countries to be empowered to be able to make an impact on these important issues.

As we move into the next century we must focus more on matters of economic development and sustainable growth. We must all follow the example of the Pugwash Movement which has shown itself well able to adapt to the new world order and unequivocal support for disarmament and dismantling of various lethal weapons.

At another level a distressing reality is that humanity has tended to use technological advances against itself. Guns play a major role in the perpetration and perpetuation of criminal activity in all societies. Information technology, which has had such an important impact in all spheres of our lives, has left us almost defenceless against hackers and other criminals. All companies, big or small, are vulnerable to this type of criminality. Once again, we see positive advances being turned into our downfall.

As we close the chapter on the 20th century, moving into the next millennium, we look forward to an era of upliftment for the poor of the world, non-proliferation of nuclear weapons, the renaissance of Africa, and the sustained progress of all nations. I am pleased that forums like the Pugwash Conference on Science and World Affairs bring together committed people and groupings from all over the world to exchange ideas and discuss matters of world importance.

It is true that the future of our continent depends on the countries of Africa working together. I believe it is equally true that the future of the world depends on committed organizations, committed people, and nations of the world coming together on matters of global concern. For this is the only way we can truly say we are building a better world, ensuring a higher quality of life for the citizens of the world. I would like to pay tribute to the Pugwash Movement for the work it has done and all the dedicated organizers and participants in this Conference. May you and we - through you, and in partnership with other similar groupings - grow from strength to strength.

DEVELOPMENT THROUGH SCIENCE

B.S. Ngubane

Deputy-President Zuma has set before you the thinking and the position of the South African government with regard to world affairs (this volume pp. 91-93). My task is to try to focus on South Africa *vis-à-vis* movements such as Pugwash and our destiny as charted by our science and technology system within a global system.

South Africa is indeed privileged to host the first Pugwash Conference on the African continent for more than 30 years. The aim of Pugwash is to bring together scholars and public figures concerned with reducing the danger of armed conflict and seeking cooperative solutions to global problems. The theme of this Conference is science and world affairs: "Confronting the Challenges of the 21st Century". This is indeed a formidable task embracing the history as well as the future of mankind. The year 2000 appears to be a watershed in the history of the world. It may well be like any other year in the ongoing lives of ordinary people and citizens of the world. However, it is historical from the perspective that the year 2000 will bring to its conclusion a tumultuous century, characterized by astounding scientific achievements, devastating wars, and explosive population growth. It will inaugurate at once a new century and a new millennium. A future, so vast in its human and historic dimensions, that it defies prediction, while posing momentous questions about social and economic viability and human vitality and values in the face of a new era.

The theme of this paper is "Development through Science" and it is the scientific management of development in realizing the praiseworthy and formidable global objectives of Pugwash that comes into focus. But although the concept of development is often propagated as a solution to all our problems it is a word open to different interpretations and used in different contexts. For instance, we tend to look at it in terms of development of the economy, physical infrastructure, information technology, medical services, agriculture, and so on. And, as you all know, the world is making huge advances in all these fields. However, the reality is that, notwithstanding innovative progress in science and technology, the present global situation regarding conflicts, the lack of basic resources in developing countries, and the pressure created by population growth, might be the worst in human history. The reason for this, in my opinion, is that the definition of the term "development" has been attenuated to describe a shallow materialistic concept, with the result that human development, basic to and a prerequisite for balanced development, does not get proper attention.

In order to create a context for the discussion of development through science I will firstly highlight our selected global themes relating to the roles of science and technology in development; then give a short overview of my department's efforts in strengthening the role of science in addressing development problems and opportunities; and conclude with a few ideas on the development of human resources for the practice of science. There is a growing consensus that the ability to develop, diffuse and commercially exploit knowledge is among the most essential factors in increasing productivity, economic performance and sustainable development. Firms play an essential role in this regard. Moreover, economies based on knowledge require science systems that can maintain and develop their essential role of knowledge production while being more responsive to social, economic, educational, and environmental needs. Support and incentive schemes must be made more efficient to strengthen public and private research and development capabilities, to improve conditions for the effective exploitation of scientific and technological knowledge, and to encourage the entrepreneurship of researchers. Significant efforts should be devoted to scientific research on technologies for sustainable growth that can contribute to solving issues on a global scale in areas such as environment, health, food production, and energy supply. Through an appropriate combination of regulatory schemes and economic incentives, governments and business must play an active role in bringing these technologies to the market and promote efficiency.

Secondly, the increased globalization of research and development and innovative activities, the emergence of global issues transcending national boundaries, as well as the growing costs of scientific research, pose a new challenge to policymakers. Governments recognize that in order to reap the potential benefits from globalization, national science and technology policies need to be directly embedded in an international perspective, so as to increase the capacity of economies to access and exploit the results of global scientific and technological activity. Rapid scientific and technological developments, in the life sciences in particular, call for approaches to policy development that involve greater interaction between the various stakeholders without sacrificing the basic principles of scientific validation. Such approaches greatly benefit from enhanced international cooperation.

It is, in the third place, recognized that some regulatory provisions may limit the contributions of science and technology to economic growth and sustainable development, by hampering scientific advances and the generation and commercial exploitation of new technologies, particularly when they come to the biotechnology sector with genetically modified organisms. Appropriate regulatory reform should unleash innovative capacity by encouraging innovative behaviour and the diffusion of knowledge. New regulatory frameworks need to take account of social and environmental policy objectives arising from the diversity of technological choices, and command public understanding of, and confidence in, science. This is particularly the case in areas of rapid change such as biotechnology, health and agriculture. The updating of regulatory frameworks would facilitate the exploitation of public research by industry while at the same time preserving public interests.

The fourth issue of global importance I would like to refer to is the need for the development, compilation and analytical use of international comparable science and technology statistics and indicators to better inform the science and technology policy debate. There should be an international effort and the OECD has indicated that it would play a key co-ordinating role in this regard.

I will now move closer to home and give a short overview of efforts of my department to support the development of South Africa and the region through science and technology. International experience and our own country's specific needs were used to draft the policy framework for getting science and technology to work as the engine for economic growth. We have located South Africa's national system of innovation in the globalized knowledge economy. The approach is based on the assumption that the national system of innovation is a set of functioning institutions, organizations and policies which interact constructively in the pursuit of a common set of social and economic goals and objectives. It was necessary to establish what these institutions and organizations are, the purpose they serve, and their relationship with the national system of innovation. Our white paper policy process of 1996 set up the framework for establishing an effective and efficient national system of innovation. Giving further impetus to addressing basic development programmes and problems through the national system of innovation was President Mbeki's State of the Nation Address on 25 June 1999. Based on this address, government released a communication strategy document emphasizing the theme "A Nation at Work for a Better Life". Consequently, government has identified the following issues to receive immediate priority attention namely: rural development; job creation; advancement of women and youth; urban regeneration; crime prevention; governance; and faster integrated delivery of quality services. Science will have to play a strong supportive role in realizing these important objectives of government.

Flowing from this, we turn our attention to the future. As already emphasized, it is part of the government's responsibility to facilitate the process of looking ahead and giving direction to economic activity and social development. We are all familiar with the idea of learning from the past as well as the importance of analysing current trends and projecting future outcomes. Learning for the future means discovering what the future might possess by examining the past, present and emerging actions. Simply put, it means thinking ahead, developing a vision of a desirable future, and making decisions today to cause that future to happen.

By means of the national research and technology foresight exercise, we set out to establish what technologies would be important to South Africa over the next ten to fifteen years. Twelve key sectors were identified and the results of this exercise will be key inputs to the strategic planning of export promotion, support to small, medium and micro enterprises, technical training, research capacity building, and the funding of research programmes. With the adoption of the national system of innovation approach, we realize that the determination of the relevancy and effectiveness of the existing science, engineering, and technology institutions requires urgent attention.

The response to this was the Science, Engineering and Technology Institutions Review. It included a penetrating evaluation of twelve institutions and, based largely on the individual reviews, a system wide review. Global trends such as the emergence of new core technologies, biotechnology, information and communication technology, and advanced materials were factored into the assessment of the value of our existing science and technology infrastructure. The recommendations generated by this exercise were submitted to Cabinet for ratification. Two very important recommendations that are in the process of implementation include the adoption of certain key performance indicators for monitoring our institutions and a revised approach in our financial support for scientific publications. In step with global developments, as already indicated, my department is giving priority attention to the promotion and support of international cooperation in the fields of science and technology. And I would like to pay tribute to Sir Michael Atiyah for the role, as President of the Royal Society, he played in helping us come to this position. Nearly 30 agreements in this regard have already been signed with both developed and developing countries. Our policy on international cooperation in the fields of science and technology operates within a framework of nine fundamental and universal characteristics of such cooperation. This framework gives direction to science and technology cooperation and also serves as an evaluative instrument.

The characteristics of the points of departure are the following:

1. Universality. Science and technology cooperation acknowledges, promotes, and utilizes universality of science and technology.
2. Motivation. The imperative of a mutually beneficial approach is recognized.
3. Cooperative. Science and technology cooperation implies continuous interaction between scientists of cooperating countries, a supportive relationship between government and scientists, and close cooperation between the sectors of the science and technology system.
4. Comprehensive. All science and technology disciplines and all areas of human endeavour are relevant to science and technology cooperation.
5. Capacity. The contribution of science and technology cooperation is dependent on the capacity of scientists to collaborate at the frontiers of science and technology.
6. Innovative. The need for an appropriate balance between directives by government and the freedom of choice of scientists, as well as the importance of multi-disciplinary approaches are acknowledged.
7. Directive. Science and technology cooperation is appropriately guided by the identification of science and technology priority fields in the light of national socio-economic needs and opportunities, as well as through the normal operation of the market mechanism.
8. Co-ordination. Science policy should be aligned with technology policy as well as with policies in other fields.
9. Financial. All sectors of the science and technology system have a responsibility for providing financial support for science and technology cooperation and co-funding of cooperation ventures by cooperating countries is inherent to science and technology cooperation.

This framework provides a well-founded base for the management of international science and technology cooperation. The vision of an African Renaissance was first articulated by President Thabo Mbeki. The end of the Cold War and the resurgence of more open political and economic interaction on a world scale has inspired new visions of political and economic renewal in Africa. The 21st Century, if it is to become the century of African renewal, should be ushered in by all its people through a programme of action, not only to sensitize the continent about its import, but also to send a strong message to the world, that in this coming century the African continent cannot be marginalized.

The new wave for the democratization of the African continent should be broadened, deepened and sustained. Africa should prove to the world that it is a continent based on the will of its people, dedicated to total emancipation. Against this background, South Africa appreciates its responsibility to play a constructive role, and my department, responsible for national science and technology policy, has a leading role. There are numerous and increasing interactions between my department and its counterparts in north, central and southern Africa. South African scientists and technologists are active in joint science and technology projects with their African partners; projects supporting promising development initiatives.

My Ministry also appreciates that we have a special responsibility towards supporting development initiatives in the Southern African Development Community, the SADC. I will mention two of these projects. Currently, a study is being undertaken to support SADC countries in updating their science and technology policies. This should strengthen their capacity to manage science and technology in support of socio-economic development.

Another key area in the drive for development is the promotion of the health status of the people of Africa. The greatest health impediment to development right now is the presence of HIV AIDS. Much has been written about the HIV AIDS epidemic. In South Africa, the Medical Research Council, amongst others, is involved in the search for an effective vaccine. Recent developments in Uganda, and evidence that the HIV rates are declining in that country, prompted the Minister of Health to send and assemble a multi-sectoral delegation to visit Uganda and study the situation at first hand. The challenge to promote the development of human resources by improving the health status of the people of Africa is urgent. We must chart a common course. The success of such an initiative will depend upon commitment and action at every level of our society.

I recently attended the World Science Conference in Budapest. This meeting attempted to pave the way for science into the next millennium. It was made clear that the gap between the rich and the poor has widened. My department is committed through our programmes to bridge this gap. Another major conclusion of the meeting was the necessity to bring the social sciences within the so-called mainstream sciences. Part of the reason for the gap between the rich and the poor was the lack of understanding of the social dynamics of change, especially technological change. This is a confirmation of the need for efforts to integrate perspectives from the natural and social sciences in addressing human capacity building as well as development programmes and opportunities.

I have tried to give an indication of issues to be addressed to promote the role of science in finding solutions to problems of global importance as well as of my department's initiatives to promote the role of science in supporting development in South Africa and the southern African region. This brings me back to a prerequisite for successful human development; namely, high quality scientists and technologists. To find solutions for problems and to exploit new opportunities for development through science and innovative ideas, we need well-trained and innovative scientists; and these scientists will have to be also indigenous people, African-bred. We need people who identify with Africa and who will be able to lead the inhabitants of Africa to a better quality of life. If Africa should have to identify only one priority basic to the realization of the African Renaissance it can only be quality of education. Such education is a prerequisite for producing competent, moral, and informed leaders, and in particular, scientists in sufficient numbers. To achieve this we need the support of the international community. The involvement of international communities in focussed educational programmes will not only facilitate education, but will also inspire the learners to be part of the ever-developing knowledge society.

The efforts of Pugwash in identifying such challenges and addressing them are commendable. However, much more is needed and Pugwash should look at the challenges of assisting, in a systematic way, in promoting human development on the African continent. A sound and well-functioning science and technology system is based on our investment in human capital. With respect to human capital, Africa has a lot to offer.

PROCEEDINGS OF THE FOURTH PLENARY SESSION

Held on Friday, 10 September 1999 at 11.45

Chair: Dr. Jeffrey Boutwell

Light Weapons and Conflict in Africa

1. **The Chairman** introduced the subject (text on p. 101).

2. The four panellists presented their views:

   a. **Mr. Jackie Selebi** (South Africa) (pp. 102-105)
   b. **Dr. Martinho Chachiua** (Mozambique) (pp. 106-108)
   c. **Lt. Col. Jacob Kamenju** (Kenya) (pp. 109-111)
   d. **Dr. Peter Lock** (Germany) (pp. 112-116)

3. **Discussion.** The following took part: Prof. Ulrich Albrecht, Prof. Mwesiga Baregu, Mr. David Fischer, Prof Luis Pinguelli Rosa, Dr. Walter Scheuer.

The session concluded at 13.50.

LIGHT WEAPONS AND CONFLICT IN AFRICA INTRODUCTION

Jeffrey Boutwell

In March 1996 a flame of peace consumed some 3,000 small arms and light weapons at a ceremony in Timbuktu, Mali, signalling the end of the threat of a potentially devastating civil war in Mali and the start of a process that ultimately led to the so-called Mali Moratorium. Adopted at the ECOWAS Summit in Nigeria in October 1998, the Mali Moratorium calls on all the states of ECOWAS to refrain from the production, import and transfer of small arms and light weapons except those specified for the military. Although not a legally binding document, the Mali Moratorium has been symbolically important as a mechanism by which neighbouring countries can begin to address the devastating effects of the uncontrolled proliferation of light weapons within states and across national borders.

In the last five years many other positive steps have been taken by a wide variety of international and regional organizations, and national governments and NGOs to confront the dangers to both states and to societies posed by the easy availability of light weapons that today are just as commonly found in the civilian population as they are in national militaries. Very briefly, these accomplishments include a wide variety of UN efforts including the work of a panel of governmental experts on small arms, the work of UNIDIR and UNDP and other branches of the UN. Also there is the 1997 Convention on the Illicit Manufacture and Trafficking of Firearms adopted by the Organization of American States. In Southern Africa, there is the work of the Southern African Development Community and the Southern African Regional Police Chiefs Cooperation Organization. There are also attempts by the European Union to tighten up export policies from many of the major weapons producers in Western Europe on small arms and light weapons. It is the work of many different national governments who have given this issue great prominence both through imposed stricter export controls, tightened border controls, and other measures, not the least of which is financial contributions to work in other parts of the world. Of special note are Japan, South Africa, Canada, Belgium, Norway, Sweden and even, to some extent, the USA. And I would just give one example of where the USA, I believe, is taking a more serious look at the problem of weapons flowing outside of the USA, and that example is a recent change in US law whereby American arms brokers are now subject to American national law even if they operate overseas.

Then there were the efforts of large multilateral organizations, such as the World Bank, which is taking a very strong interest in how its funding assistance can be used to collect weapons in those societies where World Bank civil reconstruction projects are underway. And, of course, there is the work of the International Red Cross, which continues to look at the international humanitarian law aspects of ever more sophisticated and lethal small arms and light weapons.

Finally, there is the mobilization of the international NGO community. In October 1998, at a conference held in Brussels, more than 200, and perhaps it was 300, NGOs from around the world representing every continent, came together to create the International Action Network on Small Arms (IANSA) which now has a secretariat in London and is helping to co-ordinate the efforts of many diverse NGOs working on different aspects of what is a very complex problem. Admittedly, all of these measures have only begun to scratch the surface of a horrendous threat to civil society and states around the world. Estimates are that 500 million small arms and light weapons are in circulation both legally and illegally. And yet, I think we can take great hope that in the short space of time, in five years, the international community, through work at the international, regional, national and local level, has begun to address this problem seriously.

LIGHT WEAPONS AND CONFLICT IN AFRICA

Jackie Selebi

Pugwash found its origins in the concern about the development of nuclear weapons and the threat of annihilation which these posed. The focus has, however, and importantly, not remained limited to nuclear weapons. Pugwash's role in the achievement of international peace and security has extended to the threat of devastation wrought by other weapons including, in this context, by the indiscriminate use of small arms and light weapons. While weapons of mass destruction threaten the peoples of the world with horrendous devastation through the use of a single weapon, it is small arms and light weapons which have been the primary instruments of death and destruction while the use of weapons of mass destruction has mainly been confined to certain conflict areas during the First and the Second World Wars, and in certain regional conflicts.

Small arms and light weapons have been the daily weapons of choice in conflicts and criminal actions around the world. Globally, it is estimated that there are more than 500 million small arms and light weapons in circulation. These weapons include revolvers, self-loading pistols, rifles, assault rifles, light machine guns, mortars, portable anti-tank missiles and landmines. Governments have been toppled and untold suffering has been inflicted on innocent civilians through their use. This has especially manifested itself in regional conflicts in Africa, such as in Angola, where we have a massive humanitarian crisis on our hands. In the Democratic Republic of Congo there are serious problems and a massive humanitarian conflict that we have to deal with. Conflicts of the past in Liberia and elsewhere have left a legacy of uncounted thousands of people who have been victims of small arms and light weapons.

Confronted with this reality, the international community has realized the devastation wrought by the proliferation of small arms and light weapons on socio-economic development generally, and specifically in the reconstruction of post-conflict societies. This problem can no longer be ignored. Within the UN, the Organization of American States, the Organization of African Unity, the European Union, and the SADC, important initiatives have been undertaken to address the problems associated with small arms proliferation.

Nationally, the South African Government is gravely concerned with the proliferation of any illicit trafficking in small arms, not only as it affects us within our own country in terms of the high levels of crime, but also as it impacts on the achievements of the African Renaissance. Amongst the leaders of Africa in Southern Africa there is also a concern that these weapons endanger democratization initiatives which are presently being consolidated in certain parts of our continent.

Small arms are generally associated with conflict but the reality has also shown that they are closely related to crimes; crimes such as drug trafficking, money laundering, stock theft, and robberies. These negatively affect socio-economic growth and inhibit foreign investment. The South African Government has therefore declared war against small arms, and it has declared war against the proliferation of these arms, and has made that war a priority for the South African police service. I am sure during your stay here you must have seen in our media the great debate in this country on what legislation needs to be put in place in order to control these weapons.

The government strategy to combat small arms proliferation nationally represents an integral and holistic approach. It does not only focus on the introduction of stricter control measures, but also on the eventual removal of causal factors such as unemployment, lack of education, poverty and many others. Our objective is to reduce the number of illegal arms in circulation as well as to reduce the flow of illegal weapons into South Africa. It also concentrates efforts to ensure the lawful and proper use of licenses for arms. South Africa was one of the first states to announce a landmark decision in February 1999 to destroy the stockpiles of surplus small arms rather than sell them on the open market, as was done in the past. In this regard, approximately 260,000 small rifles were destroyed in one day. We intend to continue the destruction of these surplus weapons to make sure that we do not gain profit from the existence of these.

In the context of the Department of Foreign Affairs goal of promoting South Africa'a national interests and values abroad, the issue of small arms has attained a high priority in South Africa's foreign policy. After all, what is foreign policy except the translation of domestic policy into relations with other countries. If it is important for the South African Government to contain the spread of small arms in South Africa, it would be the task of the Department such as I am heading to ensure that in our relations we make this issue an issue that we collectively in our region, in our sub-region, and internationally would have to address. In the same way that we dealt with the issue of landmines, we would need to mobilize, not only in our own individual countries, but in the region as a whole.

In order to do this, government has to look for partners. In this country we believe in partnership with civil society. But it is only when government and civil society get into partnership that effective work can be done in any endeavour that we are involved in. Many of you will know about the Conference on Landmines that took place in Norway, which I fortunately had the honour of chairing, that brought about what is called the Ottawa Treaty. At that conference it was because there was collaboration and co-ordination between governments and non-governmental organizations that a conference that was supposed to take three weeks to come out with a treaty was able to do this in two weeks. This is testimony to the effectiveness of the relationship between governments and non-governmental organizations in this instance. We have to do that in our own country and in our region to make sure that we strengthen the partnership between government and non-governmental civil society to ensure that all of us address this meeting in a manner that would bring benefits to our people.

Our Department's objective is to promote realistic and genuine endeavours to secure an international security environment in which the peoples of the world will not need to live with the threats of destruction and annihilation which have plagued them in the past. To reinforce national action on small arms the Department of Foreign Affairs came to the conclusion that it was imperative that a regional approach should be formulated so as to address the problem of small arms and light weapons; an approach which would be based on regional consents and which adopts an incremental approach. Internationally, such an approach would ensure that as each region of the world develops an indigenous approach, the building blocks will be put in place to effectively deal with this issue globally.

To give effect to this policy conclusion, South Africa's immediate international objective was to ensure that due recognition is afforded to this proliferation problem. At the level of the United Nations, South Africa has been supportive and has actively participated in various UN initiatives on small arms. South Africa was a member of the UN panel of experts on small arms which wrote the first authoritative international report on small arms proliferation. South Africa also actively participated in the follow up on the UN group of governmental experts on small arms; and participated in the UN study on ammunition and explosives. At a UN Assembly Session in 1998, South Africa initiated a resolution entitled: "Illicit Trafficking in Small Arms" which achieved consensus support between all the member states of the United Nations, and has prepared the foundations for the urgent co-ordination of international efforts to combat illicit trafficking of small arms.

On a regional level the Department has launched an initiative to build political momentum to address the proliferation of small arms at the level of the Organization of African Unity. This resulted in the OAU adopting at its summit meeting of June 1998 its first ever decision on small arms. This mandated the Secretary-General of the OAU to gather from the member states information on the extent of the problem. At the initiative of South Africa the OAU and the assembly of the Heads of State and Government which was held from 12-14 July 1999 in Algiers called for a co-ordinated South African approach to the problem posed by illicit trafficking, circulation and proliferation of small arms. Taking into account the regional experiences and activities in this matter, the OAU mandated the holding of an African experts meeting on small arms. At the UN and at the OAU the building blocks are in place to ensure that this, the issue of small arms proliferation, and the human suffering which this causes, is properly addressed. The challenge now is to utilize these opportunities to come up with practical solutions to this proliferation problem. This is to ensure that the threat of, and violence of, these weapons, no longer inhibits or prevents sustainable economic growth, prosperity and the achievement of the African Renaissance.

We would have to do a number of things to ensure that this happens. Nationally, we would have to put into place legislation; regionally we would have to harmonize the legislation that would combat the proliferation of these weapons. We would need to find ways of finding agreement on the destruction of any of the surplus of these arms. We need to promote uniformity and strengthen legal provisions relating to licensing, possession of, and trading in small arms by civilians. We would have to put in place adequate provisions that provide for the confiscation and forfeiture by states of small arms and ammunition held, or conveyed into, the country without, or in contravention of, legal requirements. We would need to establish controls over the accumulation of small arms and light weapons and associated military equipment, especially as this relates to illicit trafficking. We need to address the issue of regional transparency, making sure that mechanisms are in place to ensure that we are transparent when dealing with questions of small arms.

Amongst other things that we need to do, is to provide our neighbours with the information about acquisition of any weapons, including the forfeiture of, or confiscation of, light weapons that have been transferred to government control. At a regional level the achievement of measures such as those that I have outlined would also be expected to help to deal with this issue even though it might be slightly more complex. Our initial focus would therefore be on a wide consultative process. Now that Africa has recognized the problem the ground has been laid to achieve real progress.

However, we should not only concentrate on the demand side of the problem as being experienced in Africa. It is also essential that the suppliers of these weapons should ensure that the transfers take place in a responsible fashion and that they do not contribute to the conflicts which we are struggling to address. These countries, many of which are out of Africa - not many of us in Africa can manufacture these weapons - must take their responsibility. It creates untold misery for me to learn that there are countries in Europe who would be ready to send weapons into Angola simply because they will get diamonds. It worries me that there are countries in Europe and elsewhere who are ready to send tonnes and tonnes of these weapons of destruction into the Democratic Republic of Congo because in exchange they will get titanium; they will get minerals from those countries so they can sit comfortably in air-conditioned offices in Europe, and decide to contribute to the destruction of African lives simply because, for them, profit is more important than preserving the lives of these creatures called Africans.

We think it is important that the supply side should be addressed in a vigorous way. It is in this area that organizations such as yours could help us to work together in order to address these questions. Of the weapons that are killing Angolan children, none of them are made in Somalia; none of them are made in Gabon; none of them are made in Egypt; none of them are made in Tanzania: they are made in some very lucrative looking developed country. Figures provided in the 1996 publication on World Military and Social Expenditure state that, in the conflicts which took place around the world between the years 1900 and 1995, a total of 109 million people were killed; this is between 1900 and 1995. Of these, in the instances where separate figures were available, 62 million were civilians and 43 million were military personnel. In sub-Saharan Africa the figures for the same period of time are given as 9,121,500 in total, with 5 million of these being civilians and 1 million being military personnel. It is, however, important to note when considering these figures, that in the same publication it is made clear that the figures are not an accurate description of the situation we are confronting. The situation has changed to where in 1995 more than 90 per cent of all casualties in conflict situations are non-combatants. The same publication stated in 1995, some 15.3 million people were displaced as refugees, and the number of people in need of food and water and medical care is said to have increased by over 50 per cent when compared with the preceding decade. This is a dismal picture of the millennium which we are about to leave.

What is it that we can do to ensure that, as we enter the new millennium, we do so with the confidence and the hope that Africa, amongst others, might achieve a rebirth; that none of us would continue to support dictators; and agree that violations of people's rights should continue simply because in terms of mineral wealth we can address the problems of our people in our own regions.

Light weapons as the primary weapons of choice in conflict in Africa, and in other parts of the world, which are delivering us the figures which I have just quoted, are therefore clearly a problem which can no longer be endured and ignored. This is not only because of the atrocious images which confront us on our television screens from time to time; it is because of the unimaginable horror and suffering which is being faced by thousands of faceless people who are confronted on a daily basis by the reality and not the image.

We are duty bound; we are obliged to reach out to these people who are the victims and not the cause of their suffering. Our commitment must be absolute and consistent. It must not be a passing concern which is stimulated only when we see the images of war on television. It must be a concern that all of us are ready to address as effectively as we can.

LIGHT WEAPONS AND CONFLICT IN AFRICA

Martinho Chachiua

There is no way I can start other than by complimenting Pugwash for the simple reason that it has understood that convening this meeting in Africa, there was no other way but to talk about small arms, because while they have no capacity to destroy the world in one single day, or five minutes, the figures Jackie Selebi has given us (this volume, pp. 102-105) show that the world can be destroyed. I don't know how long it would take, but it can happen. What I am trying to do now is to share with you our conclusions within the Institute for Security Studies in Pretoria, which has been running a programme on small arms management in the region. What I am trying to do is just to give a picture of the extent of the problem. The effects have been already dealt with by Jackie Selebi. I will not touch on that, but I will discuss what we think should be the solutions.

Let me start by saying that the legacy of war complicates democratization and development in complex ways. Most of the states in Southern Africa are facing an explosion of gang violence, while struggling to reinvent themselves as responsible states for ensuring human security for all their citizens after decades of war. To me, this is partly because, at the end of the armed struggles, seldom, if ever, are all weapons collected. The conditions of insecurity that prevail in countries in transition are fertile grounds for the maintenance and acquisition of light weapons and small arms by the community at large. Physical security and economic needs are the fuel which keeps the trade in small arms moving. And it is the trade that, for the first time in history, does not require new influx of weapons to be destabilizing. Unfortunately, we are still catching weapons coming into our region but what we have is enough to destabilize the region.

Thus in countries which are in a downwards spiral towards anarchy, or have already failed: the weak, the strong, and the corrupt, have a common need to possess arms so as to ensure self-defence and maintain an advantage or exploit situations. In addition, in the nations emerging from conflict, the ownership of a weapon has not only a security motivation, but also one based on economic imperatives. Impoverished groups of people, insecure about their own potential for economic development and survival, utilize weapons as if they were cheque books, robbing to cover basic needs and/or exchanging guns for money or goods. I should add here that it is not only the poor.

It is in this light that we must look at the problem of small arms proliferation in the region. What I am trying to do is to give, not the figures, but the picture of arms availability in this region. There is no accurate measure of the numbers of weapons circulating in the region. Just to give you a few examples: in Mozambique alone, estimates of the weapons imported during the civil war range from 1.5 million to 6 million. During the United Nations peacekeeping operation, 1990-1995, about 200,000 weapons were collected. However, most of those weapons were not destroyed and soon were on the streets of Maputo again or being moved across the borders. In Angola, it is literally impossible to estimate the number of weapons. It was reported that in 1992 about 700,000 weapons were distributed to civilians by the government following the renewal of fighting. During the demobilization process, which failed anyway, they only managed to collect less than 35,000. This, combined with the lower number of police and soldiers who had been demobilized at that time, indicate that most weapons were still in the country. And I am just thinking, should peace ever come and if the example set by Mozambique remains true for Angola, once there is a peace, and people get rid of their weapons, the impact on the countries in the region could be at least as severe, if not more severe, than that of Mozambique.

Furthermore, besides the arms and stockpiles, during the 1970s and 1980s Angola has continuously received weapons since 1992, and we are talking of a variety of countries where these weapons are from. I will not mention them. Despite the sanctions imposed by the United Nations, we know that UNITA managed to get alternative routes of supply. In South Africa, which is an example which I should not leave out, in addition to more than 4.1 million licensed firearms, the estimates of illegal weapons range from 400,000 to 8 million. This discrepancy in numbers shows that we do not know how many weapons are circulating in our countries.

In Zimbabwe, disarmament and independence has been described as having been projected and thus concealment of weapons was justified as a precaution. In the post-independence period in that country weapons inflows from Mozambique and South Africa have been reported.

In Namibia, while demobilization and reintegration was relatively successful, the connection with Angola, Zambia, and South Africa makes it the country for potential growth of arms proliferation. That can be said to be equally true of Swaziland. In Zambia the major source of weapons were arms caches left by liberation movements.

Now, this picture of arms availability shows us that the region has clearly enough weapons to be destabilized, and this becomes more complicated when we look at the circulation patterns. The movement of small arms in the region occurs at two levels – intra-state and inter-state. Internally, there are two ways in which arms move from legal to illegal possession. One is when corrupt police and army officers deliberately rent their weapons to criminals. The other is when weapons are lost. The figures in South Africa are quite alarming.

The cross-border movement of illegal weapons, however, is quite commonplace. The cross-border arms trafficking is broadly facilitated by, on the one hand, the existence of increasingly well-organized transnational criminal organizations; and on the other hand, the existence of well-established covert supply networks across the region, and this is where, I believe, it is very important to look if we are really serious about solving the problem. In fact, one interesting phenomenon on cross-border arms rustling in this region is that networks and routes utilized by organized criminals seem to resemble old pipelines.

And what do I mean by old pipelines? Firstly, our liberation movements, some of which are our governments today, being unable to get weapons legally – with good intentions they had to get weapons for the liberation war – devised undercover and covert routes to get weapons. These networks linked themselves in the region and their suppliers elsewhere in the world.

Another pipeline, which is open to debate, is what I call old African roots of smuggling of prohibited goods which have been reinforced by massive population movements. What do I mean by that? Refugees are a common feature of our region. Millions of people cross the borders because of war. They stay there sometimes more than five years, and they not only get to know the societies at large, but they have to survive by doing something. Unfortunately, some of these survival issues sometimes mean weapons smuggling.

The connections between alternate networks in use for new small arms construction is very clear. When the Cold War conflicts ended and minority regimes were defeated, because their arms supply networks were already in place, the only thing which happened was that these networks changed from political motivations to personal and criminal motivations. And to me, the knowledge and the political and social connections sustaining arms smuggling are very well routed, and new understaffed, under-resourced security agencies seem incapable of tracing the business, to the extent that one could argue that wherever in Southern Africa a conflict emerges, weapons will reach it in a very short space of time.

What are the solutions? The problem is very complex. It is regional and international in scope. It has a very fluid nature. It might be political; the next phase become socio-economic; and then cultural and developmental. It is a legacy of our recent past in the region, as well as a challenge for current governors. One could go on and on with different characterizations of the problem, but what is certain is the fact that unless we engage, manage and control it, human security will remain a far away ideal to attain.

Taking into account the complicating factors, the solutions should be necessarily multi-levelled and multi-dimensional. Multi-levelled in the sense that national, regional and international initiatives have to be considered. Multi-dimensional because the factors beneath arms possession and use are not at all static. Indeed, one may start by acquiring a weapon for security and political motivations, but in the process he or she may develop a gun culture, and later the weapon may acquire an economic value. This means that a new political and security dispensation which may result from a peace process may not necessarily be sufficient for him or her to dispose of their weapon, and indeed small arms proliferation, as I have pointed out in Southern Africa, has proven this argument.

Let me conclude by saying that I have tried to show how serious and complex the problem of small arms is in Southern Africa. I have also indicated that the solution has to be an integrated package containing political and security measures, as well as social, economic and cultural components. The expectation here is that while political measures give no excuse for resort to violence, dialogue to settle political differences, employment opportunities and general economic development invalidates the weapon. Peace culture education, formal as well as informal, on the other hand, may uproot the gun culture, which may have developed in some areas of this region. I recognize, though, that to take the sort of measures that I have mentioned as a solution is wishful thinking because crime and violence are social phenomena that cannot be wished away altogether. Therefore, these measures need to be backed up by enhanced security agencies to enforce laws.

LIGHT WEAPONS AND CONFLICT IN AFRICA

Jacob Kamenju

First and foremost I will bore you a little more with a bit of historical background because the root causes of the conflicts that we have at the moment have a historical background. I will be talking mainly about the region of the Greater Horn that includes Somalia, Djibouti, Eritrea, Ethiopia, Sudan, Uganda and Kenya, and also the Region of the Great Lakes; that is Rwanda, Burundi, Tanzania and touch a little on the Democratic Republic of Congo. The first thing I would like to say is that the traditional African set up before colonialism was very simple because there were few people and they had to manage their affairs, and the conflicts were based mainly on property, and that property consisted of land and animals. In the conflicts of those days, even when they went to war, when they fought, they only took animals, and women were treated with due respect. The fights of those days targeted warriors only. It was wrong to injure or kill the aged, women and children. It was a harmonious society: there was a way they were able to live together. The transfer of power, at that time, was through generations. The generations agreed that when one generation had finished its time the next generation took over.

During the colonial period conflicts again were at low levels, mainly because the security of the African communities in these areas was in colonial hands, except for Ethiopia, which was under a sort of monarchy so to speak. One major problem that the colonial regime brought was the division between the ethnic groups, because although the ethnic groups were fighting each other, they had no ill feelings about each other. But during colonialism, the same colonizers found it easy to identify different ethnic groups so that you can separate them for ease of ruling, so you have different entities. Therefore education in some communities was compulsory and some communities became more educated than others; there are those who at this time joined the colonial military, and so forth. So, one of the root causes of conflict in Africa today is ethnicity. This was created by the "divide and rule" principle used by the colonizers to enable them to have a good grip and maintain the separation between the groupings. I mention this because I think it is important.

Another root cause of the problem during the colonial era was the Bahrain Conference, which divided Africa by the ruler. They took a ruler and said, "This is the boundary". They drew it. That is why you find different groups: one ethnic group in one country, two countries, three countries, and if this had been resolved at that time the Ogaden War probably would not have happened because the Somalis perhaps would have been together.

In the period mid-1950s to mid-1960s, most African states became independent, and, of course, they were immediately followed by Zimbabwe and later South Africa. These new nations had a lot of hope; they had a lot of promises of freedom and prosperity, but as we all know, in this Conference here, this was not to be. There were many destabilizing factors and the nations that looked to prosperity did not prosper and therefore there was a lot of creation of animosity and mistrust among the same people who were looking for independence.

Then we had the Cold War – the rivalry between the superpowers. Again the superpowers divided people even more by urging countries to be friendly not for those people's interests but for the interests of the superpowers. In itself it may not have been wrong, but it included a lot of exchange or distribution of firepower to nations which perhaps did not need it at that time. The arming of governments in some cases, like in Somalia or later in Ethiopia, and insurgents in Angola, in Mozambique, and so forth, was an ingredient for future strife. Even now, it is very difficult to define what is illicit and what is legal in terms of weapons in the region, because today you have one government in power, they get all the arms to fight insurgents and the insurgents then take over, so, what was illicit becomes licit.

Then there was the break-up of the end of the Cold War. What happened was the Russian republics and the European manufacturers continued making the arms because they needed employment; they needed their industry; and they had other reasons; but for Africa it was the planting of a time bomb that is still going on, even today. And, of course, in later years, Egypt and South Africa also became arms suppliers in this region.

So, what are we seeing? We are seeing a lot of our region with a lot of weapons. And because weapons cannot be used without ammunition, we are seeing a growth of ammunition manufacturing factories, for example, in Kenya, in Uganda, in Tanzania, in Sudan. Again, this is another destabilizing factor.

So, what is our problem? Our problem in the region is too many arms. Too many arms in the region and very little control, very little knowledge of their dynamics, their movements, their routes, their sources, and so forth. I am glad to note that Southern Africa has done quite a bit of work in this field, and we will be trying to catch up as we go on. Our borders are porous – there are no borders in some cases. The closure, for example, of the Somali/Kenyan border will have very little effect because all we are saying is no crossing but you have nearly a 1,000 km boundary and the people who live there are the same people so they will cross at the place of their choice, and the effect on controlling firearms and illicit trade becomes reduced.

The problem again starts with a country like Somalia. Somalia initially welcomed the Soviet Union for many years before they changed over to the United States and the Russians at the same time switched over to Ethiopia and brought Mengistu to power. The reason for mentioning this is because there were arms and when superpowers get involved, when they change their position in the countries, they also change their equipment. So there were American weapons in Somalia and in plenty when the Soviets went to Ethiopia and *vice versa*. Those weapons are still perhaps some of the things that are making the problem so big. Mengistu had an army of about 380,000 when he was out of power. All those people were no longer going to join the military and they disappeared with those weapons.

In the Sudan there has been 17 years of war. Again, we ask what weapons are illicit in the Sudan? Is it the government-acquired weapons to fight the war in the south, or is it the weapons which are supplied to the Sudanese Peoples Liberation Army (SPLA) by the Ugandan government for example? Again, it depends on how you look at it. But at the end of the day, the end product is that there are lots and lots of arms in the area.

Ethiopia/Eritrea - this is another example of a war which is going on. There is modern weaponry there. I don't know what Pugwash has seen about Ethiopia and Eritrea going to war because after independence the relationships between the two countries was extremely good, but there were some territorial differences which were more personal than national between the two heads of state. When in 1986 Museveni took over the Ugandan government, we estimated that the northern Ugandans who formed the bulk of the military at that time had at least 120,000 small arms amongst themselves. That army was disbanded; some of it went back into their villages and started looking after cattle, with their weapons; some of them joined the guerrilla movement, and therefore those weapons are still there, and nobody can account for them except the people who hold them.

Of course, the Great Lakes Region has got the same problem. Again, as I mentioned, it becomes very difficult to say which side is illicit and which is not. When the Hutus are in power in Rwanda or Burundi the weapons they take are licit. When they change over we have a new government, and the earlier weapons become illicit; so how can we control their weapons in this region?

We have other peculiar sources that maybe do not apply elsewhere. For example, I mentioned earlier about the government supporting the rebels in neighbouring countries. I will not go into that. But we also have some governments which have got homeguards. Home-guards are armed by government to protect their own property and their own people, but we know now that those home guards become a threat to the next community, so cattle rustling becomes a very big issue. But, unlike earlier, when cattle rustling was a livelihood, a sort of a game so to speak, the bows and arrows and the spears that were used, have now changed. They have become automatic weapons. The essence of the trade has also changed: cattle rustling has become a business. So, when you have a raid, you end up with a lot of people killed, and this time again, it is not possible to discriminate, as we did earlier, as Africans, between the old, the women, and the children, because these raids would be carried out at night; there would be a lot of people trained in weapon handling; they will burn houses, so at the end of the day the people who should be protected within the community are the people who suffer. And that is why this problem of small arms is so great, and we feel very strongly about it.

The other thing is that the presence of small arms itself, as has been said many times, is not a conflict, but the fact that an individual is holding a gun, changes the methodology of negotiation and discussion because the individual feels powerful. So, instead of negotiating, he will say, "I have what I want, or I will shoot you." So there is a turn around where people want to use force more than they want to negotiate. We get into a vicious circle whereby this insecurity or this new methodology causes the need for increased personal security. So you will find that a government in power at this stage in the region there are lots of people who will want to buy, or to acquire, weapons legally through their government licensing system, and because of the political perspectives, where you are politically right, you get a licence, you get a gun, and to me, any single weapon put down there by the government, whether for home guards or individual security, is an increase to the fuelling of the conflict, and therefore this should be limited. And by the same token, the weapons in this region held by individuals have become a sort of status symbol. You own a gun, you must be powerful. Again we must have a way of approaching this problem and restricting the government to issue weapons to the very needy cases, and, at the end of the day, we must aim at weapons going only to security forces. How far we can go in that is debatable.

The other cause, or the other problem, that we face is a vicious circle again because the conflicts and instability in those areas are causing a reduction in productivity, the use of agriculture is becoming more complex because of the raiders. The working times you find in some countries in the region are decreasing. I will give you an example of Nairobi. People close business at dusk; as the sun goes down, because there is too much insecurity. There are too many armed robberies and if you have a big industry, or you are a distributor of a commodity that is capital intensive, then you cannot trade in the night because there is a problem and the effect of that is that people waste time having to leave offices and their places of work to do shopping when they would be best doing that on their way home. We have instances again in the region where the bigger companies are not willing to pay their people in the normal manner through a payroll because of their pay packages being robbed on their way from the bank to the pay points. This puts us again into the vicious circle of reduced productivity. In the region we have very big business in private security. It is estimated in Kenya alone that the turnover for the private security sector is in excess of two billion dollars per year. Now, this money could be used in a better way if we had better security. Car hijackings in the region, perhaps the South Africans and Zimbabweans will know this, because there is a very big circle of vehicles being stolen and being taken to the unstable areas, and it is a cycle that runs from South Africa through to Central Africa and Eastern Africa. So if you are in the business of four-wheel drive vehicles you find that you have less business to do because they are being stolen. So again this inhibits development.

We have a problem of closed schools because of these conflicts, which will mean that we end up with a lot of unemployed teachers; we end up with an uneducated people; and to me this is again a very destabilizing factor that we should look into and see how we can resolve conflicts to enable people to live their normal lives. And as was mentioned earlier, this problem of small arms is taken even further by the refugees, because the refugees in most cases in the region are combatant people who have been removed from whatever they were doing because of politics, and therefore they have weapons, they have the know how, and they are much better placed to survive on their own more than the host residents.

To conclude, I want to say only two things; that Pugwash is a great movement, and I ask that you assist us to get into the study of the conflicts in the Horn of Africa in any form so that we can build human capacity for research into the conflicts. We can borrow the expertise within this grouping and I would request you also to focus much more on this region because, from what I have seen from the representations and discussions in the working groups, there is not much information coming out of this region.

LIGHT WEAPONS AND CONFLICT IN AFRICA

Peter Lock

Addressing the issue from outside Africa it seems appropriate to start by highlighting some general features of the global context in which the African security crisis has emerged.

- Globalization is reinforcing inequality. The present course of development is marked by continued polarization of income, both between nations (internationally) as well as within nations (intra-societal).

As a result, social segregation dominates the changes of the social topography of societies all over the world. Most visibly, formerly public spaces are increasingly being converted into privatized and socially selective malls. But spatial segregation is pervasive, particularly in what sociologists have dubbed "glocal cities", the forerunners of globalization and modernization.

- Increasing levels of unemployment in the regular economy mark the situation at the global level, allowing for positive deviations in some well-off countries.

As a result informal economic activities increasingly become the backbone of societal reproduction. This evolution is not restricted to developing nations and countries in transition. Particularly affected are the new entrants to the segment of the economically active population. Unemployment rates of over 50 per cent for the age group 14-24 years are common.

- A general trend to curtail state functions following the logic of the neo-liberal paradigm and profound crises of state finances reinforce each other.

As a result, privatization of security is a pervasive trend. The private security industry has turned into a vibrant economic sector whose labour force already exceeds the police forces in most countries. A few globally operating companies dominate this "crisis-proof" economic sector. To guarantee the security of its citizens is apparently not any longer the "*raison d'être*" of the state. Security and particularly protection from violence, a basic human right, are in the process of becoming a traded commodity. The adage "My home is my (proactively protected) castle" is only now in our post-modern age developing its full meaning. Those who cannot afford protected castles and the proactive commodification of their security are forced to organize their self-defence outside the legal parameters. As a result, we are observing the evolution of private security into a mutually reinforcing system of multi-polar societal "rearmament", cascading down the social ladder where it amounts to an informal militianization of the society at the lower end of the social pyramid.

The underlying structure of these trends is the globalizing economic system. A closer look at the global economy reveals that it is composed of at least three different spaces. Asymmetrical interaction between the three sectors is a constitutive element of their respective existence.

The regular economy: the rule of law is a dominant feature at its core; taxes are generally paid, though the level of effective taxation is generally declining. The unrelenting competition to attract inward investment has led to lower tax rates and excessive tax exemptions, above all, in special economic zones like the famous "*maquiladora*" border industries in Mexico. The meagre growth rates of the regular economy do not keep pace with the increase of the economically active population seeking an entrance to this space.

- *The oceans of informal economies*: The rule of law applies but marginally; its rules reflect asymmetrical power structures often resorting to violence for arbitration. Informal economies are globally the fastest growing economic space in terms of participants. Almost by definition the informal sector does not pay taxes, value-added taxes, when (rarely) interacting as consumers with the regular economy, being an exception. Though often welfare levels generated in the informal sector hardly exceed the most basic subsistence, the sector contributes to the welfare in the regular economy through interaction mainly in form of cheap services of all kind. Another part of its produce is siphoned off into the coffers of criminal actors who impose their terms wherever the rule of law is not effectively supported by the State and people do not have the means to protect themselves.
- *The globally organized criminal economy*: By definition the rule of law does not apply. There is ample evidence that the criminal economy is the fastest growing space in the global economy in terms of turnover and profits. The criminal economy constitutes a parasitic link between the spaces of the regular and the informal sectors of the economy. Conservative estimates presented at an IMF- and EBRD-sponsored seminar put the global gross criminal product (GCP) at $1000 billion, this compares with a GNP for sub-Saharan Africa of $330 billion, of which South Africa alone accounts for more than one-third.

At the level of nation-states the three spaces of the global economy interact in different proportions according to the relative strength of the State. But in spite of enormous differences in the respective configuration of the economies at the level of nation-states, one sad feature they have presently in common is unacceptably high rates of youth unemployment. This pervasive feature seems to confirm the existence of interdependent patterns in the global economy; no economy can extricate oneself from under the prevailing machinations of the global economy.

I have termed this feature "intergenerational apartheid", though this evolution is not the result of a deliberate strategy arrived at by ideological consensus, as was the case with the apartheid practiced in South Africa. It manifests itself in unacceptably high levels of youth unemployment which run as high as 60 per cent of unemployed in the age group of 15-24 years in many countries at the end of this century. There are so far no indications that this pervasive trend will be reversed in the foreseeable future. The term "intergenerational apartheid" draws attention to the exclusion of the current generation aiming to enter the pool of the economically active population from the intergenerational continuity of economic roles which prevailed in many traditional societies until recently. Due to the unmerciful processes of economic modernization the traditional succession of professional roles is replaced by migration towards the urban spheres which cannot cope with the onslaught of masses of unqualified labour. Most embarrassingly, in many failing states the level of education achieved after decolonization is not any longer being maintained. This intergenerational decline of educational levels further entrenches the observed "intergenerational apartheid". On the African continent the situation is further exacerbated by the sociological situation. Africa is the youngest continent; half of its population are under the age of 18.

Moreover, the constant migration of jobs throughout the global market place towards ever cheaper locations of production tends to reduce job offers for less qualified labour in many advanced countries as well. Certain vulnerable groups like minorities and above all young people tend to become excluded from the regular economy and provide a reserve army easily attracted by the jobs the criminal sector of the economy is offering. This is by no means restricted to drug trafficking. The sociological composition of the prison population clearly reflects the specific pattern of exclusion in each country.

What do these global trends preordain for Africa? Above all, the trends described so far manifest themselves in extreme forms throughout the continent. Violence and armed conflicts in Africa are conditioned by an advanced level of "trifurcation" of the socio-economic tissue into regular, informal and criminal economic spheres.

The absolute preponderance of either the informal or the criminal sector in association with the respective other leaves many African states without the resources needed to function even at minimum levels. As a result, the state is of often drawn partially, and in extreme cases fully, into the criminal sector. In a number of countries, tax paying economic actors and state-owned industries as a resource of the government have largely disappeared. The latter were either appropriated or "privatized" by the incumbent, though corrupt, rulers, or lost out in the context of global competition. Furthermore, it is estimated that at least 50 per cent of cross-border trade in Africa is not accounted for by any government. One part of this trade reflects traditional trade patterns which survived colonial and post-colonial regimes, while the rest is accounted for by downright criminal activity. Whatever the causes are for this state of affairs, the result is unambiguous. The state as regulator of social relations is seriously weakened and proves unable to provide proper policing on its territory. The ensuing development can be adequately described as an intra-societal arms race resulting eventually in a process of competitive militianization of the entire society.

In the initial phases of social fragmentation virtually everybody feels sheltered by adopting an explicit social identity; later pressures emerge to develop discriminating group identities because they facilitate some protection through a wide range of mechanisms, from collective political leverage and informal spatial relocation to armed self-defence. Once set in motion, the politics of collective identities are mutually reinforcing and provide a fertile playing field for ethnic, religious and political manipulators. In contrast to the seemingly virtuous causes they preach, these manipulators are generally deeply entangled in the criminal economy and can be described as "political entrepreneurs". They regularly promise a bright future provided other groups are excluded politically, economically and eventually even physically. These ideologies based on exclusion foster the militianization of the society. A latent temptation to employ violent means to resolve even minor conflicts goes with this development. Since ideologies of collective identity unfailingly portray any act of violence committed in pursuit of the "common cause" as legitimate self-defence, the dangers of escalation into a civil war are obvious. At this stage, the huge reserve armies of young men generated by the pervasive intergenerational apartheid enter the stage and see fighting as their (only) chance of inclusion, not entirely without justification.

The cynical appropriation of an ideology and the ruthless application of violence in pursuit of societal inclusion by the so-called "hittistes" (young men who lean against the wall) are empirically documented in the context of the on-going Algerian civil war [Luis Martinez, *La guerre civile en Algérie*, Paris 1998]. Young men turned into territorial warlords in suburban areas under the guise of fundamentalist fighters who became known as "emirs". Ruthless violence allows the "emirs" to position themselves in society. Contrary to outside perceptions they do not aim at a fundamentalist society. While the forms vary considerably, a large part of violence can be explained by the exclusion-inclusion syndrome.

The protracted civil war in Sierra Leone is also sustained by the large-scale exclusion of the plentiful youth in the ontext of a failing state and the underlying economic disaster [Paul Richards, *Fighting for the Rainforest*, London 1996]. In the course of the armed rebellion, largely uneducated footloose young men and children formed a social enclave with norms of its own. By now Sierra Leone's war has turned into the most palpable manifestation of a "youth war". It reflects an evolutionary process where the initial "guerrilla" leadership has long since taken a back seat. It also explains why the outside world fails to understand the seemingly irrational acts of violence in this war. Traditional procedures of mediation have failed for lack of an effective leadership in control of the fighting party.

In the sometimes emotional debate about armed conflicts, the atrocious acts of violence committed therein and the participation of child soldiers, the perception of the perpetrators is rarely taken into account. As the stage of such events is generally conveniently distant, it is bracketed with brutish, pre-modern cultures. This impedes the recognition of similarities with patterns, at much smaller scales though, in the context of industrialized countries. Sociologists report, for example, that the formation of youth gangs in metropolitan areas aiming to control criminal activities within a specific territory provide marginalized young males with more attractive roles than are offered within the regular or informal sectors. More generally we have to accept the fact that displaying an armed disposition to act violently, be it in the context of criminal endeavours or military-type fighting, offers a concrete chance for the excluded groups of the society to develop an identity which warrants a feeling of self-respect. A gun can transmute a nobody into a man whom others (are forced to) respect. Entire generations of young males pushed into economic apartheid are prone to take their chance if offered to them in the circumstances and are disposed to consider armed violence as a means to enforce their inclusion in one way or another. It is for this reason that the pervasive illicit availability of small arms in the post-Cold War period has such a destabilizing effect throughout the world. Either as tools of entrepreneurs of violence or, on their own account, young men form the recipient end of the illegal proliferation and diffusion of small arms which create havoc and untold suffering throughout the world.

In the context of the continuing social fragmentation and ensuing conflicts, the illicit availability of small arms operates as a horrific force multiplier. Whether recently smuggled into the African continent or sent in vast quantities during the Cold War to proxy governments, virtually all small arms originated in industrialized countries. It is for this reason that the industrialized countries, and Europe in particular, have a responsibility to actively join the combat against the illicit small arms availability in Africa. While many African governments have clearly recognized the absolute need to bring the black markets for small arms under control, they do not have the necessary means and are not likely to succeed if Europe does not support these endeavours.

Apart from better policing of international transactions, there is evidence that the black market for small arms is sensitive to prices and displays a high price elasticity. Proactive interference aiming at drying off certain known outlets would be a non-conventional measure worth testing. It is well known, for example, that large quantities of small arms are procured by the criminal groups from South Africa in the border regions of Angola and Mozambique, where huge surpluses from the civil war are being marketed. Investing a few million dollars in proactive undercover purchases of small arms in the marketplaces of the border towns looks cheap compared to the costs of armed criminality in South Africa. One could expect prices on the black market to soar. Additionally, the insight in black market operations gained would contribute to improved and internationally co-ordinated controls.

Many armed conflicts in Africa are entrenched and have developed their own social life and, most importantly, economies. The most annoying aspect of this development is that these wars have transmuted into a mode of (re)production for large groups of combatants, as well as traders and smugglers involved in the logistics of resupply. These war economies are based on criminal transactions, theft, deception, extortion and only rarely on productive activities. As a result these war economies generate disproportionately large numbers of refugees and internally displaced persons.

The lavish support fighting parties in Africa, as proxies of one or the other side of the Cold War antagonism, were receiving is not any longer the backbone of the war economies. The fighting parties have successfully expanded other sources of revenue. It is a bitter irony that the misappropriation of humanitarian aid has become an important element in the reproduction and sometimes even in the military logistics of the warring parties. Africa's wars have converted the continent into a major turntable of global drug trafficking. It is well established that the networks of drug trafficking also engage in the illegal trade of small arms. The illegal export of raw materials, most prominently gems, diamonds, ivory and so on. constitute yet another pillar of war economies in Africa. Finally, expatriate communities working in Europe, North America or in oil-exporting Arab countries, whose ethnic kin are involved in conflict, are tapped to send money and to organize arms transfers on the black market.

The common denominator of all resources mobilized to sustain a war economy is that they are linked to the global economy. Since the end of the Cold War the war economies have to generate their own resources and most importantly a significant share of these resources must be convertible into foreign exchange, as arms can only be procured through cash transactions on the black market. This inevitably leads to trade links with the criminal sphere of the global economy. In turn, countries weakened by an on-going armed conflict are deliberately targeted by powerful criminal actors who seek opportunities to launder money and to recruit cheap personnel prepared to take unreasonable risks as is needed in drug trafficking operations.

The indispensable global outreach of the war economies in Africa shifts the responsibility in part to the major economic powers which have the responsibility to clean up the shady segments of international trade, where the proceeds of the war economies enter global circulation and eventually the regular economy. Contrary to the perceived images of wars in Africa, they are not just local affairs. And it is not only the famous merchants of death who secure the supply of basically small arms which keep the wars going. The flip side of arms trafficking at the receiving end is the generation of convertible income which depends on criminal entry points into the global circulation of disposable commodities. The constant refuelling of Africa's war economies is brokered by shady middlemen who, for an extra-profit, clear the exports of war economies for regular circulation. Neither WTO nor OECD, for example, is known to be taking meaningful steps to improve the policing of critical commodity exports. In this sense, Africa's wars are global affairs and fall within the responsibility of dominant nations in the global economy.

To the extent that states weaken and cannot any longer sustain the rule of law, economic transactions are steered by a mix of communitarian consensus and violence, or rather the threat thereof. With the excessive illicit availability of small arms throughout Africa, often at prices which are below the costs of production, arbitration of conflicts often escalates into violence. The ensuing dynamics of violence prepare the ground for open civil war, in which the stakeholders of government positions are usually part of the problem, who abuse their position for exclusionist strategies. From an analytical point of view it might be argued that the violence and wars are no longer discrete events in Africa; the underlying structural causes cannot be distinguished.

From an individual perspective, an automatic gun is such a terrifying force multiplier that one would rather have one than not, if the environment is perceived as dangerous. The distinction between a civilian and a fighter becomes flimsy, except for the fact that being a combatant in Africa's civil wars reduces the risk of being killed. The abundance of illicit weapons constitutes indeed an onerous obstacle to shifting the direction of governance in Africa towards democratic conflict resolution and a suppression of the mushrooming criminal economic spheres. Under the hypothesis that these are essential preconditions to opening a path of development in Africa, combating the large-scale illicit small arms availability should become a priority on the political agenda of international fora.

The picture of Africa drawn in this sketch is, of course, rather selective and does not do full justice to the strengths African societies display. The gloomy side of Africa's future is the overarching weight of the criminal economy whose grip seems to tighten even further. This pressure can only be alleviated by resolute action of the major economic powers aiming at cleaning the dirty backyards of the global economy. Otherwise, African societies have strong resources of their own to manage their development.

The resilience and ingenuity of the informal sector of the African economy, which resisted colonial rule and the comprador regimes which took over at independence, are not part of the public image of the continent. But the fact, already mentioned, that 50 per cent of all cross-border trade is not accounted for by either state, is proof of strong autonomous regional organization of production, and trade which thrives in spite of the largely failed system of nation-states. If developed and protected, the informal economy has a potential to absorb the excluded masses, while the regular economy under the spell of worldwide competition does not have the potential of integrating the marginalized masses of young people. It is for this reason that the continent has to invent its path of development; the human and organizational resources exist.

PROCEEDINGS OF THE FIFTH PLENARY SESSION

Held on Sunday, 12 September 1999 at 10.30

Reports on Working Groups

Part A: Chairman: Professor Michiji Konuma

1. **Dr. Jozef Goldblat** presented the Report on Working Group 1 (text on pp. 27-29). The following took part in the discussion: Dr. Esmat Ezz, Dr. Venance Journé, Dr. Martin Kaplan, Prof. Keun Park, Prof. Luiz Pinguelli Rosa, Prof. George Rathjens, Dr. Aharon Zohar.

2. **Mr. André Mechelynck** presented the Report on Working Group 4 (text on pp. 39-44). The following took part in the discussion: Prof. Ulrich Albrecht, Prof. Mwesiga Baregu, Mr. Paul Guinnessy, Mr. Lucius Mendis, Prof. Morris Miller, Prof. Luiz Pinguelli Rosa, Mr. Carsten Rohr, Mr. Motomisi Tawana.

3. **Professor John Avery** presented the Report on Working Group 5 (text on pp. 45-47). The following took part in the discussion: Mr. Muhammad Afzal, Prof. Gabriel Baramki, Amb. Omran El-Shafei, Dr. William Epstein, Dr. Esmat Ezz, Mr. David Fischer, Prof. Allen Isaacman, Dr. Catherine Kelleher, Prof. Onaney Muñiz Gutiérrez, Mr. Hussein Solomon, Mr. Motomisi Tawana, Prof. Bhalchandra Udgaonkar.

The session ended at 13.20.

Part B: Held at 14.15

Chairman: Professor Ulbrich Albrecht

4. **Professor Morris Miller** presented the Report on Working Group 3 (text on pp. 34-38). The following took part in the discussion: Sir Michael Atiyah, Prof. Mwesiga Baregu, Prof Luis Masperi.

5. **Dr. Jo Husbands** presented the Report on Working Group 2 (text on 30-33). The following took part in the discussion: Mr. David Fischer, Ms. Sandra Ionno, Dr. Martin Kaplan, Mr. Lucius Mendis, Mr. Hussein Solomon, Mr. Motomisi Tawana.

The session ended at 15.30.

PROCEEDINGS OF THE FINAL PLENARY SESSION

Held on Monday, 13 September 1999 at 09.00

Chair: Dr. Catherine Kelleher

1. Panel Discussion: **"Sovereignty and Intervention: International Perspectives"**

2. The four panellists presented their views:

   a. Dr. Alexander Nikitin (Russia) (text pp. 121-124)
   b. Air Commodore Jasjit Singh (India) (text pp. 125-127)
   c. Professor George Rathjens (USA) (text pp. 128-130)
   d. Mr. Derek Powell (South Africa) (text pp. 131-136)

3. The following took part in the general discussion: Professor Liu Min, Professor Luiz Pinguelli Rosa, Dr. Walter Scheuer, Mr. Hussein Solomon.

4. **Presidential Address**
   The President, Sir Michael Atiyah, gave his address (text pp. 137-142).

The Conference came to an end at 12.30.

SOVEREIGNTY AND INTERVENTION: INTERNATIONAL PERSPECTIVES

Alexander Nikitin

This is a timely occasion to be discussing sovereignty, because just last night another bomb was exploded in Moscow killing many people, under the name of sovereignty; sovereignty of some part of the country which is seeking self-governance. The problem is how can we deal with issues like sovereignty, self-determination, territorial integrity, in the modern world? As a matter of fact, the notion of sovereignty is often misused and misunderstood. Some people say that sovereignty is non-recognition of some external authorities, a kind of self-determination of a group of people, territory or geostrategic space which pretends to be on its own. People who are for sovereignty and against external intervention sometimes support "freedom fighters" – but, of course, there are different interpretations, and in other parts of the world they may be called separatists, nationalists, and thus this combination of overlapping notions brings us to different concepts of intervention and sovereignty.

It was Lenin who introduced the notion of proletarian internationalism, a theory which suggests that sovereignty is obsolete; that groups of people from different nations united by economic interests should work and unite across transnational borders. Stalin brought the notion of centralization to the extreme, by domination of the central authority of the country over the provinces. In practice, Stalin converted the multinational Soviet Union into a unitary State. Brezhnev is known for Brezhnev's doctrine, which says that if some country attempts to leave the ideological camp then the central authority has the right to strike back against the sovereign rights of, say, the Hungarians, Czechoslovaks or Poles, when they were trying to break with socialism. It was Gorbachev who introduced the end of empire concepts. In the modern world we have this opposition between notions of sovereignty and intervention also in the form of conflict between self-determination and territorial integrity. The central national authority of a sovereign country contends that all the territory under their sovereignty should be integral (and cannot be broken into pieces). But, look what Boris Yeltsin said when he was a revolutionary in 1991, when he was the one behind the break up of the Soviet Union. He addressed the leaders of the Republics and suggested that they, leaders of future independent states, eat as much sovereignty as they could swallow! Several years later, when Yeltsin himself became the leader of a large country, he was saying that nobody can violate the sovereignty of the central authorities – addressing the parts of the Russian Federation that are now looking for secession or independence.

This is why in discussing contradictions between sovereignty and intervention, we very often name these contradictions differently: republics versus federal power; provinces versus centre; colonies versus metropolis.

Who has the right to interfere on behalf of the international community? No-one doubts that the UN has such a right, but also we have in Chapter 8 of the UN Charter the absolutely clear statement that regional security organizations have the right to perform peacekeeping in the area for which they have geostrategic responsibility. What kind of regional organizations are recognized by the UN? The Organization for Security and Cooperation in Europe (OSCE) is one example. In the Asian area we have a conglomerate around the ASEAN grouping, in particular, the Asian Regional Forum (ARF), an organization created specifically for Asian conflict resolution. We have the Organization of American States (OAS). We have the Organization of African Unity (OAU). Here in the Southern part of Africa there are special arrangements such as the South African Development Community (SADC). And we have the Commonwealth of Independent States (CIS) as the most recently recognized under Chapter 8 regional security organization. It was recognized in 1994 by a special resolution of the United Nations.

It is very interesting that in different parts of the world there are instances where regional organizations have intervened in conflicts without a UN mandate. They either had some special separate mandate or sent military forces in support of UN authority. In Africa, there were cases in Liberia, Congo and Rwanda, where the OAU took special legal decisions to send international African troops to the conflicts. The OAS took a special decision on the conflict in Grenada; the action in Grenada was not mandated by the UN but rather legally supported by the OAS. The OSCE in 1993 got a mandate for separate OSCE peacekeeping operations in Nagorno-Karabakh. This did not happen. The OSCE had neither the troops nor the financial resources to carry this out, but still the fact remains that the European security organization got a mandate, not from the UN, to interfere in a regional conflict. Between 1992 and today, the CIS twice, in Tajikistan and Georgia, undertook international interventions with large forces on behalf of the regional security organization.

In 1993, large CIS collective peace support forces were created to intervene in Tajikistan, consisting not only of Russian military troops but also troops sent by Uzbekistan, Kazakhstan, Kirgizstan and Russia. And the size of the intervention force was very large, more than 6,000 troops, 190 tanks and so on. It was really a war, which was waged for several years, against different local gangs of rebels. This war was finished in 1997 by the very important international arrangements which brought to the negotiating table different forces such as the Tajik government, Tajik united opposition, and the other States of the region. What is as important is that three international organizations were invited to the negotiations as guarantors of the peace accords. Not only the UN, but also the OSCE and the Organization of the Islamic Conference. The Organization of the Islamic Conference played a very interesting and unexpectedly positive role in the peace process. The regional political leader of Tajikistan was able to sign a peace accord with the leader of the opposition and this was guaranteed by the signature of UN representatives – but again, without any formal mandate for UN intervention in this part of the world.

Ossetians are split into two parts, one on the territory of Georgia and one on the territory of Russia. When, in 1992, they wanted independence, self-determination and creation of a separate Ossetian State, both countries – Georgia and Russia – were interested in thwarting this effort for independence. They created bilateral military forces to intervene in the conflict and to disengage warring parties. In the western part of Georgia we have another area called Abkhazia, which is on the border with Russia, and in 1993 also attempted to achieve an independent State of Abkhazia. Again, local regional peacekeeping forces were introduced, and quite large manpower of 1500 men with helicopters were brought into the area by a mandate of the CIS; nine Presidents of nine States gathered and signed a thick document clarifying the mandate for this international military force to stop the conflict between Abkhazia and Georgia.

In the state of Moldova there is the area called Transmistria between Ukraine and Moldova. It is populated by Russians, and again, in recent years, there have been calls for an independent Transmistrian State which would secede from Moldova. Again, the UN was able only to send a small group of observers to monitor the events, but was not able to reach any decision about whether the international community should interfere and send any international troops to the area. Instead, the presidents of Moldova and Russia co-signed a bilateral interstate agreement to intervene in the conflict and separate the warring sides. This was co-signed by the president of the self-proclaimed Transmistrian Republic, because he also was interested in stopping violence. This was quite a successful operation and violence was stopped very quickly.

Right now we have a very difficult situation in the area of Dagestan. Dagestan is a part of the Russian Federation and borders Chechnya. In geopolitical terms this area is important because it is close to the Caspian oilfields, and therefore very important to the authorities in Moscow. This is why when self-determination hostilities strike this area, Moscow is very severe in using military force to try to ensure that this part of the territory stays with Russia and remains a normal member of the Russian Federation. At the same time, the constant interference from Chechnya, and attempts to create a unified Chechnya-Dagestan Islamic State, is a clear challenge not only to Moscow but also to the world community because a decision is needed as to whether separatism or nationalism is always supported as an expression of self-determination of a people, or whether in some cases, when it happens on the territory of a well-established State, it should be considered the internal affair of the country whether to allow the group of people to secede.

Let us come back to the question of who has the right to interfere. Among security organizations, NATO is an exception. NATO is not a regional security organization. Not because the UN does not want NATO to be such a regional organization, but because this is the clear decision of NATO itself. It is not operating under Chapter 8 of the UN and stays in a separate category. This is why when 11 weeks of air strikes started on the territory of Yugoslavia, some countries, first of all Russia and China, raised the case that it is not appropriate to intervene without a UN mandate or regional organization like the OSCE. India, which is not a member of the Security Council, in general supported such a view. So three large States, which represent almost half of mankind, were against the actions which were undertaken by another group of States. And guess why Russia, China and India are against support of separatism in Yugoslavia and against unmandated intervention? I have three possible options: 1/ they don't like Muslims; 2/ they don't like Madeleine Albright!; 3/ they are afraid of separatism in their own countries. I would rule out the first option as politically incorrect and, to my mind, the truth is some way between the second and third option!

What Moscow was and is expecting from NATO is a transformation. The real transformation that was promised by NATO, from preparation for massive war towards peacekeeping. Also, Moscow is even ready to support the transformation of NATO from collective defence to collective security – for becoming a kind of OSCE with muscles. But that is not what is happening, because in the perception of Russians, NATO expanded its function by intervention, and together with political steps undertaken in the last three years has changed the attitude in many countries, including Russia, to the political and strategic future of Europe, including the role of NATO in it.

As a matter of fact, the use of military force against a sovereign state without UN mandate, which was done in the case of Kosovo, in the first stage when there was no mandate, was first of all an out-of-area operation for NATO (not a collective defence action), and was in contradiction to the UN Charter, which requires a decision from a political authority, be it the UN or the OSCE, for such an intervention. It caused a crisis in NATO-Russia relations. And it was a change to the post-Cold War order – a change to the rules of the game.

But the rules of the game had already been changed in Bosnia when the UN was marginalized by negotiations conducted directly by the G-7 who eliminated the Contact Group and circumvented the UN and OSCE. Russia, by the way, was quite happy to try to act through the G-7. We even hoped that the G-7 was becoming, if not the G-8, at least the G-7½. At the same time, the United States suggested some different arrangements for Yugoslavia, starting the Dayton process. This was not a UN process. This is why Russia and the USA have two different approaches to arrangements for Bosnia. Russia was suggesting going through international organizations like the UN and the OSCE; the United States arranged a special, unprecedented political dialogue which was orchestrated aside from international organizations.

It has been said at a previous Pugwash workshop that NATO's intervention was not legitimate but just. Others replied, that the government of Yugoslavia is not just but legitimate. We have a contradiction here. The international community should decide whether, if something is legal, you should never touch it, or whether human rights issues are serious enough that the international community should interfere even if this is illegal by international standards.

Amongst lessons to be learned from Kosovo, one could say that intervention without mandate damages the very notion of legitimate international humanitarian intervention. The example of a group of strong states who interfere by their own decision does not work in Africa or Asia. You see that in East Timor, the government of Indonesia turned to the UN to supply a legitimate international peacekeeping force and not just a combination of the forces of the strong regional states.

Pugwash should think not only about international relations, but also about architecture, especially the architecture of security. The Cold War security architecture was based on two pillars – the USA and USSR were very dominant powers. They based their domination in the world on a foundation of nuclear deterrence and balance of power. This is what I call a classical international security architecture. Then, what we have in Asia are arrangements of ASEAN. The Asian Regional Forum was very helpful in the conflicts with Vietnam, Kampuchea, Burma, and the economic sanctions applied by APEC were also very important. Russia is now looking towards participation in the Asian security system if it is pushed out from the European one. In Europe, we have a security architecture resembling a palace in the style of Rococo – an architectural style characterized by an abundance of unnecessary decorative details. In Europe we have a lot of security organizations that pretend to help in the case of conflicts between their members. We have OSCE – a good forum for talk but not so good for action. We have the CIS – the new arrival in the family of regional organizations, which try to show teeth and try to act in Tajikistan and Georgia. We have also a group of "soft" European organizations, like the Council of Europe, the European Parliament, and the European Union (still dreaming of creating an ESDI). The word is that the European Union should create its own military arm which would be different from NATO. In practice, NATO has for now become the ESDI for Europe. NATO occupies the niche in the European security architecture which before was intended not for NATO but for some international political organization attached to the European Union or, say, to the OSCE. But neither the EU nor the OSCE has been able to create its own conflict resolution mechanism.

What trends do we see in the area of the international struggle between sovereignty and intervention? First of all, the international community in the 1990s definitely interferes more than in the 1980s or 1970s, and is beginning to interfere in internal conflicts within sovereign States. That is not legally well thought through. Peacekeeping becomes a field for conflict of great power interests. We see this in Yugoslavia, for example, where Russia and NATO are vying with one another. The countries project their national interests into peacekeeping. They do not use peacekeeping as a clean international humanitarian intervention.

Another trend is that international force is being used on a greater scale, and not only for the self-defence of peacekeepers. Interference is also becoming more and more violent, more and more bloody. Much more heavy weaponry is used – aviation, reconnaissance, which is all prohibited, by the way, in the traditional UN approach to peacekeeping.

The UN has been unable to create standing peacekeeping and peace-enforcing forces. Canada and some other countries are still backing the proposal to create a strong international peacekeeping force, which would be donated by nations and operated by the UN. But in the absence of such a standing force on the eve of the twenty-first century, the role of regional peacekeeping and regional organizations grows. We do not have any formula for good relations between these regional organizations and the UN. Currently, the interactions between them are inadequate. What is certain is that humanitarian intervention cannot be equated with military intervention only. It requires much broader and more comprehensive action.

Erosion of sovereignty is, in any case, taking place in the interdependent world, by radio broadcasting, CNN, satellite television, Internet – all of that is an assault against the concept of sovereignty. You are not as sovereign as you were before, if you are the government of a small state. Everybody can operate on your territory, through radio, through television, through economic sanctions – which means that intervention, interference, becomes multidimensional and inevitable. The international community is responsible for conflict resolution among its members. But we still do not have a clear legal formula for it in Europe, Asia, Africa. We are looking for this formula in the twenty-first century. Sovereignty and intervention will be among the major problems of the coming century and hopefully Pugwash will play a role in the resolution of this issue.

SOVEREIGNTY AND INTERVENTION: INTERNATIONAL PERSPECTIVES

Jasjit Singh

We can get tied up with immediate issues and forget the fundamentals. So I will focus on the fundamentals: what is sovereignty and how should we define intervention?

The sovereign State is the only organizing principle of nation states and the international system. We do not have anything else and until such time as the world can define a new system, like world government, or something else, we are living in a world of sovereign states. And as long as sovereign States exist, certain rights and obligations go with that. Sovereignty was never absolute. I think in mythical India there is a concept of absolute sovereignty, but that is more absolute suzerainty than sovereignty. An argument, therefore, that sovereignty is not absolute is a non-starter – the question is: what happens when it is not absolute? Either it gets weakened by external pressures or internal weaknesses, or it is voluntarily relinquished. The difference with the latter is that this is the exercising of sovereignty to give up part of sovereignty. It is a sovereign decision of NATO countries to join NATO. They are not being forced into it by an external factor in any case.

The most interesting aspect of sovereignty is the sensitivity of the large and powerful states, such as the United States, China and India. The United States will not accept, most of the time, a non-US commander of a UN peacekeeping force. Nor would many of the others accept taking orders from the other ones, except perhaps an ally.

The question that comes to my mind, therefore, is do we have an alternative to sovereignty without getting into anarchy in the international system? I do not have one. And if Pugwash could find one then we would, I think, be well on our way to making a revolution in international affairs.

Should sovereignty be treated in unequal terms? Is the smaller and weaker State's sovereignty less important than that of the bigger State? If that is so, we return to the law of the jungle. What is clear is that the UN is an organization of sovereign nation states. The organization is based on the principle of sovereign equality, and all nations, according to the UN Charter, shall refrain from the use of force against the territorial integrity or political independence of any State.

One can argue that the Charter is unable, today, to meet specific situations and requirements that have arisen because of changing circumstances. In which case, the answer is to amend the Charter, not to violate the Charter at every step.

It is also written that nothing contained in the present Charter shall authorize the UN to intervene in matters which are essentially within the domestic jurisdiction of any State. This is very clear and categorical. And this is where the problem comes in again – that in some areas, the Charter, because it was defined in a certain time and place, did not visualize situations of this nature and, therefore, upheld the right of a sovereign State to manage its affairs, to pursue its own socio-economic and political goals (except where people like Brezhnev or various US Presidents tried to introduce their own doctrines in pursuit of their own influence). So, if you go by the UN Charter, even the UN cannot carry out intervention unless there is a clear understanding and mandate.

So, how should we look at intervention? It is a complex subject. At times, merely sending humanitarian aid is seen as intervening in another country. Satellite channels, communication systems, trading systems and financial transactions are all forms of intervention. So we must be clear about the context in which we are discussing intervention.

The definition that I am using for the purposes of this discussion is: an action by a state or organization inside another sovereign state without the sovereign State's concurrence. Because if you have intervened with the concurrence of the State, this is not intervention. If the President of Indonesia has accepted a UN force, should we call it intervention? In the general use of the language, yes; but in real terms, when discussing sovereignty and intervention, it is not, because Indonesia has made a sovereign choice in accepting UN forces; the UN is operating within the broad framework of the Charter even if not with specific elements of the Charter.

Let me look at the nature of intervention within that definition. I find great complexity. What types of intervention can take place? Economic, political-military, diplomatic, military? Who does it? Unilateral or multilateral? UN-authorized or UN-conducted? Each one has its own implications and must be looked at in a specific way.

Why is there a need for intervention? I could go back to the interventions of the 18th and 19th centuries. Historical accounts will tell us that some of those interventions were necessary for the sake of "civilizing the natives." A very noble idea which is not very different from the interpretation of some people today, that they are trying to protect human rights – and then they bomb a country and create about 100 times more refugees.

When an intervention is carried out to support and protect the State, one would assume that it would have the concurrence of the legitimate government of that State. There are also interventions that take place in pursuit of national and geopolitical goals of the country that intervenes. But the case that I think we are concerned most about, is the humanitarian intervention. This occurs in two broad situations: 1/ the State is weakening, some reaching levels (and I do not like the term) of imploding - some people call this a failing State, where the State institutions are no longer able to function in any coherent manner, and, therefore, on humanitarian grounds, and possibly to maintain international peace and security, there would be a need or reason for intervention.

There is also the case of oppressive regimes. Would there then be a necessity for intervention? In many ways, this will be decided, in real life, only if you come across an oppressive regime of a small country. If you have an oppressive regime in a country of a billion-odd people, I do not think that anybody is going to intervene.

On the other hand, interventions are not necessarily negative every time. There are good reasons. For example, what about the security of small and weak States? There are many circumstances where, whether or not you get the concurrence of the legitimate government, it may be necessary to intervene to try to restore security, peace, stability and order.

Unfortunately, the Charter does not have provision to cope with the problem of weakening States, oppressive regimes, and many other aspects. And that is the real problem. The problem, in my judgement, is not so much of interventions; there are certain norms and rules for intervention, and these themselves have undergone change. If you recall, in 1961 the UN mandated an intervention in Congo for suppressing a separatist movement – not to allow Congo to break up.

The UN Secretary-General has stated that peacekeeping can rightly be called the invention of the United Nations. It has been invented, but we have yet to formulate the rules, and how to manage it properly under all possible circumstances.

How have interventions in fact been carried out? There is a unilateral element. Countries do it in a variety of ways. In fact a new dimension has arisen where a formal intervention, especially of the military type, is being replaced by the non-formal semi-military intervention by militants, terrorists and Taliban-type phenomena.

In the case of multilateral actions, some are UN-conducted, some UN-sanctioned, and some are simply carried out by the regional organizations by themselves.

What we need is to finally arrive at a respectable, workable, efficient balance between sovereignty and intervention - on the assumption that sovereignty and sovereign States are not likely to disappear, even in the twenty-first century. We get examples of Europe and the European Union. I think it is forgotten that the European Union is coming together slowly, some would say painfully, but through voluntary, sovereign decisions of each of the States involved. I can give you one more example, and that is of India, which came together, barring the portion that became Pakistan. On 14 August 1947, that territory had 447 sovereign states. The Republic of India came into being on 26 February 1950. We have seen a bit of the reverse in the former Soviet Union, but we are not sure as yet what the future will hold.

On the other hand, we know that the realities of humanitarian concerns are such that intervention may be necessary. What we need is to decide how to create and implement norms. In my view, UN authority should be mandatory for any intervention. And to repeat, my definition of intervention is that an act is carried out, especially the use of force, within a sovereign country without the concurrence of that country's legitimate government, regardless of whether or not you like that government.

This is where we come up against a problem. The risks and dangers of veto. So while the veto itself may not be touched, can we think at least of modifying the rules for intervention and how the veto should play under those circumstances? And to strengthen the democratic principle, the endorsement of the UN General Assembly must be necessary for any military intervention. That then reflects the will of the international community, not merely the will of the Security Council. It is something that ought to have happened by now but has not.

The USSR used to veto a lot but in recent years has done so very little. In contrast, the USA, UK and France have all increased the frequency of veto. China once put a veto but since then it has probably believed that mere threat of veto is enough. It is this situation that is, in a way, forcing the situation to be shifted from the UN Security Council to action outside the Security Council. While the UN Security Council certainly is the primary instrument for international peace and security, if it is not allowed to function then other ways will, of course, be found. NATO can argue that it cannot accept this thing going on in its own backyard and must, therefore, intervene; Milosevic is certainly not our – and I can speak for most Indians, I think – model of a ruler or a regime. But, on the other hand, it should become part of the rules of business of the United Nations that, where there is likely to be any difficulty, the matter be referred for voting to the General Assembly – and then if the majority of the world say something, I think we can proceed.

SOVEREIGNTY AND INTERVENTION: INTERNATIONAL PERSPECTIVES

George Rathjens

I am going to start by commenting on the events of yesterday, when we seem to have got out of a terrible situation in East Timor. It seems to me that we faced, there, a really fundamental question which probably is not widely understood. I think it was a very tough test case, particularly after Kosovo. It looked, for a while, that there would be no satisfactory resolution. If we stood by, without the government of Indonesia being willing to accept intervention, and if there were severe displacements of people and massacres, such as we have seen in Kosovo, it would have been an enormous blow to the credibility of the international community, and particularly to that of the United Nations. On the other hand, if we went in without the concurrence of the Security Council (I say "we," the Western world), and intervened forcefully, that would have been an equally severe blow to the UN. It seemed to me that we had been very likely to see one of these two solutions, both of which would have devastating effects on the future utility of the UN and on its credibility as an instrument for dealing with these problems. I just hope now that if it proves necessary to put troops in, it will be done with the acquiescence, and at the invitation, of the Indonesian government.

Let me say a few words about why I have proposed that Pugwash takes up the question of intervention and sovereignty, and why I think there are very serious questions about what the world community should do. I got involved several years ago in looking at what seemed to me at that time to be the most interesting cases of intervention that had occurred in recent years, and came to the conclusion that, notwithstanding some contrary views, every case was almost an unmitigated failure.

I went back as far as the Congo case. UN intervention occurred, when the country was threatening to fall apart. It was supported in the misbelief that the UN Secretary-General at the time would take the position that it was part of the UN responsibility to help facilitate the removal of the Belgians, who were the ruling authority at the time, and would act to hold the country together. The Secretary-General, Dag Hammarskjold, instead took the position of neutrality; UN forces were not going to take sides; they were there for peacekeeping. It was a very complex case. At one time or another different members of the Security Council all disagreed with what was going on, with the sole exception of the USA. I think the USA was the only country that essentially went along, without serious objection, with what was happening. The Russians, British and French all objected. But finally it came to an end, and when all is said and done – perhaps people in this continent will say otherwise – my own superficial judgement was that if you look at it ten years later, the Congo was no better off than had that intervention not occurred. And if you look at it from the perspective of today, I would say that is true.

Of every other case since then, I do not think that any have worked out satisfactorily. (Most have arisen since the end of the Cold War, because until that time American and Western and Soviet interests were so strong that they pretty much suppressed any kinds of activities in areas of influence that might require intervention. Both the USA and Soviet Union supported outrageous governments, because they believed it expedient to do so to keep them on their side.) In each case, when the international community has gone in, there have been arguments presented in order to gain support (in some instances domestic support, in some cases international support) that reflected objectives that seemed laudatory at the time, but which were not realized at all – and in that sense, the missions failed.

We have the special case of inter-state war in the case of the Gulf, where at one time or another in order to generate support from the international community, and then to generate support in the United States, the Administration at the time had to raise every kind of argument and even then just barely got Congressional support for that American-led intervention. We were going to prevent the acquisition of weapons of mass destruction there. We were going to restore the government in Kuwait. Well, this is about the only thing that happened. We are still not sure about the weapons of mass destruction, because of the failure of the verification and inspection. It was even promised, at one time, that it would help American jobs.

It has been this way in other cases too. There was a unilateral American intervention – albeit with support of others – in Haiti because of the issue of a duly elected government not being allowed to take over by the regime there. Now that the USA has essential withdrawn, is Haiti any better off? I doubt it. And then there is Somalia. Somalia was perhaps the most extreme case, and the one where it is most easy to agree that intervention was a failure. Is Somalia any better off than before the intervention occurred? I find it very hard to believe that anybody could claim that that it is. It is true that some lives were saved, but in a broad sense it was a failure. In the case of Bosnia, the idea that we could establish a heterogeneous state, with equal rights for everybody, where the ethnic conflict would disappear, was simply not on. There has been no success. It has been promised time and again that troops would be removed quickly. That was a condition for getting support in the USA from the Congress. Troops are still there. Everybody understands that there is a *de facto* partition contrary to the understandings that were expressed at the time. And if the troops are removed, there is every expectation that conflict will break out again.

I would mention two or three things about the Somalia case which illustrate some of the problems. Somalia had a little of everything in it. The UN was there to start with but could not deal with the problems as the clan warfare broke out. We had a situation where, finally, the USA intervened, in a pretty big way, not because of Presidential decisions but because of television. There were pictures on the TV every night which led to the public believing that they just could not tolerate starving children and the chaos and disorder that were prevailing. And so the US marines went in, and there immediately developed tension between the UN people and the US government. The USA made it very clear that they were there to ensure the delivery of humanitarian relief and to establish public order, and they were not there to help rebuild a nation and to establish a reasonable civil infrastructure. And the UN authorities wanted it otherwise. But the troops were ours, we were paying the costs, we did it our way. The marines were pulled out. The mandate of the UN force to replace them was enlarged to get involved in nation-building, if you like, and it could not deliver, and in that respect failed. The last straw was when the Americans withdrew following the loss of 18 people. That illustrates, more dramatically than almost any other example, that it is going to be very hard to get nations involved in these sorts of activities if they are going to take losses and if they do not have important national interests at stake. The USA had no real interest in Somalia. This was a humanitarian-driven intervention. And for that reason, the motivation for staying was minimal. And where is Somalia now? It is the classic example of a failed State.

In some cases there is an ambiguity over the motivation for the intervention. Haiti is a case in point. It is widely argued that we went in there because we wanted to preserve the legitimacy of a government that had been elected. In fact, of course, we really went in, in large measure, because the politicians in the southeastern part of the USA could not contemplate the possibility of a massive flow of refugees.

The questions then are: why intervene, and should there be intervention, under what auspices, and with what kind of decision-making processes? I guess that everybody in this room, myself included, believes that it had best be done under UN auspices. The UN is the best instrument we can see so far in terms of political legitimacy, if not in terms of military capability. But doing so is clearly going to be very difficult and we are going to be very uneven in what we do. It is hard to see how one would get action, in response to concerns about human rights, in, say, China, or India, or any major power – especially with the veto being there. Yet we will intervene, and have intervened, in smaller States, where the costs of intervention are smaller. And this can hardly be seen as equitable or fair or non-discriminatory.

With these failures, and with it being increasingly likely that these intra-state problems are going to continue, it seems to me that this is going to be the predominant form of conflict for quite some time. I do not expect to see any more major wars. I hope we will not see any use of nuclear weapons anywhere. But we are going to see these kinds of problems, and we do have somehow to straighten them out. And if it is to be under UN auspices, clearly some reforms have to be made, and the process of decision-making has to be changed, and the capacity to intervene has to be dealt with.

In looking at these past instances - I ran a study for several years with a colleague at MIT, looking at the military requirements and how events might have been different had there been a UN force - a very strong case can be made that if this is going to be done it ought to be done under UN auspices. But it is hard to see how it can get done under UN auspices as presently organized.

There are two fundamental problems. One is the casualty problem. If the world community is going to intervene on behalf of human rights in a disinterested way, without narrow national interests being the criterion, it is going to be impossible, for domestic political reasons, to engage in intervention where there are going to be serious casualties. One State after another has demonstrated that. In the case of the USA, the Somali case is the clearest example; but we have pulled out of other places too, as soon as there was any threat of loss of life. And so the argument is made that a UN force might be able to be effective in such situations; if instead of sending draftees to areas where their lives might be at risk, you would be sending professional soldiers who undertook the risks voluntarily, then perhaps there would be less of a problem.

But severe problems in command and control have arisen in past UN interventions. Forces have been seconded to the UN and placed under control of field commanders to carry out their operations. Time and again the troops sent there have failed to respond to orders because of domestic considerations. Field commanders from a given country feel obliged, understandably, to go back to their national authorities to get authority to carry out operations that have been dictated by the overall field commander. So the command and control problem has been very difficult. And the decision-making processes are very difficult. This will have to be changed if there is going to be effective intervention in many cases. And if one takes a long time in deciding to go in, the costs, in many instances, will rise greatly. The argument is that if one is going to intervene, then it is best to do so early in the game and to do it preventively.

There is also the issue of criteria for intervention. In another study that I did, I tried to understand what consensus there might be on a legitimate basis for intervention, and I reluctantly concluded that the only thing on which I thought that there would be near universal concurrence was in cases of absolute genocide. But we have many other cases of less clear-cut abuses of human rights. Ethnic cleansing is perhaps a case in point. There is widespread public aversion to the idea. People ought to be able to live together, but in fact it is proving not to be the case. We failed in Bosnia. There were massive movements of people from ethnic cleansing at the end of World War Two and we accept that. So we have a problem deciding when intervention is going to be legitimate; what kind of military forces, if any, are used; and appropriate command and control procedures.

It seems to me that Pugwash cannot avoid getting involved in some of these questions. This is particularly so because we could not get any consensus in our group about the events in Kosovo. There were very strong feelings on the part of some members of this group that the kind of abuses that were occurring in Kosovo to the Albanian Kosovars demanded intervention by the world community. There were probably equally strong feelings on the part of others that it was intolerable to go in as was done, with the decisions being made, to be blunt, largely by the United States, with strong concurrence from the UK, and dragging the rest of NATO along, but without reference to the UN, and particularly without adequate consultation with and involvement of some other powers, notably Russia. And then on the tactics, the idea that if you are going to intervene militarily, you allow your aversion to taking any casualties to be so dominant that you end up inflicting massive damage on innocents (Serbs, in this case), that were not supporters of the regime, through bombing, which starts out ostensibly as being narrowly focused on military problems, but ends, as it has in almost every war, being close to strategic bombing, where you go after the whole infrastructure. It is a severe problem and we have to try to achieve some clarification on these points.

COOPERATIVE GOVERNANCE, PARTNERSHIP AND SOVEREIGNTY IN A FUTURE DEMOCRATIC SOCIETY

Derek Powell

I want to come at this problem from the perspective of constitutional and institutional design. What can a country that is in crisis, political turmoil, with a history such as ours, in fact do about its own stability and prosperity? I want to pick up on some of the themes that have played themselves out in the South African context over the last decade or longer and identify one or two issues which I think are quite useful as learning experiences for other countries.

I think that you will see from my presentation why the subject I am addressing, Cooperative Governance, Partnership and Sovereignty in a Future Democratic Society, is specifically relevant to the South African context, given our commitment, as a country, to the notion of cooperative governance.

Before I get into that, I want to say a few words about the global environment into which the South African State was born. It is increasingly clear that the developments in the global economy, coupled with developments in communications technology, have unleashed forces which are placing pressure on the stature of the sovereign State as the nodal point for economic, and increasingly political, organization.

It is perhaps a bit too soon to speculate, or to say with any certainty, whether the sovereign State will disappear. But one thing is clear, which is that for new democracies such as ours, the questions around the assignment of powers within the State are terribly important and terribly influenced by these developments globally. Increasingly, sub-national government is becoming more and more involved in economic and cross-border political activities, and, as a consequence, has had greater and greater powers devolved from the centre or the State.

I want to say a word about globalization, that is to say, the economic and political pressures which constitute globalization. The effect of globalization is not a neutral effect. The distribution of benefits and burdens through the processes that are unfolding in the global economy are not neutral. That is quite important, because one's perspective on the value of globalization, or the effect of globalization, will, in my view, differ if you speak from the perspective of a developed, as opposed to a developing, country.

What is important for countries such as South Africa, and continents such as Africa, and for the developing world in particular, is to mobilize its leadership and its resources to have a strong voice in these processes, so that the benefits of globalization, albeit that they are objective, historical facts, can be shaped in a way that the benefits are distributed to the developing world. And this is one of the keynotes that our President has struck in recent meetings of the OAU and the non-aligned countries.

It is within the context of globalization that the South African State was born in 1994. And it was born in a context where three characteristics define the global political landscape. The first is the influence and power of international institutions such as the UN, the IMF, the World Bank and others. Increasingly, these institutions have an effect on the relations between States, but also on the activities within a particular State. One can think, for example, of the security arrangements which exist on a regional or international level through the United Nations and other organizations – our own recent involvement in the Democratic Republic of Congo, for example, and Lesotho. But also, perhaps more intangibly, the influence of monetary institutions, such as the IMF, with structural adjustment programmes, and the like. These institutions are able to affect, in some way, the policy dimensions within a particular country.

The second characteristic of this global landscape is that many States have sought cooperative mechanisms in the form of communities of States, regional trading blocs, and other looser forms of association, with greater or lesser degree of political accommodation. These modes of reorganizing State power have raised the question of whether certain functions should be conducted at the supranational level, as in the case of the European Union, for example, where many functions of the State, which were traditionally key to the notion of the sovereign State, have graduated upwards to the supranational level. Paralleling this development is a similar drive to decentralize power to subnational level, and this development in institutional design in the global arena is captured in the term "glocalization", which refers both to the graduation upstairs of functions and the devolution of functions down to subnational government.

The third important factor is that national aspirations are still a common, if not the dominant, feature of political conflicts in many parts of the world. So whether we choose to fall into the camp that sees the world developing into a borderless environment, or with borders different to those that we know, or whether we fall into the camp of those who would argue that the nation state is still around, the one fact that is certain is that many conflicts today lead to claims to national sovereignty around a particular spatial territory.

These are the pillars of the global environment, and it was into this environment that the South African constitution was born. South Africa has a sovereign constitution that is supreme law. It also has a multi-party democracy, and it has common citizenship. Values such as human dignity, equality, and the importance of redressing a past which was divided and ravaged by both colonialism and Apartheid, are core values of the South African State. Not only are they contained in the Bill of Rights to the constitution, but they also ground some of the more important institutions which we have created constitutionally, and I will allude to some of these later. Importantly, the courts in South Africa are independent, and we have a constitutional court which is the final guardian of the constitution.

The system of government that has been created is very unusual and very much a product of our own history. We have three spheres of government, at national, provincial and local level. Each of these spheres has legislative and executive powers of their own. Most of the powers, which are distributed between the spheres, are in fact concurrent powers between national and provincial government or they are powers which are devolved to local government. What we have, then, is a system of interlocking legislative and executive powers. There are very few constitutions in the world, which have, from the outset, gone for a fairly complex system of government such as ours. And I will allude to the reasons for this complexity in a short while.

Importantly, these spheres of government are not independent; they are not autonomous. They are interdependent; they might be functionally distinct, but they are part of a single system of government. What the constitution does further is to direct that each sphere of government in its activities is bound to cooperate. The principle of cooperation is intrinsic to the South African constitution. It finds its form in the notion of cooperative government. What that means is the spheres of government must preserve the peace and national unity and the indivisibility of the country. Moreover, the fiscal regime of the country ensures an equitable allocation of finances to each sphere of government within a framework of tight fiscal discipline.

So that is the structural arrangement for government in this country. Some have called it quasi-federalism. As I have said, it goes by the name of cooperative government, and this particular formula is very much a product of the political forces, history and aspirations of South Africans. What it reflects is that, given the challenge of reconstructing this country, which has been divided for hundreds of years on the basis of race, ethnicity, ideology and wealth, the blueprint for government necessarily had to balance decentralization and political competition with the need for common citizenship, national unity and a single government leading the development of the country.

Given the history of the country, it was important to the development of the country as a whole that political competition did not play into the historical fragmentation of the country; it was necessary to use national unity as an organizing concept for a vision of development for the country as a whole. What cooperative government seeks, therefore, as an approach to government, is unity of vision, destiny and purpose, even as it recognizes diversity of interest and opinion.

Cooperation, it can be said, is one of the central pillars of the South African constitutional order. Without the will to cooperate initially, it is very unlikely that the South Africa of today would have been delivered from the political hostility and stagnation into which it was born. And it is arguable, too, that cooperation characterizes much of the way in which the government of today governs, given that it exercises its power in a world, and in a country, that is interdependent rather than independent.

Cooperation does not mean agreement on everything, or absence of disagreement or different valid viewpoints, nor is it a counterpoint to strong political and economic leadership. All of these institutions are accommodated within the constitution, with structures and institutions such as the independent courts, rights such as the freedom of political activity and expression, independent institutions such as the Human Rights Commission and Public Protector, all of which place a check on State power, and on the political trade-offs that are made on a daily basis in order to govern a country.

The best strategy for the country as a whole, irrespective of political persuasion, irrespective of particular sectional interests, is for the political, societal and economic leadership to cooperate with each other to find the best way to direct the peaceful development of the country and its people. That is the challenge for any government.

It is perhaps too soon in the history of the development of this country, post 1994, to clearly unpack and demarcate the phases of transition. But I want to highlight what I see as four trajectories where the principle of cooperation has played itself out as a strategy for governance.

The first of these is what we might call cooperation towards peace, stability and reconciliation. Without the will to cooperation, there would have been no political settlement in South Africa. What transpired in the years 1993 to 1996, was a constitution-making process that sought, and arguably achieved, an end to the conflict, democratic government, a foundation for national reconciliation, the establishment of a system of democratic government, and a process for managing the transition from an authoritarian State to a democratic one. So that was the first phase of South African constitution making where cooperation became relevant.

The constitution-making process itself was, in a sense, two phased. It was a finely tuned constitution-making process. The first phase was a negotiated settlement. During this phase we saw the establishment of a transitional executive council to manage the process of transition, and to pave the way for democratic elections and constitution making to follow. The outcome of various multiparty talks was an interim constitution. The interim constitution in South Africa has often been called the cease-fire constitution. It is the basic document which brought the parties together in a basic framework of understanding, whilst it set in motion a process for developing the constitution.

What this did was to establish a multiparty government of national unity between the National Party and the ANC, and it created certain basic structures of government, such as a constitutional court and a limited Bill of Rights. What was very interesting about this interim constitution, is that it adopted the notion of an elected constitutional assembly, to draft South Africa's constitution, after the first democratic elections in 1994. More than that, this constitutional assembly, which would do the work of constitution making was bound by 34 fundamental constitutional principles, which were set out in the interim constitution and had been agreed to by all the political role players.

It then took an unprecedented step. It said that the constitutional assembly would adopt a constitution for the country, after negotiations. It had a two-year period in which to do that. But once it had adopted this constitutional text, before that text could become law, the constitutional court of this country would have to certify that it complied with the 34 fundamental principles. In my experience, in the history of constitution making in the world, this is perhaps one of the most unique ways in which to develop a constitution, because it carefully balances the forces that shape the final product.

The second phase, between 1994 and 1996, was in a sense the birth of South Africa as a nation state. This phase culminated in the adoption and certification of the South African constitution. Reflecting back on the constitution-making process in the constitutional assembly, what is interesting to note is that there was never any formal deadlock-breaking mechanism between the parties. Bearing in mind that the negotiations involved forces who were involved in the liberation struggle in this country – not yet political parties, but certainly political movements evolving into political parties – and the National Party and other parties that had for a long period of time ruled this country in an authoritarian government, a government that had excluded the majority of citizens of this country, the fact that this process did not have a formal deadlock-breaking mechanism is significant. It is significant because in practical terms, the parties were continually required to refine those areas where they agreed and where they disagreed, and it allowed parties the opportunity of using the constitutional court certification process as a way of dealing with particular constitutional issue which, in their view, were not resolved in the constitution-making process, without undermining the basic vision of the process.

Many would argue that this was a successful way to develop a constitution, because it managed the forces of political competition constructively. It made a series of trade-offs; it balanced those forces in a way that allowed national reconciliation to take place, and for a government to be instituted which could take the process of developing the country further.

An aspect of this settlement was the Truth and Reconciliation Commission, which was South Africa's way of dealing with the havoc of its past. As we all know, it was a past where the grossest of human rights violations were perpetrated on the citizens of the country. The idea of the TRC was to persuade the perpetrators of the human rights violations to admit to their role in those events, for the truth to come out, with the prospect of amnesty. And on the other hand, to give the victims of human rights abuse the opportunity to give testimony and to receive reparations. The TRC was intended as part of the healing process towards national reconciliation.

But it was by no means an uncontroversial institution. In particular, it raised questions around whether the victims of human rights abuse could, at the end of the day, say that justice had been done. The court is still out on that question. Further, it risked criminalizing a war of liberation which was legitimate under international law. This is one of the points that the ANC has stressed in its statements on the TRC report. That was a significant question in the minds of many people in this country.

Whatever the answers to those questions, the TRC was nevertheless a remarkable, and politically mature, step towards confronting the past of the country in a way that could lead to a normal society in the long run.

The second area in which cooperation as a strategy is relevant to South Africa, is cooperation towards the reconstruction and development of the country. The South African macroeconomic strategy is known as GER – growth, employment and redistribution. What it seeks to achieve are the objectives set out in an earlier policy document on reconstruction and development – the RDP document. GER seeks to transform South Africa into a competitive, fast growing economy, and through growth to achieve social and economic development and poverty alleviation. A core aspect of this policy is fiscal discipline and a reduction of State expenditure as a percentage of GDP. The budget and budget processes are, as part of the overall economic strategy, undergoing fundamental reform.

Interestingly, GER is, in many ways, an acknowledgement that development within South Africa is not development within a hermetically sealed State. It is development in a State that is fundamentally part of a global environment and a global economy. The budget reforms in this country have been quite remarkable. We have a medium-term expenditure framework. What this seeks to do is to bring greater efficiency to fiscal allocations, and greater synergy between planning goals and budgets.

There has also been a fundamental drive to reform the public service away from a command and control type administration as under the National Party government and, in its place, to create an institution that is representative, professional and able to serve the needs of the public. Many of the transformation processes that are taking place in the public service are ahead of their time in some ways, and are processes that countries such as New Zealand, Australia and, to some extent, the UK are beginning to experiment with.

Policy reforms of this nature require a high level of convergence, agreement and common vision among key stakeholders in all sectors of society – even where there are differences of emphasis and approach. This, in turn, has required of leadership a level of maturity in building and managing relations of mutual trust, across all sectors, labour, private, civic and the community at large. The leadership in this country, in order to take forward these reforms, has necessarily had to build cooperative relations with key stakeholders. Recently, the President announced, for example, the creation of an Investment Council to pull together the talent of the country to begin to look at ways in which South Africa could solicit effective investment, and we have a forum that brings together labour, government, business and the civics into an institutional arrangement which has produced, amongst other things, the labour legislation of the country. That is a remarkable development in its own right.

Thirdly, we have what might be called cooperation towards good governance. Cooperation in governance takes on diverse institutional forms in this country. There are numerous intergovernmental structures. These intergovernmental structures are situated at both the legislative and executive levels of government and through these structures the various spheres of government cooperate with each other to plan and implement policy and other activities. We have, for example, at the legislative level, the National Council of Provinces. The idea of this institution is to give regional government a stake in national law-making processes. At the local government level, we have what is known as the South African Local Government Association. It is an association of local government with a view to ensuring effective local government participation in decision-making processes at national and provincial level.

What is interesting about South Africa is that the constitutional regime places an enormous emphasis on local government as the site of service delivery, but also as the site of economic and social development. Hence, we find considerable representation for local government in policy processes at national and provincial levels of government.

Other ways in which partnership or cooperation find institutional expression in the country are through public-private partnerships, which are mechanisms for service delivery which play themselves out at local government level, where private sectors and the communities and government have formed partnerships to deliver services to the communities. There are numerous institutions that operate at the community level, such as community policing forums. You have campaigns aimed at building civic responsibility, and the notion that government is not only about government in the traditional sense, it is about the country as a whole taking responsibility for its own destiny.

So what we have in South Africa is a range of institutional mechanisms which encourage participation from the different sectors in our society in those processes which fundamentally shape the direction that the country has taken. Whether they work well or not depends, to a greater or lesser extent, on one's perspective, but on the whole the participation of the core sectors of society is seen as a fundamental objective of government in this country.

Finally, to talk about cooperation in the broader context, we could say that South Africa's strategy is cooperation towards regional and continental development and international relations. South Africa plays a leadership role in many institutions which it shares with other countries in Africa. In institutions such as the OAU and SADC, and through initiatives such as the African Renaissance, which is an attempt to build a developmental focus for the continent as a whole and to stimulate and mobilize the leadership of the continent to play an effective role in international institutions and processes.

What is important about our policy in this regard is that great emphasis is placed on South-South relations and working within the developing world to shift power relations in a way that allows for the distribution of benefits to developing countries.

PRESIDENTIAL ADDRESS

Michael Atiyah

Last year I concentrated on the traditional concerns of Pugwash – weapons of mass destruction – and I also alluded to the high-profile issue of terrorism. These must still remain high on the Pugwash agenda, they are the areas where Pugwash has much expertise and credibility. Unfortunately, progress in these matters is slow and international attention tends to focus on the latest crisis (for example, Kosovo, or East Timor), so that nuclear disarmament and control of biological weapons do not get the public attention they deserve, and without public interest, politicians tend to drag their feet.

So, while recognizing the continuing importance of the whole disarmament process, I will this time focus on issues of peace rather than war. After all, peace is much more difficult. In war the military take over, national dissension is forgotten and everything is simplified. That is why politicians frequently prefer war, or the threat of war, to the more difficult task of governing in peace.

If we take a long view of the way the world is governed we can discern trends, we can discuss the utopia to which we aspire and then we can try to help the process along. This is the goal of Pugwash and many other organizations with broadly similar aims. Where are we now, where do we hope to get to and how are we going to get there?

Conventional wisdom says that democratic government and the market economy are winning the battle. Ever since the collapse of the Soviet Union things seem to be going in this direction. Perhaps South America is the clearest example of the triumph of democracy. A continent traditionally run by military cliques and dictators is now in democratic hands. Eastern Europe, the former Soviet bloc, is joining up with the more mature democracies of the Western Europe. Asia is in the process of converting economic development into political change.

Naturally there is a wide variation in the style and strength of these various democracies. Some are well established and stable, others are very new and fragile. For some the development of democracy was a long and natural part of their history and culture, for others it represented a sharp change and a foreign import. Nevertheless, we can take reasonable hope from the rapid enlargement of the democratic world and look on this as a good augury for the future.

If indeed democracy is the way of the future then we need to look carefully at the particular problems that this gives rise to and prepare for them in advance. It is not clear that a world of democratic governments is automatically a world living in peace and harmony.

First of all, we must recognize that there is more to true democracy than an elected parliament. We need a free press, free radio and television. We need a fair and independent legal system, an honest and impartial police force. We also need the military establishment, and preferably also the religious establishment, to be kept at arms length. We need a respect for human rights and the protection of minorities. Finally, and most crucially, we need a sound decision-making process – one which responds to an informed public opinion. I will follow up this point later, but first let me look in detail at the tricky question of minorities and human rights – highly pertinent in view of the situation in the Balkans.

The basic premise of democracy is that, after free discussion, votes are taken and (within limits) the view of the majority prevails. But this begs the basic question: what is the size and extent of the unit which votes? In the aftermath of the First World War, Woodrow Wilson proclaimed the right of self-determination. Cynics will say that, as the United States did not have an Empire, this idealism was at the expense of the former European Imperial powers, notably Austria, Britain and France. Essentially it meant that colonies should become independent, and this process, accelerated by the Second World War, effectively saw a proliferation of new nation-states, all exercising "self-determination."

But what determines the boundaries of a "nation-state", the area in which the democratic system is supposed to operate? A quick look round the world, or at the membership of the United Nations, shows that there is often little rhyme or reason in what constitutes a nation-state. History is the only common feature, and since history has been so varied, this is reflected in the variety we now have. Here in Africa we have been reminded that the colonial powers drew lines on the map that did not respect traditional communities. I shall return later to the problems this produces for the UN at the international level, but my present point is simply the arbitrariness in the frontiers within which the democratic process is supposed to operate. Take for example the highly controversial one of Ireland. The demography of the situation is simple. In Ireland as a whole the Catholics are in a majority, within Northern Ireland the Protestants predominate, and the division of Ireland reflected this. A somewhat similar situation prevails between Lebanon and Syria. Taken together (which they were in former times) the Moslems are the majority, whereas Lebanon, as carved out by the French after the first world war, had a Christian majority. I mention these two examples just because I am familiar with them, and because the problems that these divisions have given rise to are still with us. I am sure all of you can cite parallel instances in other parts of the world. Many of the most intractable problems of the world at present arise from the historical origins of nation-states.

But this problem is not just an accident of history, or the aftermath of Imperial rule. It is more fundamental and is inherent in the ideas of democratic government. Given the complex history of mankind and the intermingling of peoples that has occurred over the centuries, no perfect boundaries can be drawn which make each nation-state entirely homogenous with respect to race, religion, language and culture. The world is a polyglot mixture and becoming more so every day.

The potential conflicts between different groups inside a nation can sometimes be kept in check by a forceful non-democratic government. Yugoslavia, under Tito, and the diverse republics and nationalities of the old Soviet Union, under Communism, provide recent examples. British rule in the days of the Empire kept tribal conflict in Africa at bay. Independence and democracy of sorts has unleashed a Pandora's box of hostility and violence. This is not an argument in favour of empires, but it does illustrate the deep nature of the problem. How do we keep peace among different groups in a free environment?

Even in Western Europe, one of the more peaceful and stable parts of the world, internal conflicts are still with us. I have already mentioned Ireland. There are also the Basques in Spain and the more complex relation between Flemings and Walloons in Belgium. On the other hand, the example of Switzerland shows that harmonious coexistence between different peoples is certainly possible, provided there is a high degree of local autonomy and adequate constitutional safeguards to protect minorities.

A close look at Western Europe suggests a few thoughts on the possible utopian future where all such conflicts are eliminated.

What we see emerging, in a truly democratic context, is a hierarchy of levels of authority interwoven with each other. We have the European Union, under which are the various nations; many of these are themselves federal, with autonomous provinces, and further down the line we may have smaller regions and then cities and towns. At each stage we have democratic control and, in principle, each level of authority supervises the one beneath it to ensure that the rules are kept. Disagreements or potential conflicts at one level can be passed up for adjudication to the next level. In so far as the European Union has developed a policy for the component parts of the former Yugoslavia, this seems to be the sort of model it has in mind. In other words, within the larger family of European nations, the rivalries in the Balkans can hopefully be kept on a leash.

I think it is recognized that the recent progress in Northern Ireland has been made easier by the fact that both Britain and Ireland are now parts of a larger entity, namely the European Union. Perhaps even more crucial has been the involvement of the United States as an outside player, and here perhaps there is an important lesson to be learnt. In the past there would certainly have been many British politicians who would have objected strenuously against the United States being brought into the Irish problem, on the grounds that this was an internal problem, and that we did not want any infringement of our sovereignty. One hears similar statements in other parts of the world. But there are in the end only two ways to solve a conflict, either by brute force (and this is rarely a permanent solution) or by some negotiated agreement. Such an agreement is difficult to achieve without the involvement of one or more outside parties.

Some 15 years ago I was having a dinner discussion with a government minister who had academic pretensions. We were discussing the nation-state in the context of Germany (at that time still divided). He was arguing that the nation-state was a 19th century invention and that its time was past. I was emphasizing the strong emotional attachment to German reunification that I had always found on my frequent visits to Germany. History has vindicated me, but the minister had a point. The increasing complexity of the modern world, mainly as a result of technological change, has meant that national sovereignty is now highly circumscribed. Our national politicians still talk as though they exercised great power, but in real fact their freedom of manoeuvre is strictly limited. Actually, when things go wrong, our politicians usually ascribe this to world economic forces beyond their control, but they are happy to accept the credit in good times!

One of our problems is that we have not yet made the necessary adjustment in our ideas and our discussions to the new reality of interdependence. Let me illustrate this from the scene in Scotland where I now live. As you may know, after more than 300 years, Scotland now has its own parliament with very considerable powers. This reflects the strong sense of identity of the Scottish people and the existence in Scotland of separate legal and educational systems and a separate church. So far so good and autonomy is now accepted, but the big question around which the political parties circle is whether Scotland should acquire total independence or not. It is argued that, within the European Union, an independent Scotland could be a viable unit. Now, although I live in Scotland and sympathize with the need for Scottish autonomy, I see very little difference in practice between the alternatives ahead. As part of the European Union and as a neighbour of England, it is not clear that a separate army and foreign policy would make much difference to the situation. Words like independence or national sovereignty no longer have much real meaning and their continued use by politicians is window dressing intended to enhance their own importance.

So let us agree that interdependence between countries will gradually lead to a system of interlocking relationships which may help to prevent outright conflict, and let us return to consider the operation of democracy within each country. As I mentioned earlier, democracies are not necessarily peaceful. It is enough to recall that many Germans in the 1930s voted for Adolf Hitler. Unfortunately, banging the nationalist or military drum has always been popular. Is then democracy together with human weakness an inevitable recipe for the triumph of dangerous demagogues? That would be a dismal prospect. I am a perpetual optimist, but there are I believe rational grounds for optimism. The 21st century, even in its first decades, is going to be transformed by the internet and other forms of electronic communication. This will mean that the natural arena for the spread of information and ideas is the whole world. It will be much harder to build up the misinformation, self-delusion and hatred that demagogues thrive on. The most dramatic recent example of the effect of new methods of communication was the collapse of Communism in Eastern Europe. It is widely accepted that it was the inability of the communist authorities to control the free flow of information which contributed to the speed of the collapse.

But there are other even more striking ways in which electronic communication may affect the workings of democracy. In ancient Athens, the birth-place of democracy, the whole citizenship could in practice turn up to discuss and decide. But on the large scale of modern states, this is no longer possible and we have elected representatives instead, although Switzerland with its constant referenda is an exception. In recent times, with the use of increasingly sophisticated public opinion polls, perhaps the views of citizens can be tested more frequently. But electronic communication offers a much more effective way of consulting the public. In a sense, we can revert almost to the Athenian model of direct involvement. As yet this has not happened, but it is technologically possible and there seems every reason to believe that something like this will happen. After all, why bother to take a small sample of the population when you can consult the lot? The phrase teledemocracy has already been coined to describe this future scenario. In essence we are now on the verge of the second phase of the communication revolution. The first stage, the era of radio and television, was passive: the ability of the public to receive information rapidly. The second active stage is when the public can respond in a two-way process.

I imagine this teledemocracy will start in the United States, since this is technologically the most advanced country, and since American politics is already very much dominated by opinion polls. The question is: how will it work in practice and what will the consequences be? As we all know, the way a question is put frequently determines the nature of the answer. For example, if the British public were asked: are you in favour of abolishing the pound and becoming subservient to Brussels? the answer will surely be negative. However, if the question is: are you in favour of strengthening the British economy by joining the Euro? the answer might be positive. So, if we move to a new style of democracy in the next century where the public is consulted, as matter of course, by electronic means, the key point will be the management of the process. Who asks the questions?

If this becomes a world-wide phenomenon, even in countries where democracy is still in its infancy, the dangers are obvious. George Orwell would have recognized them. Where internal objectivity cannot be guaranteed or inspire confidence there will have to be external monitors, rather the way the United Nations monitors elections in critical areas.

You may think that I am describing a very distant world, and that we have more pressing problems, but the speed of technological change usually exceeds expectations and is far in advance of the controls that need to be exercised. It is, I think, significant that our student participants were much more aware of the opportunities and problems opened up by the internet. It pays to look ahead and plan for the future. The world has been late in coming to grips with scientific revolutions in nuclear physics, chemistry and biology. We should not make the same mistake again.

I have attempted to outline a future of interlocking nations with mechanisms for adjudicating conflicts. At some level and for important issues this may need the intervention of the United Nations or one of its subsidiary bodies. Unfortunately, at the present time, the United Nations is not nearly as powerful and effective a body as the world needs. Of course, it is much better than nothing, and it has had many successes, but it has obvious limitations. There are manifold reasons.

It is widely recognized that, after 50 years, the UN should be drastically reformed to reflect the realities at the end of the 20th century. But vested interests have effectively blocked any serious movement, and the major powers are perhaps not unhappy to see the UN remain a weak body, thus leaving effective control in their hands.

Perhaps attempting to reform the UN from within, by delicate negotiation, is a hopeless task. Instead, some impartial body or bodies, preferably non-governmental, should try to think the problem through from scratch. An attempt should be made to produce a new structure for the UN which would be equitable, effective and realistic. This could then be handed to the UN to provide at least a new basis for discussion. Just tinkering with the present set up is perhaps inadequate, more radical thinking may be necessary.

I should emphasize that I have no clear idea of what a reformed UN should be like. There are many anomalies which should be removed, and the experience of inadequacies over the last 50 years should be fully utilized. The problem is extremely important, since the present international scene is an untidy mismatch of activities by the UN, by NATO, by independent states, and this is not a recipe for stability in the long run.

One weakness of the present UN is its poor financial position. This reflects other aspects, such as the lack of confidence in it by the United States. There is also a general distrust of large bureaucracies, which are seen as black holes that suck up money. The European Union suffers from a similar syndrome, with Brussels seen as an expensive and over-luxurious bureaucracy. So a reform of the UN has to address both issues. How to provide more resources and how to see that these are properly spent.

One idea which has been floated, that would provide the UN with an adequate financial base, would be to impose a small tax on international financial transactions. This is proposed by James Tobin, a distinguished economist, for two quite separate reasons. In addition to solving the financial problems of the UN it would also dampen down the dizzying scale and speed of money flows round the world. The argument is that such dampening would remove some of the froth without doing serious financial damage, and the result would be to add stability to an otherwise dangerously volatile scene.

Since Pugwash elected a mathematician as its President, I am afraid you have to pay the price of listening to a little arithmetic! But, do not worry too much, the sums will be easy since I will not attempt to be too precise. But, following good astronomical practice, I will just deal with orders of magnitude – counting the number of zeros.

First, a little history. In the 19th century, when Britain, as the first country to industrialize, was the leading economy, about 50 per cent of British savings were invested abroad. This helped the economic development of many parts of the world, not only in the British Empire. You might think, with the vast amount of money that is moving around the world, a similar process is now taking place, with the developed countries investing a large fraction of their savings in the developing countries. You might be confirmed in this view when you hear that international financial transactions approach 2 trillion dollars daily and are rising fast. But you would be wrong. In fact the fraction of savings of rich countries that is invested in poorer countries is more like 5–10 per cent, rather than 50 per cent. Of the 2 trillion dollars that I mentioned, about 90 per cent consists of very short-term movement, money that moves back and forth over a matter of a few days. This is speculative money that has nothing to do with investment. To be blunt, this is gambling on a global scale.

The motive behind the free movement of money was the encouragement of trade and investment – not the support of gambling. It is for this reason that serious economists have discussed ways of reducing short-term speculative money movements, with its destabilizing effects, such as we saw during the recent Asian financial crisis, without reverting to old-fashioned currency controls which inhibit genuine commerce. This is the main purpose of the "Tobin tax": the support of the UN is, for economists, an incidental if beneficial by-product. Moreover, other good causes could also be supported.

Now let us look at the sums, the orders of magnitude involved. A tax of say 0·01 per cent (and this figure is just illustrative) on international financial transactions of 2 trillion dollars daily would yield 200 million dollars daily. For the genuine business man 0·01 per cent, added to his other costs would be negligible. But, for the short-term speculator, the effect would be very discouraging. If I borrow 1 billion US dollars for two days, say at 5 per cent annual rate, it will cost me about a quarter of a million dollars. A transaction tax of 0.01 per cent, on the other hand, would cost me 100,000 dollars, a significant addition. This might make me think twice.

So the tax would be a severe discouragement to speculators but it would not eliminate them all. Suppose they were reduced by a factor of 2. Thus instead of collecting 200 million dollars daily the Tobin tax would only yield half that amount namely 100 million daily, approximately 35 billion annually. If only one half of this amount went to the UN, this would be over 15 billion which is about three times its present budget. This is perhaps a realistic minimum of what a truly effective UN would need.

So the sums make sense in a broad sort of way. Many of the parameters could be altered without changing the overall picture. The Tobin tax would be beneficial economically and it would provide a firm basis for the UN. The objections are mainly political, since the United States, in particular, is not keen on seeing a UN with real power. There may also be more legitimate concerns about so much money being handed over to a single body with inadequate controls. There would be opportunity for massive corruption and extravagance. These dangers could be mitigated by channelling the Tobin tax through some other international organization, such as the IMF, which could be responsible for monitoring the way the UN spent the money, and could withhold it if it was not satisfied. Such a division of responsibility is the most effective way of ensuring probity and efficiency. It is encouraging that the Canadian Government has put the Tobin tax on the international agenda and asked that a serious feasibility study should be made.

Let me move now to a point I made earlier, the arbitrariness of the boundaries of many nation states, and let me relate this to the UN. The UN consists of individual states, with their status essentially frozen by historical accident as of 1945 (though subsequent changes have occurred, sometimes peacefully but often as a result of force). Moreover, the moral legitimacy of the governments of many states can be questioned. As we have heard, here in Africa there have been, and still are, a number of regimes which are corrupt, undemocratic and disregard human rights. But Burma reminds us that this is not solely an African problem. In this respect, the UN is like a gentleman's club where gangsters are accepted, provided only that they pay their dues and wear jackets and ties!

Admittedly, to follow this analogy, some of the respectable members are increasingly unhappy to have such undesirables in the club and they try to bend the rules to find an excuse for expelling them. Perhaps, the rules should even be changed to allow bank accounts and tax returns to be examined?

Well, you get the idea. Reform of the UN, either by explicit changes, or by more flexible interpretation of the Charter should allow the UN to interest itself in the internal affairs of its members. We should no longer pretend that, within its borders, a government is entitled to do entirely what it likes. Human rights violations or gross corruption should not be tolerated, just because they are committed by a member of our club.

Current events in East Timor show that there has been some positive movement in this direction, but we should remember that, for more than 20 years, the world ignored the repression in that area.

Notice that I have not said outright that the UN should acquire the right to interfere in the internal affairs of its member states. The word "intervention" has, as our South American colleagues have pointed out, unfortunate connotations. I am not going to spell out in detail how, and under what circumstances , the UN should get involved with the internal affairs of member states. I am not an international lawyer (and I realize we have several distinguished lawyers among us), nor am I an expert on the UN Charter, but I do think this whole question lies at the heart of many of our current problems and Pugwash will shortly be undertaking a serious study of it. We all want an international order where we operate within the law, but this requires a better legal framework, one that harmonizes more with morality. *A world based on injustice can never be at peace.*

PART III

STATEMENT AND REPORTS FROM THE CONFERENCE

WORKING GROUP 1

A NUCLEAR-WEAPON-FREE WORLD

NUCLEAR DISARMAMENT: A LOST CAUSE?

Bart van der Sijde

The Present Situation - Sketch of the Major Developments

Without any doubt, nuclear weapons will be one of the many burdens which humanity will have to carry when it enters the next century and the next millennium. On the threshold of a new era it is at the very least extremely disappointing, indeed dramatic, to have to observe that the chances of definite, general nuclear disarmament are so low, that it is even questionable whether we will ever be rid of this type of weapon of mass destruction. This is all the more disappointing because the end of the Cold War has removed the main reason for the failure to eliminate nuclear weapons immediately after the end of World War II, when the 1946 Baruch Plan provided a way of doing so. This failure was caused at the time by the increased hostility between East and West. Now, after a disappointingly brief period of surprising, even exuberant nuclear disarmament between 1987 and 1993, it has become clear that the importance attached to nuclear weapons by the USA and Russia, and perhaps other states as well, is so deeply entrenched that it is at odds with the execution of Article VI of the Non-Proliferation Treaty. Indeed, if present policies were to be continued they would abrogate it.

This development, which has surfaced during the last two or three years, does not just involve maintaining nuclear weapons as such. That is in itself an extremely grave and disappointing fact. It also involves keeping large numbers of weapons. For the USA alone this comes down to maintaining a force of at least 10.000 nuclear warheads[1]. We must expect that Russia will more or less follow this trend. Many individuals and organizations are extremely worried about the total stagnation of the nuclear disarmament process. An example of that is an official statement by the Pugwash Council in October 1998[2].

The Threat Remains

All this means that nuclear weapons will essentially remain, along with the threat of the destruction of civilization, including the phenomenon of the nuclear winter discovered in the 1980s by the chemist Paul Crutzen. The effects of reductions of nuclear weapons cannot, after all, be compared to those of conventional weapons. A country which reduces its conventional armoury by 75% has usually chosen a defensive posture. But the reduction in nuclear weapons from 70,000 (in 1986) to 20,000 in the future cannot be described as a decisive improvement. An immense and absurd inflation of numbers had already taken place during the Cold War. A situation threatens in which this will essentially not be corrected. In a report by the American Academy of Sciences a figure of 20 warheads is described as sufficient for inflicting decisive damage on Russia by the United States[3]. Only at levels far below 20,000, at around 100, can it be said that the destructive capacity, though still immense, is no longer sufficient to wipe out all of humanity[4]. With 1,000 weapons, this catastrophe is still a realistic possibility.

This means that in the coming century the extent of the threat of nuclear weapons will be closely tied to the political climate, mainly that between the United States and Russia. If this were to deteriorate, we would return to a situation of deterrence comparable to the Cold War. This also means that we will be going into the twenty-first century with an exceptionally heavy, negative moral burden. As was stated in the Advisory Opinion of the International Court of Justice of July 1996, nuclear weapons are not in accordance with international humanitarian legislation and the international law of war. By stubbornly ignoring these statements, all the nuclear weapon states (officially seven and in fact eight) have placed themselves above the law. This also undermines the credibility of five of these nations in their role of peacemakers and mediators as permanent members of the Security Council.

The Threat of Inequality

There is also the threat of an undesirable and unacceptable difference of approach to chemical and biological weapons on the one hand, and nuclear weapons on the other. Chemical weapons were first used on a massive scale during the First World War and a Protocol for the prohibition of the use of these weapons was formulated as early as 1925, signed by many countries. The Chemical Weapons Convention was ultimately signed in January 1993, providing an international framework for the elimination of these weapons. The strength of the Convention can of course be undermined by countries refusing to sign. The Biological Weapons Convention dates from 1972 and provides a similar framework for this type of weapon.

Despite the series of treaties somewhat limiting the nuclear weapons phenomenon it is regrettable and disappointing to have to observe that not even the use of this type of weapon, fifty years after its creation, has been banned. On the contrary, attempts to achieve "no first use" declarations are ignored. The differences in approach between nuclear weapons on the one hand and biological and chemical weapons on the other, can no longer be explained and defended by the time differential of thirty years. Characteristic and shocking as far as this is concerned are the attempts by the United States especially to have chemical and biological weapons defined as illegal in the framework of the International Criminal Court and to exempt nuclear weapons from this status. This is a clear attempt at countering the statements by the International Court of Justice by legalizing nuclear weapons through a different process.[5]

Non-Proliferation and Disarmament are Inseparable

The two large nuclear weapon states seem determined to monopolize the nuclear arms phenomenon, thus providing an alibi for the other recognized and non-recognized nuclear weapon states and creating the risk that new nuclear weapon states will present themselves one by one, as happened with India and Pakistan in May 1998. This despite their well-intentioned dedication to the non-proliferation of nuclear weapons. This dedication is however being structurally undermined by the attempt to turn the temporary division between nuclear and non-nuclear weapon states into a permanent one. More than 160 states have forsworn nuclear weapons and are largely part of nuclear-weapon-free zones. It will be clear to every independent and thinking individual that this current division cannot be a foundation for the international legal order. At the so-called Prepcom of April 1998, a preparatory conference for the NPT conference to be held in 2000, the non-aligned countries expressed similar criticism.[6]

Despite this situation the nuclear weapon states and NATO emphasize the proliferation problems of weapons of mass destruction and long-range missiles while ignoring the issue of their own nuclear arms. On the issue of nuclear non-proliferation there are in my opinion hardly any controversies. Nobody applauds an increase in the number of nuclear weapons states. Outside India and Pakistan there was no applause for the nuclear tests, only understanding for the shared responsibility: that of the countries concerned[7] and the nuclear weapon states.[8] And every interested party is closely monitoring events in North Korea, Iraq and Iran.

Changes in the Rationale for Nuclear Weapons

Attempts are nevertheless proceeding to broaden the rationale for the application of nuclear weapons within the framework of counter-proliferation, where these weapons no longer serve just to deter nuclear weapons but to counter weapons of mass destruction, including biological and chemical weapons and even the threat of missile attacks in general.[9]. This last is a typical example of the adaptation of new arguments rationalizing the use of nuclear weapons. It started as a counter to a possible nuclear threat from Nazi Germany, followed by use against Japan to achieve a swift end to the war in the Far East. After that it found a new application as a deterrent weapon during half a century of Cold War and now a new aim has been found: counter-proliferation. This while even American insiders are of the opinion that the United States is powerful enough to counter the new threats with conventional means.[10] There is continuity in these arguments: namely that nuclear weapons should not be abolished.

The Present Situation - A More Detailed Assessment

Spokespersons of nuclear weapon states and NATO allies generally paint a far more positive picture of the present situation than that presented here. They usually point at the substantial nuclear disarmament which has taken place since the end of the Cold War. If problems concerning further steps are acknowledged at all, these are generally put down to the position of Russia, more specifically the Duma, which is accused of disrupting the process of nuclear disarmament through not ratifying the START II agreement for the last six years. The sombre outlook for the so-called "cut-off treaty", which attempts to stop the production of fissile materials, are blamed on the non-official nuclear weapons states and any Third World countries which are blocking this initiative. The indefinite extension of the NPT (1995) and the coming of the Test Ban Treaty (1996) are also pointed at as proving progress has been made. Such comments are in themselves correct, but only part of the truth.

Disarmament

Let us start with the disarmament process itself. In linear terms this is indeed substantial. As of this moment the number of nuclear warheads is 50% of the maximum of 1986, which is true for both the United States and Russia (the former Soviet Union). As a result of present policy this total will in future possibly be reduced to 30% of the 1986 figure. The number of strategic warheads ready to be launched has also been reduced by some 50% for the two major powers and if START II is implemented will drop to 25% and under START III even to 15% of the 1986 levels. The number of American nuclear weapons in Europe has already been reduced to about 20%.[11] In themselves these are quite impressive reductions but some essential facts should be noted. All these reductions or proposed reductions are the result of decisions taken in the golden era of nuclear disarmament around 1990: after that no new steps were taken, except for START III. Typical for the reigning optimism around 1991/92 are statements made by William Arkin[12], among others. He stated that the substantial number of measures taken by Bush and Gorbachev in September and October 1991 must have consequences for nuclear disarmament. The first steps towards abolition had been taken! In fact, the present situation in Europe (with the presence of 100-200 American tactical nuclear weapons) indefinitely freezes a situation in which tactical nuclear weapons were not completely abolished. The present problem is that if START III, allowing for the deployment of 2,000 strategic nuclear weapons by both superpowers, is the end of disarmament (which is not at all unlikely), then we will be stuck with a somewhat reduced nuclear force (including reserves) of some 20,000 warheads.[13] If it were indeed the final stage of nuclear disarmament then the situation would hardly be an improvement on 1986, aside from the improved political relations.

New Consciousness Nuclear Weapon States

Besides limiting the significance of the present disarmament achieved until now the increased confidence with which the Big Two have exhibited their nuclear armaments is of importance. In the USA a series of documents made public through the Freedom of Information Act show that a re-evaluation has taken place. A presidential document (PDD-60) clarifies that nuclear arms will remain the cornerstone of American defence policy. The National Stockpile Stewardship Management Program (NSMMP) lays down the careful custody of all nuclear knowledge, at least until 2065, as well as maintaining the number of warheads (operational and in the stockpile) at a level of about 10,000. The fissile material of thousands of semi-dismantled nuclear warheads, known as "pits", has also been stored, and is not being transferred to the civilian sector[14]. Regarding Russia, the situation is even more complicated and alarming because of the poor level of maintenance of many nuclear systems. This implies risks not only for those directly concerned, but also for many countries in the northern hemisphere. These risks cannot be calculated accurately but should not be underestimated. There is a great contradiction between the numbers which Russia wants to maintain for political reasons and the numbers it can maintain technically and financially. Only a fraction of the strategic systems are ready to be launched at any given moment.[15] Not ratifying the START II agreement is an act of extreme short-sightedness on the part of the Duma in relation to the country's interests. It is in fact an irrational attempt to compensate for weakness in conventional forces. From 1985 onwards the nuclear arsenal of Russia has been decreasing by 2,000 warheads a year but, as already noted, this decrease will probably stop when it reaches the level of 10,000. In Russia, in a decision similar to that taken by the President of the United States, nuclear arms have again been declared to be a fundamental component of national security.[16] There has even been a rediscovery of the tactical nuclear weapon, of which the government wants to maintain a stockpile at a level of 5,000, which could even be used in civil wars on Russian soil.[17] In the light of these developments it should be clear that the implementation of Article VI of the NPT will not happen any time soon.

Of course the world's nuclear weapons problem is not solely defined by the two major powers. Besides the self-evident fact that the other three nuclear weapons recognized in the NPT have arsenals consisting of several hundred warheads, the political position taken by these states is also of great importance. Until now this has not been encouraging and the question of whether they will ultimately show a greater willingness than the major powers to give up their nuclear arms, remains unanswered.

Besides this, one must sadly acknowledge that the United Nations Conference on Disarmament does not at present have any nuclear items on its agenda. This is partially because of the blocking of the cut-off initiative and partially because the nuclear weapons states want to negotiate on their mutual affairs outside the CD forum.

Success With a Catch

A critical note should also be made regarding recent disarmament successes like the extension of the NPT and the Comprehensive Test Ban Treaty. Both are very valuable, but especially in the framework of non-proliferation and (no longer) in the framework of the nuclear disarmament of the nuclear weapons states. During the extension conference of the NPT in 1995, eighteen Principles and Objectives were accepted concerning further steps towards nuclear disarmament. That is why a great number of states were ultimately prepared to agree to the permanent extension of the NPT. The annual Prepcoms held to evaluate this process - including the last one in April 1998 - were however a failure because the nuclear weapons states and their allies blocked every cautious step in the direction of nuclear disarmament.

The historical aim of the CTBT has been weakened especially by the United States which has made itself almost independent of classic nuclear tests through computer simulations and sub-critical tests.[18]

The Great Silence

A further point indicating the existence of a firm policy to maintain nuclear weapons concerns the way in which proposals and statements made during the last three years by various individuals and organizations have been treated or ignored.

In the first place, the Advisory Opinions of the International Court of Justice in July 1996. I know of no official reaction by the nuclear weapons states to these Opinions. At the time they opposed requesting an Advisory Opinion from the Court and they now ignore these opinions. This is true for the illegality of using or threatening to use nuclear weapons as well as the "directive" to negotiate with an eye on a clearly defined final goal: total nuclear disarmament. More commonly these states will emphasize the part of the Opinion which states that it cannot decide on the legitimacy of the use of nuclear weapons in those cases where there is a threat to the continued existence of the state, in order to undermine as much of the other parts of the Advisory Opinion as possible.

Almost at the same time, in August 1996, the Canberra Report sponsored by the Australian government was published. The innovative part of this report consisted especially of the argument for taking a number of steps to reduce the dangers of nuclear weapons, before ultimate abolition. Three measures were suggested: the dropping of the alert status, separation of the nuclear warheads and missiles and the introduction of the "no first use" principle. In Dutch politics there is cautious support for the first and perhaps for the second measure. But the third runs into difficulties. The appeal for a "no first use" policy by the German minister of foreign affairs Fischer ran into a storm of criticism from most nuclear weapon states. And the proposal for a staged total abolition of nuclear weapons runs into even more opposition. A process of nuclear disarmament in stages is simply not open for discussion. Neither of the two major states has shown the least support for the suggested measures.

Besides these two initiatives we should also mention the statements by generals (December 1996) and politicians (February 1998), circulated by former American Senator Cranston. These statements generally support the Canberra report. The identity of the individuals prepared to take a public stand on disarmament is important. Prominent amongst the generals is Lee Butler, until 1994 supreme commander of the nuclear forces of the United States. Amongst the Dutch politicians who signed, ex-prime ministers Van Agt and Lubbers stand out.

In the NGO world the Abolition 2000 appeal is the best-known initiative. It dates from the NPT extension conference in 1995 and strives for the official acceptance by that year of a plan for total nuclear disarmament. Various groups of countries have also been active in the framework of the United Nations. The most relevant of these is the initiative taken by eight countries, among them Brazil, Sweden and Ireland and known as the New Agenda Coalition.[19] The original statement, published in June 1998, proposed a process of nuclear disarmament which would include a no-first-use declaration and the withdrawal of tactical nuclear weapons to the territory of the countries which owned them. In the autumn of 1998 a somewhat diluted resolution based on this statement was passed in the United Nations General Assembly. The great majority of the NATO states (twelve out of sixteen) abstained and did not follow the three nuclear weapons states in their "no" vote. Perhaps a cautious act of rebellion?

References

1. See the report "The future of U.S. Nuclear Weapons Policy", p.35,36; published by the US National Academy of Sciences 1997 and the article "US Nuclear Stockpile" by Robert S. Norris and William M.Arkin, The Bulletin of the Atomic Scientists, July 1998, p 69-71.

2. See the declaration "The Impasse in Nuclear Disarmament", a special statement by the Council of the Pugwash Conferences on Science and World Affairs, Pugwash Newsletter November 1998, p. 2-3.

3. See the report of the National Academy of Sciences referred to in note 1, p.43.

4. The number 100 was mentioned in this connection by Paul Doty, director of the International Security Program, Center for Science and International Affairs, Harvard University, Cambridge, Mass. in the US during the

symposium commemorating the ninetieth birthday of Joseph Rotblat. In his opinion 1000 nuclear weapons could still destroy all of humanity.

5. This point arose during the negotiations on the "Rome Statute of the International Criminal Court", 17 July 1998. At the end of 1997 IALANA published a communiqué on this issue entitled: The International Criminal Court: a crime to poison but not to nuke?

6. Working Paper presented by the Members of the Movement of Non-Aligned Countries Parties to the Treaty, Prepcom document 28 April 1998.

7. See the article "India ontmaskert zichzelf", Bart van der Sijde in the Dutch daily NRC Handelsblad 13 May 1998.

8. See for example the articles "Veroordeling van Indiase kernproeven is eenzijdig" by Karel Koster and Dirk Jan Dullemond, in the Dutch daily Trouw 14 May 1998; "Kernwapenstaten na de proeven van India: wat je zegt dat ben je zelf", by Jan Marinus Wiersma in Trouw 26 May 1998. See also the article by John P.Holdren (until recently chairman of the Executive Committee of the Pugwash Council): "Nuclear proliferation and United States Responsibilities". THe Chicago Tribune, 2 June 1998.

9. See "Doctrine for Joint Theater Nuclear Operations" of the US Joint Chiefs of Staff, 9 February 1996.

10. Position taken by former US Ambassador Thomas Graham Jr., leader of the American delegation during the NPT extension negotiations in 1995, during a combined session of the Defence and Foreign Affairs Committees of the Dutch Parliament, 9 February 1999. He was supported in this by General Lee Butler, former commander of the US nuclear strike force.

11. See "Taking Stock" (NRDC), Washington DC, March 1998. In a recent (end of 1998) statement in Parliament by the Dutch minister of foreign affairs van Aartsen a similar percentage is mentioned.

12. See the article "Nuclear Weapons headed for the Thrash" by William Arkin, Damian Durrant and Hans Kristensen in The Bulletin of the Atomic Scientists, December 1991, p.15-19. Their reaction is also discussed in the book "Kernwapens op hun retour?", Cahier 61 of a series published by the Study Centre for Peace Affairs, Catholic University of Nijmegen, p.14 by Bart van der Sijde and others.

13. The possible existence of a "resistance level" at 2000 strategic nuclear warheads was mentioned earlier in the article "Kernbewapening moet uit het slop" by Bart van der Sijde and R. Kruisinga in the daily NRC Handelsblad 16 June 1998. Dr. Steven Miller, director of the International Security Program, Center for Science and International Affairs, Harvard University, Cambridge, Mass. USA, made similar allusions during the 48th Pugwash Conference (29 September - 4 October 1998). Former US Air Force General Lee Butler also indicated that such a "resistance level" exists (The Hague, 9 February 1999).

14. See article by Arkin et.al. referred to in note 1.

15. See the article "Taking Nuclear Weapons off Hair-Trigger Alert" by Bruce G. Blair, Harold A. Feiveson and Frank N. von Hippel in Scientific American, November 1997, p 42-49.

16. The Russian parallel is the Russian Federation President's Decree nr. 1300, 17 December 1997.

17. See Report 27th Workshop of the Pugwash Study Group on Nuclear Forces, 21-22 February 1998 by Gotz Neumeck, Pugwash Newsletter volume 35, nr.2, November 1998.

18 Quoted in a press release dated 21 April 1998: Disarmament Groups force DOE to release portions of secret "Stewardship" plan for nukes; Agency now admits it will 'replace' nuclear weapons in stockpile and develop 'new nuclear options for emergent threats'.

19. Eight countries took the original initiative, including Sweden and Ireland (Slovenia withdrew later). See "Towards a Nuclear Free World: The Need for a New Agenda", Draft Resolution for the UN, 26 October 1998.

A ROAD TO NUCLEAR DISARMAMENT

Bart van der Sijde and Karel Koster

This paper looks at the process of nuclear disarmament. The central thesis advocated here is that in time nuclear weapons should be abolished completely. In the recent past numerous measures in that direction have been suggested. Four approaches to nuclear disarmament may be discerned: they strengthen each other and together should lead to nuclear disarmament and non-proliferation: 1) The political delegitimization of nuclear weapons; 2) the establishment of nuclear-weapon-free zones; 3) securing nuclear weapons against unauthorized or hasty use; 4) gradual decrease in the number of nuclear weapons.

The Political Delegitimization

Under this heading a number of policy issues will be discussed. These policies will be very important as far as the credibility of nuclear weapons states and their allies (especially NATO) is concerned. Do they really want to be rid of the nuclear weapons, or will the non-nuclear weapon states have to conclude that they do not? This especially concerns measures and approaches which, if adopted, could strengthen non-proliferation and disarmament, but if not could result in initiatives like the cut-off treaty not getting underway.

Negative Security Assurances

A central point in this connection is clear negative security assurances for non-nuclear weapon states - the promise not to use nuclear weapons - which should not be qualified by unnecessary limitations. These have existed for a long time but were clearly defined in 1995 and in effect allow only one exception: that concerning a country which acts in concert with a nuclear weapon state which realistically is a nuclear threat. This limitation is understandable and is in fact more relevant for NATO member states with US nuclear weapons on their territory than for other countries. It has been the political practice of the last few years which has undermined this principle. Both American and Russian policy documents do not exclude the possibility that nuclear weapons will be used for non-nuclear purposes, not only against other weapons of mass destruction but against missile attacks in general. As a result of this process of undermining, the desire of a number of countries with sufficient infrastructure, political aspirations and enemies to acquire nuclear weapons will be encouraged. Although other motives also played a role, India is a good example of this.[1] Perhaps clear transgressions of the NPT could be punished by withdrawal of the negative security assurances. The nuclear paragraph of the NATO strategic concept, of which a new version will be adopted in April 1999, should at least contain a statement that nuclear weapons are weapons of last resort. Every suggestion that NATO nuclear weapons could play a role in the framework of counter-proliferation policy should be removed.

We argue for a strong restoration in daily political practise of sharp and precise negative security assurances which are not encroached on by the nuclear doctrines of the nuclear weapon states.

No-First-Use Principle

A second aspect of delegitimization is defined by the introduction of a no-first-use principle. From recent discussions following on the bombshell dropped by the German minister of foreign affairs Fischer, it is clear that the nuclear weapon states are at present not prepared to engage in this debate and it is therefore questionable whether one or more NATO member states would succeed in really launching this debate. Precisely because no first use automatically entails a firm security guarantee, it is a matter of some importance whether it will actually be introduced: both supporters and opponents are in fact in agreement on this.[2] In our opinion the principle lends itself more for inclusion in a treaty than as a unilateral declaration. That provides mutual protection and commitments. The most far-reaching and desirable option is for all seven *de facto* nuclear weapon states (and preferably Israel as well) to conclude such a treaty. Even an agreement between Russia and the United States would in itself be a valuable contribution. At present only China maintains the principle unilaterally.

We argue for a strong effort to introduce the "no-first-use" principle by treaty in the near future.

Use and Threat of Use

A third possibility is to make the use and threat of use of nuclear weapons a crime. This could be done in accordance with the Advisory Opinions of the International Court of Justice and International Law and within the framework of the International Criminal Court. The attempt by the Western nuclear weapon states to create an exception for nuclear weapons must be strongly rejected and prevented. Such a success would crown the efforts by the nuclear weapon states to legitimize their status, as defined in the NPT, indefinitely but would ultimately form a grave, irresponsible undermining of the Treaty.

Within the framework of the creation of the International Criminal Court, the use and threat to use weapons of mass destruction should be described as being contrary to International Law.

Continuation of Negotiations

In the opinions of the International Court of Justice no concrete proposals can be found on the process of nuclear disarmament. Apparently this is seen as a process which should take place in the political sphere. But there are indications of the outlines within which this process should take place. Without any doubt the three steps described above fit seamlessly into the advisory opinions of the Court. The American refusal to negotiate START III before START II is ratified in the Russian Duma, is at the very least at odds with the Advisory Opinions of the Court. The refusal by the Russian Duma to ratify is clearly in contravention with the advice of the Court to negotiate for total disarmament.

Both the refusal by the Duma to ratify START II and that by the American Senate to negotiate START III contradict to the Opinions of the International Court.

Nuclear-Weapon-Free Zones and Non-Proliferation

The concept of nuclear-weapon-free zones is a fairly old one, but only gained real meaning after the end of the Cold War.[3] The Treaty of Tlatelolco for South and Central America dates from 1967, but only became politically viable after differences between Argentina and Brazil were resolved. It covers all of South and Central America, except for Cuba. The Treaty of Rarotonga, covering the Pacific and signed in 1985 also had a problem: France continued its nuclear testing on Muroroa. After the last nuclear test in 1996 that country rapidly cooperated with the implementation of the Treaty and it is now in force without any limitations. Australia and New Zealand are both partners.

In the 1990s two new nuclear-weapon-free zones were created: one covering southeast Asia from Burma to Indonesia (in 1995) and the other covering all of Africa (in 1996). The first is not quite completed (there are some territorial problems with the nuclear weapon states) and the second still awaits a number of ratifications, but is supported by the nuclear weapons states. A fifth treaty is being prepared and will include a number of states of the former Soviet Union (Kazakhstan and others). Mongolia also regards itself as a nuclear-weapon-free state.

A Nuclear-Weapon-Free Zone in Europe?

These five treaties cover all of the easy regions. Through the introduction of the zones described above the nuclear weapons problem has become a matter of the Northern Hemisphere, concentrated around the five recognized nuclear weapon states, Europe, the Middle East, South Asia and the Far East (China, North Korea). There are discussions now and again on the creation of nuclear-weapon-free zones in some of these regions. But existing relations in a number of these regions make the idea of a nuclear-weapon-free zone an illusion. Now that the Cold War is over, the possibility of a nuclear-weapon-free zone in Europe or at least a part of it, should also become a subject of debate, as should other regions. It is hypocritical to support nuclear-weapon-free zones for other difficult parts of the world but to reject one for Europe, as is now the case. No one denies that a nuclear-weapon-free zone in Europe, with three nuclear weapon states as well as six NATO member states with American nuclear weapons on their territory, is a complicated matter. But a decision not to exclude this possibility beforehand could definitely contribute to the further lessening of tensions between East and West. There are enough European states who want to remain politically independent of Russia, and do not want to become a member of NATO, to initiate the discussion. It is, however, clear that the chances of a relevant nuclear-weapon-free zone in Europe are tied to the chances of further nuclear disarmament. As long as NATO regards nuclear weapons as an essential part of its security and certainly if the US weapons remain in Europe, a potential nuclear-weapon-free zone here will be of little significance.

*A policy by the nuclear weapon states and NATO which promotes the installation of nuclear-weapon-free zones elsewhere in the perspective of non-proliferation, but blocks it in Europe, is not credible and needs to be changed radica*lly.

Non-proliferation and a good system of safeguards are indivisible. Although after bitter experiences with Iraq following the Gulf War, the climate within the International Atomic Energy Agency (IAEA) has improved, it remains a relevant question whether that organization has at present sufficient power to be able to act flexibly in countries like Iran and North Korea. Special inspections still need to be initiated by the IAEA itself and cannot, as in the Chemical Weapons Convention, be enforced by countries.

Reducing the Danger of Nuclear Weapons

A number of measures were suggested in 1996 in the Canberra Commission Report to make the current stock of nuclear weapons less unsafe. Included in this framework are the so-called "de-targeting" of nuclear warheads, introducing the "off-alert" status, physical separation of warheads and missiles, and introducing flexible, physical barriers against rapid launch. In fact the existing direct communications between the responsible nuclear commander and the president (or prime minister) and the famous "hot line" between the presidents of the Great Powers can also be counted among them. The information collection by numerous systems which are independently capable of giving early warning of an attack also contributes. This gives a greater degree of confidence to ignore signals coming from one specific system and which may well prove false.

Barriers Against Use

Unfortunately the debate about the proposals in the Canberra report has not visibly started yet. As far as we know, "de-targeting" has been introduced but it is commonly assumed that this involves a delay of mere minutes before launch and is therefore essentially a symbolic start. Such a delay can be fairly easily absorbed within the normal warning time. The term "off-alert" status is usually taken to refer to a delay lasting hours. The British authorities have already introduced this and even refer to a readiness stage lasting days.[4] We need to realize that we are entirely dependent on the authorities for the information needed to judge the meaning of these measures. But in general taking up an "off-alert" posture is regarded as an essential contribution to reducing the danger of an early or accidental nuclear launch. In view of the concerns about the Russian nuclear weapon arsenal this is a very important measure. Nevertheless, this measure encounters a practical difficulty and one of principle. In our opinion it can only work if there is strict reciprocity between the two great powers. Furthermore, it also requires extensive verification, otherwise partial evasion is too easy. Introduction of such a regime requires the permanent stationing of experts on both sides, who would be allowed into the very innards of the competing nuclear power. Truly an innovation! A measure which could be introduced with fewer far-reaching consequences would be for the Americans to adjust the number of nuclear weapons at ready-to-launch status to that of the Russians: probably less than 1,000.

An issue for debate is whether - given sufficient political will to increase the security of the nuclear weapons systems - the separation of nuclear warheads and missiles and/or the addition of physical barriers would not be easier from the point of view of verification. The deployment at sea of submarines without nuclear warheads on board, and placing huge covers on top of intercontinental missile silos is easier to verify by a group of observers on the spot or (in the latter case) by satellites.[5] The last measure is not only more inspiring but also has advantages as far as verification is concerned.

It is strongly recommended that the measures for reducing nuclear dangers proposed in the Canberra Report should be supported, especially the separation of nuclear warheads and missiles, as well as the placing of physical barriers.

Nuclear Disarmament

Besides non-proliferation, nuclear disarmament is the ultimate central issue. The three previous measures are meant to reduce the role of nuclear weapons, but will not result in abolition. A central issue is that the absurd destructive potential of the Cold War has only been halved (to 35,000 warheads) and that therefore the enormously inflated numbers are still an issue. As has already been noted, if a new policy is not formulated which differs drastically from the existing one so as to continue the fruitful period of disarmament measures taken around 1990, then we are heading for a total of 20,000, almost completely in the hands of the two great powers.

The START Process

There is some prospect of a START II agreement with a ceiling of 3,500 warheads and START III with a ceiling of 2,000 to 2,500, to be attained by 2008. The realization of START II will however be delayed to 2003 (as far as the warheads are concerned) and 2007 (for the launchers), *provided* the Treaty is ratified *in time* by the Russian Duma. The execution of START III is again tied to Start II, so that even the direct prospects of the disarmament measures proposed so far are very uncertain. Of course, most of the blame must be apportioned to Russia, but the possibility of attaining a more balanced package from the viewpoint of the Russians through Start III, is being blocked by the American Congress. Of course this situation has to be solved first. Unfortunately practical politics have constantly created new arguments for the Duma not to agree to ratification: among them NATO enlargement and the Iraq and Kosovo questions. The nationalist and communist parties in the Duma are virtually unapproachable politically by the West. They nevertheless are taking a great responsibility upon themselves by blocking a policy which is vitally necessary for Russia: a rapid, controlled and drastic reduction of their nuclear forces. It would be wise for the Americans to propose as well as take every imaginable measure to convince the Russians of this. Blocking START III negotiations is not a part of such a policy.

Tactical Nuclear Weapons

The present situation is a sombre one, because we have only taken a few steps on the path to disarmament. The START III negotiations should - as has been urged by others - include far more than strategic nuclear weapons, that is the number of operational missiles with a certain number of nuclear warheads. Tactical nuclear weapons should also be included in the negotiations for the first time. This includes the nuclear weapons deployed in the so-called co-user NATO member states. These are weapons deployed in the Netherlands, Belgium, Germany, Italy, Greece and Turkey. In a nuclear war situation these states would become *de facto* nuclear weapon states and can for this reason - among others - be regarded as semi- nuclear weapon states. This is in fact not in accordance with the Non-Proliferation Treaty and this situation has - understandably - come under increasing attack by the signatories of the NPT.[6] A first step could be the withdrawal of these nuclear weapons from Europe and the removal of their infrastructure. In view of the measures already taken in September and October 1991 there is in fact only one satisfactory solution in this area: the complete abolition of these weapons. The land and sea components of these weapons were, after all, abolished in 1991 and the air component strongly reduced. As far as the tactical weapons are concerned there is in fact just the choice between the current status quo (if need be including withdrawal from Europe) and complete abolition. Through abolition the above criterion - no deployment of nuclear weapons on the territory of other states - has automatically been met. However, two huge political barriers have to be surmounted before tactical nuclear weapons can be abolished: in the first place the Cold War taboo surrounding the last American nuclear weapons in Europe; in the second place the Russians will have to undo the rediscovery of tactical nuclear weapons, including their eventual use in internal conflict.

Tactical nuclear weapons must be abolished within the framework of the START III negotiations and therefore also the deployment of American nuclear weapons in Europe; Russia therefore needs to re-evaluate the role of tactical nuclear weapons.

Large Reserves

Another point which can no longer be ignored is the enormous number of nuclear weapons kept in storage besides the immediately deployed weapons. The Americans are known to maintain three categories of some 5,000 warheads each in storage as part of the Stockpile Stewardship Management Program (one of them consisting of the disassembled "pits" which include the intact nuclear core). These last are therefore not usable in a crisis situation, but the plutonium and uranium is not being taken out of commission. Although it is difficult to make an outsider's calculation on the amount of extra reserves which is needed or necessary from a technical standpoint, one cannot be satisfied with the information that two reserve warheads are needed for every operational one. This could only be justified from a need for absolute security, as if the future of the nation were to depend on the malfunctioning of a single warhead.

Reserve stores also need to be drastically limited in the framework of the START III negotiations and only a limited number of extra warheads should be maintained above the ceilings of deployed weapons agreed upon in the Treaty.

On the Way to One Hundred

If developments went our way then, somewhere between 2007 and 2010 a situation would be created in which about 2,000 warheads would be deployed by the USA and Russia, tactical warheads would have disappeared and only a marginal extra reserve be maintained. The small nuclear weapons states could be left out of this stage, because their numbers vary between 200 and 400, while Israel, Pakistan and India would have even fewer.

There are signs, certainly on the part of the USA,[7] of stubborn resistance to reductions below 2,000. If nuclear disarmament is to mean anything, however, than this barrier will also have to be crossed. There is little point in making detailed proposals for the contents of a START IV, V, and so on in 1999. However, there need to be continuing negotiations about the next stage, even if the last one has not yet been implemented or ratified.

We propose that *real* nuclear disarmament only starts at a level of *100* per great power and adjusted numbers of a few dozen for the other nuclear powers. The total number of nuclear weapons should be at most 300 to 400. We would define the road there as the *preliminary* to *real* nuclear disarmament. The 100 number is simply a symbol for a number which is bigger than one and amounts to about one per cent of present numbers. It is still true that the Second World War in the Far East was ended by two nuclear weapons with a relatively low yield. The conclusion of the US National Academy of Sciences, that a mere twenty nuclear weapons represent a decisive destructive potential against Russia, still stands.

Our position is that real nuclear disarmament only begins at a level of about 100 and that all prior reductions simply amount to fighting inflation. We also propose that this stage can be reached in a period, no longer than 20 years, which should be agreed upon now by all the nuclear weapon states involved (for instance at the fiftieth anniversary of the Non-Proliferation Treaty in 2020).

The question of the survival of the Anti-Ballistic Missile Treaty is also again relevant. In March 1999 the American Congress decided that the government should make preparations for an extensive anti-rocket shield, the construction of which could be started soon. This would almost certainly destroy the ABM Treaty and result in a period of political confusion. If the Russians oppose it, START II ratification will almost certainly be delayed once again.

The Final Stage

What can be said about the final stage? First, that *complete nuclear disarmament*, to be pursued in a more or less defined period, can be the only politically legitimate *goal* as defined in Article VI of the NPT. But there are reasons for distinguishing between the preliminary stage to 100 and the final stage. Right until today the nuclear weapons states have refused to commit themselves to a fixed timetable for complete nuclear disarmament. The most likely explanation is - as has been postulated - that they do not want to get rid of nuclear weapons and even want to maintain large numbers of them. However, they could also argue that there are too many uncertain variables in the process for them to decide on any kind of final date. The proposal for 100 is also meant to force them to take up a position as regards their real motives. A time-bound reduction to 100 meets the justified problems connected to a time-bound reduction to "zero".

In the literature various comments have been made about the uncertainties of the final stage and the studies needed in various areas to clarify the possibilities and dangers.[8] Naturally, the motivation for making these studies will increase when it becomes clear that reaching the final stage is no longer utopian. We foresee that solutions must be found for a number of problems during the final stage of disarmament:

1) Complete transparency regarding nuclear weapons, fissile materials and nuclear installations, between the five recognized nuclear weapon states, to avoid abuse during further disarmament. Naturally, this must be accompanied by an atmosphere of trust, in which the parties sense that the others can be trusted. It is, after all, impossible to gain absolute certainty about the presence or absence of the last kilogram of plutonium. When this stage is reached, it should be possible for the five permanent members of the Security Council to act in concert, as guardians of the world community, instead of competitors on a knife's edge of mutual nuclear threat.

2) Cooperation by the three nuclear weapon states not recognized as such by the NPT, at the same level as between the official nuclear weapon states. This also means reducing arsenals to levels comparable to those of the recognized nuclear weapon states.

When these two conditions are met, the size of the nuclear weapons arsenal can be reduced further, to a few dozen. There would then be only the existing, new or suspected threats of a few countries.

3) The situation regarding the cooperation or refusal of countries like North Korea, Iran and Iraq and perhaps others in future. There is little point in speculating about possible situations, but it seems quite clear that the Security Council will not allow those countries to be in possession of the last nuclear weapons. Depending on voluntary cooperation or else that attained through pressure, and the actual level of threat (suspected or nearly certain ability to make nuclear weapons, suspected or nearly certain stockpile of ten to twenty weapons), a solution can hopefully be found here as well.

4) The situation regarding terrorist organisations. In view of the effort made by for example Iraq in the 1980s we do not regard it as plausible that a terrorist organization which does not have *de facto* control over a large piece of territory, will be capable of developing a nuclear weapon. Greater anxiety should be reserved for the possible disappearance of nuclear weapons from Russian arsenals. Numerous unconfirmed stories circulate about the disappearance of several small nuclear weapons. If no serious threat from that kind of weapons arises in the coming 20 or 30 years, such rumours should be discounted. Besides, there is little point to maintaining a nuclear deterrent, no matter how small, against a terrorist organization. There is probably never a target which can be attacked: such organizations should be fought through internal security measures.

Some analysts, including Charles Glaser,[9] suggest that the total abolition of nuclear weapons is a dangerous matter. As the knowledge to make nuclear weapons will not be lost, they will again be produced whenever new conflicts occur. The country which succeeds first would then have a dangerous monopoly. Glaser even estimates that the danger of using nuclear weapons will then be greater, because there will be no deterrence. We regard this reasoning as very speculative, because such confrontations between nuclear weapon states and non-nuclear weapon states already occur without nuclear weapons being used. The complete abolition of nuclear weapons will in any case only be achieved when all parties are sufficiently convinced that mutual trust will allow it.

The most important argument for continuing to pursue complete nuclear disarmament remains that it is very unlikely that any stable world peace can be achieved in which a number of countries still possess nuclear weapons. The maintaining of nuclear weapons as weapon of mass destruction is furthermore a sign of a serious lack of morality in world society.

References

1. The desire on the part of the new nationalist Hindu government to create a degree of national unity was one important reason for the nuclear test of May 1998. Others which are always mentioned include national security

regarding China (with Pakistan hardly being mentioned). The Indian Pugwash member Jasjit Singh, however, started his speech defending the test at the annual Pugwash conference in Sept-Oct 1998 by noting the general arguments described in this article.

2. See the article by Bart van der Sijde "Kernwapen kan overbodig worden" in the Dutch daily NRC Handelsblad 28 December 1998.

3. See the article "Kernwapenvrije zones anno 1998" by Bart van der Sijde in the Dutch periodical *Transaktie*, vol.27, nr 1 1998, p.69-85.

4. The British step is noted in the Pugwash declaration of October 1998.

5. See the article "Taking Nuclear Weapons off Hair-trigger Alert" by Bruce Blair, Harold A. Feiveson and Frank N. von Hippel, *Scientific American*, November 1997, p.42-49.

6. See the article "Troubled Treaties: Is the NPT tottering?" by Rebecca Johnson, The Bulletin of the Atomic Scientists, vol.55, nr.2, March-April 1999.

7. See the comments and note on this in the introductory article "Nuclear Disarmament by the year 2000: A lost cause?"

8. See the interview with McNamara, p.45-53 in the book *The Gift of Time* by Jonathan Schell, 1998, Metropolitan Books; and the article by Charles L. Glaser "The Flawed Case for Nuclear Disarmament", *Survival,* vol. 40, nr. 1, p.112-128. A first approach to this can also be found in the report "Laying the Foundations for getting to Zero: Verifying the Transition to low levels of Nuclear Weapons" by Patricia M. Lewis, VERTIC research report no.1

9. Glaser, "The Flawed Case for Nuclear Disarmament."

MOVEMENTS TOWARDS COMPREHENSIVE NUCLEAR DISARMAMENT

Liu Gongliang

The Basic Goal of Nuclear Disarmament

For more than 40 years, the international community and the people of the world have worked hard to push forward the process of comprehensive nuclear disarmament, but progress has been very slow. Up to now, the nuclear reductions between the USA and Russia have primarily been limited to the deployment levels of their strategic offensive forces. More than 20,000 warheads are still stored in their arsenals, and it is easy to couple these warheads with delivery systems. India and Pakistan conducted nuclear tests in May 1998, which has focused renewed attention on the threat of a nuclear arms race in South Asia. Several months from now, the world people will welcome in the new century, and it is time for us to learn the lessons from the past 40 years, and promote the process of comprehensive nuclear disarmament.

Firstly, we should make clear the purpose of nuclear disarmament. There is no doubt that the final goal of comprehensive nuclear disarmament is to eliminate nuclear weapons in the world. However, the practice of arms control shows that it is a long-term goal. A step-by-step approach is needed to reach the basic goal, and great care should be taken at every stage. The first step is reducing and eliminating the danger of nuclear war, including by preventing the danger created by the quantity of nuclear weapons and fissile material. This is an important step for long-term international security; it also is an urgent step in the current international situation. Russia's Duma has not ratified START II yet, after the Iraq and Yugoslavia crises, and confidence between the USA and Russia has dropped. Russia's military has declared that its nuclear missiles would retarget NATO. According to START I and START II, the numbers of deployed nuclear strategic weapons are falling, but the danger of nuclear war is increasing.

Nuclear disarmament is not an independent movement. It depends on processes of arms control, including nuclear arms control and non-nuclear arms control. Nuclear disarmament and the NPT regime complement each other. If nuclear states did not take active measures to promote the process of nuclear disarmament, an effective NPT-regime would not be achieved. But, it should be emphasized that nuclear disarmament cannot be advanced by a failed non-proliferation regime or by a regional nuclear arms race. It is the same for nuclear arms control and conventional arms control. The basic purpose of arms control is creating a peaceful environment for the human race. Nuclear weapons have a massively destructive effect, so we should pay more attention to preventing the danger of nuclear war, but in the past two decades some new high-tech weapons, such as cruise missiles, depleted-uranium bombs and graphite bombs, have been developed by some developed countries. Those new weapons can also inflict terrible casualties. We also should control or prohibit those new high-tech weapons.

Some Technical Issues Must Be Solved

Comprehensive nuclear disarmament would meet some complex technical issues. The USA and Russia have solved some technical issues for their bilateral nuclear reduction agreements. Compared with future multilateral deep nuclear disarmament, this is very straightforward. All nuclear weapon states should research those technical issues which must be solved in multilateral disarmament collaboratively.

Effective Verification Techniques

A real nuclear disarmament process should be conducted under monitoring of an international organization, and using effective verification techniques. Those techniques should meet the following demands:

- We need to develop a technical system that could identify effectively reduced warheads and reserved warheads allowed by treaty.
- Highly precise accounting techniques. Because there are more than 20,000 warheads stored in the US and Russia's stockpiles, how to account for the large number of weapons, especially the "pits", is a very important and sensitive problem. Great improvements to the accounting precision is a key technique; the uncertainty would have to be less than 0.1%. If the uncertainty were larger than several per cent, several hundreds warheads could be miscounted, which is not acceptable. The United Nations inspection team reached no conclusion on Iraq's nuclear weapons plans after their strong intrusive inspection during more than five years. Nobody has confidence to ensure that they can completely account for all nuclear weapons in all nuclear weapon states and all *de facto* nuclear weapon states.
- How to account for the means of delivery and how to monitor the illegal production of all kinds of delivery vehicles also need to be researched.

Reliable and Safe Reduction Techniques

There are different reduction techniques in deep nuclear disarmament. According to various proposals on future deep nuclear disarmament, there are three kinds of techniques that would be used. These are:

- Dismantling nuclear weapon systems including separating warhead and vehicle, and dismantling or destruction of launchers and silos. Every step could be opened to monitoring but sensitive information must not be revealed.
- Dismantling nuclear warheads, including separating ignition and high explosive, separating high explosive and pit, and dismantling the pit. The pit is a core component of a nuclear warhead. Up to now, the USA and Russia have not dismantled their pits in implementing START-I or START-II. Separating plutonium from the pit is difficult and dangerous work. It needs safe and reliable supporting techniques.
- Disposing of the plutonium dismantled from nuclear warheads. This is a critical measure for irreversible nuclear disarmament. Though some technical proposals have been researched, nevertheless it depends on political decisions, energy economy, verification approach, accounting methods and environment impact, and no approach has been accepted by all sides yet. We need to pay more attention to this key issue.

I believe that more technical problems will be met as nuclear disarmament deepens, so to strengthen research on these topics and to improve them continually are important responsibilities for scientists. Scholarly exchange or collaboration between scientists under the non-proliferation principle benefits the development of verification techniques.

Multilateral Nuclear Disarmament and Confidence-Building Measures

Multilateral Deep Nuclear Disarmament

If the USA and Russia reduced their nuclear weapons by a sufficiently big margin, it would bring other nuclear states into the nuclear disarmament process. It would turn the bilateral process into a multilateral process. It can be said with certainty that the verification regime for multilateral disarmament will be very different from those measures used in USA-Russia bilateral agreements. For START I and START II national technical means (NTM) are used to verify compliance with the agreement. It is well known that both USA and Russia have strong NTM with which to monitor the other. For a multilateral arms control treaty, it is necessary to build a transparent international monitoring system that can be shared by all participants, and this needs an international verification organization.

Another sensitive question is who are the natural participants in multilateral nuclear disarmament negotiations and treaty? One viewpoint is all the CD's members should participate in the negotiations and join the treaty. They think that nuclear weapons threaten world security, and every country has a responsibility to promote and to monitor the process of nuclear disarmament of nuclear weapon states. An opposing point of view is that only the five declared nuclear weapon states would need to negotiate the treaty, because every step of nuclear disarmament would touch upon nuclear secrets, and there would be a risk of nuclear proliferation if non-nuclear weapon states joined the negotiations and future verification of the treaty. Another knotty problem is how to treat some *de facto* nuclear weapon states such as India, Pakistan and Israel. Some people demand insistently that India and Pakistan sign the CTBT and join future nuclear disarmament, but another possibility is that India, Pakistan and Israel should sign the NPT and abolish all nuclear weapons as a non-nuclear weapon state like South Africa. If we treated those states as a nuclear weapon state, the non-proliferation regime in the world would be broken, and the process of nuclear disarmament would be hindered. We should balance non-proliferation and disarmament, since showing any bias towards one of two related issues will harm international security.

Confidence-Building Measures

From the previous analysis it becomes clear that the verifying of comprehensive nuclear disarmament would not only rely on technical measures, but also depend on political openings. There is a close relationship between political demands and technical capabilities. If political demands increase, the technical solution may become more difficult or more expensive. If technical advances or improvements in effectiveness and less intrusiveness are accomplished, political demands may ease.

It would take a long time to reach the above conditions. Before solving all technical and political problems, confidence-building measures are needed. In recent years, a lot of scholars have advocated "de-alerting", "de-activation", " de-targeting", and "de-mating" as confidence-building measures. Bilateral discussion between the USA and Russia were reported. We well know that in both the USA and Russia more than 4,000 nuclear warheads are still on 15-to-30-minute alert. They keep their weapon systems on hair-trigger alert and maintain a launch on warning policy. So de-alerting may reduce the risk of nuclear war and especially reduce the danger of accidental nuclear war. I think that a no-first-use policy is a better measure than de-alerting. It implies the highest level of de-alerting. I think that people who support de-alerting have no reason not to support the no-first-use policy.

Theatre Missile Defence, High-Tech Conventional Weapons, and Nuclear Disarmament

The US House of Representatives has approved building of a nationwide defence against ballistic missiles (NMD) system. The USA and Japan also have decided to establish a Theatre Missile Defence (TMD) system cooperatively, this so-called "defensive area" stretching as far as Taiwan. China and Russia strongly oppose the policy. It will certainly throw an obstacle in nuclear disarmament's way and may even trigger a new arms race. Taiwan is a part of China's territory; the US-Japan TMD system extends to China; nobody should believe that China would willingly reduce its nuclear weapons in these circumstances.

We should treat arms control integrally in the world. True international security will never be reached by separating nuclear and conventional arms control. If we do this, those countries who have no advantage on advanced conventional weapons or who have no money to develop advanced conventional weapons would take the nuclear option or strengthen their nuclear reliance – it is a very dangerous trend. We should oppose arms races in nuclear forces, in outer space, in advanced conventional weapons which have terrible destructive effectiveness, and in all fields.

VERIFYING THE TRANSITION FROM LOW LEVELS OF NUCLEAR WEAPONS TO A NUCLEAR-WEAPON-FREE WORLD

Tom Milne

This paper looks beyond the CTBT and the NPT to the requirements for verifying the transition from low levels of nuclear weapons, defined as being in the hundreds per nuclear weapon state, to a nuclear-weapon-free world. Since the United States and Russia currently have more than 10,000 warheads each, deployed, stockpiled or awaiting dismantlement, the paper is concerned with a stage of disarmament that can occur only some years in the future.·

Whatever one's views on the prospects for nuclear disarmament, there is little reason to anticipate that nuclear weapons will be eliminated in less than 20 years. Every nuclear warhead in existence could be dealerted and/or demilitarised in a matter of months· and dismantled in a period of years,· but enduring reservations held by most policy-makers, and exhibited by public opinion in the nuclear weapon states, mean that progress is likely to remain cautious and incremental.

So any discussion about verifying the final stages of nuclear disarmament must try to take into account the ways in which verification possibilities could alter on a timescale of two or more decades.

The Role of Verification in a Nuclear Weapons Convention

If and when states decide to relinquish their last remaining nuclear weapons, it is likely that they will do so under an international treaty. The content and form of a future nuclear weapons convention remains to be negotiated,· but such a convention is almost certain to include a verification regime.

Verification is the system whereby parties to an arms control or disarmament agreement can assure themselves about the compliance of others and demonstrate their own compliance. It is a means of building confidence in a disarmament regime and is generally complemented by other safeguards and confidence-building measures.

It is important to be realistic about the role that verification can play in a disarmament process. Confidence in disarmament derives from a variety of interdependent measures. In the case of verification, technologies capable of detecting treaty violations will help to deter potential violators by making cheating more difficult, more expensive, and perhaps more likely to be discovered. Verification techniques also provide assurance that if cheating does occur it will be detected above a certain threshold, providing the opportunity for enforcement measures to be set in motion, or for states to react in some other way. Other confidence-building measures include export control regimes, transparency surrounding wider defence activities, declaratory statements, trade agreements, and consultative mechanisms that allow for disputes to be resolved at low levels and before they get out of hand.

Verification is thus one of the methods by which states can be assured of the desirability and feasibility of a disarmament regime. In the case of nuclear disarmament, verification is a highly important means of providing such assurance. Many of the relevant means of verifying nuclear disarmament are well understood and known to be effective. And they are not impeded by the presence of a pervasive civilian industry, in the way that verifying chemical disarmament, for example, is complicated by the widespread presence of precursors of chemical warfare agents.·

It is sometimes argued, however, that the viability of a prospective or existing treaty is equivalent to its verifiability – that unless treaty compliance can be verified with complete certainty the treaty is worthless. This is too narrow a view. Even though many aspects of verifying nuclear disarmament are likely to be effective, uncertainties in verifying overall compliance with a nuclear weapons convention are inevitable, due to limitations of technologies and practical constraints imposed by negotiated treaty provisions.

Verification is thus just one of many possible interlocking measures in an international treaty and should not be seen as the most important. While verification regimes have an important role to play in encouraging the participation of states in international disarmament treaties, and in ensuring that these treaties are effective, they are only one element in an array of confidence-building measures built into a disarmament process.

Getting to Low Levels of Nuclear Weapons

The prospects for the total elimination of nuclear weapons some years hence will be affected by the experience with disarming between now and then. It is conceivable that the process of reducing stockpiles from current to low levels will build confidence in the disarmament regime, demonstrating that nuclear disarmament can be a peaceful and cost-effective means of maintaining national and international security. But equally, reductions to low levels might provoke a "backlash" against disarmament, reinforcing doubts about the desirability and feasibility of eliminating nuclear weapons. This might occur if, for example, agreements are not effectively implemented, they cost too much, or the verification system fails to detect a treaty violation.·

On the optimistic side, it is entirely possible that the process of going from current to low levels of nuclear weapons could "put in place a substantial part of the technical, institutional, and political ingredients that would be needed for an NWFW."· National nuclear weapons programmes might be made more transparent, for example, with greater openness about their activities and increasing involvement of weapons lab employees in open science. International safeguards on the nuclear fuel cycle might be extended to cover all relevant facilities in all states. There might be much greater openness among the nuclear weapon states about the size, form and location of remaining stockpiles of fissile material in military programmes. A "pattern of knowledge" · might be established among the nuclear weapon states, based on regular contacts between scientists from the weapons laboratories, regular on-site inspections, and so forth, giving confidence that no systematic deception was taking place in any state. And the chemical and biological weapons conventions might establish effective international bans on possession of these classes of weapon.

In addition, arrangements for verifying the elimination of the warheads themselves are likely to be introduced into disarmament agreements as stockpiles are reduced to low levels. (To date, nuclear reduction agreements have been concerned only with warhead delivery vehicles.) A first step could be to monitor the accumulation of warhead pits at a storage facility to await final disposition. While not necessarily providing assurance that a particular type or class of warhead was being dismantled, this would, if nothing else, give confidence that the total stockpile was at least being reduced in size. At a certain point, however, more rigorous verification of warhead dismantlement is likely to be required. Considerable research has already been carried out in this area, mostly in the USA, and a number of reports are available in the open literature.·

The first part of this paper outlines the process of verifying the elimination of warheads under a prospective international agreement, drawing on the U.S. studies. The purpose is to summarize what is already possible using existing verification techniques and to point to areas where further

R&D is needed. Some or even most of the verification measures discussed would, ideally speaking, be put in place before the final stages of nuclear disarmament.

The second part of the paper is concerned with the issue of detecting clandestine stockpiles of weapons or weapons-grade materials. The discussion is centred on the implications for verification of increasing openness in the world and improving technological verification capabilities.

Verifying the Elimination of Nuclear Arsenals

For the sake of clarity, the process of verifying the elimination of nuclear warheads can be divided, albeit somewhat arbitrarily, into four parts: first, the comprehensive declarations of warhead and fissile material inventories; second, consolidation of warheads and fissile material in "bonded store;" third, demilitarisation measures; and fourth, dismantlement of warheads and secure disposition of the contained fissile material.·

Comprehensive Data Exchange

In the process of moving from current to low levels of nuclear weapons, the nuclear weapon states should agree, at some point, on an integrated data exchange, including details such as type, serial number and operational status of every nuclear warhead, and all relevant aspects of their military nuclear programmes. This declaration, continually updated, would be the starting point for a comprehensive approach to verifying nuclear disarmament to zero.

As yet there is no system of integrated data exchange and verification that any of the nuclear weapon states is willing to endorse. In June 1995 the U.S. proposed a detailed data exchange with Russia, covering existing inventories of warheads and fissile materials, dismantlement of nuclear weapons since 1980 and production of fissile material since 1970. But this was rejected by Russia as too "comprehensive" and inconsistent with a "step by step" approach to transparency.·

The only existing transparency surrounding warhead and fissile material inventories has come from partial declarations unilaterally made by the U.S. (which has revealed the most among the nuclear weapon states about both warheads and fissile materials), Russia (which has done little more than declare quantities of plutonium and highly-enriched uranium surplus to military requirements) and the U.K. (which has promised greater transparency about fissile material stocks and is consulting with academic bodies and non-governmental organisations (NGOs) about declassification of information on its nuclear weapons programme).·

Such partial transparency could be the start of a series of incremental measures to build confidence, but as yet it does little to satisfy formal confidence-building and disarmament objectives.· The sooner that comprehensive data exchanges are made, and either the nuclear weapon states or an international body can begin to monitor all nuclear warheads and material, the better.

Introducing at an early stage reciprocal inspections among the nuclear weapon states (or inspections by an international body) of warhead storage facilities, production plants, and research and development sites, will create greater confidence as arsenals are being reduced to low levels and when they eventually are reduced to zero. There is less incentive to cheat on a disarmament agreement while numbers of warheads are relatively high (as is the case in the U.S. and Russia today), if only because of the period that would necessarily elapse before the cheating could have any possible "value" – longer than most governments could expect to survive.

It is likely that, initially, any such inspections would be performed mutually among the nuclear weapon states, as for example in the START process, rather than by an international organisation such as the International Atomic Energy Agency (IAEA). At some stage in the disarmament process, however, inspectors from non-nuclear weapon states should become involved, to provide international scrutiny. A balance has to be struck between creating international confidence in disarmament and putting unnecessary impediments in the way of progress. There are obvious difficulties, for instance, in international inspection of the dismantlement of warheads. The nuclear weapon states would not want to divulge certain information on nuclear weapon design and indeed are obliged not to do so by the Nuclear Non-Proliferation Treaty (NPT). But at the same time, and as discussed at later points in this paper, there are steps that could be taken – the construction of dedicated dismantlement facilities, for example, designed to accommodate international verification – that might make it possible for the nuclear weapon states to provide sufficient transparency to international bodies. In planning and preparing for nuclear disarmament, each of the nuclear weapon states might usefully review exactly which details of warhead design and construction have to remain secret, in the first place from other nuclear weapon states, and secondly from the IAEA, a body which has much experience of working with and protecting sensitive information.·

Bonded Store

The starting point for consideration of the final step to zero can be taken as a situation in which all remaining warheads and weapons-grade fissile material are held in "bonded store." That is to say, each nation in possession of nuclear warheads and fissile material produced outside safeguards has made available a detailed inventory of its remaining warheads and materials. These warheads and materials are stored in sealed containers, monitored either by permanent representatives appointed according to an international agreement or using a range of sensors to transmit data directly to an external site.· Inspectors will have access to the storage sites to audit the inventories and verify that the objects and materials held are nuclear warheads and weapons-grade material as declared.

Research is underway on warhead authentication techniques, based on measurements of radiation emissions and other characteristic signatures. But it is expected that authentication of nuclear warheads will be one of the most problematic aspects of verifying the disarmament process. The requirement is to authenticate an object inside a sealed vessel, declared to be a warhead, without revealing sensitive design information.· Examples of objects that might be substituted for real warheads include warheads in which natural or depleted uranium has been substituted for plutonium and highly-enriched uranium (HEU); objects closely resembling real warheads but fabricated to less demanding specifications; or genuine warheads containing much less fissile material than the warheads supposedly being eliminated.· Likewise, but more straightforwardly, techniques are being developed for confirming that weapons-grade plutonium, for example, is contained inside a sealed canister, again without disclosing classified information.

If nuclear weapons are still deployed when the decision is taken to go to zero, arrangements will need to be made for their secure transportation to the warhead storage or dismantlement facility. Chain-of-custody procedures are already used by the nuclear weapon states for the routine transportation of warheads, but these would have to be opened to verification. Richard Garwin describes, for example, how a fibre optic net could be placed over each weapon and a tag attached that would be invalidated if the net were broken.·

When secured in bonded store, a unique tag should be affixed to each of the warhead canisters and containers of fissile material. If an untagged warhead canister or container of fissile material were then to be discovered subsequently, this would be direct evidence of cheating on the disarmament agreement. Tagging would also allow an inspector to verify the authenticity of a particular warhead or container of fissile material selected from the declared inventory at random. In this way high confidence of detecting cheating could be obtained by statistical sampling techniques (in place of exhaustive checks), simplifying the verification procedure.·

Demilitarization

As nuclear arsenals are reduced, concerns are likely to be raised about the potential for breakout, when a nation openly or covertly withdraws from a disarmament agreement and reconstitutes a military nuclear force. Whether breakout poses a serious threat is controversial.· If breakout is perceived to be a danger, however, technical measures that provide reassurance should be considered.·

One means of reassurance would be to eliminate, under verification, all ballistic missiles capable of carrying nuclear weapons. Although the absence of nuclear warheads is the basic yardstick for verifying a nuclear-weapon-free world (because nuclear warheads can be delivered in many ways other than by missile), the elimination of strategic delivery vehicles would remove an important part of a nation's strategic nuclear capability.·

The INF and START agreements provide proven models for the verified destruction of missiles, which could be replicated in a multilateral treaty. And although INF and START are bilateral US-Russian arrangements, the elimination of nuclear delivery vehicles has been studied by other nations.·

A second method of demilitarisation would be to make an irreversible physical alteration to the warheads held in bonded store, rendering them unusable without disassembly and refabrication. U.S. warheads can be demilitarised, in this sense, using a technique that is quick to carry out, inexpensive and verifiable.· The same might be true for warheads produced by the other nuclear weapon states.

Dismantlement of Warheads and Disposition of Fissile Material

After warheads have been authenticated, they can remain in bonded store in containers protected with tamper-indicating seals until they are transported, employing chain-of-custody procedures, to the dismantlement facility.

In the U.S., where techniques for verifying warhead dismantlement have been studied in some depth, the term "transparency" is often used for the monitoring of warhead dismantlement, rather than "verification."· This is because of the widespread concern that it would be impossible to "verify" that an object being dismantled is a warhead without revealing more about design features than would be acceptable from a security standpoint, or legal under current national regulations.· The expectation is that, over time, mutual transparency will provide sufficient confidence for dismantlement agreements to work without rigorous verification.

It is conceptually simple to consider a dismantling facility as having one entrance and one exit. The passage of warheads entering the facility should be monitored and correlated with warhead components, in forms that have no security classification, passing through the exit. The fissile material would then be placed under international safeguards and taken to storage facilities to await final disposition and the other warhead components destroyed.· The dismantlement facility should be enclosed within patrolled boundaries, equipped with sensors and other technical safeguards. All objects leaving the facility should be inspected (with radiation detectors, for example) to prevent warheads or materials being smuggled out.

Citizens of the country that owns the warheads should dismantle them. The dismantlement process need not be observed, since the facility could be inspected periodically (in theory after each warhead is dismantled) to ensure that neither warheads nor materials were being illegally accumulated. To give additional confidence, chain-of-custody procedures could be employed within sections of the dismantling areas. It is also possible that either remote observation of the dismantling process via video camera or direct observation by disarmament inspectors could be arranged after appropriate shielding was in place.·

The need for intrusive verification of the dismantling process could be considerably reduced if it proves possible to use radiation "signatures" to correlate warheads entering the dismantlement facility with warhead components exiting. U.S. laboratories are currently investigating this technique. The main uncertainties concern disclosure of sensitive information, the uniqueness of the radiation signatures, and the alteration of signatures during the dismantlement process.·

Arrangements for the verified dismantlement of nuclear warheads might be more complicated if dismantlement takes place in a facility still used for warhead production and maintenance. A minimum level of remanufacturing and reconditioning of warheads will continue in each of the nuclear weapon states until a decision is taken to disarm completely. The U.S. Pantex Plant, near Amarillo, Texas, for instance, might be used for dismantlement and maintenance operations simultaneously, as might the U.K.'s nuclear weapons research laboratory, Atomic Weapons Establishment (AWE) Aldermaston, if the U.K.'s nuclear arsenal is dismantled in stages.

The criteria against which potential dismantlement schemes will be evaluated include the confidence the scheme gives that warheads are being dismantled as stated, the danger of inadvertent loss of classified information, and the scheme's international negotiability.· More mundane considerations will be the cost of inspections and, while the facility continues to be used for routine maintenance of warheads, the impact on operations.· The U.S. Department of Energy has made a detailed study of options for dismantling warheads with differing levels of verification at the Pantex Plant, which is currently the only facility in the U.S. authorised to disassemble nuclear warheads.· The other nuclear weapon states should commission similar studies.

To minimise the risk of disclosing sensitive information, and possibly to demonstrate a commitment to disarmament, the nuclear weapon states might also wish to evaluate the option of constructing new facilities to accommodate the verified dismantlement of warheads. The U.S. is again leading the way. The DOE has already carried out a study of the suitability of a newly constructed facility at the Nevada Test Site, known as the Device Assembly Facility, for use as a dedicated dismantling plant.

Lastly, there is some debate over whether it would be worthwhile verifying the destruction of non-nuclear components of warheads. On the one hand, it is argued that these parts could easily be replaced clandestinely, while on the other, that monitoring of the destruction of non-nuclear components would "add to the preponderance of evidence that functional nuclear warheads are being dismantled."·

Summing up, many of the technologies and techniques needed for verifying the elimination of nuclear warheads have been worked out at a general level. For the most part, research has been carried out in official and semi-official studies by U.S. organisations. While it is essential that these techniques be refined and improved, what is most important now, if disarmament is to proceed expeditiously, is for each of the nuclear weapon states themselves to study the central verification problems and requirements in order to identify particular techniques and approaches that they are willing to endorse.· They should study the generic techniques for verifying warhead dismantlement, such as warhead authentication and chain-of-custody arrangements, and give detailed consideration to the logistics of dismantling the warheads in their respective stockpiles, including, for example, the practicalities of accommodating international verification at their potential dismantlement facilities.

Whenever possible, verification research should involve international collaboration, both to perfect verification techniques and to build confidence in the disarmament process. The arrangements for verifying the transition to a nuclear-weapon-free world will have to be negotiated; it would be useful for nations to develop verification techniques co-operatively so that the eventual treaty negotiations are not hampered by technical disagreements. ·

As verification and transparency measures are introduced into international nuclear disarmament treaties, starting with START III, the states involved will have to carry out R&D and practical preparations, such as have been begun in the U.S.. Introducing transparency and verification into national nuclear weapons programmes might have a significant financial cost. Research and ingenuity could reduce this cost, however, and early investments in these fields and sharing of the results would be very useful.

Detecting Cheating: Trends and Possibilities

By far the hardest verification problem facing nuclear disarmament is the detection of illicit acquisition or retention of nuclear weapons or weapons-grade fissile material. This is an area that would benefit from further study by each of the nuclear weapon states and as many other nations as are willing to devote resources to supporting nuclear disarmament.

With respect to detecting potential cheating by the current nuclear weapon states, the main verification problem will be establishing that neither warheads nor materials have been secreted away before the disarmament agreement enters into force. As far back as the early 1960s, when stockpiles were much smaller than today, a British analysis of controls on fissile material, based on a study of the U.K.'s nuclear resources as they were then, concluded:·

> As far as the major nuclear powers are concerned, the control problem of the secret plant would be small compared with the problem of clandestine retention of fissile material.

A JASON study published in 1993 draws the conclusion that "one can never count on finding clandestine warheads."· And Steve Fetter goes further, saying that:·

> from a purely technical point of view, it would not be difficult to hide the existence of a few dozen (or perhaps even a few hundred) nuclear devices from inspectors.

Some reassurance might be gained from attempts to reconstruct the complete history of a nation's nuclear programme from the declarations it makes, in order to check the warhead and fissile material declarations for internal consistency and to reconcile the declarations with intelligence data. Yet such an enterprise could turn out to be a mixed blessing, for it would assuredly reveal uncertainties and inconsistencies in the declaration. It might not matter that, in all probability, innocent explanations exist for such discrepancies; they could still be used to stall progress in disarmament. Investigative techniques used for "nuclear archaeology," whereby physical evidence is collected to verify declarations (samples taken from nuclear reactors, for example, in order to determine, within certain bounds, the total quantity of plutonium produced in the reactor), might also be of doubtful value, for the same reason.· On the positive side, if the nuclear weapon states approach the matter of establishing initial fissile material and warhead inventories in an open and co-operative manner, this would be likely to have a generally helpful impact on the disarmament process.

The British analysis referred to above sets out in some detail the multiple difficulties that would have faced any nation attempting to establish, with a useful degree of certainty, the quantity of fissile material produced in the U.K. by 1962. The inherent uncertainties have multiplied many times with the passage of nearly forty years, both on account of the increasing scale of the verification task, and because many of the senior managers in the early years of the nuclear programme, without whom it might be impossible to resolve particular controversies, are no longer alive.

South Africa's nuclear disarmament has provided the only practical experience of attempting to verify the quantity of fissile material produced in a programme operating outside international safeguards. In 1991 the IAEA was given 20 years of records to analyse. Although tests were made to establish that production records, for example, were authentic, it proved difficult to confirm the quantity of highly-enriched uranium produced at South Africa's enrichment facility.· Hans Blix, then Director General of the IAEA, concluded in a speech to the 1995 NPT Review and Extension Conference that:·

> [E]ven in the case where the Agency has been shown the most extensive co-operation and openness and has conducted the most extensive inspections, it cannot positively affirm that a declaration presented is correct and complete. It can only report that after very thorough verification, nothing has been found suggesting the contrary.

It is worth considering that the uncertainty exists despite the fact that the South African programme was small scale, producing only six warheads. About 1,000 persons worked on the project, with roughly 250 involved at any one time. Waldo Stumpf, Director of the Atomic Energy Corporation of South Africa, makes an upper estimate of the total cost of the programme as about $US500m (at historical exchange rates) – a tiny fraction of comparable expenditure in the U.K., France or China, let alone Russia or the USA.

With respect to a nation attempting to violate a nuclear weapons convention by building and operating a secret production plant, the existence of any large plant would be obvious to national technical means (NTM),· so the problem is detecting a small-scale, clandestine nuclear programme developing one or a few nuclear weapons. Currently, this issue is associated mainly with the objective of deterring further horizontal proliferation of nuclear weapons. Like the illegal concealment of warheads or materials, this is a difficult problem, and one that might become even harder as technology, such as laser enrichment of uranium, advances.

The potential development of small clandestine arsenals will probably be seen as an increasingly menacing threat when the current nuclear weapon states reduce their arsenals to low levels. As disarmament continues, there will be a need for technologies that can be used in international monitoring by organisations such as the IAEA, and for technologies that provide the best possible information-gathering capability for national intelligence services. In this connection, the nuclear weapon states have probably all studied the potential for detecting clandestine nuclear activities, based on extrapolations of their own industrial experience.·

Thus it is not in question that it would be possible – that is to say, possible based solely on technical considerations – for an industrialised country to circumvent any international nuclear disarmament regime. It might do so either by sequestering a number of warheads or a quantity of fissile material before the agreement enters into force or by the clandestine manufacture of nuclear warheads after the agreement is in place.

While there is no getting away from this unfortunate fact, it is, of course, quite wrong-headed to reason that pursuing nuclear disarmament has, therefore, to await general and complete disarmament, and developing imperfect means of verification is a waste of time. Nuclear disarmament cannot produce complete safety from nuclear weapons. Nobody is saying that it can. It might, however, prove safer than any alternative, as well as preferable in all sorts of other ways.

More than this, nuclear disarmament is likely to take decades to accomplish – as already discussed – and verification possibilities could change significantly in this period. Two of the factors most likely to affect verification possibilities on this timescale are trends towards greater openness in the world and towards improving technological verification capabilities.

This paper is based on Tom Milne and Henrietta Wilson, *Verifying the Transition from Low Levels of Nuclear Weapons to a Nuclear-Weapon-Free World,* VERTIC, 1999. For reason of space, the above text is abridged and references have been omitted. A full version of the paper is available, on request, from the London Pugwash Office.

INDIAN NUCLEAR DOCTRINE

Jasjit Singh

The contours of Indian thinking that would finally culminate in an accepted nuclear doctrine have been clear for a long time even if the issues were not debated in detail earlier. At the broader level of strategic thinking Indians have sought peace so that human development could be pursued as the highest priority and in more effective perspective. The decade of wars starting with 1962 further convinced people that peace was a pre-condition for human development and hence defence would need to be seen in conjunction with development and not as alternate choices. The role of military power in this context remains focused on defence of India. The philosophy of defence thus has been embedded in the concept of defensive-defence.[1] The roots of this philosophy lie in the traditional cultural and political Indian approach to conflict and its prevention/resolution.[2] This has also implied a minimalist approach to political and military aims of defence (restricted to India's territorial integrity), investment of resources, and approach to war (and its prevention as a key objective). The wars that India had to fight were not started by India; it sought to keep them limited and circumscribed; and it sought termination at the earliest possible opportunity even if some of them (like the 1947-48 war) left unresolved problems that were to later generate conflict, and others (like the 1962 war) left disputes unresolved. It was inevitable, therefore, that the approach to nuclear weapons also would be guided by the same philosophy that has shaped the strategic culture.

The existence and proliferation of nuclear weapons has posed a serious security, political and moral dilemma to India from the very beginning. Mahatma Gandhi perhaps was one person who realized the political impact of nuclear weapons clearly from the time the bomb was dropped on Hiroshima. India from that time made persistent efforts to get the international community to abolish nuclear weapons since that would resolve the nuclear dilemma. Global nuclear disarmament, therefore, was a central objective of Indian policy from the very beginning. Arms control measures including those that sought non-proliferation could at best only be interim steps towards the final goal. Hence the emphasis on the balance of obligations between those who possess nuclear weapons and those who do not; and nuclear disarmament was seen as the only effective and durable measure to the challenges posed by the existence and proliferation of nuclear weapons. This principle had guided India's approach to the NPT except that when the treaty was negotiated into its final shape the commitment to disarmament was very weak. Even then India sought alternate solutions to the nuclear dilemma. More recently de-classified US documents clearly indicate the unsuccessful attempts to seek security guarantees from the United States and the Soviet Union which might enable India to sign the NPT as a non-nuclear weapon state.

India's policy, therefore, was constructed on seeking nuclear disarmament as a solution to its security, political and moral challenges. In the meantime, it sought to keep its options open and this offered the optimum choice.[3] But this was not seen to be what it was: a policy of restraint. India's demands for nuclear disarmament were interpreted as unrealistic and moralizing by countries that sought to retain nuclear weapons for their own security. Unfortunately, India's struggle for nuclear disarmament failed to gather the support of the dominant powers even after the end of Cold War. On the other hand, we witnessed the focus shifting almost entirely from disarmament to non-proliferation during the 1990s. It is against this background that India nuclearized overtly in 1998.[4] Given the policy of seeking restraint and pursuit of a minimalist policy it was logical to expect that India would adopt a nuclear doctrine in harmony with its strategic culture and approach to national security and defence. Restraint would now operate at a somewhat higher level, but restraint would still be the central parameter of the principles governing nuclear policy and strategy.

The Doctrine

India became the sixth state with nuclear weapons in May 1998. The broad parameters of the nuclear policy were spelt out in the following months. On 17 August 1999 the government released the draft nuclear doctrine for India (see Annex) to encourage public debate. The document would be considered by the government in due course for a final decision.

The draft has naturally attracted a great deal of comment which was the stated purpose of releasing it to the public. Two basic points need attention. Firstly, this is not the doctrine accepted or endorsed by the Government of India as yet. It is, as the document says, a draft proposal of the National Security Advisory Board. Secondly, many comments in India and abroad have confused the very concept of a "doctrine" in spite of a clear expression in the Preamble of the paper where in para 1.6 it is stated that "this document *outlines the broad principles* for the development, deployment and employment of India's nuclear forces." (Emphasis added) A doctrine is expected to define the principles and concepts for the guidance of policy-makers after it has been approved by duly constituted authorities. It must not be confused with policy, strategy or even posture which no doubt may be expected to flow from it. This elementary fact seems to have been missed by most critics. For example, many people have complained that the doctrine does not indicate the "cost" of India's proposed doctrine. The issue of costs is very important. But should a doctrine paper go into costs? If it does, would it remain a document that outlines "broad principles" or become a policy planning paper? On the other hand, the document consistently states that India would pursue a doctrine of credible "minimum" deterrence. The concept of minimum is not merely related to a single aspect of, say, numbers, but also to costs and posture.

Similarly there has been a view that the doctrine does not spell out the size of the arsenal, or for that matter, the likely threat. This was to be expected since the doctrine would hardly be expected to define these issues. The strategic and security environment is not static and keeps changing consequent to a range of factors depending upon the actions and attitudes of other states. Advances in technology impact on strategic capabilities and military power in various ways. While national security interests would normally remain constant the framework under which they have to be pursued is subject to change. Thus the minimum level of capability required cannot remain static. For example, needs of deterrence are likely to decay with passage of time. But it would be difficult, and even imprudent to follow a policy based on making a firm judgement. The doctrine, therefore, provides a clear enunciation of the principle that the concept of credible minimum deterrence is a dynamic one where the actual size, components, deployment and employment of nuclear forces would be related to the changing realities of the strategic environment, technological imperatives and the needs of national security.

Much of the critique of the document has arisen from two factors. The first is the rather unique approach of making the proposal public for discussion and debate. Unlike conventional weapons whose development, deployment and employment would remain restricted to military and defence establishments, nuclear weapons concern all citizens. In an established democracy the people have a right to know what principles will govern the nuclear policy. The second is even more important. The doctrine does not adopt the conventional wisdom of other nuclear weapon states. To that extent this is not only in contrast to the acknowledged wisdom of the main nuclear powers but seeks to chart a new path.

The Central Logic

The central logic of the doctrine is built on five essential principles which are discussed in the following paragraphs.

Firstly, India's nuclear weapons are meant only to *deter nuclear weapons threat/use*. It is obvious that nuclear weapons cannot possibly provide deterrence in all circumstances. For example, they are unlikely to deter terrorism or even conventional wars. In fact limited conventional wars remain a distinct possibility even where nuclear weapons are present. At best nuclear weapons may deter full-scale conventional war unless one side is willing to risk the possible escalation to nuclear threat. On the other hand, the experience of the nuclear age bears witness to the conclusion that limited war is still possible under the nuclear overhang. The Kargil war was not the first such war between two nuclear neighbours. What is often forgotten is that thirty years earlier China and the Soviet Union fought an intense local limited war across the border on the Ussuri River. This is why the doctrine does not expect that nuclear weapons would deter all forms of war and armed conflict. But unlike most other nuclear weapon states, India's nuclear weapons are not meant to deter the use and threat of use of conventional weapons, chemical weapons, biological weapons or a generalized formulation of protecting national interests anytime anywhere. The doctrine requires that the nuclear policy should seek to deter rather than fight a war with nuclear weapons. The orientation of the doctrine is thus quite narrowly focussed on the need to defend India through deterrence against any nuclear weapons threat. India's use of its nuclear weapons would be predicated on failure of deterrence, that is, if and when an adversary uses nuclear weapons against India. This severely limits the potential development, deployment and employment of the nuclear arsenal.

Secondly, the doctrine rests on the principle of *"No-First-Use"* even against nuclear threat. It commits Indian nuclear policy to one of *retaliation only*. Para 2.4 of the draft paper expresses it in the following terms:

> The fundamental purpose of Indian nuclear weapons is to deter the use and threat of use of nuclear weapons by any State or entity against India and its forces. India will not be the first to initiate a nuclear strike, but will respond with punitive retaliation should deterrence fail.

The only country to adopt a similar commitment is China although uncertainties have always existed whether China's commitment constitutes a political statement or is a doctrinal precept. A political statement would be reassuring but cannot be relied upon to be sustained in times of crisis. But a doctrine of no first use would also require a strategy and posture consistent with the doctrine. While the doctrine itself cannot be verified (as much as a political statement cannot be accepted as binding) the posture can certainly be judged as to its consistency with the doctrine. And any change in the posture would provide an indication of change in strategy.

A retaliation only doctrine accepts the fact that India will not initiate a nuclear strike under any circumstances. On the other hand, in order to reduce the risk of an attack on India it promises in unequivocal terms that sure and certain retaliation will follow such an attack. Assuming that rationality operates, as indeed it is expected to in any deterrence equation, this prospect, which would lead to unacceptable punishment to the aggressor, is expected to function as a deterrent.

Thirdly, as a defensive doctrine of no-first-use and limiting the use of nuclear weapons to retaliation only, India's draft doctrine is in total harmony with the UN Charter. Article 51 of the Charter under Chapter VII clearly endorses "the inherent right of individual or collective self-defence if an armed attack occurs against a Member of the United Nations." Requirements of self-defence against nuclear weapon threat can either be met through global nuclear disarmament (which would eliminate the means of threat generation) or achieved through possession of capabilities that could deter the use of nuclear weapons (there being no other credible means of defence against nuclear weapons). In fact, UN Security Council Resolution 1172 violates the UN Charter since it seeks to deny India the right enshrined in the Charter and eliminate capabilities that the permanent members of the Security Council themselves possess and jealously guard by disregarding their treaty obligations of nuclear disarmament.

Fourthly, the doctrine emphasizes that global, verifiable and non-discriminatory *nuclear disarmament is a national security objective*. This is to be expected from a country that has pursued the goal of complete abolition of nuclear weapons for decades in order to remove its own security dilemma, as well as ensure greater prospects for international peace and security. In fact, the doctrine clearly states that the absence of nuclear disarmament has created the need for India to acquire its nuclear deterrent. It also makes it clear that India's nuclearization has emerged out of the failure of the international community to institute disarmament half a century after nuclear weapons came into use and more than three decades after the five nuclear weapons states were bound by treaty obligations to do so.

And last but not least, the doctrine is based on the concept that the nuclear policy, strategy and posture would be based on the *minimalist principle*. The emphasis on "minimum" deterrence clearly defines this principle in relation to the capability sought, the size of the arsenal, costs involved, the level of retaliation required and the nuclear posture in peace time and in times of crisis and active threat. And hence the emphasis on "minimum." It is unfortunate that in spite of repeated emphasis on minimum deterrence, even simple statements of intent to respond to a nuclear attack with punitive retaliation are interpreted as "massive retaliation" by reputed organizations well versed in the lexicon and concepts of nuclear strategy.[5]

There has been a view that the doctrine promises an open-ended build up of the nuclear arsenal by India. The most often cited reason in support of this contention is that the doctrine is similar to that of the superpowers because it seeks to build an arsenal based on a "triad of aircraft, mobile land-based missiles and sea-based assets." But the reason for the superpowers' massive build-up was not because of the principle of the triad, but because of the overall nuclear doctrine to deter war itself and conventional as well as other forces equipped with weapons of mass destruction. Reference to a sea-based component of the triad has been interpreted as an indication of a large if not a huge arsenal, but it is clear to anyone who examines the issues objectively that with a no-first-use doctrine and strategy, survivability is the key to the success of deterrence strategy. And survivability is what gets enhanced with a triad strategy. In fact, a sea-based arsenal would be more survivable with a smaller arsenal. On the other hand, it would take quite some time for India to possess an operational sea-based (essentially submarine carrying nuclear ballistic missiles) deterrent. For the foreseeable future the operational force is likely to be limited to the other two components of the triad.

Some Issues

The draft doctrine makes it clear that India will not follow the disastrous route of the first tier nuclear weapon states who built up tens of thousands of nuclear weapons in a mindless arms race which they find difficult to even reverse. On the other hand, some important issues have been raised related to the doctrine which need to be addressed.

Apprehensions have been expressed that this doctrine would set off a nuclear arms race. But when a country decides that it would not be the first to use force and tailors its doctrine accordingly, the prospects of any arms race are intrinsically reduced. In fact, the doctrine promises that India will work for an international treaty banning the first use of nuclear weapons. It is indeed surprising that a doctrine that repeatedly affirms the principle of no-first-use and use in retaliation only is considered by some people to trigger an arms race! Any objective examination of the concept would reveal the reverse to be the case. For a moment let us assume that all nuclear weapon states follow the principle of no-first-use enshrined in the draft doctrine. In that case no country will initiate a nuclear strike. Logically this should trigger an arms race in reverse reducing arsenals to minimum levels and under-cut the necessity for deterrence if the doctrine is implemented honestly.

The concerns about a potential arms race reflect a perception arising out of the prism of Cold War mind-sets where first strike (and even "launch on warning") was the order of the day. This necessitated large arsenals and potentially pre-emptive strikes with nuclear weapons. The result was the peak of 67,000 nuclear warheads. But China did not follow the same route and relied on minimum deterrence concept with a no-first-use commitment. There is no evidence that China's nuclear posture led to any arms race. Even in future China's nuclear arsenal is likely to be dictated by the size of US/Russian arsenals and issues like ballistic missile defences and so on, rather than the defensive doctrine of India.

There is also a concern that a doctrine of no-first-use would expose India (and its population) to a first strike by an adversary. This is true to a large extent. But what needs to be seen clearly is that this danger has existed all along, and certainly ever since the Indian security environment was nuclearized. The choice of use of nuclear weapons against India by China has existed since 1964 and by Pakistan since 1987. Pakistan even held out nuclear threats a number of times since then. Any objective analysis of the US despatch of nuclear-armed Task Force 74 headed by USS Enterprise would indicate that the action carried an implicit nuclear weapons threat.[6] If no such actual use took place it was entirely due to their own choice rather than any deterrent factor. The issue, therefore, is that a nuclear deterrent would reduce the incentive for possible threat or use of nuclear weapons against India by any country. But the doctrine of no-first-use does place India at risk of a first strike by another country.

The crux of the issue here, therefore, is not that risk of a first strike against India would exist but what would be India's capability to reduce if not eliminate that risk. Without possession of nuclear weapons there would be no capability at all to deter a nuclear weapons threat. With nuclear weapons, there would be two broad options: that of a first pre-emptive strike against an adversary, or a no-first-use strategy. India has opted for the latter. The risks and dangers in the former case are much higher in spite of the fact that a much larger arsenal on hair-trigger alert would be needed. The potential for miscalculation and accidents increase exponentially under these circumstances. In fact this was the condition under which the world has existed during the East-West confrontation which also led to the build up of such huge arsenals. This was also the major reason for the extensive deployments of a variety of weapons and delivery systems, extended deterrence, and extensive elaborate global command and control systems. This strategy has also led to concepts and capabilities of launch-on-warning and so on. Obviously this approach would demand tremendous resources without any certainty that it would protect India from nuclear attack.

It is in this context that the concept of no-first-use must be seen. The adversary would be able to undertake a first strike; and this is more likely to seek decapitation of the arsenal rather than attempt a counter-value strike. But this is exactly why survivability of the nuclear arsenal is so critical. Sufficient force must survive to undertake punitive retaliation which, in the words of the doctrine, would inflict "unacceptable damage" on the aggressor. This would aim to provide the necessary deterrent against a first use by the adversary.

It also needs to be recognized that the risks of nuclear threat and possible use increase with the prospects of an inter-state war where one or both sides possess nuclear weapons. A conflict situation, therefore, lowers the nuclear threshold by increasing the risk of possible use. India's interests are better served if there is no nuclear threat. This is why abolition of nuclear weapons is perceived as a national security objective. In the absence of nuclear disarmament there is a need to keep the nuclear threshold as high as possible so that the nuclear threat factor does not enter into India's operational security situation. India's interests are prevention and avoidance of war which would naturally raise the nuclear threshold. It would also be necessary to keep conventional military capability sufficiently high to provide a credible effective deterrent against conventional war. This would help to raise the nuclear threshold. It can, of course, be argued that if India maintains a higher level of conventional military capability than before this would increase the threat perceptions in Pakistan which may then tend to lean even more towards early reliance on nuclear weapons. This only enhances the importance of the credibility of Indian nuclear deterrence. There is no objective reason for threat perceptions in Pakistan to increase since all wars since 1947, including the latest one in the Ladakh sector of Jammu & Kashmir state, were initiated by Pakistan; and all that Pakistan has to do is not to start another war. By the same logic the reverse would be true in the case of China which in any case is modernizing and upgrading its military power in a significant way.

There is no doubt that India's nuclear doctrine poses a serious challenge to the prevailing doctrines of offensive orientation and first strike strategic doctrines of the US/NATO and Russia.[7] China has been demanding a treaty on no-first-use among the weapon states. The Chinese and Indian doctrines now indicate a counter view to the traditional aggressive doctrines of other nuclear weapon states who visualize use of nuclear weapons against non-nuclear threats. It is not surprising, therefore, that the USA responded to the draft doctrine by saying that India is "moving in the wrong direction."[8] But since the State Department spokesman acknowledged that the USA had yet to examine the document, this appears to have been a reflex action to a doctrine which reflected thinking in total contrast to the US doctrine and belief system.

Conclusion

In conclusion it can be stated with a high level of confidence that the proposed doctrine denotes a serious effort at restraint in the principles for the development, deployment and employment of nuclear weapons. India exercised restraint for decades through its policy of keeping the nuclear option open. Unfortunately this was not adequately recognized and the noose of non-proliferation without disarmament was pursued by the nuclear weapon states and their allies who themselves are nuclear weapon states in security terms if not in legal terms. The draft doctrine promises restraint within the new framework of a state with nuclear weapons that only seek to deter a nuclear attack rather than exercise nuclear hegemony.

In the final analysis we have to ask a question of our own. There remain many in this world who for some reason or the other are opposed to India acquiring nuclear weapons. Their views must be respected. But now that India has declared itself to be a state with nuclear weapons and possesses an efficient nuclear arsenal, the proposed doctrine must be judged from the standpoint of whether a defensive doctrine of credible minimum deterrence which relies on no first use and use in retaliation only enhances peace and stability as compared with an aggressive doctrine of first use seeking to use nuclear weapons.

References

[1] For example see the then Defence Minister Shri KC Pant, "Philosophy of Indian Defence," in Jasjit Singh and Vatroslav Vekaric (eds), *Non-Provocative Defence: the Search for Equal Security*, Lancer International, New Delhi, 1989, pp.215-233

[2] Jasjit Singh, "Conflict Prevention and Management: The Indian Way," *Asian Strategic Review 1995-96*, IDSA, New Delhi, 1996, pp.9-26.

[3] See Jasjit Singh, "India's Nuclear Policy: A Perspective" Strategic Analysis, vol. XIII, no. VIII, November 1989, pp.781-802

[4] Jasjit Singh, "Why Nuclear Weapons" in Jasjit Singh (ed) *Nuclear India*, Knowledge World, New Delhi, 1998, pp. 9-25.

[5] Introduction to "India's Nuclear Doctrine" 18 August 1999 in *BASIC Publications*, http://www.basicint.org/indiadoct.htm dated September 3, 1999 as distributed as Background Document at 49th Pugwash Annual Conference, 7-13 September 1999, Rustenburg, South Africa.

[6] In fact the then US president Richard Nixon later stated that he had considered using nuclear weapons in 1971. See *Time Magazine*, 29 July 1985.

[7] The (NATO) *Alliance Strategic Concept Press*, approved by the Heads of State and Government, Washington DC 23-24 April 1999, Press Release (NAC-S(99)65 24 April 1999. Para 62 states that "The fundamental purpose of the nuclear forces of the Allies is political: to preserve peace and prevent coercion and *any kind of war*." (Emphasis added). The new Russian doctrine states that Russia would use its nuclear weapons "in response to the use of nuclear or other mass destruction weapons against it or its allies, and also in response to large-scale aggression involving conventional arms in situations critical for the national security of Russia and its allies" (FBIS-SOV-1999-1007 dated 7 October 1999).

[8] "Clinton writes to Vajpayee, US calls N-doctrine a wrong step" by Aziz Haniffa, *India Abroad News Service*, 19 August 1999.

Annex

Draft Report of the National Security Advisory Board on Indian Nuclear Doctrine
17 August 1999

Preamble

1.1. The use of nuclear weapons in particular as well as other weapons of mass destruction constitutes the gravest threat to humanity and to peace and stability in the international system. Unlike the other two categories of weapons of mass destruction, biological and chemical weapons which have been outlawed by international treaties, nuclear weapons remain instruments for national and collective security, the possession of which on a selective basis has been sought to be legitimized through permanent extension of the Nuclear Non-proliferation Treaty (NPT) in May 1995. Nuclear weapon states have asserted that they will continue to rely on nuclear weapons with some of them adopting policies to use them even in a non-nuclear context. These developments amount to virtual abandonment of nuclear disarmament. This is a serious setback to the struggle of the international community to abolish weapons of mass destruction.

1.2. India's primary objective is to achieve economic, political, social, scientific and technological development within a peaceful and democratic framework. This requires an environment of durable peace and insurance against potential risks to peace and stability. It will be India's endeavour to proceed towards this overall objective in cooperation with the global democratic trends and to play a constructive role in advancing the international system toward a just, peaceful and equitable order.

1.3. Autonomy of decision making in the developmental process and in strategic matters is an inalienable democratic right of the Indian people. India will strenuously guard this right in a world where nuclear weapons for a select few are sought to be legitimized for an indefinite future, and where there is growing complexity and frequency in the use of force for political purposes.

1.4. India's security is an integral component of its development process. India continuously aims at promoting an ever-expanding area of peace and stability around it so that developmental priorities can be pursued without disruption.

1.5. However, the very existence of offensive doctrine pertaining to the first use of nuclear weapons and the insistence of some nuclear weapons states on the legitimacy of their use even against non-nuclear weapon countries constitute a threat to peace, stability and sovereignty of states.

1.6. This document outlines the broad principles for the development, deployment and employment of India's nuclear forces. Details of policy and strategy concerning force structures, deployment and employment of nuclear forces will flow from this framework and will be laid down separately and kept under constant review.

2. Objectives

2.1. In the absence of global nuclear disarmament India's strategic interests require effective, credible nuclear deterrence and adequate retaliatory capability should deterrence fail. This is consistent with the UN Charter, which sanctions the right of self-defence.

2.2. The requirements of deterrence should be carefully weighed in the design of Indian nuclear forces and in the strategy to provide for a level of capability consistent with maximum credibility, survivability, effectiveness, safety and security.

2.3. India shall pursue a doctrine of credible minimum nuclear deterrence. In this policy of "retaliation only", the survivability of our arsenal is critical. This is a dynamic concept related to the strategic environment, technological imperatives and the needs of national security. The actual size components, deployment and employment of nuclear forces will be decided in the light of these factors. India's peacetime posture aims at convincing any potential aggressor that:

(a) any threat of use of nuclear weapons against India shall invoke measures to counter the threat: and
(b) any nuclear attack on India and its forces shall result in punitive retaliation with nuclear weapons to inflict damage unacceptable to the aggressor.

2.4. The fundamental purpose of Indian nuclear weapons is to deter the use and threat of use of nuclear weapons by any State or entity against India and its forces. India will not be the first to initiate a nuclear strike, but will respond with punitive retaliation should deterrence fail.

2.5. India will not resort to the use or threat of use of nuclear weapons against States which do not possess nuclear weapons, or are not aligned with nuclear weapon powers.

2.6. Deterrence requires that India maintain:
(a) Sufficient, survivable and operationally prepared nuclear forces,
(b) a robust command and control system,
(c) effective intelligence and early warning capabilities, and
(d) comprehensive planning and training for operations in line with the strategy, and
(e) the will to employ nuclear forces and weapons

2.7. Highly effective conventional military capabilities shall be maintained to raise the threshold of outbreak both of conventional military conflict as well as that of threat or use of nuclear weapons.

3. Nuclear Forces

3.1. India's nuclear forces will be effective, enduring, diverse, flexible, and responsive to the requirements in accordance with the concept of credible minimum deterrence. These forces will be based on a triad of aircraft, mobile land-based missiles and sea-based assets in keeping with the objectives outlined above. Survivability of the forces will be enhanced by a combination of multiple redundant systems, mobility, dispersion and deception.

3.2. The doctrine envisages assured capability to shift from peacetime deployment to fully employable forces in the shortest possible time, and the ability to retaliate effectively even in a case of significant degradation by hostile strikes.

4. Credibility and Survivability

The following principles are central to India's nuclear deterrent:

4.1. *Credibility*: Any adversary must know that India can and will retaliate with sufficient nuclear weapons to inflict destruction and punishment that the aggressor will find unacceptable if nuclear weapons are used against India and its forces.

4.2. *Effectiveness*: The efficacy of India's nuclear deterrent be maximized through synergy among all elements involving reliability, timeliness, accuracy and weight of the attack.

4.3 *Survivability*:

(i) India's nuclear forces and their command and control shall be organized for very high survivability against surprise attacks and for rapid punitive response. They shall be designed and deployed to ensure survival against a first strike and to endure repetitive attrition attempts with adequate retaliatory capabilities for a punishing strike which would be unacceptable to the aggressor.

(ii) Procedures for the continuity of nuclear command and control shall ensure a continuing capability to effectively employ nuclear weapons.

5. Command and Control

5.1. Nuclear weapons shall be tightly controlled and released for use at the highest political level. The authority to release nuclear weapons for use resides in the person of the Prime Minister of India, or the designated successor(s).

5.2. An effective and survivable command and control system with requisite flexibility and responsiveness shall be in place. An integrated operational plan, or a series of sequential plans, predicated on strategic objectives and a targeting policy shall form part of the system.

5.3. For effective employment the unity of command and control of nuclear forces including dual capable delivery systems shall be ensured.

5.4. The survivability of the nuclear arsenal and effective command, control, communications, computing, intelligence and information (C412) systems shall be assured.

5.5. The Indian defence forces shall be in a position to, execute operations in an NBC environment with minimal degradation.

5.6. Space based and other assets shall be created to provide early warning, communications, damage/detonation assessment.

6. Security and Safety

6.1. *Security*: Extraordinary precautions shall be taken to ensure that nuclear weapons, their manufacture, transportation and storage are fully guarded against possible theft, loss, sabotage, damage or unauthorized access or use.

6.2. *Safety* is an absolute requirement and tamper proof procedures and systems shall be instituted to ensure that unauthorized or inadvertent activation/use of nuclear weapons does not take place and risks of accident are avoided.

6.3. *Disaster control:* India shall develop an appropriate disaster control system capable of handling the unique requirements of potential incidents involving nuclear weapons and materials.

7. Research and Development

7.1. India should step up efforts in research and development to keep up with technological advances in this field.

7.2. While India is committed to maintain the deployment of a deterrent which is both minimum and credible, it will not accept any restraints on building its R&D capability.

8. Disarmament and Arms Control

8.1. Global, verifiable and non-discriminatory nuclear disarmament is a national security objective. India shall continue its efforts to achieve the goal of a nuclear weapon-free world at an early date.

8.2. Since no-first use of nuclear weapons is India's basic commitment, every effort shall be made to persuade other States possessing nuclear weapons to join an international treaty banning first use.

8.3. Having provided unqualified negative security assurances, India shall work for internationally binding unconditional negative security assurances by nuclear weapon states to non-nuclear weapon states.

8.4. Nuclear arms control measures shall be sought as part of national security policy to reduce potential threats and to protect our own capability and its effectiveness.

8.5. In view of the very high destructive potential of nuclear weapons, appropriate nuclear risk reduction and confidence building measures shall be sought, negotiated and instituted.

A NUCLEAR-WEAPON-FREE ZONE IN THE TAIWAN STRAIT

Shu Yuan Hsieh

When Jiang Zemin visited the United States in October 1997, he made a scary historical analogy. In discussing the Taiwan issue, he declared that the Americans had fought a civil war to reunify the nation. Jiang wished he could push our current world trend of amicable resolution back a century or two in order to resolve conflict with war. It seemed that he had not experienced true democracy and prosperity, otherwise he would have understood that people today with higher education and more wealth would prefer peaceful negotiation, rather than the utilization of arms and soldiers to settle disputes.

China has never renounced the use of force against Taiwan, and over the past years China has been threatening to take military action against Taiwan in various scenarios. Ever since the mid-1950s the United States has tried to induce China to agree formally to the renunciation of force in the Taiwan Strait. So far, it has failed. In fact, Chinese missiles were fired into Taiwan's waters in 1995 and 1996, which prompted the Clinton administration to send the USS Nimitz to the Taiwan Strait where hundreds of commercial vessels pass daily. Not only does Taiwan depend on uninterrupted sea lane communication in the Taiwan Strait for its economic lifeline but Japan, the United States, South Korea, and southeast Asian countries also need a stable shipping route for their international trade through the Taiwan Strait.

Certainly, one can calculate that the potential theatre of flash point to ignite a fire between China and its "designated" rival will be the Taiwan Strait. How do we manage to prevent the possibility of a military clash sparked around the Asia-Pacific region? How can Asia-Pacific nations come to agree on confidence-building measures for the promotion of stability? What kind of strategy do we pursue to maintain peace in the region?

As one has witnessed, since 1949 tensions have always surrounded the Taiwan Strait because of the political and military hostility across it. The medium-range ballistic missile tests conducted by China near Taiwan's coast in the mid-1990s certainly created a temptation for Taiwan to reconsider whether it needs to acquire nuclear weapons for its security. As India and Pakistan have just raised a great anxiety (in 1998) by nuclear testing for more countries wishing to step in the nuclear club, there have been many voices urging the Taiwan government to redevelop nuclear weapons capability. So far, Taiwan has remained as a non-nuclear weapon state, although it is not a signatory of the Nuclear Non-Proliferation Treaty (NPT).

The major reasons for Taiwan not to adopt a nuclear weapons policy are as follows:

- nuclear weapons are such unusable arms that they just have no utility;
- nuclear weapons do provide some security benefits by deterring hostile behaviour, but the costs of nuclear acquisition exceed the benefits; and
- in the past, none of the states that possess, or are thought to possess, nuclear weapons have had their security problems solved by the fact of nuclear acquisition.

As to the costs of nuclear weapons development, the advantages of having them do not outweigh political, military, and economic costs. We must have close relations with our neighbours, friends and allies, such as Japan, the United States, Korea, southeast Asia, and the European Union. We could lose their diplomatic and all other support, if we pursue our ambition of acquiring weapons of mass destruction. The reality is that having friends is more important than having nuclear hardware.

Since China continues to proclaim the use of force against Taiwan (the most recent statement was made in Tokyo on 27 November 1998 by Jiang Zemin) and the fact that China plans to test again Tong-fong-31 intercontinental ballistic missiles towards Taiwan, we are convinced that something has to be done to promote the stability and peace of not only the Taiwan Strait but also the Asia-Pacific region. The mere tests of missiles by a hostile state should not spoil our democracy, freedom, and prosperity in Taiwan.

Of course, for Taiwan to join the Theatre Missile Defence (TMD) programme with the United States, Japan and Korea, and to purchase research and develop high-tech weapons, are favourable and even vital to its defence. However, for the long-term objectives of regional prosperity and security, Taiwan should pursue further a wide variety of bilateral and multilateral agreements relating to measures such as confidence-building, pollution fighting, drug traffic checking, pirate and refugee controlling, and last but not least, establishing a nuclear-weapon-free zone (NWFZ) in the Taiwan Strait.

Considering China as a major nuclear power in Asia, we need to bring it into our arrangement for security in the region. Although China has often stated that the purpose of its nuclear weapons is solely self-defence, its launching of nuclear-capable missiles during the military manoeuvre in the Taiwan Strait in 1995 and 1996 casts doubt on this policy. Especially, when a high ranking member of the Chinese armed forces threatened an American ambassador during the Taiwan crisis that, if an intervention was carried out by the American fleet, China would conduct a nuclear attack on Californian cities. It seemed not exactly in accord with Chinese nuclear doctrine – a no-first-use policy.

Later in August 1996, China's top arms negotiator at the United Nations, Sha Zukang, announced that China's no-first-use nuclear weapons policy does not apply to Taiwan. Even though Beijing's regime sought to correct such a statement, it is an indication of inconsistency among the high Chinese Communist Party members, which poses a direct danger to peace in the Asia-Pacific region and the world.

When the USS Nimitz was sent to pass through the Taiwan Strait during the missile crisis, the United States assumed the right of passage as freedom of the high seas, while China considered it as the right of innocent passage through territorial sea. Based on the law of the sea, the Taiwan Strait should be used for international navigation. We, therefore, need to engage China more in the international community in order to let it comprehend international law. To include China in establishing an NWFZ in the Taiwan Strait is one of the best ways to have China observe international law.

The concept and practice of the NWFZ began in 1959 when 42 nations signed the Antarctic Treaty, followed by the Treaty for the Prohibition of Nuclear Weapons in Latin America and the Caribbean (Tlatelolco Treaty) in 1967 and the South Pacific Nuclear-Free Zone Treaty (Rarotonga Treaty) in 1985. The conclusion of the Southeast Asia NWFZ treaty in 1995 and African NWFZ treaty (Pelindaba Treaty) in 1996 has given virtually the entire land area of the Southern Hemisphere the status of a zone free of nuclear weapons. By contrast, the Northern Hemisphere is rather tardy on this matter. The Taiwan Strait could lead the NWFZ movement over the North. The signatories to an NWFZ treaty for the Taiwan Strait could include Korea, Japan, Taiwan, southeast Asian countries, as well as Russia, China and the United States, the nuclear weapon powers in the region.

The idea of creating an NWFZ was reinforced by the decision on the indefinite extension of the Non-Proliferation Treaty (NPT) in 1995. At that time, two documents were also adopted by the Review and Extension Conference of the Parties to the NPT. One of the documents was Principles and Objectives for Nuclear Non-Proliferation and Disarmament, in which the Conference encouraged the establishment of NWFZs, in the belief that such zones would enhance global and regional peace and security.

The incentive for creation of an NWFZ around the Taiwan Strait is the desire by the regional nations, which utilize the Strait as a major trade route, to pressure China to abandon its use of force in the area and to prevent an arms competition in the region. As China is, and will likely continue to be, one factor that has always preoccupied the countries of the Asia-Pacific, its threat to use military force challenges everyone, trying not to risk the dangers of antagonizing China. Because China has rather weak naval and air forces, its leverage is missiles, especially ballistic missiles. It has advanced to the forefront of missile technology in short- to medium-range ballistic missiles like the M-9, M-11, M-13, and so on. These missiles threaten not only Taiwan's security but Korea's and Japan's sea lanes as well.

Moreover, southeast Asian countries need the same international waterway and air space for navigation. Historically, China and India have been the traditional regional powers shaping the socio-economic and even political destiny of southeast Asia, much before Western influences. Since Taiwan is located on the pivotal situation of the sea lanes between northeast and southeast Asia and between the Indian and Pacific oceans, the Taiwan Strait is an indispensable route to economic development and military security for southeast Asian nations. The three nuclear weapon powers in the region, the United States, Russia and China, might have to provide negative security assurances to guarantee the complete prohibition of nuclear acquisition, manufacture, testing, and stationing of nuclear weapons in the Taiwan Strait.

The main purposes for an NWFZ treaty are to contain the spread of nuclear weapons and, at the same time, to compel nuclear weapon states such as China, Russia and America to eventually abandon their nuclear arsenals. In general, primary measures of an NWFZ treaty cover:

- the non-possession of nuclear weapons by zonal states;
- the non-stationing of nuclear weapons within the zone by any state; and
- the non-use or non-threat of use of nuclear weapons against targets in the zone.

Such arrangements clearly indicate that nuclear testing and use or threat of use of nuclear weapons are banned in a designated zone.

Thus far, the five existing NWFZ treaties have not prohibited port calls and transit by ships and aircraft carrying nuclear weapons. Therefore, the nuclear weapon states have no objection to the treaties. We can have similar rules to cover as many participating states as possible to the region.

By having an NWFZ around the Taiwan Strait, furthermore, we can develop more transparency and confidence between China and other big powers. It will also extend arms control and help avoid a new arms race in the region. Meanwhile, a part of relations between Taiwan and China can be solved.

We should promote the idea of building the Taiwan Strait NWFZ treaty in order to strengthen regional cooperation and to ensure stability and peace.

RATIONALE FOR NORTH KOREA'S DEVELOPMENT OF A NUCLEAR PROGRAMME

Young Hoon Kang

It was around 1989 when the media first invited world attention to North Korea's construction of huge nuclear facilities in the area of Yongbyon, North Korea. But the tensions appeared to ease away when the two Koreas agreed to the Inter-Koreas Joint Declaration for Denuclearization of the Korean Peninsula at the end of 1991.

In reality, neither North nor South Korea need to develop an independent nuclear weapons capability, since their national security has each been guaranteed by the world's strong nuclear powers, China and the United States respectively, on the basis of military alliances. Therefore, a lingering question to many people seems to be North Korea's motivation for developing its own not tactical but strategic nuclear weapons capability, despite the near collapse of its national economy. In this regard, I would like to discuss something about the rationale for North Korea's ambitions to develop its own nuclear programme. I think the bye-laws of the Korean Workers Party (that is, the Korean Communist Party), the highest authority in North Korea, appear to provide a clear suggestion of the rationale for North Korea's ambition for developing its own nuclear weapon capabilities.

According to the bye-laws, last revised in October 1980, the present task of the North Korean Workers Party is to achieve complete victory of socialism in the North and to complete national liberation throughout the Korean nation. The last national task is to build a Communist Society for the entire people in the Korean peninsula under the principle of the Juche (that is, self-reliance) ideology. In order to fulfil its mission, the North Korea Workers Party has repeatedly made its position clear that it must expel "American imperialistic military forces" from South Korea and terminate "the Colonial status" of the South.

Under these policy objectives and strategic guidance for the national unification under the Red Flag, North Korea has been developing its nuclear weapons system, that is, nuclear bombs and strategic missiles. North Korea test-fired its Daepodong-1 strategic missile with a range of 2,500km on 31 August 1998. Presently, North Korea is preparing to test-fire its Daepodong-2, which has an estimated range of 4,000km or more. With regard to a nuclear weapons capability, Admiral Dennis C. Blair, Commander-in-Chief of the US Pacific Command, testified before the US Congress on 3 March 1999 that the Kumchangri facility could be completed in four to six years, which could give North Korea the hidden and secure capabilities to produce materials for nuclear weapons, to say nothing of some nuclear experts' suggestions that North Korea already has enough materials for several nuclear bombs.

In any case, there is a danger that the North Korean leaders might overestimate their nuclear capabilities and conclude that, with the threatened use of their weapons, they could deter the US forces stationed in South Korea from taking any retaliatory actions against the actions of North Korea. Thus, if North Korea gains a nuclear capability, the North Korean leaders might attempt to use their nuclear weapons capability not only as an instrument of foreign policy but also as a shield under which to wage their war of national liberation against South Korea in the form of localized warfare and guerrilla tactics without the fear of US military retaliation. North Korea already has well-trained guerrilla forces, numbering over 100,000. With their well-prepared military transport capabilities, one must remember that they can be ready for a military mission at any time.

Even though the real purpose of North Korea's nuclear programme may be to acquire a military means for communist revolution in South Korea, it has further complicated the whole security situation in East Asia. In reaction, the Japanese raised at once serious questions about the adequacy of their anti-missile defence system, and decided to launch an independent intelligence satellite. At the same time, these developments, together with the 1997 adoption of new guidelines for US-Japan defence cooperation, have strongly provoked China's hostile reactions.

In Washington, North Korea's test firing of the Daepodong-1 missile, and preparations for the test-firing of the Daepodong-2, have also sparked renewed anxieties and prompted an agreement with Japan in September 1998 on conducting joint research on theatre missile defence (TMD), which might stimulate an arms race between China and Japan.

Thus, North Korea's development of a nuclear programme will not only absolutely nullify the inter-Korean Joint Declaration for Denuclearization of the Korean Peninsula agreed by the North and the South in December 1991, but will also threaten to destroy the existing latent nuclear-weapon-free zone in the region of northeast Asia, which includes the Korean Peninsula and Japan, to say nothing of foreclosing the possibility of further expanding the scope of the zone in the region.

The Korean Peninsula is the only place in the world where the national interests of four major powers (the USA, China, Japan and Russia) intersect. That is why the development of nuclear weapons in a poor, small country like North Korea has caused a chain reaction among the surrounding four major powers. Of these four major powers, three are strategic nuclear powers and one has a virtual nuclear capability that could be developed in a short period of time into an operational nuclear weapons capability. If, through multilateral collaboration, these four major powers could not persuade a small developing country like North Korea to refrain from developing its own nuclear weapons capability, the future of mankind looks indeed very gloomy and unstable.

This situation surely deserves the close attention of the Pugwash Conferences from the point of view of nuclear proliferation, arms control and stability and peace in the Asia-Pacific region. It is to be hoped that the Pugwash Conferences could provide the international community with its expertise and wisdom to help alleviate the tension created by North Korea's nuclear and missile programme.

DEVELOPMENT OF THEATRE MISSILE DEFENCE AND ITS EFFECT ON INTERNATIONAL STABILITY AND ARMS CONTROL

Liu Min

Since 1994, some American arms control experts have conducted extensive analyses and discussions on the potential effect of theatre missile defence systems in terms of technology, economics and politics. The analyses led to another heated debate on arms control and disarmament issue in the USA and international community since the opposing movement against the "Strategic Defence Initiative" (SDI) began in 1983. The "SDI" was suspended in 1990 for technological and economic reasons, when the "Theater Missile Defense Program" began. The former Strategic Defense Initiative Organization was restructured as Ballistic Missile Defense Organization (BMDO) to develop advanced anti-ballistic missile technology. The proposed missile defence system would be deployed on the territory of the USA and its allies under the excuse of countering the attack of theatre ballistic missiles for which the vehicle re-entry velocity is no more than 5 km/sec and which have a range below 3,500 km. Nevertheless, the analyses conducted respectively by arms control experts from USA, Russia and China have repeatedly demonstrated that this kind of TMD systems (e.g. Theater High Altitude Area Defense (THAAD) system) would counter the 10,000 km range ICBM with a re-entry velocity exceeding 7 km. This would violate the Treaty between the United States of America and the Union of Soviet Socialist Republics on the Limitation of Anti-ballistic Missile systems (or the ABM Treaty) signed in 1972. The ABM Treaty has restricted both sides from developing ABM systems and played a stabilizing role in international arms control.

In March 1997, the Presidents of Russia and the United States of America held a summit meeting in Helsinki, and the issue of the ABM Treaty was part of their discussions, which resulted in some joint statements.

In September 1997, Russia and the United States reached agreement on the limitation on the test of higher-velocity TMD system in the Standing Consultative Commission (SCC). Under this agreement, during tests of higher-velocity TMD systems:

- The velocity of the ballistic target-missile will not exceed 5.0 kilometres/ second
- The range of the ballistic target-missile will not exceed 3,500 kilometres.

What is the purpose of the US theatre ballistic missile defence programme? Is it for countering enemy's attack of theatre missile from neighbouring areas? If you look into it, you will find that none of the neighbouring countries of the United States are capable of developing 3,500 km-range ballistic missile and are not hostile towards it. People have to consider why the United States has been spending so much money and energy to develop such ballistic missile defence systems. What is its real intention?

Some people are pushing the US government to gradually direct its TMD system program toward developing a more sophisticated Ballistic Missile Defense (BMD) system, which is also called National Missile Defense (NMD). The current development of theatre missile defence system in the USA would legally and technologically lead to the development of ABM systems. This will violate the ABM Treaty. The relevant reports issued by the USA, the joint statements by the Presidents of both countries at the Helsinki summit meeting in March 1997, and the agreements signed by the State

Secretary of the USA and the Foreign Minister of Russia in September 1997, all show that the USA wanted to amend the ABM Treaty and pave the way for the development of the new anti-missile weapon systems.

The USA has so for devoted great resources for technological preparation. A lot of effort has been put into the launch test, laboratory modeling and design of land-based, sea-based and air-based anti- missile systems. For instance, the BMDO which has been in charge of overall research for anti-missile systems, and the Aircraft Ltd. of Lockheed Missile and Space Corporation , which has been in charge of design and development of the THAAD interceptor, had finished overall design and production of the THAAD test type before July 1995 and conducted missile flight tests during the period from 1996 to 1999. In the meantime, other sea-based and air-based interceptor designs and development have also been conducted.

This indicates that the USA is going further and further on the way to substantially develop ABM systems.

The Demarcation Agreement for higher velocity theatre missile defence systems signed by Russia and the United States in September 1997 has amended the ABM Treaty signed in 1972 and eliminated the legal obstacle for development of a theatre missile defence system. The United States want to develop new ballistic missile defense systems and turn them into advanced ABM systems. In addition, it wants to deploy those systems on its own territory and Japan, South Korea and Taiwan region as well. This intention has incurred strong protest from the international arms control community. Chinese, Russian and French government publicly criticize such an idea, and there is also a strong voice inside the United States against the plan. The East Asia region deployment plan has made the Chinese people sceptical of American intentions: Is the US missile defence program really based on its own security concerns? The Chinese people would have to worry that the arms control stability in North-East Asia region would be destroyed and the countries in this region will be prompted to reconsider their own security and defence programme. They may, on one hand, vigorously develop and upgrade their own strategic weapons to ensure sufficient and effective counterattack when facing attack from outside; on the other hand, they may develop similar TMD systems to achieve effective interception when facing missile attack. This obviously would result in a new wave of arms racing and thus damage regional stability and international security.

Based on the analysis above, we believe:

1. In today's world, the trend is towards peace, development and the relaxation of tensions. The Cold War ended several years ago. The development and research of TMD does not go with the trend of the times. Nor is it conducive to international disarmament efforts, and it will exert negative impacts on the global and the Asia-Pacific regional strategic balance and the stability in the next century.

2. The international arms control experts should pay attention to the effect of BMD development on international security and protest the deployment of TMD systems in Asia or any other sensitive region to assure the security and arms control stability in these regions and prevent a new wave of arms racing.

3. The USA should stop the research, development and testing of TMD and NMD systems to lay a mutual-confidence foundation for the implementation of START-II and the negotiation of START-III, thus promoting comprehensive nuclear disarmament.

THE RISK OF THE MISUSE OF HIGHLY-ENRICHED URANIUM (HEU) BY TERRORISTS

Francesco Calogero and Giancarlo Tenaglia

Terrorism

Maybe there exist experts on terrorism: we certainly do not belong to that group. But let us nevertheless venture a few elementary considerations.

There exist individuals and groups in the world who are prepared to perpetrate acts of ultimate violence. Their motivations might be embedded in some kind of rational political project (for instance, the fight for a "free and independent X", be it Ireland, Tibet, Chechnya, Daghestan, Kurdistan, Sud Tyrol, Israel, Palestine,...; or, the struggle for a "more democratic" government: in Africa, in Latin America, in Asia...), or they might be rather crazy (as, for instance, in the case of certain "religious sects"), or they might fall somewhere "in between" (as, for instance, might be the case of those American citizens who consider their Federal Government to be an occupying colonial oppressor; moreover, all religions are indeed "in between" rationality and faith, although fortunately most of them - but unfortunately not all of them - now decry violence). Their motivations may originate from grievances having some legitimacy and based on great sufferings by individuals or nations (Zionists, Palestinians, Kurds, Irishmen,...), or from purely personal affairs (perhaps at the origin of the deviant behaviour of the so-called Unabomber - who killed and maimed several individuals in the United States via clever explosive devices delivered by mail, and who turned out to be a man of above-average intelligence and education – and had an aborted love-affair). But the assertion highlighted above is hard to deny, indeed its validity has been proved time and again in recent times by terrorist acts (in Japan, in the United States, in the United Kingdom, in India, in Russia, in Latin America...).

Any realistic prospect of using a nuclear explosive device for terrorist purposes is likely to appear very attractive to terrorists. Not to all such groups and individuals - perhaps those whose motivations are more political might be turned off by the excessive, and excessively indiscriminate, effects of any nuclear explosion - but on the other hand they might be greatly attracted by the hope of using a nuclear explosive capability for blackmail and negotiation. As for the crazies, the mystique of the "ultimate weapon" is likely to be quite attractive.

A standard objection to the concern about nuclear terrorism is that anybody bent on producing cataclysmic large-scale damage - to property and to people - can use other, "easier," means (for instance, biological agents). We do not find such arguments convincing, firstly for the reason indicated above (attractiveness of the "ultimate weapon" and of its "big bang", for psychological and propagandistic reasons, and/or in the hope of using it for blackmail and negotiation), but also because the dichotomy "easy/difficult" is largely subjective: what is easy and what is difficult depends on a number of individual circumstances - it is quite conceivable that for certain individuals, due to their backgrounds and contacts, the nuclear route might actually appear as the *easiest* option to produce a major effect.

In conclusion, we believe it is quite justified to be highly concerned about the risk of a terrorist use of nuclear explosive devices - the more so because, as we explain below, this is a realistic threat, indeed a threat likely to materialize soon. This concern is shared by most experts: indeed lip service is amply paid by world political leaders to this threat. Yet we are convinced that this risk is in fact badly underestimated - thereby allowing it to linger, even when obvious steps to lessen it are available. The main purpose of this paper is to explain the motivations of our concern, and to indicate what, in our opinion, should be urgently done.

Technological Facts

Out of every 1,000 atoms of natural uranium (NatU), 993 have nuclei composed of 92 protons and 146 neutrons (isotope U-238: 238=92+146 is the total mass in nucleon units: *up to small corrections*, the mass of an atom coincides with the mass of its nucleus, the mass of the nucleus with the sum of the, approximately equal, masses of its constituents, protons and neutrons), while 7 have nuclei composed of 92 protons and 143 neutrons (isotope U-235: 235=92+143): hence NatU is 99.3% U-238 and (only) 0.7% U-235. Only U-235 is suitable for the liberation of nuclear energy. Hence NatU has to be *enriched* in the isotope U-235 to become a source of nuclear energy. Most nuclear reactors for the production of electric energy utilize as fuel uranium enriched to between 3% and 5% U-235 (from the 0.7% of NatU): this is generally referred to as low-enriched uranium (LEU). To manufacture a nuclear explosive device a much higher proportion of U-235 is needed, at least above 80% (the higher the proportion of U-235, the easier it is to manufacture a bomb). Uranium containing such high percentages of U-235 is called highly-enriched uranium (HEU - actually any uranium containing more than 20% U-235 is called HEU, but in the context of this paper we mean by HEU only uranium containing such a high proportion of U-235 to be *easily* usable as raw material for a nuclear explosion - indeed most of the HEU we will be discussing comes from the dismantlement of nuclear weapons).

The enrichment of uranium in the isotope U-235 is a difficult and costly job, because one must separate atoms which are chemically identical and only differ very slightly in their mass (only by marginally more than 1%; 238-235=3, divided by 238). Sophisticated technologies are needed, and a great consumption of energy. To enrich 1,000 units (for instance, kilograms) of NatU to 3.5% one must separate out 800 units of U-238: so that the 7 original units of U-235 are then indeed 3.5% of the remaining 200 (=1,000-800) units. Less than 200 units remain to be separated out, even to produce pure U-235; obviously less than 200 units must be additionally separated out, to produce HEU from LEU. It is thus clear that, starting from a given quantity of NatU, much more "separative work" is required to turn it into LEU, than to subsequently turn that LEU into HEU: say, at least 4 times more separative work (800/200 = 4) is needed to produce 200 units of LEU from 1,000 units of NatU, than to produce 7 units of HEU from 200 units of LEU (this is not quite right: it is easier to separate out the U-238 atoms when there are a lot of them than when there are relatively few[1]). Note moreover that roughly 30 times (200/7 ≈ 30) more LEU than HEU can be obtained from the same quantity of NatU. The cost of the operation is substantial, mainly due to the large quantities of energy required by it, and it is roughly proportional to the "separative work" needed. NatU is rather abundantly available in nature, but it is not very valuable since it is LEU, not NatU, that is used in the vast majority of reactors to produce electrical energy. LEU is instead a quite valuable commodity, because it is widely used to commercially produce (nuclear) energy, and its production from NatU is both *difficult and expensive* (for a discussion of actual prices, see below; very roughly, the cost of LEU is one-third from NatU, two-thirds from the separative work, and the order of magnitude of its market value is $1,000 per kilogram - or perhaps a little less). HEU is, of course, a much more valuable material (very roughly, 30 times more expensive to produce, since - as we just saw - from the same amount of NatU, and with almost the same amount of separative work, one can get about 30 times more LEU than HEU: so one might very roughly estimate its value at, say, *$25,000 per kilogram*, corresponding to a value a bit below $1,000 for LEU). However, in contrast to LEU, HEU is not commercially available, because of its potential as basic material with which to manufacture nuclear explosive devices. On the other hand LEU is not only *unusable* for manufacturing nuclear explosive devices, but also very "proliferation resistant," inasmuch as it impossible to produce HEU from LEU unless one has mastered the technology of uranium enrichment. This technology is only possessed by certain countries: of course by the five nuclear-weapon countries - USA, Russia, United Kingdom, France, China - and by a few others (by South Africa, by Pakistan, probably by Israel and by India, perhaps by Ukraine, perhaps by some other technologically advanced countries, perhaps very marginally - a capability to produce grams in a laboratory rather than kilograms or tonnes in a factory - by Iraq and by North Korea). It is certainly way beyond what any terrorist group or organization might muster, now or in the near and middle-term future (decades).

Manufacturing a nuclear explosive device is on the other hand an *easy* task, if a sufficient quantity of HEU is available, since to produce a nuclear explosion it is essentially enough to press together, sufficiently quickly (fractions of a millisecond), a sufficiently large mass of HEU (see below). Most experts agree that even a very small group (possibly even a single person), working clandestinely, say, in a garage, and using only technologies and materials that are easily available commercially and/or via easily accessible black markets (for instance, high quality conventional explosives), could manufacture a nuclear explosive device that would have a high chance of working, however imperfectly - provided an adequate quantity of HEU were available. The order of magnitude of the required quantity is *100 kilograms HEU* (for a sophisticated design, entailing high compression of the material - achieved by spherically symmetrical implosion - before the explosive release of nuclear energy, much less suffices: a typical modern nuclear warhead contains about 20 kilograms of HEU, see below). The likely explosive yield of such a primitive device would be in the kiloton range, and it might be one order of magnitude larger, namely of the order of magnitude of the Hiroshima bomb. In any case, a one-kiloton nuclear explosion - namely an explosion releasing, in a fraction of a second, an energy equal to that produced by one thousand tonnes (one million kilograms) of conventional high explosive (TNT) - if set off in the centre of a major, densely populated, city - would represent a catastrophe of horrendous proportions. It might kill promptly over 100,000, cause as many or more delayed deaths due to injuries, burns and radioactivity, produce property damage amounting to tens or hundreds of billions of dollars, including the possible transformation of the area around the explosion into a wasteland uninhabitable for years due to radioactive contamination.

And let us emphasize that, contrary to what many might believe, no great competence (for instance, no mastery of nuclear physics) is required to manufacture such a nuclear explosive device, *provided an adequate quantity (say, 100 kilograms) of HEU is available*. All the required information, and even blueprints, are easily available in the open literature. Nor are other nuclear materials needed. For instance, the six uranium bombs manufactured by South Africa before it decided to give up the military nuclear option did not have any neutron initiator, and they were never tested - yet nobody has any doubt that they would have performed to specification, with a yield probably larger than that of the Hiroshima bomb. Indeed that bomb design, which was indeed a uranium device, was never tested before it destroyed – on 6 August 1945 - the city of Hiroshima. (The explosive device tested at Alamogordo on 16 July 1945 was a plutonium bomb, such as the one that destroyed Nagasaki - 3 days after Hiroshima.)

One should also keep in mind that it is more difficult to manufacture a transportable weapon - meant to be delivered, by aircraft or, even worse (due to the higher accelerations it would have to withstand), by rocket, far away from where it has been assembled - than a device "in a garage" - an explosive device which could then be easily set off where it has been manufactured via a delayed trigger, allowing ample time for getaway. Clearly this is instead the natural option for a terrorist gang or individual. The space requirements for such a project are perfectly consistent with the size of a private garage, or a basement cellar, or for that matter an apartment, in any city: indeed, the total volume of the required quantity of HEU (100 kilograms) would fit easily in a single small suitcase (the specific weight of uranium is very large). This also provides an indication of the ease with which such a material could be transported; although of course, to avoid any problem with criticality, one would be well advised to divide it into several different bags, and make sure to keep them apart.

Finally, let us also emphasize that the handling of uranium - in the context of a project to build clandestinely a nuclear explosive device - would entail no significant radiation hazards.

To summarize: *the only effective barrier to terrorist use of nuclear explosive devices is the difficulty, for the terrorists, of getting hold of about 100 kilograms of HEU.*

These facts indicate the tremendous risk entailed by the possibility that one or more terrorists get hold of 100 kilograms or more of HEU. This is further discussed below. But let us immediately emphasize that this risk is uniquely associated with HEU. Indeed, the only other material which can be used to manufacture a nuclear explosive device is plutonium, but in that case the task is much more difficult, not only because plutonium is potentially much more dangerous than uranium to handle, but especially because, in order to produce a nuclear explosion with plutonium, the assembly of the material must be achieved so quickly, and under such geometrical constraints ("spherically symmetrical implosion"), to be rather certainly beyond the capabilities of any group of terrorists: it would require a sophisticated technological project and much experimentation, it certainly could not be done "in a garage." Nevertheless plutonium does present a substantial risk of *governmental* nuclear-weapon proliferation, because what is beyond the capabilities of a terrorist group working clandestinely, may instead be quite feasible for a state (acting openly or clandestinely). For this reason much concern has been focused, in recent years, on the question of how to dispose, in proliferation-resistant manners, of the substantial quantities of plutonium made available by nuclear disarmament, as well as of that routinely produced in nuclear reactors (which is however essentially unusable before the spent fuel rods are reprocessed, to recover the plutonium from them, by separating it from the highly radioactive waste). This is a rather complex issue, which we need not discuss here. Let us just note in passing that the quantities of plutonium now available are roughly one order of magnitude less than the available HEU (see below); however plutonium continues to be produced in nuclear reactors, as a by-product of the nuclear production of electricity, while presumably no more HEU is now produced (at least by the major producers: not by the USA, presumably not by Russia, not by the United Kingdom nor France, hopefully not by China; probably still, marginally, by Pakistan and perhaps by India). Hence, as regards HEU, the main problem relates to the existing stocks, which are enormous (see below). Moreover, in contrast to the case of plutonium, for HEU - to which attention may be confined in the context of a discussion, like the present one, focused on the *sub-governmental* terrorist threat - the simple technological facts outlined above offer an obvious strategy to dispose of the HEU made redundant by the process of nuclear disarmament, which neatly eliminates the risk of (governmental or sub-governmental) nuclear-weapon proliferation: one should de-enrich ("blend down") the HEU by mixing it with NatU, producing thereby LEU which can then be stored for future use as a (non-explosive!) energy source, or even sold (worldwide) to electrical utilities for use as nuclear fuel. The process of de-enrichment (contrary to enrichment) is *easy*, and it does not require any major consumption of energy, hence it is also relatively *inexpensive*: and it yields a material, LEU, which is a valuable commodity. For every kilogram of HEU one can thus produce, say, 30 kilograms of LEU (see above: this is obviously a very approximate value, the precise figure depends on the exact grades of the original HEU and of the LEU being produced). Of course, the process of de-enrichment entails the waste of some of the separative work that went into the production of HEU from NatU; but, as explained above, only a minor fraction of that work gets lost, if the final product is LEU. Let us also mention that, to deal with the potential risk inherent in the existence of HEU, an immediate measure that could be taken, without prejudice as to what final kind of LEU one might want to produce and find some customer for, would be to de-enrich the HEU enough to exclude any future explosive use, without necessarily going all the way down to LEU. For instance one could de-enrich the HEU down to 20% - that is, so that the remaining U-235 component does not exceed 20%.

Commercial/Economic/Political Facts

HEU is not available for sale, and indeed it has no commercial uses, although there do exist a few research reactors in the world that employ it as fuel, and most of the reactors used for naval nuclear propulsion - mainly in nuclear-powered submarines (possessed only by the USA, by Russia, by the United Kingdom and by France) - are fueled by HEU.

On the other hand, large quantities of HEU are available, especially in Russia and in the USA. Russia possesses the largest stock, whose precise size has not been officially announced, although it is known that the quantity made available by nuclear disarmament (already decided: indeed, largely completed and/or in progress: roughly, half and half) exceeds *1,000 tonnes, namely one million kilograms* (an estimate often made mentions the figure 1,200 tonnes: see below). Hence, diversion of one per cent of one per cent of the HEU available in Russia would be sufficient to provide enough material for a terrorist nuclear explosive device. Or, equivalently, *the stock of HEU available in Russia is sufficient for the manufacture of at least 10,000 easily realizable, primitive nuclear explosive devices* (and perhaps for five times as many modern nuclear warheads).

In principle this material is securely guarded. But it is well known that the reality is otherwise.

All this material is under the control of the Russian Ministry of Atomic Energy (MinAtom), the Russian institution that manages all nuclear matters, civilian and military, in Russia. It is estimated that as many as one million people depend for their living on MinAtom. The majority of them live now under dire conditions. Monthly salaries for most of them are in the low tens of dollars, and their payment (of course, in roubles, not dollars) is often delayed by months. Among the least well-treated are those guards and accountants, at the lower end of the MinAtom staff, who are ultimately responsible for the day-to-day security of the storage sites - hundreds of them widely dispersed in Russia - where nuclear materials are stocked. As a telling symptom of how bad the situation is for the MinAtom personnel one may recall that one of those at the very upper end of it - the director of one of the two main nuclear-weapon laboratories in Russia - committed suicide recently, clearly because he was overwhelmed by the responsibility he felt towards his underpaid staff.

It should, moreover, be kept in mind that a substantial portion of the HEU is stored as oxide - in powder form, hence very easily diverted and stolen in small quantities at a time. Moreover, while most of it is by definition weapon-material, some of it is earmarked as fuel for naval propulsion, hence possibly subject to less careful security.

In any case let us re-emphasize that the main reason of concern is *the enormous disproportion between the quantity of HEU available in Russia, and what would suffice for an easily manufactured explosive nuclear device: a ratio higher than 10,000.* Even if the Russian HEU were, as a rule, well guarded (and there are reasons to doubt that this rule applies widely), one might reasonably guess that some fraction of it, say a few per cent, is not so well cared for; is it not likely that, of that fraction badly cared for, say, a few per cent is will be stolen? But such a quantity is still enough to manufacture -easily - several nuclear explosive devices, each of which could destroy a city.

Indeed a number of episodes of nuclear smuggling are known to have occurred. It is plausible to conjecture that many more have occurred which are not known. For one thing, almost all known episodes involved smuggling via Europe, whereas it is well known that the most porous frontiers of Russia are those towards the South. (A glimpse at this kind of developments: "On the border between Bulgaria and Romania police have seized 2 kg of enriched uranium in the possession of a Kurd traveling from Turkey to Moldova."[2]

The incentive for stealing and then selling HEU is significant: it is an extremely valuable material. As we indicated above, its market value, after transformation into LEU, can be (very roughly!) estimated to be of the order of *$25,000 per kilogram.* Its value as HEU is of course inestimable. For instance, Saddam Hussein's Iraq tried to produce HEU clandestinely, and succeeded in producing a few grams, at the cost of several billion dollars. Clearly these figures entail a significant temptation for individuals who starve while their salary of a few tens of dollars per month does not get paid! On the other hand, the material which might eventually trickle into the hands of terrorists need not cost them such large sums (say, $2.5 million for 100 kilograms - although there do exist well-funded terrorist organizations, for whom such sums of money are quite manageable).

These figures are also likely to motivate a keen interest by criminal organizations to get hold of this material - because of its value. But criminal organizations might also have other interests: for instance many believe that the Chechen Mafia has already made sure it possesses an adequate supply of HEU, to use its terrorist potential as an ultimate blackmailing tool against the Russian government, if need be.

An obvious conclusion to be drawn from all this is the need to reinforce the security of all nuclear materials, and in particular of HEU, wherever it is stored, and in particular in the former Soviet Union, where the breakdown of the previously existing societal structure (which, plagued as it was by heinous aspects, did however function) and the dire economical conditions have created a particularly (potentially) dangerous environment. This is, incidentally, our reason for focusing on the situation in that part of the world, and in particular in Russia, where by far the largest stock of HEU is stored: HEU, of course, represents a risk wherever it is, and we are not certain that the apparently more efficient custodial systems in Western countries, and especially in the USA where the second largest supply of HEU exists, are themselves completely foolproof against diversion and stealing, especially by insiders. Much has been done and is being done to improve the situation in Russia, via collaborative activities by Western, and especially American, experts with MinAtom and its installations. Yet concern remains justified, as for instance indicated by a recent assessment of these initiatives and of their prospects, which we report *verbatim* and *in toto* below, copied from the July 1999 issue of *Physics Today*, the monthly bulletin of the American Physical Society, signed by Washington Editor Irwin Goodwin:

> *Protecting Russia's Nukes*. In his State of the Union speech last January, President Clinton implored Congress to substantially increase US technical and financial aid to Russia to make sure that its nuclear materials for bombs and warheads are safeguarded and fully accounted for at all times. He proposes that the part of the Department of Energy's anti-proliferation program that helps Russia protect its weapons-usable plutonium and highly enriched uranium from theft and diversion by terrorist groups and "rogue" countries should go up by a modest $5 million in fiscal 2000 above the $140 million already allocated for the current fiscal year.
>
> The sum seems inadequate to deal with the size and scope of the situation. Still, it's a hefty increase on the $3 million that Congress provided for that activity in 1993, the first year of the Clinton administration.
>
> With the collapse of the Soviet Union, its cold-war system of controls over nuclear facilities and materials has become dysfunctional. During the Soviet era, security depended on the discipline and loyalty of managers, workers, and guards in the nuclear weapons complex. They were all well paid and well respected within Soviet society, and, like all Soviet citizens, they were subject to surveillance by the KGB and other police-state agencies. But the breakdown of society and the meltdown of the economy have changed all that. US officials recently confirmed periodic rumors and press accounts of at least seven smuggling incidents in the early 1990s involving small amounts of weapons-usable materials, which they suspected were stolen from sites in Russia and the former Soviet Union.
>
> According to US government estimates, Russia now has about 150 metric tonnes of plutonium and 1,200 metric tonnes of highly-enriched uranium. About half of each of the inventories is in weapons and the other half in such forms as metals, oxides, solutions, and scrap at hundreds of institutes, warehouses, and various enterprises throughout the country.
>
> Since 1997, when the National Academies' National Research Council examined the condition of Russia's storage and safety of nuclear materials, the situation has worsened. A report by another research council panel, led by Richard A. Meserve, a partner at Covington and Burling, a prestigious Washington law firm, revisited the problems. The four-member panel concluded that the US should continue funding the program, in partnership with Russia, for the next decade, putting up at least $145 million for each of the next five years. The panel also recommended that the program train more Russians to manage and guard nuclear materials and set up better accounting systems to keep track of

> the materials, consolidate hundreds of kilograms of material at fewer sites, and step up to protecting radioactive cores in nuclear submarines. 'Although joint efforts by Russia and the United States have strengthened security at many sites, we believe that terrorist groups or rogue nations have more opportunity to gain access to Russian plutonium and highly enriched uranium than previously estimated,' said Meserve. 'Given the current situation in Russia, reducing the risk of illicit transfer of nuclear material will require years of steady work. Controlling the spread of these materials should be a high priority of US national security.

We agree with this assessment, although we note that it fails to emphasize the special character of HEU in creating a terrorist threat. Strangely and regrettably, this is a common flaw of such analyses.

We also note that, important as it is to emphasize the need to improve the *safekeeping* of these materials, a more promising option to address this problematique is to work towards their *elimination*. We already saw that, for HEU, a natural and convenient technological route towards this goal is to transform it into LEU. Indeed, as we now report, major progress in that direction has been undertaken, but unfortunately - and for bad reasons - much less has been achieved so far than it might have been possible.

The HEU Deal

The HEU deal is an agreement between Russia and the United States, signed in 1993 by the Presidents of these two countries, whose outcome should be the elimination of a large fraction of the HEU possessed by Russia: 500 tonnes, almost one half of the total stock possessed by Russia.[3] The envisaged procedure is for Russia to blend down the HEU to LEU and to sell this, and transfer it, to the United States. The total amount of money initially envisaged in the deal is of the order of 12 billion dollars (note incidentally that, in order of magnitude terms, this corresponds to the estimate of the value of HEU mentioned above: indeed $25,000 per kilogram corresponds to $12.5 billion for 500 tonnes, namely for 500,000 kilograms). The total time over which the deal should have been implemented is 20 years.

In fact, according to an official release by the United States Enrichment Corporation (USEC, see below), at the end of 1998, more than five years after the agreement had been signed, the total transfer amounted only to "1,487 metric tonnes of LEU derived from 50.5 metric tonnes of warhead HEU. Equivalent to about 2,297 nuclear warheads."

Before proceeding any further, let us recall again the rough estimates presented above, as to the quantity of LEU obtainable from a given quantity of HEU - 30 times more - and the amount of HEU contained in nuclear warheads - 20 kilograms. These are again reasonably consistent with the precise data given in this USEC release: 1,487 / 50.5 = 29.45 ; 50,500 / 2297 = 21.98 (although the specificity of the number of "[e]quivalent ...nuclear warheads" quoted by USEC hardly makes sense). On the other hand, over more than five years at least one-quarter of the total amount of HEU covered by the deal should have been processed, namely at least 125 tonnes (125 = 500 / 4); but less than half that quantity, indeed only 50.5 tonnes, was dealt with. Moreover a 20-year period of implementation was excessively long to begin with, if gauged against the tremendous risk entailed by any delay in eliminating HEU.

It is indeed of interest to compare the approach taken in the HEU deal with that taken in a qualitatively analogous, if quantitatively much smaller, case. At some point it became known to American authorities that 600 kilograms of HEU were somewhere in Kazakhstan, entailing the risk that they be diverted or stolen. An agreement was then struck with the relevant Russian and Kazakh authorities, appropriate amounts of money were paid, an American aircraft was secretly flown to Kazakhstan and all the HEU was whisked to the United States, deposited in a safe place and earmarked to be blended down to LEU ("Operation Sapphire"). The affair was managed, indeed appropriately so, as a matter affecting the security - indeed, the survival - of the United States and of the rest of the world, and it was dealt with as quickly and as decisively as such a (correct) assessment implied. The amount of money involved in the operation was probably of the other of magnitude of one thousandth of that envisaged for the HEU deal (some $10 million, as against $12 billion), as indeed the amount of HEU involved was about one thousandth of that covered by the HEU deal (600 kilograms against 500 metric tonnes, namely 500,000 kilograms).

We suggest that the elimination, or at least a drastic reduction, of the (much larger, roughly by three orders of magnitude) quantity of HEU available in Russia should have been approached, to the extent possible, in the same spirit. From a purely technical point of view it is obviously much easier to transfer out of Russia 500 tonnes of HEU, than the roughly 15,000 tons of LEU into which this quantity of HEU is supposed to be blended down before being shipped out of Russia: indeed, this is a job that could be accomplished in a week, rather than 20 years. It is objected that Russia would never sell such a strategically important material as HEU, which could be used to manufacture nuclear weapons, to the United States. But this argument is flawed - or at least it should have been put to a *serious* test. The fact is, as everybody knows - including everybody in Russia, both in MinAtom and in the Duma - that the United States, as indeed Russia, is awash in HEU, which is now being recovered from the reduction of nuclear-weapon arsenals whose dimensions, both in Russia and in the United States, had grown over the Cold War period to a size ridiculously (or rather, tragically) excessive, by any rational estimation. Nobody can doubt that, whatever the future has in store, the United States possesses an excessive quantity of HEU - which it is indeed in the process of eliminating, by blending it down to LEU: over 15 tonnes of HEU have already been so processed, a contract to blend down 50 more tonnes by March 2005 has already been signed. *The acquisition of additional HEU has no military relevance for the United States*. Moreover, Russia could have asked and obtained the right to verify that the HEU transferred out of Russia would indeed be blended down to LEU, and it could also have asked for and obtained assurances that this be done quickly.

On the other hand, it is certain that the Russian side - be it MinAtom, be it the government, be it the Duma - was and permanently is in dire need of cash and, therefore, certainly very likely to be positively responsive to any offer which might entail the immediate acquisition of a very substantial amount of money - rather than its payment over a 20 year period. Part of these funds could then be earmarked to strengthen the security of the remaining HEU, and to pay the MinAtom personnel, including those who might have been employed in the blending down of the HEU if that operation were performed in Russia. But let us emphasize that, whatever might be the final fate of those funds - even if a sizable part of them were to end up in numbered accounts in Swiss banks - they would have been well-spent - probably more cost-efficiently than any other outlay ever allocated by the United States to matters of national and international security - inasmuch as they would have eliminated a tremendous potential threat, many thousand easily manufactured nuclear explosive devices, each of them capable of annihilating a city via a terrorist act of wanton destruction - at a cost comparable to that of a few B-2 bombers - a cost to be paid once and for all, to solve a problem in a manner entailing no future expenditures (compare with the several times larger amount of money already spent in the vain pursuit of ballistic missile defence).

But things unfortunately went differently, the HEU deal, after having been concluded at the highest political level, presumably on the basis of the kind of security considerations we have highlighted, was then transformed by the United States into a commercial affair, in which financial interests - indeed, the financial interests of some individuals and some corporations - have become uppermost in determining the course of events. It is because of this that the HEU deal has been implemented slowly, and indeed has been and still is on the verge of collapsing altogether. To quote Thomas L. Neff, who was, incidentally, the first one to put forward the idea of the HEU deal: "Washington has consistently put domestic commercial and financial interests, most notably those involved in the recent privatization of the U. S. Enrichment Corporation, ahead of US national security interests."

The initial fateful mistake was to introduce the notion that the HEU deal could be accomplished at no cost to the American taxpayer, on the grounds that the subsequent commercial sale of the LEU was supposed to compensate, or perhaps to overcompensate, for the payments made to Russia. While such a happy outcome might have been possible, and indeed might still be possible, it put the entire operation on the wrong track, causing it to be driven by money rather than security concerns. The next mistake which was then made, consistent with this wrong approach, was to assign the implementation of the deal to the United States Enrichment Corporation (USEC), which was the American agency dealing with uranium enrichment in the United States (rather obviously, the institution least likely to want to see the acquisition of enriched uranium from overseas). To make matters much worse, USEC, which was originally fully owned by the Federal Government, was in the process of being privatized, and it has in fact now become a private corporation, whose motivation is therefore obviously profit. The privatization seems to have been patterned after the fashion of what we have seen happening in the former Soviet Union recently, with state assets ending up in private hands at a huge profit for some fortunate individuals. For more details, including the intricacies of commercial considerations that have caused the slowing down at best, and the eventual total collapse at worst, of the HEU deal, we refer to Neff's article quoted above.

If a nuclear explosion destroys an American city (or, for that matter, any other city in the world) and the HEU used to produce it can be traced to have come from Russia, justifying the suspicion that such a terrible event might have been prevented by a different management of the HEU deal, yielding a much quicker elimination than is now in progress of a large part of the Russian HEU, the question will be raised of whose responsibility it was to let private profit motives prevail over national and international security considerations, on a matter of such potential risk. Or maybe the matter should be raised even before such an explosion?

In fact, perhaps a total collapse of the HEU deal might be desirable, or at least a complete end to USEC's involvement, a solution USEC itself might eventually opt for, if it becomes clear that the prospects of making additional money out of this affair are doomed. Nor, indeed, can it be expected that a private corporation act on the basis of considerations other than profit, for shareholders and management alike. The need would then become, hopefully, inescapably clear for a fresh approach to the problem, to be dealt with as a matter in which security considerations are uppermost, and financial considerations only secondary. The goal, of course, is crystal clear: get rid as quickly as possible of as much HEU as possible, by blending it down to a sufficiently low concentration of U-235 (say, 20% or less), so as to render it unusable as a raw material for nuclear explosions.

What Should Be Done

There are several courses of actions which should be pursued in parallel, and which are so obvious as to require no elaboration.

Every effort should be made to strengthen the safekeeping (accounting and physical security) of HEU, wherever it is, and especially where there are reasons to suspect that such safekeeping is in need of improvement.

The appropriate security agencies should try to intercept any leakages of HEU, before it gets into dangerous hands. They should also try to infiltrate terrorist groups, to try to get tipped off about dangerous projects. But probably the danger comes mainly from unexpected quarters, perhaps from the crazies rather than from the politically motivated.

Every person and institution who has financial resources and cares for world security - from George Soros to the European Community, from Bill Gates to OPEC - should offer to pay to have HEU quickly de-enriched to a non-explosively-usable composition, containing no more than 20% U-235. In exchange for such payment, which should of course go to the owner of the HEU, some preferential title could be negotiated, to be redeemed (possibly, who knows, at a profit - but this should not be the motivation for entering the deal !) when that material is eventually remanufactured into nuclear fuel and sold to electrical utilities.

But, given the present state of the world, a key responsibility to intervene in this matter rests with the United States government, of course to the extent Congress allows it to act sensibly: but the matter is so clear cut - in the importance of its security implications - that it would provide a wonderful opportunity for the President to display leadership - and for the advisors to the President to push him to so behave - and for the arms community in the United States to focus on this problem more effectively than it has done so far.

The overriding priority should be clear: *to render unusable for explosive purposes, by blending it down, as much HEU as possible, as quickly as possible.*

References

[1] Realistic figures put the Separative Working Units (SWU) needed to produce 1 Kg 3.5% LEU at 5.4, and that needed to produce 1 Kg of 80% HEU at 200. Hence, to produce 200 Kg of 3.5% LEU, 1080 (=200 times 5.4) SWU are needed, while to produce 7Kg 80% HEU, 1400 (=7 times 200) SWU are needed. Hence, 320 (=1400-1080) *extra* SWU are needed to produce 7Kg HEU than 200Kg LEU, indeed much less than the 1080 SWU needed to produce 200 Kg LEU, although the ratio of these two figures, 1080/320 = 3.4, is closer to 3 than to 4. See *Nuclear Weapons Databook, Volume II, U.S. Nuclear Warhead Production,* by Thomas B. Cochran, William M. Arkin, Robert S. Norris, and Milton M. Hoenig, NRDC, Inc., Ballinger Publishing Co., Cambridge, MA 1987.

[2] Program for Promoting Nuclear Non-Proliferation, Newsbrief, Second Quarter 1999, p. 20: the first source given is *Reuters,* 22.04.99.

[3] For a more detailed description of the HEU deal than given here see, for instance: Thomas L. Neff, "Privatizing U. S. National Security: The U. S.-Russian HEU Deal At Risk," *Arms Control Today*, August/September 1998, pp. 8-14; for up-to-date information on the implementation of the HEU deal see: http://www.usec.com.

YET ANOTHER LESSON TO BE LEARNED FROM THE NATO-YUGOSLAV WAR

Jozef Goldblat

According to available sources, about 60kg of highly-enriched uranium, supplied by the Russians, are stored at the Vincha Institute of Nuclear Sciences near Belgrade. This amount of uranium is sufficient to build at least one nuclear explosive device. For both political and technical reasons, the Yugoslav government did not make such use of the material in question. (Some advocates of nuclear deterrence argue that NATO would not have "dared" bombing Yugoslavia, if the latter had possessed a nuclear weapon.)

If, however, the Yugoslav Institute had been hit by NATO bombs or missiles, deliberately or by accident, the uranium and the spent nuclear fuel stored there would have spread and caused radioactive contamination of the surrounding area. Radionuclide exposure, in sufficiently high doses may result in radiation sickness, thyroid damage, early death, genetic defects in future generations, and psychological trauma. The civilian population would be most severely affected. Any yet, there is no explicit prohibition against the use of radiological means of warfare.

According to the 1977 Protocol relating to the protection of victims of international armed conflicts, "nuclear electrical generating stations" may not be attacked, but zero-power reactors – such as those possessed by Yugoslavia – which are used exclusively for research purposes, are exempt from legal protection. Even nuclear power stations may cease to be protected if they provide electric power in significant support of military operations. Nor do the Protocol prohibitions cover other installations containing large quantities of radioactivity and posing serious hazards, such as cooling ponds with fuel elements removed from the reactor, reprocessing plants, where spent reactor fuel is chemically treated to separate uranium and plutonium, or storage tanks with high-level radioactive wastes. It is, therefore, noteworthy that the Protocol urges the parties to conclude agreements among themselves to provide additional protection for objects containing dangerous materials.

So far, only India and Pakistan – countries having a comparable level of nuclear development – have reached an agreement banning the destruction or damaging of their nuclear installations. In addition, the Treaty of Pelindaba, which established a nuclear-weapon-free zone in Africa, banned such acts. However, bilateral or regional agreements would hardly suffice. To prevent radiological warfare which could harm both belligerents and non-belligerents, a global ban on attacking nuclear facilities is needed. As long as war cannot be prevented, all possible measures must be taken to reduce the losses of human life and property to a minimum.

A nuclear site can also be destroyed by accident. Indeed, it happened more than once that NATO "surgical" strikes in Yugoslavia missed the target. To minimize such occurrences, the bombing or shelling of objectives, whether military or civilian, which are situated in the vicinity of nuclear installations, should be prohibited as well.

Another problem is the safety of nuclear materials under the conditions of armed conflict, when the materials may be less protected than in time of peace and, therefore, more vulnerable to abuse or theft. This danger cannot be removed by an international treaty. The country, on the territory of which nuclear materials are stored, must be held responsible for their diversion to any illicit ends.

WORKING GROUP 2

EMERGING SECURITY THREATS: AFRICA AND GLOBALLY

COMPREHENSIVE MEASURES TO REDUCE ILLICIT SMALL ARMS AVAILABILITY

Peter Lock

The Problem

The enormous stock of small arms accumulated during the Cold War has turned into a dangerous time bomb which the international community appears incapable of controlling. Illicit small arms availability has become a global threat to the sovereignty of states at macro- and micro levels. The full impact of the unbound small arms arsenals will only be felt in the years to come because of the magnitude of the volumes circulating already beyond effective controls. Present levels of disturbances, violence and systematic violations of human rights are but the harbinger of the societal fragmentation enabled and reinforced by the pervasive availability of illicit small arms. However, illicit small arms availability is neither an independent variable in the social matrix nor the root cause of fragmentation. It is a dependent variable, though regularly mutating into an active virus in the context of certain levels of social fragmentation which brings about a widespread weakening of the state. The ensuing security vacuum, in turn, constitutes a dynamic pull factor of illicit small arms flows.

Formal and informal privatization of security and criminalization of economic transactions are indicators of a general trend towards violence and the threat thereof becoming a dominant factor in the regulation of social relations. As a result, a dynamic process of internal rearmament takes root in an increasing number of states. Their economies are burdened with escalating transaction costs and their societies with intensifying levels of violence. A projection of present socio-economic trends in many regions suggests that widespread disruptive consequences of the pervasive availability of illicit arms are to be expected, if the international community fails to find a fast track to effectively decommission the surplus stocks. In addition, it is necessary to create a potent body tasked with vigorous intervention in black markets aiming at drying them up.

Presently various entities within the United Nations, intergovernmental bodies at the regional level and a myriad of non-governmental organizations have taken the initiative to combat illicit trade and availability of small arms. Unfortunately, the widespread political attention and the number of institutions involved does not yet correlate with measurable positive changes in the field of controlling illicit small arms availability. This enduring and unrelenting situation can be attributed to at least three underlying factors. Firstly, the knowledge available in the public domain of how small arms are illegally circulating is still rudimentary, which excludes customized countermeasures; secondly, most initiatives address the problem from the supply side only, possibly because the dynamics of the demand side are often politically sensitive and controversial and therefore inhibit co-ordinated countermeasures. Finally, rather traditional measures like registers, for example, originating in the arms control debate in the extinct bi-polar world are extended to cover small arms which were largely disregarded as an issue during the Cold War. Instead, innovative concepts are needed to respond to the largely intra-societal dynamics of the pervasive low-level destructive violence with its latent potential to escalate into civil war, the dominant form of current armed conflicts which are dominated by small arms.

Towards a New Paradigm of Armed Violence

Societal violence, crime, organized crime, armed conflicts, intra-state wars, and even international wars are no longer discrete events. A profound paradigmatic change of the concept of war is in the offing. The situation can be compared with the fundamental shift in development economics in the late fifties when Albert O. Hirschman opened a new area of thinking with his concept of "unbalanced growth". Successful conflict resolution must be based on the new paradigm of armed conflicts. With the transformation of the character of wars, their causes and the logic for their continuation are also changing.

After the Cold War armed conflicts are more directly constrained by the economic resources the fighting parties can mobilize locally. Therefore, a proper understanding of the underlying economics in on-going wars is a precondition to design efficient sanctions and other interventions intended to end the fighting. In the context of current civil wars the roles of politician, war commander, monopoly trader and organized crime leader blend gradually. Radical ideologies manipulating often spurious identities serve as a cynical cover for the shrewd personal interests of such leaders. Yet the ideological grip they hold over the population on the basis of constructed identities endows them with negotiating power *vis-à-vis* the international community. Understanding this often complex power equation is essential for peace negotiators.

Presently non-requited transfers in support of fighting parties are rarely available, diaspora contributions being sometimes an exception. The required resources must be accumulated or collected locally, often by force and blackmail. Fighting parties turn into complex economic conglomerates, often commanded by wealthy traders turned warlords. In particular, the non-state parties in conflict must generate foreign currency to pay for their military resupply on the black market. Four major sources have been identified in empirical studies of war economies: illegal export of raw materials and minerals; the drug trade; the appropriation and misuse of humanitarian aid; diaspora communities through voluntary contributions and blackmailing.

Illegal transactions are a common denominator of "war economies". Globally criminal economic activities appear to benefit more from higher growth rates than the regular economy. Experts have estimated that the global GCP (gross criminal product) has reached $1000 billion, of which $400 billion accrue to the drug sector alone. Global networking of criminal traders facilitates easy access to illegal markets. These "markets", however, are characterized by high risk premiums and violence. They serve as a force multiplier for warring parties and create a class of war profiteers, both inside and outside the conflict zone.

However, the underlying logic of every illegal economic activity aims at legalizing the proceeds which requires safe entry points into the legal economic circulation. This indispensable exchange between illegal black market activities and the regular commodity and monetary circulation makes "war economies" vulnerable to well-targeted interventions by the international community. In other words, the lack of efficient controls of criminal economic activities and the frequent collusion of sovereign states (not only of tax havens) are the fertile ground where "war economies" take easy roots. Exports of UNITA's diamonds and the Kmer Rouge's tropical wood and gems are just two conspicuous examples of illegal trade tolerated by the international community.

Without real resolve to combat the criminal economic circulation little progress is to be expected in constraining parties involved in armed conflicts and in luring them to consider a non-violent conflict resolution. In particular, it should at last be taken into account that most traditional international embargoes have reinforced the income concentration and the criminal control of the embargoed economies in the hands of the incumbent rulers who were the target of the embargo. An ill-conceived embargo has the potential to initiate a self-reproducing pattern where the power of a criminalized elite depends on the continuation of the fighting as well as the embargo. Milosevic and the nomenclature of his regime exemplify such a pattern.

While the killings and human suffering in ongoing wars appear to be absolutely irrational from an outside perspective - therefore the outside world presses for an end of the "irrational" fighting- it is important to recognize that conceivable post-conflict scenarios are likely to have many losers to whom an end of the fighting appears undesirable. Once ended without comprehensive proactive outside support fighting is likely to resume or else criminal elements maintain a tight grip on the economy and stifle any dynamic market-oriented development. The present post-conflict situation in El Salvador highlights this danger. More people are being killed annually with firearms than during the civil war. Many economic transactions are not regulated by the market, instead they are regulated by violence or the threat thereof.

Intergenerational Apartheid

During the nineties one destabilizing structural variable has emerged almost universally. The incapacity of most economies to integrate the younger generations into the regular labour market, as they reach the age of normal entry, provides a teeming yeast to the illegal and criminal economic circuits and an unlimited reserve army for intra-state armed conflicts. Unemployment rates of 50 per cent and even up to 70 per cent are reported for many countries in the age group 15–24 years. The radical modernization of most Third World and transition economies in the course of the current globalization has to find the means and strategies to overcome the tense and expanding intergenerational economic divide, lest societal violence will prevail. The latter is likely to take on an ideological guise and eventually escalate into often amorphous armed conflict. In this process the state slowly disappears as a legitimate actor and war fighting becomes a surrogate mode of production for many actors, which explains why the civilian population is the most endangered species in current wars.

Illicit small arms availability empowers the malignant forces of societal polarization and confrontation. In the absence of other chances to develop self-respect, the ownership of an illegal firearm provides idle young males with a surrogate for an accepted role in society. Thus, the combination of the independent variable (large numbers of idle young males) and the dependent variable (copious availability of illicit small arms) make up a terrifying time bomb the international community appears unable or rather unwilling to defuse for the time being.

Unilateral and Asymmetric Strategies

National "gun" cultures and the presently prevailing conditions among nation-states are so diverse that an efficient universal set of regulations to control small arms is unlikely to emerge rapidly. Measures taken without further delay at local, national and regional levels must serve as precursors for an eventual set of international conventions. It therefore remains important to encourage national governments to take unilateral steps to control the illicit availability of small arms. Any delay of national measures while waiting for international harmonization will cause additional harm impossible to repair at later stages. Unilateral initiatives also serve as important role models, which will help to draw hesitant governments into a world-wide coalition focusing on the elimination of illicit small arms availability. It is important that presently safe countries do not take their status for granted, because in times of globalization no economy is immune against producing social fragmentation, "inner cities", "bidonvilles", "no go areas for the police" and other pull factors.

No single measure will provide an encompassing solution, but an array of improved controls and proactive measures will eventually curb the pervasive armed violence which puts the very existence of many states in danger. Against this background, selective measures designed to reduce the illicit availability of small arms and their malignant role in fostering societal fragmentation will be discussed in the second section. It does not enumerate the various proposals presently discussed in the major small arms fora, such as tightening export controls and so on, this presentation seeks instead to complement this discussion with partially innovative measures.

Proposals

Black market networks appear to be able to deliver small arms, if necessary in large quantity, whenever and wherever effective demand articulates itself. Proactive intervention to halt illegal flows across borders by custom services and police forces is presently not feasible because too little is known about these networks, their warehouses, the operation of their financial settlements and, not least, their collaborators within governmental agencies around the world. While the intelligence agencies of certain states ought to have a fair understanding of the black market, because some of these actors were used as proxies during the Cold War to perform deniable supplies, this information is not shared internationally with control agencies nor is it available in the public domain. Thus, improving the control of small arms trade requires reliable information on the basis of worldwide field observation and data collection.

Surprisingly, up to now no institution collects continuously data on the illicit proliferation of small arms. There is urgent need to create an independent source of information in this field. SIPRI (Stockholm International Peace Research Institute) and OGD (Observatoire Géopolitique des Drogues, Paris) are excellent role models for an efficient institutional form to collect reliable information on the illicit proliferation and societal diffusion of small arms. SIPRI, in the field of major arms transfers, and OGD, in the case of drugs trade, have proven that a modest, but targeted research effort unfettered by political intervention can highlight sensitive transactions, the actors themselves are not disclosed, by screening open sources only. It is to be hoped that a smaller country like the Netherlands or Switzerland will fill the this obvious gap in the landscape of peace research in the near future and support an independent research facility serving as a *global small arms observatory*.

An important element to improve our knowledge about the illegal flows of small arms is comprehensive and reliable information from the places where they are being used in armed conflict and crime. From journalists to humanitarian aid workers there are many yet untapped, potential sources of much improved information, provided these observers are qualified to report in correct detail what weapons they observe during conflicts. One way of qualifying them would be the widespread availability of a concise small arms identification booklet which would allow them, among others, to tell the difference between an Uzi, Galil, Kalashnikov, G-3 or FN without looking for a label.

Since humanitarian organizations are severely obstructed in their work as a result of the almost effortless availability of small arms in many conflict zones, it should be in their interest to improve the information on the black market networks which supply regions of conflict with arms and ammunition. Anticipating the creation of a global small arms observatory, humanitarian organizations should sponsor the production and global distribution of a simple *small arms identification booklet*. It would certainly improve the quality of reporting on armed conflicts and societal violence and cumulatively provide information which would allow the reconstruction of illicit flows of small arms, a precondition to targeted intervention.

Supply Side

Improved policing of illicit transactions is the most obvious step to take, in order to curb the black market. All present international, mostly regional, initiatives pledge to improve the standards of controls. But this may not prove to be sufficient to curb the illicit transfers. In some cases governments do not have the infrastructure necessary to effectively control the transactions originating on their territory. An efficient and internationally co-ordinated plan of action must include a broad range of measures, some of which are geared towards structural improvements producing substantial results only in medium and long term. Some steps could be taken instantly to discourage fraudulent exports.

The case for a transaction deposit. An important source of illegal flows is the falsification of export licenses or export consignments never reaching their documented destination. Some traditional exporters of small arms, like the Czech Republic, simply do not have the diplomatic network which is indispensable to corroborate suspicious export applications. In order to facilitate an adequate control of licensed exports without burdening the tax payer with the costs which might outweigh the economic value of the transaction, the introduction of a transaction deposit might considerably improve the effectiveness of governmental controls. In order to avoid the creation of competitive disadvantages for smaller producing countries, less capable of certifying the correct execution of export transactions, all arms exporting countries should agree to charge an export deposit for licensed small arms exports.

Export control bureaucracies and custom services are regularly confronted with fraudulent documentation which are not easy to detect. It is not known how often licensed arms exports are converted into an illicit commodity during the transaction and end up in the wrong hands. The increasing number of cases of attempted fraud may reflect improved scrutiny of the controlling institutions, but it is also possible that fraud is generally on the increase.

While major weapon systems have a certain visibility which makes an eventual detection of a fraudulent transaction likely, this condition does not pertain to small arms, ammunition and spare parts. The case of smaller arms is different. For a variety of reasons SIPRI never attempted to cover small arms: firstly, data were difficult to come by; secondly, in the context of the military confrontation between West and East the issue was not of high priority; and thirdly, it was argued that small arms would constitute a certain proportion of the trade in major arms and could thus be assessed *ceteris paribus*. While the latter has certainly not been the case, given the fact that most internal wars are predominantly fought with small arms, and generate demand for a specific range of infantry weapons, it remains a sad reality that even today very little is known about the trade in small arms.

Additionally, since the mid-eighties the enduring economic and financial crisis throughout large parts of the Third World has weakened governmental bureaucracies in a significant number of smaller countries to such an extent that it has apparently become easy to corrupt them and obtain forged documents which facilitate illegal consignments of small arms. Black market traders are thus able to generate perfect looking documents even without exploiting the sophistication of modern computers and copy machines. In short, false destinations seem to be internationally on sale; moreover, the number of ports where customs offices are inexpensive to bribe is apparently numerous, facilitating convenient re-routing of shipments to their illegal destination.

But a similar observation points to some of the smaller transition countries where the capacities of the export control administration are limited. The administration does neither have the personnel nor the logistical means to thoroughly check the applications for export licenses and to supervise the correct execution of the respective shipments. The smaller transition countries in Central and Eastern Europe have typically considerable manufacturing capacities in the field of small arms, which makes them an attractive target for black market actors.

The known cases of fraud using a third country as an *entrepôt* suggest that a certain class of countries is emerging where the agreed rules of international relations do not apply any longer. Several fraudulent transactions were reported in the literature where Paraguay has been implicated. This must have serious implications for the procedures of surveying arms transfers at the international level. These countries must be identified by consultative procedures among exporting countries. At the same time, all suppliers have to institute measures which make sure that the controlling body in the exporting country maintains an efficient leverage of intervention throughout the entire transaction. Presently, hardly any sanctions are at hand to effectively sanction the fraudulent re-routing of an export consignment if the mishandling occurs beyond the territorial borders of the exporting country and the exporter can not be directly implicated. Virtually no leverage exists to undo the illegal delivery.

Given the continued pressure from the demand side and the plentiful existence of traders who dispose of the criminal energy to enact fraudulent transfers, governmental controls must be considerably intensified lest illicit transactions will spread further. However, some supplier countries do not have the resources required to effectively tighten controls. More generally, it is difficult to justify that the taxpayer should shoulder the considerable costs which accrue from the indispensable control procedures. The trade of this commodity reflects above all the legitimate entrepreneurial interest of the arms manufacturer or trading company to make a profit. The transaction costs should be fully part of the entrepreneurial cost calculation and be reimbursed to the government. However, the black sheep among the exporters are responsible for the escalation of the costs of the control system. Charging exporters with a standard fee would punish the law-abiding exporters. Making a transaction deposit compulsory for every licensed transaction instead would allow for an automatic punishment if the transaction is not carried out as stipulated in the license. In addition, the interest earned during the deposit period would help to cover the costs of controlling arms exports.

Making a substantial transaction deposit compulsory for every licensed transaction would give the government a financial leverage to enforce a correct execution. A transaction deposit of 25 or 50 per cent of the value of the respective contract would serve as a powerful economic disincentive to all contracting parties to engage in fraudulent transactions. Because the deposit will only be returned after an independent source certifies the full and correct execution of the licensed export, a careful scrutiny of potential customers lies in the self-interest of the exporting company. Even under the worst-case assumption that a ruthless exporter counts on losing the deposit in his calculation, the weapons would arrive at the black market at least 50 per cent more expensive than the going market rate, leaving the government with additional resources to further improve controls.

The introduction of a *transaction deposit* allows governments to act unilaterally and shift the burden of proof to the exporting party. It is conceivable to consider deposit rates which decrease with the visibility of the respective weaponry, dispensing major weapon systems from any deposit. The lower the visibility the more important substantial deposits would be, because it is this end of the illicit arms trade which relates to on-going wars, societal violence and fragmentation of states.

Avoiding the Misuse of Legal Small Arms

Virtually all mass-produced small arms are manufactured and traded in abidance with existing laws. However, at some point in their respective long life-cycles, large quantities of small arms pass into the sphere of illegal circulation and misuse. Present levels of illicit small arms availability is the cumulative result of a large variety of schemes which permanently re-supply the illegal circulation of small arms and ammunition. The various critical points during the life-cycle of a weapon must be addressed with specific control mechanisms and measures of general prevention. A generally underestimated source of illegal circulation is the legal private ownership of firearms in countries with little restrictions attached to private gun ownership. Regulative schemes used in other fields provide an orientation of how firearms and ammunition can be safeguarded from entering illicit circulation and misuse.

The case for an obligatory third party liability insurance. The transformation of legally possessed firearms (manufacturers, traders, armed forces, law enforcement bodies, and private licensees) into instruments of crime, violence and internal warfare takes many routes. It is an obvious truth that the criminal energy in this process is not restricted to the demand side. Greed, extra profit and passive corruption, covert support and interference (mainly by governments), fraud and negligence, are among the facilitating factors at the supply end, while criminal energy and corruption are the main tools on the demand side. By no means all facilitating factors are punishable, nor are the sentences always commensurate with respect to the often appalling consequences of the ensuing criminal offences and violations of humanitarian rights. Generally, the original owners are never held liable on the basis of their negligence or active participation for the damages the respective firearms provoke during the illicit phase of their life cycle. With respect to the actual perpetrators, claims of victims are often of no avail because only on rare occasions are there sufficient assets that can be forfeited.

Therefore, the liability of legitimate owners of small arms, in case their weapons enter for whatever reasons into illicit circulation and are eventually identified to be implicated in damages of various kind, urgently requires greater attention. As a first step, it is proposed to introduce a compulsory third party liability insurance for every small arms item produced or traded irrespective of the status of the ultimate legitimate owner. The logic of this proposal draws among others upon the regulation of car ownership in many countries, as well as on already existing international conventions regulating and securing the liability in the case of transborder risks like transport of oil and fissile materials at sea. In the latter cases the international community perceived the need to react to emerging large-scale dangers and established an obligatory third party liability coverage.

Given the recognition that the indiscriminate diffusion of small arms puts the existence of entire states into danger and causes human suffering and enormous economic damages almost everywhere, the situation is ripe for the international community to agree on joint measures to reverse present trends. It would lend support to countries presently most affected by the scourge of illicit arms availability. A convention on small arms liability would be one such measure.

Cars and guns are both commodities which can cause major damage to third parties, particularly when misused. The damaging act itself may be perpetrated by a third person following a theft or just by accident. A car owner is obligated to take out a third party liability insurance, because the state rightly wants to protect the victimized third parties. Damages are covered irrespective of the financial capability of either the owner or the actual perpetrator. The regulation gives priority to the victim. It also protects the tax payer from being shouldered with the bill as a debtor of last resort.

It is striking that in the case of firearms third parties are not protected in a similar manner against accidentally or intentionally caused damages when misused. Absolutely no liability protecting third parties takes effect in case a firearm is stolen. This is a serious lacuna which should be corrected, particularly since the general taxpayer is usually charged with the costs provoked by the misuse of firearms; alternatively victims are left without compensation. The changes required can be initiated unilaterally at the national level. They should, however, rapidly expand into an international convention, because illicit circulation of firearms is transnational. And more importantly, the international community has to find ways of coping with the heritage of millions of small arms already circulating illicitly and not covered by any insurance.

When discussing third party liability it is important to take into account that no institution or person entitled to possess firearms is aloof from the risk of theft and eventual misuse of its inventory. This applies to the armed forces as well as to the police, where the custody of firearms is not always as tight as one would expect. The need to introduce the concept of liability is best illustrated by the enormous damages caused by the huge stock of small arms recklessly left behind by the US-forces at the end of the Vietnam war. These weapons were marketed off-shore by the exhausted Vietnamese government, entered the web of global illicit circulation, and resurfaced in the hands of insurgents and criminals all over the world engendering untold suffering and enormous destruction. It would offer an interesting exercise for insurance mathematicians to calculate the rate that they would have had to have charged the US-army for each American gun left behind in Vietnam to cover the (economic) risk of their illegitimate use off the Vietnamese battlefield.

Every firearm carries the potential of creating havoc, when misused. This risk must be translated into costs which are to be borne by the legitimate holder. In order to implement this logic, every firearm should be produced with a unique mark (number) to allow for tracing of the original legitimate owner. The only feasible moments in the life-cycle of a firearm for collecting an insurance fee or a tax to cover potential and statistically predictable damages are the production itself and every legitimate trade transaction. For practicable purposes the total fee would have to be collected with the first transaction after production.

It seems evident that the private insurance industry would collect a considerable fee against a policy covering all possible liabilities, taking into account the long life-cycles and the enormous damage any firearm can cause. The principle to charge the owner of small arms with the costs of potential liabilities would considerably increase the costs of gun ownership. It would make the true (societal) costs of gun possession finally transparent. It might also reduce the effective demand for firearms as life-cycle costs increase. Finally, a healthy pressure to improve the safekeeping of firearms would come from the insurance industry trying to protect its stakes. Technical innovation, like personalized trigger locks, would rapidly enter the market, because such firearms would in the end allow legitimate gun owners to reduce the insurance premium.

As a first step it would be important to establish the principle of compulsory third party liability insurance to be levied on the ownership of guns. It must apply to the military and the police alike, because arms stolen from their stock are as likely to become the means of destructive acts as any other firearm. (Different rates are conceivable, but this is a technicality which the insurance industry will easily solve at a later stage.) Such a regulation can be introduced unilaterally, but the nature of illicit circulation and international diffusion of small arms strongly suggests to pursue in parallel the implementation of an international convention. For example, a sniper gun stolen from the stores of the FBI might end up in some remote country and be used in terrorist activities. Thus, such a risk insurance (or tax) would have eventually to be operated as an international trust fund charged to regulate cross-border liabilities. Enormous sums can be involved when small arms not legitimately possessed see action in violent conflict and civil wars.

This proposal clearly embraces market mechanisms to achieve its ends and considers this an appropriate complement to tighter government regulation of firearm ownership and trade. It is likely to provoke an outcry from the gun lobby and also from the armed forces and the police. However, the latter must understand that risk coverage is a reasonable principle. Once the problem is handled as normal insurance business it will look less threatening to legitimate holders, because additional costs will correspond to their specific risk classification.

But the important aspect of the proposed move is the translation of the risk small arms embody into monetary terms, thus making the costs of "gun cultures" transparent. The increase in effective prices would discourage overstocking at all levels and possibly reduce the demand for new production as excess stocks will be marketed first. It also ends the effective subsidization of gun ownership. Presently the taxpayer shoulders the damages ensuing in case of theft and misuse. With an obligatory insurance the gun owner would finally pay for the eventual damages his individual choice to carry a firearm purporting to enhance his personal security entails.

Of course, this short exposition note can only provide a stimulus for a debate of how *compulsory third party liability insurance* for small arms can contribute to improving the safety standards, and reduce illegal circulation. Small arms are a dangerous commodity, but lack the normal treatment of dangerous commodities. Otherwise, some obligatory insurance regime would long since have been introduced as in the case of car ownership. But as the problems are becoming more pressing and costly every day, it should finally be possible to tackle them more vigorously, in this case from the economic end, and call for an end to the effective subsidization of privately owned small arms. However, innovative approaches to old problems always need a long breath and must work towards winning wide political support.

The case for an ammunition tax. Many observers would agree that the supply of ammunition is a key factor in most ongoing armed conflicts. In many conflicts the warring parties are absolutely dependent on the efficiency of the logistics of the black market. This reliance on the black market to supply the fighting forces introduces an economic dimension into the political equation which determines the level of fighting and possibly the eventual disposition to seek a political solution to the conflict. Moreover, in some protracted conflicts fighting has virtually developed into a "mode of production", where the calculation of the "fighters" should be particularly sensitive to the prices of their “inputs”, not least because illicit trade networks operate on a cash basis in convertible currencies.

In general, however, prices are low, quite often below the cost of manufacture, as huge stocks of surplus are being cascaded in a chain reaction which increases the likelihood of leaks filtering into black markets. In contrast to the creation of new jobs in the regular economy being virtually unaffordable in the context of war-torn economies, the economic threshold of turning armed violence into a profitable business is rather low due to the pervasive availability of cheap small arms.

The long shelf-life of most small arms suggests that it would be worthwhile to additionally explore the control of ammunition supplies as a means to effectively reduce the availability of operational small arms at least in the medium term. The definitive consumption, when used, promises an impact on the availability through improved controls and significant increases in prices.

Against this background, it is proposed to lobby governments to levy a significant (consumer) tax on *all* ammunition fitting small arms used in armed conflicts. Such a tax would be absolutely consistent with the historical evolution of taxes and their legitimacy. An important element in the evolution of taxes are levies on demeritorial goods, like alcohol and tobacco. A tax on ammunition would follow in this tradition where the state acts on the basis of a consensus that a certain commodity has negative external effects and affects the common welfare. The current international discussion indicates that this consensus is rapidly maturing with respect to small arms. though the conversion of this consensus into a consumer tax on ammunition is unfortunately not yet on the political agenda.

A tax on ammunition would be a non-fiscal tax aiming at changing consumption patterns and the behaviour of the consumers. Small arms have become an increasingly important public health issue; therefore, they require a health-oriented steering of “consumption patterns”. In changing the relative price of this commodity the government explicitly intends to act against the potential (mis)use of this commodity. To make an impact a rate between 100 per cent and 200 per cent should be considered, in order to truly act against the unnecessary use and overstocking of ammunition. Being a mass-produced commodity with only a limited number of producers, an ammunition tax should be collected at the source.

The tariff must be levied for all ammunition produced. This is perfectly acceptable because the tax is neutral for the government as a legitimate customer. The higher prices its agencies have to pay are fully balanced by tax income. All private users of small arms would have to pay a premium for their hobby or for the privilege of presumed self-defence. This is perfectly justified given the risks of inevitable health-hazards due to the danger of illegitimate use of privately owned firearms in case of theft and other ensuing negative external effects.

With respect to international trade in cases of certified deliveries to legitimate governmental recipients the consumer tax would be relinquished. In all other cases of export the tax would be levied at the source. The additional tax income would help to cover the social costs of the misuse of firearms which are presently shouldered by the general tax payer. Given the fact that a tax levied indiscriminately at the source would be collected from foreign customers as well, a large part of the tax income should be dedicated to an international trust fund charged with coping with the disastrous consequences of reckless proliferation of small arms in the past.

There are, of course, also arguments against such a consumer tax. As a matter of principle, neo-liberal economists argue that any consumer tax infringes the liberty of the consumer, but this tax intervention must be weighed against the negative externalities of the diffusion of small arms. “If we don’t, others will” (sell and export cheap, duty-free ammunition) - this standard argument against national initiatives will be advanced by the different lobbies associated with the ammunition business. This argument is not valid as can easily be understood, when applied to the cultivation of marijuana and other drugs, in order to create additional jobs for farmers.

It will certainly be argued as well that the stocks already in black market circulation will earn windfall profits as a result of the institution of such a tax. This will inevitably be so, before black market stocks will begin to be crowded out. It should also be taken into consideration that the introduction of an ammunition tax will provide an environment for black market activities if neighbouring states do not co-ordinate their tax policies. The mutual history of the United States and Canada provides a number of negative examples when different tax policies created black markets and huge profits for cross-border smuggling[1].

But these hurdles are not insurmountable. The history of demeritional taxing shows a surprising convergence between different cultures and continents. Ammunition seems to fulfil the properties needed to justify an internationally co-ordinated promotion of a tax. The most important factor bringing the international community in line with a process of strictly applying a hefty consumer tax on ammunition is the emerging consensus on the dangers of small arms diffusion and the commonly felt need to counteract this tendency.

From a security perspective the control of ammunition and its taxation is far more efficient at the source than at the level of ultimate consumer. It also allows for a rigorous accountability of those legitimate international transactions which would not be subject to taxation. Higher prices for ammunition lower the chances of unlawful appropriation of ammunition, because all actors will be less inclined to overstock ammunition and the final demand is likely to display some elasticity to prices. Finally, if the major international suppliers (G-8 and NATO member states among others) could agree on a co-ordinated tax on ammunition, they would easily be able to muster sufficient pressure to make most other suppliers conform to such an international tax regime as happened with other demeritional taxes in history.

The case for a recycling deposit. Increasingly states intervene in markets and put up regulations to make sure that environmentally hazardous commodities are being properly recycled to avoid lasting negative effects if left as waste. Thus, car manufacturers will eventually be forced to recycle their cars at the end of their useful life. The costs of recycling will be charged with the sales price. To minimise the risk of illicit firearm possession and black market activities small arms should also come under an obligatory recycling regime aiming at eliminating a potential source of illicit transactions.

All small arms items are particularly sensitive commodities which constitute a potential danger if the decommissioning does not lead to the destruction and recycling of the components. To ensure a proper recycling of firearms and decommissioned ammunition, a recycling deposit should be charged with the sale of firearms and ammunition. The deposit must be a substantial sum, in order to make it attractive for everybody to return any weapon found or not any longer needed and cash in the deposit. If further investigations of ammunition supplies confirm that illicit refilling of empty cases constitutes a problem, a deposit should be levied on ammunition as well, in order to secure the safe return of the cases to the legal sphere. Aluminium soft drink tins, also a mass produced item, carry such a deposit in Denmark.

Such a compulsory deposit would have a double effect. It would raise the economic threshold to acquiring firearms, but it would also contribute to reducing the average life cycle of the existing stock of weapons and hence their availability to enter illicit circulation in general. The incentive to decommission weapons and be rewarded by a refund is likely to forestall the accidental appropriation and entry of many firearms into illicit circulation. If old stocks of ammunition can be converted into cash, they are also less likely to end up in the cross-border black markets.

A recycling deposit operates not unlike a buy-back programme, but its costs are borne by the legitimate or tolerated owners of firearms and it is made a permanent feature. Given the large existing stock which was acquired without the payment of a deposit, the start capital would have to be advanced by the government during the initial phase, while the programme would become self-financing after a certain period. Compared to the potential security gains the initial investment required is modest.

The case for an international trust fund for the destruction of surplus stocks of small arms. Large surplus stocks of small arms, originating from uncontrolled accumulation of reserves during the Cold War, remain a source of illegal circulation which requires co-ordinated international action. The cascading of surplus weapons from the arsenals of industrialized countries to the inventory of the armed forces of less wealthy nations was started in the early nineties. Unfortunately, the weaponry and the volumes involved were often not commensurate with legitimate needs of the recipient armed forces. Sometimes the suppliers were merely interested in economising on recycling costs. As a result these weapons ended up in the arms depots with diminished safety against leakage into illicit global circulation. More importantly, however, the profound economic crisis in most transition countries and a significant number of developing countries has weakened the public services, including the military to such an extent that the safety of the arms depots is no longer guaranteed. In such precarious circumstances recycling of surplus arms is not feasible.

Measured against the enormous economic damage and human suffering caused by the global availability of illicit small arms, investing in an international trust fund for the destruction of surplus stocks of small arms will produce high rates of return in the form of foregone damages. Against this background the formation of an international trust fund for the destruction of surplus stocks of small arms, possibly under the auspices of the United Nations, should receive priority on the agenda of the General Assembly or the CD in Geneva. Since many individuals would be sympathetic to the creation of this trust fund, one might conceive this fund as a foundation which receives governmental as well as individual contributions.

Provided sufficient funds can be assembled the trust fund would actively pursue a strategy of identifying critical surplus stock and offer its services which in certain cases might involve in addition to destruction and recycling a compensation for the lost market value of the weapons, in order to make it attractive to governments to contract the trust fund for surplus reduction.

Demand Side

A case for reversing the culture of violence. Reducing the availability of small arms is first and foremost a process of reconstructing mutual confidence; it calls for a consensus of putting the state back into the driving seat along the road towards improved security. The different measures proposed to improve the control of firearms and to reinforce the supply side controls are only the start of an extensive political discourse needed before any legal action can bear fruit. The ensuing discussions will already be part of the cultural and political changes required to convert declaratory policies of governments into effective measures of improved control. Bringing the diffusion of firearms under control is not merely a legal act, it requires overcoming the latent gun culture whose "virus" is more firmly established in some societies than in others. Unfortunately, the propagation of the gun culture is presently well entrenched in the global electronic media. Some non-governmental organizations, like the US-based National Rifle Association, strategically sponsor the gun culture. This "virus" does not recognize national borders and often operates as a precursor of the global network of illicit arms trade. It has been reported that the hero of the Rambo movie is perceived as a role model among young Africans in war-torn societies.

The case for restrictions on advertisements for small arms. The aggressive visibility of advertisements for military weapons in all kind of print media is but one example of many accepted practices which contradict the declared intention of many governments to fight the diffusion of small arms, including weaponry of military design. Advertisements for military weapons[2] in print media are particularly suspect, since the readers are legally not entitled to purchase the products advertised. Since procurement officers are certainly not appropriately addressed by advertisements praising the performance of automatic guns and military ammunition in a wide range of journals, it is pertinent to question the purpose of such advertisements.

To secure contracts in the legitimate procurement market, sending extensive documentation, offering competitive prices or bribing the right person would seem much more efficient and indeed practised strategies to win contracts. So, what is the logic behind the advertisements for sniper guns, automatic weapons etc.? Is it merely to entice the appetite of prospective customers, fanatics who are addicted to weapons, so that they would place their orders on the black market which eventually translates indirectly into higher demand in the legal market where the advertising manufacturer has its stakes? Or is it about cultivating a political climate where the focus is on military solutions to political conflicts which guarantees business as usual for the arms industry?

The fact is that, in general, media aggressive publicity for weapons which nobody is entitled to possess, except for the armed forces and law enforcement agencies, is tolerated worldwide. It appears to indicate the prevalence of a permissive political climate with respect to the diffusion of small arms. Just as advertisements for medicine, smoking, and drinking came under regulation in many countries, advertisement for weaponry, which individuals are under no circumstances authorized to possess legally should be restricted to the absolute minimum required for rational procurement procedures.

Displaying ammunition, as if it were lipstick in the window of an up-market perfumery, attempts to impart ammunition with an aesthetical aura and to obfuscate the inherently dangerous character of the commodity. Given the legal antecedents of restricting advertisements for certain other products, the chances to challenge the public relations practice of the international small arms industry look promising. The aim must be to curb the indiscriminate propagation of firearms, at a minimum to those legally restricted to military use only. Early parliamentary initiatives in the FRG[3] to prohibit advertisements for military weapons in general, print media should be reactivated and provide a model for an initiative within the European Union.

Children and minors around the world are exposed to the products of the dominant American entertainment industry. Armed violence is a pivotal feature in these products where product placing is a major element of the business. Thus, firearm manufacturers compete for placing their violent gadgets as long and as often as possible. The minimum regulation required would be a code of conduct aiming at reducing the open propagation of armed violence and a *prohibition of product placing related to small arms*.

The case for massive intervention in the illegal circulation of small arms. Armed violence, organized criminality and protracted civil wars are often a reflection of economic activities which are not based on a set of rules of law. In the context of intra-state armed conflicts, armed predatory acts and extortion often turn into an independent mode of production. It is therefore reasonable to assume that the different violent actors in their capacity as "*homines oeconomici*" are sensitive to the costs of their tools of production. Thus, prices charged in black market transactions influence the actual demand and hence the societal diffusion of small arms. If the argument that the illegal circulation of small arms is marked by price elasticity holds, then massive market intervention aiming at elevating prices would significantly contribute to the reduction of illegal circulation and societal diffusion of small arms.

Black market intervention requires financial resources and a trustworthy institution under the authority of the regional organizations like the OAS or OAU and possibly national agencies, as in India[4] for example. The ECOWAS initiative is probably the politically most advanced agreement at the regional level to control small arms. With strong financial and institutional backing from the EU in the context of the Lomé III agreement, the feasibility of aggressive black market intervention could be tested in this region, which is still infested with large quantities of free-floating small arms. Such a programme, which is not restricted to traditional buy-back schemes at the end of civil wars, is expected to tackle the problem of illicit small arms availability by pricing small arms and ammunition out of reach to potential black market customers.

The implementation of such a scheme has to cross many hurdles. It is often argued that it is not possible for an accountable agency to act on the black market. It must be stressed, however, that throughout the Cold War the intelligence networks of major powers purchased particularly small arms on the black market, in order to avail themselves with stocks of "deniability" to support their proxies. The design of the agency is also a sensitive issue, but given the damage caused by the illegal diffusion of small arms, the organizational hurdles should not be insurmountable. Among others one can certainly borrow from the experience in international cooperation to combat the illicit trade of drugs.

While international programmes for mine clearing are widely accepted and receive considerable support, the idea of a follow-up initiative focusing on clearing illicit small arms has not yet taken root. Like mine clearing, small arms "clearing" has important humanitarian and economic dimensions. For regions where protracted intra-state conflicts have produced elaborate networks of black market supplies specializing in small arms and the complementary illegal channels for the export of marketable products, like gems, diamonds, tropical wood and diverted humanitarian aid commodities, the availability of a *Rapid Intervention Fund* (RIF) administered by the United Nations would be helpful. Governments would be eligible to receive funds in support active intervention in the black market and to take measures aimed at controlling the illegal economy. The same would apply to countries where an early warning of an emerging conflict has been given.

But the supply of organized crime might also be directly targeted in some cases. In southern Africa the border towns in Angola are believed to operate huge black markets where the organized crime in South Africa is believed to procure its weapons from stocks spilling over from the protracted civil war. A few million dollars invested in mopping up the arms bazaars in the border region would make a considerable contribution to containing the sprawling violence in South Africa.

The financial volumes required for such a rapid intervention fund are modest if compared with the costs of conflict and post-conflict reconstruction. However, the rate of return measured in added global economic development should be high and therefore it might be appropriate to involve the World Bank in the provision of the required funds.

The case for dual standardization of small arms calibres. An important measure to improve the control of the illegal proliferation and diffusion of small arms would be internationally agreed standards for calibres of small arms which strictly distinguish between weapons designed to serve the military and law enforcement, and weapons designed for sport, hunting and legal private ownership. Such a set of norms would, in the long run, facilitate much improved controls. It should be acceptable to all nations because it does not directly interfere with national regulations of private ownership of guns. At the same time it would be an important step towards ensuring that ammunition proposed by the UN-expert group on ammunition[5], as ammunition for official use and private consumption, is unequivocally identifiable. The diffusion of arms and ammunition from official use to illegal non-state ownership and illicit circulation would become instantly transparent.

Liberal sales regulations with respect to private acquisition of firearms and ammunition would not any longer lead to illegal exports of ammunition to feed international black market networks. It would become inherently more difficult to procure ammunition for illegally possessed automatic weapons as their possession is globally restricted to the military and the police. In a transition period of ten years it should be feasible to achieve a complete separation of the two markets.

Once such a set of manufacturing norms is in place, it will become easier to exert international pressure against ammunition manufacturers supplying international black markets with non-private calibres. It may even be possible to win the consent of arms manufacturers because the introduction of such dual-standardization would mean additional business.

Small arms are a case for *common responsibility.* Most small arms and ammunition illegally circulating around the globe were legally produced in industrialized countries. At some stage these weapons were separated from the chain of legal custody. The ways and means of the transfer into illegal ownership almost always require the availability of convertible foreign currency because black markets operate only on a cash basis. Thus, the recipient party of the illegal trade must control some economic activity linked to the international (global) market generating the indispensable convertible cash. Often this linkage is also constrained to black market transactions as is the case with the notorious diamond trade which keeps the civil war in Angola going. In every case, however, the economics of the black market for arms can also be targeted from the demand side through tightening the controls on illegal economic transactions and targeted economic embargoes.

As a result, ongoing *armed conflicts*, even in remote impoverished regions, are never isolated events. To the contrary, their dynamics and logistics depend on their *continued participation in the global economy*, though often through the backdoor of criminal sectors. Without the oxygen of economic interaction with the global economy armed conflicts would suffocate or at least transform into rather low levels of violence.

Thus, tackling the small arms availability in ongoing wars requires an understanding of the underlying economics which would allow targeted and internationally co-ordinated action to close the revenue generating backdoors of the global economy through which criminal brokers organize logistical life lines for the fighting parties. More generally, illicit small arms availability and the growth of the global GCP seem to be closely linked, not only in cases of armed conflict, as increasing levels of armed violence in fragmented societies in Latin America and Africa testify.

The awareness of far-reaching consequences of tolerating criminal economic activities for the diffusion of armed violence is still underdeveloped. From an economic perspective no armed conflict is far away. In some cases, it might be necessary to sharpen the legal instruments[6], in order to stem the arms supply to ongoing wars. But there is no denying that all armed conflicts are economically on our doorsteps, and the spoils form part of our consumption pattern, which makes them our common responsibility.

References

[1] I am indebted to David Leclerq who drew my attention to this aspect.

[2] For a detailed analysis of advertisements for military weapons see: Peter Lock, *Rüstungswerbung Der Boom gegen die Krise der Rüstungsindustrie*, Militärpolitik Dokumentation No.41/42, Frankfurt 1985.

[3] In 1985 the Social Democratic Party proposed to insert a paragraph into the Law to Control Military Weaponry (*Kriegswaffenkontrollgesetz*) banning the advertisement for military weapons in general print media.

[4] Tara Kartha's 'Tools of Terror Light Weapons and India's Security' (New Delhi 1999) describes in detail a scenario on the sub-continent which clearly endorses an active market intervention.

[5] Report of the Group of Experts on the problem of ammunition and explosives, GA 29 June 1999, A/54/155.

[6] While some would argue that closing loopholes is impossible, the crippling record of the American embargo against Cuba suggests that a lot can be achieved provided the necessary political will exists.

ECONOMIC ANALYSIS OF THE ARMS TRADE AND ARMS PRODUCTION

Jean-Paul Hébert

The question of the dissemination of weapons and their production is one of the factors that conditions peace in the world. The role of weapons - especially light weapons - is exposed in a very lucid and complete manner in the paper by Peter Lock (this volume, pp. 201-214). I wish to add two proposals for future work.

Creating a Better Knowledge of Arms Transfers

The crucial light-weapon question shows gaps in information as to transfers in this domain. However, these gaps do not concern only light weapons. The knowledge that we have of some world arms transfers is also piecemeal, imprecise and sometimes contradictory.

At the present time, two types of source exist: sources in value and sources in physical volume.

Sources in Value

These are themselves of two types. There are data from SIPRI on the one hand and American data on the other.

SIPRI has published, since 1968, a yearbook including a chapter on the world transfers of major conventional weapons. It is an important source but is weak in the methodology, since the Swedish institute chose to construct an indicator that is not precisely a real financial flux measure. This indicator is a good instrument of analysis in the long term and for comparisons, but it must not be confused with a precise assessment of expenditures of arms procurement. Besides, these data do not give any detailed indications on the bilateral exchanges.

The American data are gathered mainly in two publications: on the one hand, those of the ACDA (US Arms Control and Disarmament Agency) that has published since 1970 a yearly report (World Military Expenditures and Arms Transfers) and, on the other, the research service of the American Congress (Congressional Research Service) of which the yearly report written by Richard Grimmett (Conventional Arms Transfers to Developing Nations) gives a retrospective panorama (seven years) on the world weapon transfers (agreements and deliveries) that does not limit itself, in spite of the title of the document, to developing countries. One can add that since 1997 the Military Balance, the IISS (International Institute for Strategic Studies) publication, also had data; but these are not original; they use data from ACDA and the CRS, as well as national data from some countries (such as France).

However, each of these sources has its own methodology to evaluate transfers of arms, and there exist meaningful differences between them with regard to certain countries. The difficulty is increased by the fact that the same source is carried from one year to the other to revalue its chronological sets, therefore presenting considerable differences sometimes. Finally, there is often a real conflict between international data and data of national origin. This is especially clear with regard to the case of France. For example, French sales of the year 1994 were estimated by SIPRI as $705m in the 1995 yearbook and $1021m in the 1996 yearbook. The CRS gave $700m in its 1995 edition and $1400m in the 1996 edition, and ACDA, in 1995, mentioned $800m. The discrepancy (3.5 to 7 billion francs) is very far from the official data of the French Ministry of Defence (16.8 billion francs) and even of the French customs data (12.1 billion francs).

Sources in Physical Volume

Since 1992 another source appeared: the register of the United Nations on conventional arms. This register corresponds to a desirable transparency effort. However, one knows that it has limits: it is only based on the voluntary declarations of countries (and some actors important to the arms market, such as Saudi Arabia, Kuwait and Pakistan abstained from answering). It only records a limited number of major weapon categories, to the exclusion of light weapons and small artillery. These categories themselves are very general (fighter planes) or ambiguous ("missiles and missile launchers"). Above all, movements are only recorded in physical terms and not in value.

Dependence on Sources of Information

The difficulty in reconciling different sources and obtaining accurate information is illustrated especially by the content of the 1998 report of Richard Grimmett. Indeed, this report includes a radical reappraisal of American weapon sales.

This report deals with the same categories over several years. It chapters the transfers to developing countries, and describes the synthesis on transfers towards the set of countries of the world (from which one can deduce transfers, towards the developed countries). The main sellers (United States, Russia, France, United Kingdom, China, Germany, Italy) are individualized. Data concern agreements as well as deliveries. The report gives retrospective sets over eight years (1990-1997 in the report of 1998). Some figures give quadrennial regroupings (1990-1993 and 1994-1997). From one edition to the other, it proceeds to certain adjustments on data, but until now these adjustments remain limited and could be considered like simple statistical adjustments. It is not the same in the 1998 report that proceeds to a very important revision of data concerning the United States. If one compares data concerning the seven common years (1990-1996) of reports of 1997 and 1998, one notes indeed that the total deliveries of arms of the United States in this period are shown to decrease by 25 per cent (76.9 billion dollars according to the 1997 report, 57.4 according to the 1998 report). Reappraisals concerning the other important sellers are not of the same size, even for those whose data present difficulties of statistical comprehension: numbers for China do not vary, those for Russia are increased very slightly (from 34.8 to 35.2 billion dollars), as is the case for the United Kingdom (from 35 to 35.2). Results of France are revalued by nearly five per cent (from 16.7 to 17.5 billion dollars). The reappraisal is stronger for Germany (from 10 to 11.2 billion dollars). In fact, the most appreciable variation concerns the category "other European" countries of which the valued sales 1990-1996 in 1997 to 14.8 billion dollars are reviewed to rise in 1998 to 18.5 billion dollars. But it is obvious that the movement is not of the same order of size as the one pertaining to the American sales.

This decrease of 25.2 per cent of the American sale assessments concerns sales to developing countries (-25.8 per cent: 51.7 billion dollars to 38.4) as well as sales to the developed countries (-24.2 per cent: 25.2 billion dollars to 19.1), that this decrease is due to conditions of economic crisis of the Asian countries or Latin America, or to the weakness of the price of oil.

The same drastic revision to the decrease applies to agreements passed to the American industry: the total amount, years 1990-1996 agreements, was estimated by the CRS in 1997 to 113.4 billion dollars. In 1998, for the same period, the assessment is only 80.7 billion dollars (-28.9 per cent). The decrease is very big for agreements with developing countries (73.4 to 48.7 billion dollars, a decrease of 33.6 per cent), but it is far from being negligible for the developed countries (of 40.1 to 32 billion dollars, a decrease of 20.2 per cent).

Of course, the result of these compensating revisions (strong decrease of American sales, rise for the other countries) is to reduce substantially the part of the United States in the world arms trade.

Thus, for the year 1996, the part of the United States in the world arms deliveries would not be 45.8 per cent but 31.5 per cent and for the period 1990-1996 this part would change from 35.7 per cent to 28.3 per cent.

The report does not provide any methodological indications to explain this revision, so it is impossible to decide on the real level of data.

The importance of this reappraisal in a source that became, with the passing of years, one of the main references, underlines the necessity to arrange controllable and comparable sources.

It would therefore be useful, in the setting of activities of Pugwash, that one initiates a working group, including economists and experts in particular in transfers of the main exporter arms countries (US, UK, France, Russia, China, Germany, Italy notably) in order to establish more detailed and controllable data on the financial values of arms transfers, country by country, followed by supplier (enterprises) by supplier.

The Movement of Transnationalization of Arms Businesses

To the same end (to have a better economic knowledge of phenomena connected to transfers and the production of arms), it is surely desirable to be very attentive to the evolution of the industrial structures of systems of arms production.

In these last years, the main evolutions have been characterized by three phenomena :

- privatization: most state arsenals have been transformed in societies. The national societies have been privatized (completely or partially). The movement is especially strong in Britain, France, Italy, Spain, Sweden…;
- diversification: businesses are generally less specialized in the military production. They are therefore less dependent on state orders;
- consolidation: the most visible movement is the one that produced in the United States the emergence of three gigantic groups : Boeing-McDonnell, Lockheed-Martin-Loral, and Raytheon-Hughes, but it also develops in Europe (British Aerospace-Marconi in Britain, Aérospatiale-Matra in France, Dasa in Germany).

Until now this consolidation has occurred in a national setting (in the case of the United States) or in a national or European setting (in the case of the European countries). But this movement of concentration has little chance of remaining completely outside the world economic evolution. However, what one observes in the other sectors (banks, insurance, energy, car) it is the transnational alliance constitution. It is therefore probable that, in years to come, one will see such a transnationalization of arms enterprises.

This phenomenon will evidently pose problems of political control. States risk losing a part of their restraint on enterprises in the sector. Laws and regulations themselves (in particular, as for exports) risk becoming unworkable. The enterprises policy will be influenced less and less by state-controlled decisions. Such a process can only open the way to an increased dissemination of the arms of all nature.

I propose the creation of a Pugwash working group especially dedicated to this question, in order to follow the transnational industrial alliances closely and to evaluate without delay the consequences of this evolution to capacities of State control.

TRENDS IN TERRORISM

Tara Kartha

Terrorism has undergone many shifts and changes over the past several hundred years, moving from violence sponsored by Islamic rulers against the invading Crusaders, killings for revolutionary aims, or equally for liberation from rulers perceived to be oppressive or non-representative. While terrorism was certainly a problem with human, social and potentially security dimensions, today it is the last aspect which is increasingly to the fore, even as the human costs increase.

In terms of security, it is apparent that terrorism is increasingly becoming a form of warfare, where apparently independently and loosely structured groups are able to wreak havoc and launch virtual war against well-armed security forces. Indeed, it appears that Livingstone's warning of terrorism being a manifestation of a third world war are now more credible. In terms of human costs, the rise in casualties is apparent even as the actual number of terrorist incidents goes down. Though terrorist incidents were lowest in 1996, they produced the fourth highest casualty rate since 1968.[1] This is testimony to the fact that the lethality of the terrorist has increased vastly, a phenomenon especially apparent in the developing world, where counter-terrorist forces have correspondingly few modern means available to tackle them. Clearly, terrorism has shifted gears, moving into a higher form of violence that comes close to warfare. However, it is important to note that the *objectives* of terrorism have not changed at all. These have remained political with destabilization as the primary tool.

In charting the shifts in terrorism it is as well to briefly enumerate the characteristics of terrorism prior to 1990.

- Groups were usually motivated by Marxist-Leninist/Maoist/ "capitalist", or in a sense secular ideologies. There were clearly laid down "manifestos" and objectives.
- A clear leadership was apparent and identification was possible.
- Violence was discriminate, in the sense of being carried out for maximum political benefits. Targeting was largely symbolic.
- Responsibility was claimed with considerable fanfare and press releases.
- Foreign patronage was usually well known.

While terrorism was usually seen as an option by the weaker party against a much stronger one, in the 1970s-1980s it was the primary "cheaper option" of the superpowers in destabilizing states seen as friendly to the other. Though it may be argued that states have always sponsored terrorist movements (the example of the *hashashin* of nearly 900 years ago being one) it is, however, a fact that blatant state patronage of these elements gave a near legitimacy to these movements, and more importantly gave a certain legitimacy to the use of terror to pursue the interests of states. Secondly, it also placed tremendous resources in the hands of actors who until then had to content themselves with the odd gun or stolen explosive. Third, since these terrorist movements were transnational - that is, launched from one state into another - there was little restraint on violence or targets. Fourth, and most significant, was the use of religion to justify terrorism against the atheist Soviet Union (in terms of the *jihad* in Afghanistan), which had a ripple effect. If in 1979 the jihad was launched against the Soviet Union with American money, in Iran an equally dangerous but more subtle war was declared with the first modern religious terrorist group appearing during the same period. By the end of 1987 more than 80,000 guerrilla groups had been trained in Pakistan alone. The numbers trained in Iran or Libya are unknown.

The end of the Soviet Union meant not only the withdrawal of aid (from all sides), it also meant a loss of a convenient transnational ideology. Additionally, on the ground, it meant that there were a number of groups "up for sale" to the highest bidder. These were mainly regional powers who after closely watching the Iraqi conventional conflict as compared to the covert anti-Soviet operation, were obviously capable of assessing which was the more sensible option. In the developing world - struggling to close the gap between the peoples' expectation and what governments could reasonably deliver - religion was a clarion call to action. A process of religious revival that had been afoot since the mid-1970s was harnessed by various leaders, the clergy, and various sub-state entities (like intelligence agencies and individual revolutionaries and indeed businessmen), and non-state entities (like narcotics traffickers) to replenish and re-energize various movements that seemed to be on the verge of faltering. In the Indian northeast, various church organizations provided funding for Christian rebels (who, ironically, followed an elementary "socialist" ideology); in the Punjab a revivalist Sikh religious ideology was armed; and in Afghanistan the jihad was replaced by a motley crowd of Sunni and Shia and "secular" groups, each controlled by various regional actors. In France a huge wave of bombings by the Algerian *Groupe Islamique* was followed by the assassination of the Israeli Prime Minister by a religious extremist. The USA found its citizens under attack in a wave of violence in Saudi Arabia and Pakistan.

Thus by the 1990s at least 64 of a total of 96 active terrorist groups were clearly identifiable as being religiously motivated, even if the doctrine and theology that they swear is violently extremist and far removed from actual precepts. Second, the fact emerges that 14 have ex-mojahideen among their ranks, and third, 17 are based in or closely involved with Pakistan (this last figure is a conservative estimate - if transitory links are taken this rises to more than 21). Fourth, nearly all Asian groups were found to be involved with the trafficking of narcotics. Fifth, all found havens in various "non-states" or states whose governments seemed to be under the sway of various Islamist or criminal elements under dubious control.

Apart from these were areas where states were wracked by ethnic, nationalist or regionalist regional loyalties and beliefs, whose citizens clash with state forces in a paroxysm of violence that may be called "religious extremism." While these are certainly of a lesser number than the Islamic groups, nonetheless they must be taken into account. Thus, this paper uses the term "radical extremism" to mean all kinds of violence against society, perpetrated by groups or states, on the basis of a narrow and exclusive ideology that sanctions the use of extreme violence in pursuit of political aims.

The Inputs that Shape Militant Radicalism Today

Today there are, therefore, several separate but not unrelated strands that go into the making of modern terrorism. These are briefly discussed below.

There is an unprecedented spread of the tools of terror - assault rifles, grenades, and explosives are readily available in the major hot spots around the world. A rough estimate puts the numbers of weapons on the loose at around 500 million. The reality could be anywhere around this figure. In explosives, this is even more difficult to guess. Consider that during the 1980s, Czechoslovakia sold some 1,000 tonnes of Semtex to Libya and around 40,000 tonnes to Syria, North Korea, Iran and Iraq.[2]

This has two implications. On the one hand, small countries with limited resources can wage virtual wars at a minimum cost to themselves, especially if they have little production capacity of their own. Second, and closely related to this, states can now arm discontents with little or no chance of the transaction being brought home to themselves. For instance, weapons originating in South America have been seen in the hands of jungle bandits in India, while Chinese weaponry arms nearly all groups in Southern Asia including Myanmar, Philippines, Bangladesh, Kashmir and Afghanistan. Third, the possibility of small groups operating independently (up to a point) of any authority raises the possibility of originally state-sponsored groups moving out of anyone's control. A good example is that of the Tamil Tigers who are now adept at sourcing the international black market in arms, and who may be purveyors of weapons (especially explosives) to other groups in the region.

There is apparent a slowly increasing level of technological capability with the modern terrorist. The path from the .303 to the Kalashnikov has been a bloody one and, unfortunately, the upward climb continues with the acquisition of shoulder-fired anti-aircraft weapons and chemical and biological weapons. Once the realm of fiction, the Aum Shinriyko incident served to highlight the new possibilities. Nuclear terrorism is also seen as a real possibility, with a crude nuclear weapon perhaps towed to a harbour or on a truck. Given the technology available to certain groups, this is now no longer simply decried as alarmist scenarios. Experts are uneasily aware that it is only a matter of time. This technology diffusion is also apparent within the same group which, apparently covering all ends, had thoughtfully set up a top notch ammunition and gun manufacturing factory to supplement the stocks of weapons already in their hands. The improvisation techniques of the Tamil Tigers similarly has always managed to keep the security forces unprepared. Rumours of a small submarine and nascent aircraft capability (possibly gliders) have, however, yet to be confirmed.

Narcotics have intertwined with religious and covert movements so that the two sometimes meld into one. Terrorism is now big business. It has been noted that there has been a doubling of drug production between 1986 and 1988 in Afghanistan, a period that corresponds with the first signs of Soviet attempts to draw back from the region, as well as US questioning of the aims of the war. Today, narcotics have become part and parcel of the conflicts in Afghanistan, Tajikistan, and threaten to spread instability into Kyrgystan, and further into the Caucasus. Similarly, Lebanon remains a primary centre of production and trafficking with the collusion of a number of terrorist organizations, and possibly with Syrian military officers involved. Precisely the same situation is apparent among the Kurds who use the refugees in Italy and Germany as conduits. Another case is that of Sri Lanka, where the Tamil Tigers association with narcotics trafficking is so well hidden as to almost escape notice. It is worth noting that cadres at the top level belonging to these outfits follow the lifestyle of executives of multinational corporations rather than desperadoes fighting for a homeland. In short, terrorism and militancy is today big business. As one analyst notes, if the Palestine Liberation Organization were an American corporation, it would have been on the list of Fortune 500 companies, with assets that in the mid-1980s were valued at $5 billion.[3] Given the fact that corruption occurs at almost every level due to the considerable sums of money involved and the ease of accessing this (which may simply involve a customs or a military officer turning a blind eye as a shipment goes past) the possibility that there are built-in factors operating against the ending of a conflict or the termination of a terrorist grouping needs to be kept in mind.

Even as the old generation of terrorist leaders loses support, there is a mixed floating population of armed mercenaries available for "hire". It is true that men of the reputation of Abu Nidal and even the Kurdish leader Ocalan have become "too hot to handle." In this sense their own fame, loss of backing, as well as the consistent efforts of anti-terrorist forces of the USA, as well as Israel, have led to their being sidelined. However, it is as well to note that conflict spots of the world today host entire generations of fighters who have grown and matured on nothing but war and violence. The UN Special Rapporteur, Enrique Bernales Ballesteros, in his report to the Secretary-General notes:

> a growing reliance on mercenaries, which for our purposes means seeking out foreign experts who are paid for services of destructive violence and deadly effectiveness. The Special Rapporteur has maintained and continues to maintain that mercenaries exist, that they are not a small number of individuals; that they are groups of professionals selling their skill in war and violence; that they are also criminal organizations, an international blight in their perpetration of acts of violence which ruin human lives ... and extend to terrorist attacks that have touched off or aggravated conflict, with often catastrophic results.[4]

This list includes ex-fighters from South Africa and Angola who have set up their own "private security services." Outfits like Executive Outcome make a business out of conflict, with the difference that the violence is now "privatised." Their organization and arsenal rival those of most developing countries, and their numbers are growing. The list is, however, topped by ex-Afghans who have migrated to several conflict areas. "Ex-Afghans" are known to have migrated to Algeria, Egypt, Sudan, Iran, the former Yugoslavia, Chechnya, the Indian state of Kashmir, the Philippines, and Pakistan itself. Many of these are nationals of their own states, but are sometimes under hunted by their own governments leading them to look for other markets for their talents. Several important points emerge, however, among most mercenaries.

a) Recruits to the jihad (or indeed to the various "liberation" movements of the Cold War) who joined up in their early twenties or even prior to that, are at present in their late thirties and have no other profession to fall back on.
b) They were indoctrinated with the most radical interpretations of religion, with little theological grounding or indeed any more logic than that of promised glory and benefits in the afterlife.[5]
c) Mercenaries tend to drift from one group to another, making police detection difficult.
d) Many are also "part-time terrorists" and may be hired out for a special assignment, after which they may disappear for a while.
e) The Afghanistan-Pakistan region is the nodal centre for Sunni terrorism, professing an ideology based on the narrowest of beliefs. The Taliban is only one segment of this, and indeed is less extreme when compared to the *Lashkar e-Jhangvi* in Pakistan, a militant Sunni organization that has many a murder to its credit.

The last decade has seen the largest movement of refugees and displaced persons (27 million) constituting an eleven-fold increase since 1970.[6] A refugee population almost invariably is the recruiting ground not only for fighters, but also for the carriers of narcotics and other cogs in the wheels that keep "Terrorism inc." a running proposition. At another level, the refugee areas are also those where dreaded terrorists routinely "disappear," evading capture by anti-terrorist squads. While there have been cases (for example, Ramzi Youssef) where the announcement of a huge ransom has yielded results, more often than not lesser known but equally dangerous men have moved across borders with the aid of the men who "control" the refugees camps.

There are "non-states" (Afghanistan), states in a state of near collapse (Cambodia, Congo), or those suffering from almost perpetual internal conflict (Pakistan, Lebanon, Myanmar) which offer ready shelter to all kinds of criminal and extremist elements. Collapsed states like Afghanistan have little control over who stays or transits through their territory, a situation similar to that of Cambodia, where there is a ruined economy and a corrupt police or state force tends to be ready to deal with whatever organization promises it funds. It is not surprising, therefore, that Cambodia is a hunting ground for arms dealers and the like. Other states with poor or non-existent control over their territories (outside of the national capital) also fall into this category. At another and perhaps more dangerous level are those states whose systems have become part and parcel of terrorist, narcotics or other criminal activities (Pakistan, previously Thailand, Myanmar, parts of China, southern parts of Russia, and parts of Southern Africa). These states usually neighbour conflict areas and, therefore, are vital in providing the infrastructure vital to the carrying on of terrorist or militant activities. The existence of these non-states adds to the new terrorism in the following ways.

a) It allows the patron state (not necessarily a neighbour) to pretend an almost complete ignorance of the activities of training camps, and continue to fund them. This is most applicable should NBC terrorism emerge from out of these "non-states."
b) These countries are home to a motley band of terrorists (often with completely different or even competing aims) from different countries and regions. It has been seen that much of the cooperation that is evident at a ground level has been engendered at training camps. For instance, former trainees of the Abdul Sayyaf training camp, and later the Harkat training camps, are known to have sheltered each other, while others have been provided with "contacts" in the arms bazaar.
c) These are rudimentary camps and, therefore, come up quickly even after a bombing raid, or may simply move a few kilometres to a different location.

Common roots and common objectives - a high degree of cooperation among groups. The bringing together of Shia and Sunni terrorist groups is credited to the Saudi millionaire turned terrorist leader Osama Bin Laaden. Al Queda is an important manifestation of this, and is reported to be functioning in Sudan, Afghanistan, and with links in the territory of the warlord Aideed. Apart from this is the cooperation between groups like Hamas and the Al Jama'ah a'-Islamiya of Pakistan. The 18th Convention of this Islamic group in Pakistan brought together delegates from more than 30 Islamic groups and movements. A pledge of unity of the Palestinian and Kashmiri "jihad" was made, and a huge donations campaign for the latter held. Other groups present included the head of the Islamic jihad in Bangladesh, and the Turkish Fadilah. Needless to add, every amenity was provided by the government for the smooth running of the "conference."

Other groups operating into China (based in the north of Pakistan) are also said to have links with Palestinian groups, while the Islamic Movement of Uzbekistan has close links with the Taliban. When its leader Tahir Yodasev fled Tashkent after a strong reaction from the security forces, he was given a house in Kandahar and a base for training his fighters.[7]

The Taliban phenomenon is spreading - the new "Just wars" are here. There is clear evidence that extremist groups (already called the Taliban by the press) are working in Bangladesh, Uzbekistan, Kashmir and perhaps into Dagestan and Chechnya. Note that the US bombing of terrorist training camps brought out threats of a Third World War from extremist Chechen leaders. What is important to note is that the leaders of these groups often threaten the state leadership with dire consequences if their deadly "just war" is curtailed. Equally significant is the fact that many of these outfits are funded by intelligence agencies (as in Pakistan) or outside powers (as in Syria) making them, in effect, a state within a state.

The merging of regular and irregular war. Terrorism was always difficult to define, but today in South, West and Central Asia, terrorism is increasingly merging into war. While it is true that states have regularly used non-state actors for the purpose of terrorism for strategic gains, what is now apparent is that state actors are now acting in concert with non-state actors, giving the latter a degree of lethality and capability that is quite unprecedented. Examples of this abound. The use of the Pakistani army in the attack on Kabul (weighing in on the side of the Taliban) is one, while the use of Russian air assets on the side of Abhkaz rebels is another. The most graphic and recent example is that of the conflict in Kargil (May-July 1999) where Pakistan engaged in covert action with the use of regular troops. The fact that the occupation of Indian territory across the Line of Control (LoC) in India was done by Pakistani infantry (its Northern Light Infantry) has been well-documented by international publications.[8] This is an extremely dangerous development, since it allows the state to ignore even the most basic rules of war and international behaviour. It is also a direct prod to interventionist tendencies, since the state feels it can "get away" with such adventurism.

Measures to Deal with the New Terror

- Restricting the tools of terror (deny the weapons).
- Dealing with mercenaries at the international level (deny recruits).
- Create the "stateless terrorist" (deny shelter).
- Pressure banking havens to reduce their secrecy laws where criminal funding is concerned (deny funding).
- Build up the failed states as a priority (deny territory).

Apart from this, it is as well for major aid givers to realize that Islam by itself is hardly a threat to anyone. Prominent moderate religious leaders in failed states need to be empowered to effect change. Their voice is more likely to be heard than that of the foreign-trained social workers or political activists.

It is a tragedy that states like Afghanistan, Cambodia and Mozambique have been virtually abandoned to their fate. The need to bring a semblance of order back into their systems must remain a primary concern of the so-called "international community". Indeed, neighbours would be better engaged in rehabilitation programmes, and working out regionally a plan to provide training, employment opportunities and to revive infrastructure. As one African delegate pointed out, a little money goes a long way in bringing order to areas which have been close to starvation for years. Institutions like the World Bank, which have a post conflict reconstruction unit, need to be involved in this with the full cooperation of local religious leaders aimed at reviving the traditional networks.

It is tragic that those who enthusiastically armed the Africans (for instance) are now notably reluctant to contribute effectively to the chaos in that continent – a chaos that is basically engendered by the fact that the region is awash with light weapons, explosives and small arms. Similarly, both the United States and Russia (in its former role of a superpower) have pumped in enormous amounts of weaponry of a value that far exceeded the "aid" that was once granted to the Zahir Shah regime. In short, peace has never been popular – war is not just interesting copy, but profitable business.

The United Nations is no less to blame. Of the sum total of states that are by and large adversely affected by the terrorist menace, few have been able to cooperate at the level of the General Assembly to put forward anything more than banal "calls" for action. Moreover, possible cooperation is adversely affected by the UN clinging to the phrase of "self determination" – almost a holy mantra in UN language – which is at variance with today's realities. Much more to the point would be stern action to see that states implement the international conventions that most are already party to like those that ensure the individual's political and human rights.

References

[1] Ian O. Lesser, "Countering the New Terrorism," RAND, Project Air Force, MR-989-AF, 1999.
[2] *Ibid.*
[3] Cindy C. Combs, *Terrorism in the Twenty-First Century*, Prentice Hall: New Jersey, 1997.
[4] Report on the question of the use of mercenaries as a means of violating human rights and impeding the exercise of the right of peoples to self determination, submitted by the Special Rapporteur of the Commission on Human Rights, Fifty-first session, Agenda item 109, A/51.392, 23 September 1996.
[5] The diaries and letters of killed militants in Kashmir reveals an almost complete ignorance of the basic tenets of Islam, while there almost always are exhortations to the recruit on the rewards that would accrue to him should he die in battle against the infidels. Author's review of seized literature in Kashmir, June 1998.
[6] Human Development Report, 1996, p. 26.
[7] Ahmed Rashid, "Heart of Darkness," *Far Eastern Economic Review*, 5 August 1999.
[8] See, for instance, an interview with a Pakistani soldier in *Time* 12 July 1999.

ORGANIZED CRIME AS A NEW THREAT TO SOCIETIES IN TRANSITION

Ulrich Albrecht

Crime, or deviation from orderly conduct, remains an inherent part of human experience. Organized crime, the sustained engagement in disrespect of orderly conduct, is largely a new experience, at least with regard to the recent relative importance of the phenomenon, especially in societies in transition. This paper is an effort to delineate the systemic roots of this development. It aims to assess the influence of organized crime in such societies. The intention of this presentation is to interpret organized crime in a comparative manner as a phenomenon of transitional societies by and large, ranging from former state socialist countries, to Latin American societies that turned away from military rule, and other developing countries faced with enforced social change. The basic hypothesis is the global coherence of organized crime.

To start with, key definitions are needed. *Societies in transition* encompasses a broad spectrum of societies: Latin American countries in the process of *transición* from military rule; former state socialist countries in Central Eastern Europe involved in transformation towards the goals of democracy and market economics; countries, mainly from the developing world, which are characterized by a past of authoritative regulation and a marked change in their statehood, which may have weakened them.

In Germany, where the authorities had to come to grips with the presence of organized criminal activities emanating from Central Eastern European neighbour countries, a joint task force of Federal and regional experts recently arrived at the following definition:

> Organized crime is the planned execution of criminal acts directed by the search for profit or power, which in each case or in their collectivity are of major importance, if more than two participants engage for a sustained period or for an undefined time in a division of labour
> (a) using business or business-like structures,
> (b) using force or other means of intimidation, or
> (c) cooperate in order to penetrate politics, the media, public administration, the judiciary, or the economy.[1]

The technical details of this definition will be by-passed here, in order to scrutinize the emphasis on the economic aspects ("search for profit," "business-like structures," penetrate "the economy"). The understanding is widespread that organized crime reflects, according to a top German prosecutor, "criminality of capital" (*Kapitalkriminalität*),[2] the "continuation of market economics by other means."[3]

This bridges the gap with, as Edwin H. Sutherland coined the term in 1938, *white-collar criminality*, economic activities at the fringes of the legal and civic system. Yet, the onslaught reaches presently much further. In transitional societies one observes the emergence of an immoral economy, alongside the conventional sector, also characterized as the underground economy.

The main problem when dealing with organized crime is the unreliability of available information. Investigative journalist sources and scientific assessments are scarce. There are many publications, mainly of a journalistic nature, with doubtful allegations. The most informative sources are research establishments of police headquarters, such as the Research Unit of the German Federal

Criminal Agency (*BKA*), and their outlets in other parts of the world (for example, the Middle-European Police Academy/MEPA in Budapest which is supported by the German BKA). Partly because those police bodies feel overburdened to come analytically to grips with the internal modes of operation of organized crime, partly because collaboration among scientists from different countries is still easier than among police establishments and intelligence communities, analysts from academia find doors wide open if they offer help in understanding the rise of organized crime.

Developments in the criminal sphere apparently culminate in present Russia. A systematic search in Russian newspapers about the informal sector and the role of the *mafia* (thus the adaptation of the term in this country) found that the "shadow economy" represents a share of 40 per cent of the Russian economy, and provides the main part of the "black" economy (Prime Minister Chernomyrdin is cited with an estimate of "between 20 to 50 per cent"). Representatives of the criminal sector are said to be in control of 90 per cent of companies and economic organizations (leaving mainly the biggest firms untouched). According to other accounts, up to 80 per cent of all Russian "economic subjects" are controlled by the *mafia*. The MVD (Russian Ministry of the Interior) counts "some thousand mafiotic gangs," consolidated into 150 "associations".[4] Such a rise has not been predicted in writings about *transición* or transformation and deserves special scrutiny in order to understand and to challenge this development.

The annual turnover of organized crime was estimated in 1998 at US$500 billion, half of this amount covering drugs. 50 per cent of this sum involves laundering in the international banking system.[5]

Organized Crime as Surrogate Regulation

Current assessments about regulation in transitional societies contain key notions which are borrowed from other contexts - and which indicate a requirement for novel concepts in understanding - and challenging - the phenomena. There is language about "informal sectors" in national economies, insinuating activities at the fringes of social life, carried out in the niches of the market system. This view defies the fact that so-called informal sectors today account for a dominant area of economic activity in those societies, for a significant share of the population, and that they represent one of the pillars of economic growth. Furthermore, there are statements in a moralistic tone about corruption and the requirement to combat it, without an assessment of the inherent links between corruption and societies undergoing enforced social change. Thirdly, there are complaints about old-fashioned modes of social promotion in the creation of new functional elites, called (using a term borrowed from mountaineering activities) *Seilschaften* (roped parties), with hints to effective interaction of old *nomenklatura* cadres climbing up the social ladder with mutual support, outperforming in elite formation less compromised candidates. Fourthly, and of greatest concern, is the awareness of organized crime, especially with regard to the notorious Russian *mafia*.

The four notions mentioned - informal economic activity, corruption, roped parties and organized crime - manifest different aspects of regulation in transitional societies. Seen from the perspective of informal regulation these developments can be read as fulfilling surrogate functions in a debilitating society. Blackmailing represents a sort of tax, and - again in this parallel - the taxpayer gets a return, protection, which the formal sector is unable to provide or to provide sufficiently. The supply of illegal gambling opportunities responds to demands that are not accepted in the formal system. Transfers of drugs and of arms also represent supply of goods which are denied in the sphere of formal regulation. As the New Zealand expert Sam Pope summarizes: "Organized crime is seen as the freest, least contained form of capitalism."[6]

The focus of informal economics lies on activities not regulated by the law (the role of law and regulation will preoccupy the present paper). The non-legitimized interaction between economic and state/public spheres provides the domain for corruption. Non-meritocratic promotion is the essence of the *Seilschaften* process. All these phenomena indicate, in conventional wisdom, defiance of the orderly conduct of affairs, contradicting the established patterns of transition and transformation, or political development *per se*.

Societal Ruptures as Opportunities for Irregular Activities

The sudden changes in Latin American countries which turned away from military rule, the soft revolutions in Eastern Central Europe, and the decay of statehood in a number of developing countries, opened tremendous opportunities for resort to informal processes of social regulation.

Officials who had survived purges in state socialist countries or Latin American military dictatorships developed, as a means of survival, strong connections among their likes in order to disguise the past, to protect themselves, to possibly strengthen influence, and to find a *modus vivendi* under new circumstances. In this context, Heiko Pleines sees a new "party mafia" emerging in Russia.[7] His observations about the other side of Gorbachev's reforms appear as suitable for application to circumstances in other societies in transition.

As the "second economy" was widely accepted in the past of authoritarian regimes, it provided a major opportunity to push this scheme during the heyday of the immediate post-authoritarian situation. The underground economy needed protection, and got it from the opponents to far-reaching reforms. By exploiting the split among the functional elites about reforms, the underground economy formed a strategic alliance with the party *mafia* and merged into a network of corruption and organized crime.[8] Especially in Russia, "the economic reforms, which were launched in the late 1980s, offered the corruption network and organized crime new fields of activity".[9] Gorbachev's anti-alcohol campaign pushed production and distribution of one of the products in greatest demand into illegality. Thus the campaign "did for the Soviet gangs what Prohibition had done for their US counterparts in the 1920s."[10] Laws meant to transform the underground economy, which had become vital for Soviet society, proved counterproductive and to the benefit of the black economy due to the inexperience of the lawmakers.[11] The handling of privatization in particular seems to have opened wide the doors for illegitimate profiteering - an observation which will be true also for a variety of transitional societies which experimented with privatization. And the enormous rise of speculative international capital transfers in the process of globalization contains a great share of sums representing criminal activities.[12] In 1945 there were 55 so-called extra-territorial tax oases, off-shore banking havens outside the control of financial inspection. The present number is 175 oases.[13]

There are findings from other societies which point at societal ruptures as the cause of the rise of informal regulation. Robert S. Leiken reports that severe corruption in South Korea was "the result of interaction between Korean traditional norms and modern industrial development."[14] He also informs about "a forthcoming survey by economist Edgardo Buscaglia, [that] in Argentina, Brazil, Ecuador, and Venezuela bribes and payoffs to court officials rose sharply as a proportion of court costs between 1986 and 1993."[15] The Corruption Perception Index, which is compiled annually by the Berlin-based NGO Transparency International, shows that the bottom third of the 54 countries surveyed is occupied exclusively by developing and former socialist countries.[16]

In Latin America, public office has been inseparable from patrimonial obligations to family, clan, clique, party, and the shadow economy. "While producing economic growth during the 1970s, these regimes also siphoned resources to clients in the private sector and the labour bureaucracy and to political parties and *caudillos*," states Leiken about the phase of *transición*. "In the 1980s, these profligate practices helped to push Latin American states into a debt crisis, sparking democratization and market reform."[17] The Colombian weekly *Semana* revealed detailed accounts of the mafia *caudillos*, according to which one clan alone holds one-third of all property in the country. Colombia in this view is "not only a narco-democracy but something like a great national drug cartel."[18]

Thus the enormous rise of organized crime in transitional societies does not come by accident. It is directly connected with the great social and political ruptures resulting from the change in regime. And it won't be limited to a transitional phenomenon after the status of consolidation (a key notion in the academic literature about *transición* and transformation)[19] has been reached, unless its roots are sufficiently understood and appropriate countermeasures are taken.

A Proposal

The basic idea of this contribution is to reverse the dominant approach towards organized crime and other unlawful activities, which are based on the standards set by the OECD profile, that is, the rule of law, liberal market procedures, and functioning representative democracy. Instead, informal patterns of regulating societies will be used to interpret the reproduction of the societies concerned. This means to deal with concepts which enjoy little respect in established discourse. Yet the analytical promise is to get much closer to an appropriate understanding of the role of organized crime than the defiance approach will allow.

The promise is indeed much broader. This proposal does not only offer fresh insight into processes in present societies, but, in addition, into processes of regulation in the preceding social formations in Latin America, Central Eastern Europe and elsewhere. If one thinks about it, this hypothesis should not come as a surprise. The new forms of regulation which are so decisive have their deep roots in former authoritarian modes of societal organization. Analysis of these societies in the past has been fixated on the state and formal institutions in the communist world and in Latin American dictatorships, partly because those closed societies did not not offer much more for research and assessment from the outside. A number of developing countries of relevance in this context have also not been particularly accessible to social science research. In addition, it was also a central element in national ideologies that the formal institutions effectively regulated society.

Only a few dissidents submitted before 1989 that the state socialist system was basically run in informal ways. In the German Democratic Republic, Rudolf Bahro, in 1977 perceived widespread corruption as a "somewhat higher reason for the relative functionality of our system."[20] The Soviet defector Mikhail Voslensky gave in his book about the *nomenklatura* a long list of accounts of corruption, *Seilschaften* and misconduct in the former Soviet Union. Sovietologists generally did not take these early hints as seriously as they should have done before the state socialist system collapsed. The semi-legal black market which was instrumental in exchanging scarce goods, the administrative distribution of strategic resources via authoritarian decision procedures, careers in state socialist parties or in the military establishments in Latin America, in essence followed modes of informal regulation rather than nominal formal procedures. The "administrative markets" in these authoritarian societies were in fact markets in the sense that the elites negotiated transactions. The real capitalist market regulator, the price, was of secondary importance. The status of this statement is hardly more than a hypothesis - but one which opens up a large research agenda.

The Essence of Informal Regulation

Informal regulation is essentially based on strong and continued relations within a micro-milieu which are not formally defined and which are also not formally institutionalized. Emphasis rests upon the aspect of strength of a social relationship (as it can be regularly observed within communities of organized crime).

The material contents of informal modes of regulation parallel those of formal regulation: information is transferred, resources are being exchanged, influence and authority are exerted. Furthermore, support is mobilized, and emotional and effective relations develop.

Even if informal regulation exists beyond the law it is based on rules, and disregard of those rules entails sanctions. Rules and the sanctions linked to them in informal regulation, especially in the case of organized crime, are greatly at variance with the ones regulating the formal sphere of society.

In the case of informal regulation also social interaction requires a certain amount of knowledge and trust in rules. In order to corrupt somebody the corruptor needs a minimum of information about his victim, as well as about his preparedness to be forthcoming if certain conditions are met. On the other hand, the one going to become corrupted needs advance confidence in the expectation that the corruptor will actually deliver. Informal cooperation is thus characterized by an "investive risk" unknown to players in formal relationships. The question of Robert Axelrod in his book about the evolution of cooperation, how "robustness of reciprocity" can be achieved, applies also to criminal cooperation. The starting point of Axelrod: "Under which conditions cooperation arises in a world of egoists without a central staff of government?"[21] is shared by me.

The Axelrod question applies in both directions with regard to organized crime: from the informal superiors, who have not been voted in by anybody, down to their followers (about whom those superiors have no assurances of loyality). Emrich *et al.* find in their seminal study about roped parties: "The power of the lead man in the last resort depends upon the secure expectation that association with this particular social relation benefits every member of the roped party."[22]

If rules are misread or disregarded, sanctions apply which in the case of organized crime differ greatly from control based on formal law. Formal regulation is based on the specific knowledge of the kind and extent of sanctions if the law is broken. Informal regulation sanctions are characterized by uncertainty about their actual extent. The openness of retaliation in nuclear exchanges according to the doctrine of flexible deterrence represents the apex of this kind of informal regulation - the adversary should remain principally uncertain when and where nuclear weapons were to be used. This provocative example is cited as a hint to related aspects of sanctions in informal structures, such as methods of frightening, of blackmailing and terror, again linked with uncertainty about the actual outcome.

Furthermore, sanctions in informal regulation are characterized by excess. One feature of sanctions in this area is an enormous openness towards violence, as well as rapid escalation. German investigators into *mafia* killings remain impressed by the brutalization in sanctioning ("A whole magazine is emptied onto a victim where a few bursts of fire would have sufficed"). In communities of organized crime the disregard of rules often ends with the physical annihilation of the accused, in order to remind the rest of the group of the code of honour of the organization.

Both aspects, uncertainty about the extent of sanctions and knowledge of the openness to brutal violence, apparently effectively control processes of informal regulation. This scheme of regulation is not an open one, but reflects a highly closed social subsystem. This explains in part difficulties in combating organized crime. The sociologist Mühlmann defines such groupings in general as

> 'relatively closed', exclusive, clublike associations inside larger communities, with limited and controlled access to membership, often also with a hierarchic structuring of members according to rank, and in this instance controlled and staggered possibilities of promotion from 'lower' to 'higher' grades, with growing exclusiveness and growing esoterics of the cultivated 'knowledge' ... of the goals of the association.[23]

One may honestly ask what the traditional Italian mafia and the new Russian *mafia*, or Nigerian gangs and Chinese organized crime have in common, and whether the discrepancies will not predominate. Analysts heavily contradict one another. They see the universality of organized crime as based on three features:

1. The dominant means of control is money. Lt.-Col. Juri Adashkevitch, in Germany a liaison officer from the Russian FSB (a follow-up to the KGB in charge of internal security) describes the Russian *mafia* as a "profit-maximization scheme" characterized by tight control via money. The economist L. Gatovskij agrees: the objective "is the highest possible profit at any cost."[24] There are a host of similar assessments.
2. Organized crime normally uses formal commercial activities for its dealings. According to German findings 92 per cent of all activities of international organized crime are linked with

operations of firms (in Russia 70- 80 per cent of all private companies are said to be under *mafia* control, and according to the Russian Ministry of Finance only 16.6 per cent of all juridical persons regularly pay taxes).[25] German authorities are aware of 350+ criminal organizations, registered companies, which operate throughout Europe.

3. Organized crime operates internationally and participates fully in the development towards globalization. Seventy per cent of the persons involved in organized crime on German soil are non-nationals.[26] There are clear signs of clandestine networking.

Activities of modern organized crime comprise traditional operations such as blackmailing, prostitution, meddling contracts, printing counterfeit, systematic theft including the unauthorized unloading of transport vehicles (for example, of emergency supplies provided by international organizations), armed protection services, illegal gambling, grand-scale smuggling, trafficking drugs and arms, alongside with more recent actions such as facilitating illegal immigration, the illicit international transfer of private cars, nuclear materials, human organs in transplantation medicine, rare animals, treasures from museums and private collections, and ecological dangerous waste, manipulations with credit cards, and money laundering. Obviously the sea-change in the international system and the opening of borders especially in Europe is the chief reason for the new activities which have illicit transfers as the common denominator. Sam Pope finds: "Criminality transgressing borders is by now the biggest economic activity in all of the world."[27]

General Findings

The general discourse about regulation in transitional societies and elsewhere looks at formal institutions, notably the state, and - motivated by neoliberal convictions - the reduction of these. There is a tendency to underrate a third and surrogate alternative: the rise of schemes of informal regulation outside the law.

In societies in transformation with their enforced change of institutions, limited guarantees of legality and widespread paucity of the public purse, informal regulation tends to become endemic. A functionalist perspective sees also positive aspects: the social animal develops surrogate means of control, there is at least some kind of regulation. To some extent wealth is redistributed, underpaid public servants achieve a decent income, small and medium business gets promoted. Given the international dimension of informal regulation, it contributes also to compensation in the glaring differences in standards of living.[28]

This perspective needs to be confronted with the negative impact of organized crime. The direct ones are at least fivefold. First, national economies incur higher costs for the community, because the extra costs of these activities are disseminated. Second, general welfare becomes neglected and turns into a secondary goal for the public service. Third, organized crime focuses on distribution and tends to neglect production. Fourth, in public administration norms of gratification tend to supersede norms of efficiency. And finally, institution-building and the build-up of their authority in the new democracies are hampered.

The actual conduct of societal transformation, the ways and means of how decisions were made and implemented, apparently greatly facilitated the rise of organized criminal activities to a substantial degree. The mistakes made will differ from country to country. But there is definitely a need to carry out field research about the "other" side of enforced change, the numerous faults and deceptions which contribute so much to the present disappointing state of affairs in a large number of countries.

The issue at stake is, however, much more principal in nature. Codification of the rules of regulation has passed an apex, and according to neoliberal convictions formal institutions are in the retreat. This implies an enormous loss of public political space, what is accepted by the organizers of the process (who certainly did not mean to leave part of this space to criminal actors). The Lancaster political scientist Michael Dillon speaks in this context about "liberalism's principled hostility to politics," and cites Etienne Balibar in order to illustrate the shortsighted perspective of police organizations which focus upon the criminal person, and not the structures which mainly propel organized crime: "Just as there is an economic fetishism of *things*, so there is a juridical fetishism of *persons*, and in reality these are one and the same thing because the contract is the other side of the exchange and each is presupposed by the other."[29] Yet, theoretically based insight is required in order to subdivide legality from illegal doings, to separate accidental criminal conduct from organized efforts, in order to understand the interrelationship between the normality of capitalism and the abnormality of organized crime.

It may well be that proponents of capitalism become aware of the actual costs of organized crime, for example. in transaction costs, and that in the end the very system joins in efforts to combat the organized criminals.

Political space is public space, in contrast to the private sphere, with which organized crime is connected. Sceptics foresee indeed a decay of politics:

> Politicisation ... is interminable even if it cannot and should not ever be total... each advance in politicisation obliges one to reconsider, and so to reinterpret the very foundations of law as they had provisionally been calculated or delimited... Nothing seems to me less outdated than... [this] classical emancipatory ideal.[30]

References

[1] *Gemeinsame Richtlinien der Justizminister/-senatoren und Innenminister/-senatoren der Länder über die Zusammenarbeit von Staatsanwaltschaft und Polizei bei der Bekämpfung der Organisierten Kriminalität*, 1990 (Anlage E RiStBV) [Collective Guidelines by the ministers/senators of Justice and the ministers/senators for the Interior of the German States for co-operation between Public Prosecutors and Police in combatting Organized Crime].

[2] Hans-Ludwig Zachert, Die international organisierte Kriminalität" [internationally organized crime], in: *Internationale Politik*, no.2/1995, p.6.

[3] Susanne Krasmann/Werner Lehne, "'Organisierte Kriminalität' im Windschatten der Globalisierung legaler und illegaler Märkte" ['Organized crime' in the shadow of the globalization of legal and illegal markets], in: *Vorgänge. Zeitschrift für Bürgerrechte und Gesellschaftspolitik*, no.1, March 1997 (137), p.107.

[4] Marion Kunze, *Literaturstudie zum Thema: Informelle Wirtschaft und informeller Außenhandel in der Russischen Föderation/GUS. Dimensionen, politische Strategien, theoretische Fragen* [A study of the literature about the theme: Informal economy and informal foreign trade in the Russian Federation/CIS. Dimensions, political strategies, theoretical questions], mimeo, Berlin, no date given (1997).

[5] Robert Harnischmacher, "Organisierte Kriminalität. Mißbrauch von Banken und Versicherungen als Kanäle" [Organized criminality. The misuse of business banks and ensurance companies as channels], in: *Europäische Sicherheit*, April 1998, vol. 47, no.4, p.44.

[6] Sam Pope, "Grenzüberschreitende Kriminalität" [criminality transgressing borders], in: *Politische Studien*, no.351, vol.48, January/February 1997, p.26.

[7] Heiko Pleines, Organized crime and corruption in Russia since 1987, in: *Russia & the successor states briefing service*, Harlow, Essex: Longman, October 1995, p.3 (This series has been discontinued in 1995). - Pleines' text is based on one of the best collections of writings about Russian organized crime.

[8] Pleines in his otherwise excellent contribution tends to overlook this split in the functional elites of the former Soviet Union.

[9] Ibid., p.7.

[10] Mark Galeotti, *The Age of Anxiety. Security and Politics in Soviet and Post-Soviet Russia*, Harlow: Longman, 1995, p.184.

[11] For detail see Pleines, loc.cit., p.7.

[12] For a general appraisal, see Elmar Altvater/Birgit Mahnkopf, *Grenzen der Globalisierung. Ökonomie, Ökologie und Politik in der Weltgesellschaft* [The limits to globalization. Economics, ecology and politics in the global society], Munster (Westfaelisches Dampfboot) 1996, esp.chapter 5. - Pleines (loc.cit., p.11) reports that "since 1990 up to US$100 billion have been taken out of the country without authorisation."

[13] Cf. Harnischmacher, loc.cit., p.45.
[14] Robert S. Leiken, "Controlling the Global Corruption Epidemic", in: *Foreign Policy*, Winter 1996/97, no.105, p.60.
[15] Ibid., p.61.
[16] Transparency International and the Index can found in the Internet (http://www.is.in-berlin.de/service/ti.html) or can be contacted via e-mail (ti@contrib.de)
[17] Ibid., p.65. - See also: Silvia Colazingari/Susan Rose-Ackerman, „Corruption in a Paternalistic Democracy: Lessons from Italy for Latin America", in: *Political Science Quarterly. The Journal of Public and International Affairs*, Fall 1998, vol.113, no.3, pp.447-470.
[18] Cited after Harnischmacher, loc.cit., p.46.
[19] The pace-setter is the book series by Guillermo O'Donnell/Philippe C. Schmitter/Laurence Whitehead, *Transitions from Authoritarian Rule*, 3 vol., Baltimore/London: The Johns Hopkins UP, 1986.
[20] Rudolf Bahro, *Die Alternative. Zur Kritik des realexistierenden Sozialismus* [The Alternative. A Contribution to the Critique of Real Existing Socialism], Cologne/Frankfurt 1977, p.281. - Bahro defines corruption as „this systemic bribing towards state loyality which Stalin was able to initiate from the above and which by and large never was outside the control of the top" (ibid.).
[21] Robert Axelrod, *The Evolution of Cooperation*, German ed., p.3
[22] Eike Emrich/Vassilios Papathannassiou/Werner Pitsch, Klettertechnik für Aufsteiger. Seilschaften als soziales Phänomen [Climbing techniques for ascendants.Roped parties as a social phenomenon], in: *Kölner Zeitschrift für Soziologie*, 48 (March 1996), no.1, p.150.
[23] Ernst Wilhelm Mühlmann, "Geheimbünde" [secret associations], in: Wilhelm Bernsdorf (ed.), *Wörterbuch der Soziologie*, Stuttgrat 1969, p.130
[24] Adashkevitch quote from a statement in Berlin April 1997; L. Gatovskij, „Reforma i interesy" [reform and interests], in *Voprossy Economiki*, Moscow 1995, no.6, p.123. The Russian quotes are taken from Kunze, loc.cit., footnote 17, p.12. - Kunze's survey is based upon 78 articles in the Russian Press.
[25] Cf. E. Stroev, "Korrektirovka ekonomiceskich reform i rasbota koncepcii biuzetnoj politiki", in: *Voprossy Ekonomiki*, 1997, no.1, p.77.
[26] Figure from Landeskriminalamt Berlin [Berlin Criminal Investigation Office].
[27] Pope, loc.cit., p.27.
[28] Cf. e.g. Egmont R. Koch, *Grenzenlose Geschäfte. Organisierte Wirtschaftskriminalität in Europa* [business without limits. Organized economic crime in Europe], Munich 1988, p.19, who refers to the smuggling of cigarettes as „providing jobs to thousands in Italy and Spain", and to „the trafficking by the shirt mafia (by which) customers of European mail order firms have access to inexpensive supplies."
[29] Michael Dillon, "Criminalising Social and Political Violence Internationally", in: *Millennium: Journal of International Studies*, 1998, vol. 27, no.3, p.543 and 565 (Etienne Balibar, *The Philosophy of Marx*, London: Verso, 1995, pp.71-72; italics in the original).
[30] Jacques Derrida, "Force of Law: The 'Mystical Foundations of Authority' - Deconstruction and the Possibility of Justice", trans. M Quaintance, *Cardozo Law Review*, 1990, vol.11, p.971.

STRATEGIC IMPACT OF TRANSNATIONAL CRIME

Jasjit Singh

> Organized crime poses a direct threat to national and international security and stability, and constitutes a frontal attack on political and legislative authority, and challenging the very authority of the state. It disrupts and compromises social and economic institutions, causing a loss of faith in democratic processes. It undermines development and diverts its gains. It victimizes entire populations, targeting and capitalising on human vulnerability. It co-opts, entraps and even enslaves segments of society, especially women and children, in its diverse and interrelated illegal undertakings, particularly in prostitution.
>
> *Discussion guide for the Ninth United Nations Congress on the Prevention of Crime and the Treatment of Offenders* (A/CONF.169/PM.1 and Corr.1), para 39.

The strategic impact of transnational crime (which broadly includes the crimes listed in the report of the first CSCAP Study Group) can be broadly divided into two categories: that related to the impact on civil society, and that impacting on the stability and viability of the State. Civil society and the modern nation-state are deeply inter-related in the modern world, and one cannot really exist without the other, at least in the sense of a liberal system which respects both the rule of law as well as norms for societal functioning. While crime, and its variety that transcends national borders to link with other criminal activity, has always existed in states and society, what requires a closer look are the implications of the increasing interdependence and "globalization" of human interaction within and across states. Revolution in communications and information has opened tremendous opportunities for faster growth of human development. But on the other side of the coin there is also the increased potential for misuse of the capabilities provided by these revolutions for illicit and criminal activities. Networking is becoming the key to globalized economic, social and political activity, especially across national frontiers, which often has been cited as the model of the modern civilization. The very same capabilities are also available to criminals for transnational activities.

The dynamics of transnational crime also need to be seen in the context of what may well be the strongest driving force for its growth and spread: the revolution of rising expectations which has been sweeping across the globe for some years now. Unlike a few decades ago, when formal education was a prerequisite for knowledge and awareness, the information/communication revolution has brought awareness of potential and real quality of possible life to even the uneducated in remote parts of the world. The impact of the audio-visual electronic media is powerful. As a consequence, expectations have been rising rapidly. The problem is that actuality and satisfaction level cannot rise at the same rate. The result has been an increasing divergence in the expectations-satisfaction levels. The poor were always trying to get rid of their poverty; but even the rich are looking for more riches. The resultant instability has been at the root of a great deal of turbulence and instability in society and states. The divergence between expectations and actuality leads to strong incentives for rapid acquisition of money and power. An ever-greater number of people are trying to get rich quickly. This not only leads to corruption and criminal activities, but also fuels ethno-centric and sectarian political agendas.

Impact on Civil Society

There are a number of identifiable consequences of transnational crime that impact on the functioning and viability of what may be termed civil society. While specific criminal activities have their own consequences on the social fabric, the overall effect is to undermine civil society. Some examples of the effects are briefly outlined below.

Social Problems

The most obvious impact of transnational crime is the expansion of social ills and problems. The International Commission on Peace and Food has concluded that:

> The drug menace is transnational in character with far-reaching implications for societal and international peace and security. The countries and regions which produce these drugs, and through which the trade flows, have been afflicted by endemic violence and social turbulence. Inevitably, drug trafficking is linked with illicit arms, terrorist groups and the Mafia. Criminal gangs are increasingly gaining control over the administrative and political structures of drug-producing states, where drug-related corruption permeates the military and government.[1]

Drug trafficking, for example, has a major debilitating effect on society. By its nature, drug trafficking is a transnational as well as criminal activity. Spread of narcotics usage in the country of production as well as the country of final destination is inevitable. But the passage of narcotics across national frontiers leads to a situation in which transit countries get affected. For example, significant quantities of narcotics transit through India from two (of the world's three largest) drug production areas in the world, that is, the so-called "Golden Triangle" to India's east, and the "Golden Crescent" to its west. Drug addiction has been rising as a consequence of this traffic, although the most dramatic rise has been in Pakistan itself where the number of addicts increased from a mere 2,000 in 1980 to over 3 million by 1994. The worst impact of narcotics spread is that it starts to involve the youth. This has repercussions for the future of society. The International Commission on Peace and Food has gone on to say that: "The use of and trade in narcotics represents a menace not only to the health and well-being of individuals and societies, but also to international security."[2]

Rise in Criminal Activity

Transnational crime by definition tends to increase criminal activity in the states concerned. Criminal mafiosi operate across borders to maximize the effect of their activities and to protect their own functioning from the law enforcement agencies of a particular state. One of the goals is to take advantage of the weakness in any co-ordination policies and measures existing between the countries concerned. For example, there is no universal extradition treaty covering all states although an increasing number of states are now resorting to bilateral treaties so that safe havens are not easily available to criminals. Meanwhile, transnational crime places a premium on national criminal activity which receives sustenance from the larger capacity resources of transnational criminal organizations and activities.

Nexus Between Crime and Politics

The nexus between crime and politics appears to be rising worldwide. This has a far-reaching impact on society and its civilized functioning. Politics under such circumstances has to make allowances for, and at times is driven by, the interests of the criminal groups. Transnationalization of crime expands the scope of such compromises. The rule of law, which is the central basis of civil society, starts to be seriously undermined by the crime-politics nexus.

Privatization of Law and Order

Increase in crime makes it difficult for the state to provide adequate security for its citizens. This becomes a more serious issue where security of people in the corporate sector and important personalities is concerned. In areas of high level of insecurity, crime and violence, citizens tend to organize private security forces and acquire arms for self-defence. Private armies have already emerged even in the most democratic and organized peaceful states. These are organized by the criminal organizations on one side, and the citizens on the other. The State, in the face of high levels of insecurity, tends to turn a blind eye if not actually to encourage and promote such processes. In the short term this does provide a higher level of security in society. But it also creates situations of leakage of arms and easier access to the tools of violence for criminal and militant groups. In any case, privatization of security undermines the institution of the State which otherwise carries the primary responsibility for the security and well-being of its citizens.

A typical example of a variation of privatization of law and order is the case of the former Soviet Union. Over 13,000 policemen have gone to work for the 300 criminal gangs (each with some 2,000 members). The combined effect of various types of transnational criminal activities is to degrade the rule of law. This in turn leads to erosion of faith in the legitimate institution for providing security and ensuring peace in society: the State. Weaknesses in good governance are intrinsically generated by such criminal activities. Undermining of the rule of law has serious implications for civil society and stable statehood.

Impact on Stability of the State

By all accounts, the stability of states, especially in the post-Cold War period is threatened more by the extensive use of armed violence than any other single factor. But it is the easy availability of high power high lethality light weapons that has expanded the scope and frequency of use of armed violence. Ready availability of modern small arms transforms political dissent into violent activity. It is not surprising, therefore, to observe that:

> A disturbing feature of contemporary world is the spread of a culture of violence.....In some societies, the trade in narcotics has been responsible for raising the general incidence of violence. Russia and some parts of Eastern Europe have seen a surge of violence as criminal syndicates seek to exploit the new freedoms. Widespread criminalization can threaten the very functioning of a state.[3]

Weakened Institutions

Increasing crime in a state and society has an inevitable consequence in weakening the institutions of a nation-state. Modern nation-states and their society are deeply integrated and interdependent entities. For the state to function effectively and efficiently, each component institution of the state must function effectively. In fact, the strength of institutions, which range from the political institutions, to economic and social organizations, defines the viability and stability of a state. Criminal activities, by definition, are the antithesis of the institutions of a modern nation-state. They pose a challenge to the strength of these institutions, especially when undertaken by organized transnational groups and gangs.

Transnational crime, like drug trafficking, has a debilitating effect on institutions of the State and society. As James Velde writes, "Drug mafiosi now influence large sectors of the career bureaucracy, elected political leaders and police and military forces of such countries as Colombia, Bolivia, Peru, Myanmar, Kazakhstan, and maybe soon, Russia."[4]

One of the most crucial modern institutions which gets weakened by transnational crime is that of democracy. In countries that are in the initial stage of transition to democracy, or those with fragile democratic institutions, the impact could easily lead to irreversible trends.

Undermining Economic Stability

The central objective of most transnational criminal groups is predominantly economic. Such organizations and groups are normally in position to generate large incomes. A European Parliament study estimated that over $100 billion in profits from illegal drug activity was successfully laundered throughout the world in 1995.[5] The same study concluded that laundered drug money throughout the world earned $820 billion during the decade of the 1980s. Colombia's drug traffickers operate a multi-billion dollar business and launder much, if not the most, of their profits in foreign banks. Drug money is reported to account for 25 per cent of Pakistan's GNP.[6] Transnational criminal organizations with even much lower levels of economic turnover naturally command entrepreneurial and management skills at the cost of their availability for legitimate activities. Profits generated through criminal activity are normally high, and make economic management a much more difficult task. More important, a parallel economy gets established based on the financial activities of criminal groups. Such "black" economies undermine legitimate economy and can destabilize development efforts and steps to ensure economic equilibrium.

Transnational crime poses a serious challenge to the integrity of financial and commercial institutions at the international as well as national levels. The corruption of financial institutions (much as happens in respect of political institutions) is a serious phenomenon that undermines public confidence with a serious long-term negative impact on the global confidence in the economy (and the political system's ability to rectify the ills). Money laundering at the international level, for example, erodes the capacity of governments to manage, control and regulate the global financial system.

Retarding State-building and Human Development

Many states in recent years have embarked on building liberal democratic institutions and market economies after the end of Cold War. Such a dual transition has been extremely difficult for many states, especially where the criminal justice system was weak. Organized crime has made the task and prospect of a successful transition more difficult. This has far-reaching strategic implications, not only for the countries concerned, but for the larger international community also.

Promotes Violence

> [Those] who have died in war since [9 August 1945] have, for the most part, been killed by cheap, mass-produced weapons and small-calibre ammunition, costing little more than the transistor radios and dry-cell batteries which have flooded the world in the same period. Because cheap weapons have disrupted life very little in the advanced world, outside the restricted localities where drug-dealing and political terrorism flourish, the population of the rich states have been slow to recognize the horror that this pollution has brought in its train. Little by little, though, recognition of the horror is gaining ground.[7]

Colombian guerrillas and active insurgent groups, who once received financial support from Nicaragua, Libya and Cuba, now rely on drug trafficking to sustain their operations. Drug money is believed to account for 25 per cent of Pakistan's GNP.[8] Drugs accounted for 20 per cent of

Colombia's foreign exchange, 30 per cent of Peru's foreign exchange, and 50 per cent of Bolivia's foreign exchange.[9] Such high levels of financial turnover in drug trafficking also requires protection of the assets, trade and profession of transnational crime. This is sought from non-governmental sources, subverting the State and employing violence for coercion and self-defence by the criminal organizations. Gang wars become a more frequent phenomenon adding to the overall violence levels in society.

Erosion of State

One of the biggest strategic threats emanating from transnational crime is the erosion of the State and speeding up of the unravelling of institutions that form the core of a successful state system. For example, the drug market's enormous size, combined with its illegality, has generated huge corruption, lawlessness and violence throughout Latin America and parts of Asia. The scope of the trade in narcotics has eroded the authority of the State as the symbol and enforcer of the law in many Latin American countries. Transnational crime weakens the State, and this in turn offers greater opportunities to carry out criminal activities. Increase in crime in turn further erodes the State and its institutions. Thus a vicious circle can start off.

Threat to Sovereignty

By its very nature, transnational crime implies the violation of national borders. State borders have become porous to trade, financial systems and communications. But such activities are mostly legal, and there remains an effort to regulate what moves across the frontiers. Transnational crime is trans-border and illegal by definition, and hence directly challenges the fundamentals of state sovereignty. The flow of illicit arms, drugs and narcotics, and other contraband, and international linkages and networking among organized criminal groups, negates the very concept and exercise of national sovereignty of states involved. States may retain sovereignty in a formal sense, but they would no longer be able to exercise it in the classical way.

Response Strategy

By definition, transnational crime requires international co-operation if it is to be addressed effectively. In pursuance of the mandate of the UN General Assembly resolution 49/159, the Commission on Crime Prevention and Criminal Justice proposed a set of recommendations adopted in Economic and Social Council resolution 1995/11. In fact, there is also urgency in doing so in terms of time since crime and criminals tends to spread in a cancerous growth. It has been proposed that the main focus of international co-operation should revolve around the following areas:[10]

(a) Harmonization of legislative and other countermeasures.
(b) Training and exchange of law enforcement and criminal justice personnel.
(c) Information-sharing among relevant agencies.
(d) Joint force operations or task forces.
(e) Protection of witnesses, investigators and judges.
(f) Elaboration of new measures and updating of existing ones, on the basis of periodical assessment of results achieved.
(g) Technical co-operation and assistance in drawing up such counter measures.
(h) Clear identification of co-ordinating authorities in the individual countries.

(i) Measures to encourage the adoption and improve the implementation of existing co-operation arrangements.
(j) Effective co-ordination of activities at the bilateral, regional and multilateral levels.

References

[1] *Uncommon Opportunities: An Agenda for Peace and Equitable Development*, Report of the International Commission on Peace and Food, Zed Books, London, 1994, p.37
[2] *Ibid*
[3] *Our Global Neighbourhood*, The Report of the Commission on Global Governance, Oxford University Press, New York, 1995, p.16.
[4] James R Van De Velde, "The Growth of Criminal Organizations and Insurgent Groups Abroad due to International Drug Trafficking", in *Low Intensity Conflict & Law Enforcement*, vol.5, no.3, Winter 1996, pp. 466-483.
[5] *Ibid.*, p.479.
[6] "UN Official Says Drug Money Corrupting Pakistan's Economy", United Press International, March 31, 1994.
[7] John Keegan, *A History of Warfare*, cited in *Our Global Neighbourhood*, n.3, p.130.
[8] "UN official says drug money corrupting Pakistan's economy", *United Press International*, March 31, 1994.
[9] Alfred McCoy, *War on Drugs: Studies in the Failure of US Narcotics Policy*, Westview, Boulder, 1992, p.4
[10] Phil Williams and Ernesto U. Savona (eds), *The United Nations and Transnational Organized Crime*, Frank Cass, London, 1996, p.104.

WEAPONS OF MASS DESTRUCTION, DOCTRINES FOR THEIR USE, AND THE THREAT FROM TERRORISM

Jorma Miettinen

Nuclear Weapons

In the 1990s a rearrangement of the global strategic balance has taken place. The United States has become the only superpower. Following the dissolution of the Soviet Union, Russia inherited its vast nuclear arsenal. The threat by mutual deterrence has become less marked but the disarmament spree of the beginning of the decennium has died. The enlargement of NATO to Eastern Europe increases tensions.

China

With the breakdown of the Soviet Union, the value of China as a strategic companion to the USA vanished and the first Clinton administration started taking a critical ideological line. It mellowed during Clinton's second term but the Taiwan lobby in the US Senate maintains the controversy by supporting the independence drive of Taiwan's President Lee and exaggerating the role of spying in the modernization of China's nuclear weapons. The USA is shocked by China's fast economic growth which has continued for over 10 years. If it continues for another 10 to 15 years it will make China the world's richest nation. China has recently tested several modern nuclear warheads but has not built a massive nuclear arsenal like the two other superpowers.

The United States

The USA declares, as one pillar of its defence strategy, to: "shape the international security environment in ways favourable to US interests...by maintaining significant overseas force deployments." Some countries and political, ethnic or religious groups consider this doctrine to be hegemonic. Since they are not strong enough to openly threaten the USA, they do it secretly, by methods of terrorism. The USA has had to widen its military posture to include counter-terrorism, also realizing the possibility that terrorists might use weapons of mass destruction.

As for nuclear weapons, President Clinton issued in 1995 a "negative security assurance," pledging that the United States would not use nuclear weapons against non-nuclear weapon states parties to the Non-Proliferation Treaty. However, Special Assistant to the President Robert Bell has stated that negative security assurances will not tie the hands of US decision-makers faced with a chemical or biological attack.[1] (The USA has given up these weapons and cannot retaliate "in kind".) The *Doctrine for Joint Theatre Nuclear Operation*[2] identifies non-state actors as likely targets for nuclear strikes. Thus, the strike made on 7 August 1998 against the Al-Shifa Pharmaceuticals Industries plant in North Khartoum, on the pretext that it was producing a precursor of the nerve gas VX, could have been made using a nuclear weapon, particularly if there had been an underground extension with controlled entrance (civil defence shelter, BL-4 laboratory, whole body counter, and so on).

The only evidence turned out to be a soil sample obtained by the CIA just outside the plant which contained trace amounts of an organophosphate EMPTA (o-ethylmethylphosphonothioic acid), a Schedule 2 compound under the CWC which is used as a precursor to many civilian products like fungicides, pesticides and antimicrobial agents. It was in no way a valid justification for the bombardment in which many innocent workers died.[3]

The USA is producing new modifications of nuclear weapons suitable for targeting potential proliferators, for example, the B61-11 nuclear bomb which is described by the Department of Defense as suitable for targeting Libya's alleged underground chemical weapons plant in Tarhunah.[4] What a lucrative target for this bomb the Al-Shifa plant would have been if it had had an underground shelter!

The USA has now admitted its mistake in bombing the Al-Shifa plant.[5] The Clinton administration did not challenge a lawsuit filed by the owner of the plant, businessman Saleh Idris, and agreed to release US$24m that Mr Idris had deposited in US banks. The Al-Shifa plant made both medicines and veterinary drugs and it has raised Sudan's self-sufficiency in medicine from 3 to over 50 per cent and produced enough veterinary medicine for all of Africa. The USA has not compensated the relatives of the killed workers or the owner of the plant.

The present US nuclear policy does not support non-proliferation. Instead, it proves to the developing nations that nuclear weapons *are* important for achieving military and political objectives.

In November 1997 President Clinton issued a classified Presidential Decision Directive (PDD) 60 which "permits US nuclear strikes after enemy attacks using chemical or biological weapons."[6] It names "rogues states" as possible targets. Several US officials have listed such rogue states. The list contains 6-7 states, but the following five are always on the top: Iran, Iraq, Libya, North Korea and Syria. All these are non-nuclear weapon states.

The cause of the broadening of the targeting of nuclear weapons is the assumed increase of proliferation of WMD.[4] This has caused a multimillion upgrade of the US nuclear arsenal to make possible quick shifts of targeting and changes in the nuclear doctrine. Russia and NATO are also making changes in their nuclear doctrines.

Russia

Russia is increasing its reliance on nuclear weapons. It has changed its nuclear doctrine to possible first use and emphasized the role of nuclear weapons in defence. NATO enlargement, US BMD efforts, and modernization efforts of US nuclear weapons, are mentioned as causes of this development. There is pressure to start production of a new generation of tactical nuclear weapons in spite of the fact that there exist 5,000 usable older ones widely spread and 1,200-1,500 warheads in reserve. Lack of financing has prevented the new production, so far.

Chemical Weapons

The United States has decided to destroy its whole chemical weapons arsenal and the process is well underway.

Russia has the world's largest chemical weapons stockpile, 40,000 tonnes as agents, and it has not yet even started its destruction. It delayed to the utmost the ratification of the Chemical Weapons Convention (CWC), signed in Paris in January 1993, and has not yet presented a comprehensive plan for their destruction.

The USA has financed some joint research projects in Russia regarding destruction techniques, probably mostly to get information. Russia's justification for the delay of the destruction has been the necessity to get foreign financing for it.

All parties to the CWC are requested to destroy their stockpiles in 10 years (by 2007) but there exists a possibility to get a five years extension by request. Evidently, the true reason for delaying is that the Russian military want to preserve a weapon considered to be valuable, being politically easier to use than nuclear weapons.

Other Countries

Several countries, which have not signed or ratified the CWC, evidently have chemical weapons capability. Such countries are, for instance: Iraq, Israel, Syria, North Korea, Taiwan and Egypt. The Organization for the Prohibition of Chemical Weapons (OPCW), however, is functioning satisfactorily and there is reason to hope that some day even these countries will come to the conclusion that the CWC can give them better protection than an illegal CW capability.

Biological Weapons

Together with about 140 other states, all the superpowers are parties to the Biological and Toxin Weapons Convention (BTWC) which entered into force in 1975. Unfortunately, it has not prevented the proliferation of biological weapons. It has one serious weakness: it does not have any verification mechanism.

Since 1995 the Parties have worked for a legally binding instrument to strengthen the treaty.[7] Drafts have been presented for a compliance regime having the following components:

1. Declaration of *dual use capabilities* (plants normally producing legal civilian products but quickly convertible for production of military products).
2. *Challenge investigations* on evidence of possible non-compliance.
3. *Non-challenge visits* to confirm accuracy of declarations and clarify unresolved questions.

It is assumed that most of the roughly 140 States Parties would probably declare less than 100 facilities but some may have several hundred, possibly even up to 1,000.

Drafts have been presented for an "OPBW" resembling the above-described OPCW. Some progress has been made in four years but the Protocol is still far from ready.

Most countries, including the United States, destroyed their biological weapons in the years following the signature of the treaty (1972). The Soviet Union did not. In 1973 it started a large secret project for developing offensive biological weapons, which remained essentially secret until 1979 when a release of anthrax spores took place in Sverdlovsk. The explanations of the Soviet officials that the resulting epidemic, which killed at least 66 people, was due to meat poisoning were not credible. The first hints of the truth were soon obtained when a deserter, Mark Popovskiy, came to the USA.[8] The first deserter to tell about the vast production plants of the Biopreparat Company, a biologist called Vladimir Pasechnik, who came to England in 1989, just when the Soviet Union started to dissolve, frightened the British and American intelligence organizations by telling about his work to develop the "Black Death" (*Y. pestis)* as a biological weapon. He said he had developed a genetically modified variant resistant to vaccines.[9] President Bush and Prime Minister Thatcher were informed about Pasechnik's revelations and they used pressure on Gorbachev to open the Biopreparat plants to foreign inspectors. This happened in January 1991, at the highpoint of détente. There were huge fermentors and other equipments, all empty and clean, and sullen personnel who answered questions reluctantly and evasively.

However, a comprehensive report about Biopreparat was obtained in 1992 when the First Deputy Director of the Research and Development Department of Biopreparat, Dr. Kanatjan Alibekov, a Kazak, arrived in the USA. He has abbreviated his name to Ken Alibek. He knew the organization thoroughly. When at its largest, at the end of the 1980s, it comprised 50 institutes and 100,000 workers. There were research, development and production plants for bioagents and vaccines and weaponizing plants and testing stations and fields. The greatest testing field was on the Rebirth Island in the Aral Sea. Biopreparat also produced antibiotics and other products of biopharmacy and veterinary medicine. The secret activity took place behind this public façade of civilian production.

Ken Alibek came into publicity for the first time in the USA on 2 March 1998 on the "Prime Time" television programme. On 9 March 1998 his interview was published.[9] He revealed that the Soviet Union weaponized hundreds of tonnes of anthrax spores, bubonic plague and smallpox bacteria into strategic MIRV missiles which were then targeted on the largest US cities.

Today, Biopreparat has been divided into two parts. One part is being converted to civilian production, for example cosmetics and liquors; the other part is under military command and top secret. Some Western intelligence experts believe that work on offensive weapons is continuing. The largest plant is located at Sergiev Posad, ca. 50 km to the north east of Moscow. It is next door to a monastery which is a great tourist attraction. The charming tourist guides have no notion about the nature of the institute next door.

Iraq's Bioweapons

After the Persian Gulf war, the organization founded by the UN to verify the destruction of Iraq's weapons of mass destruction, UNSCOM, confirmed that Iraq had had between 1991-1995 bioweapons under development, but they had been destroyed. The truth about the large-scale production of anthrax and botulinum toxin and their loading onto SCUD missile warheads in hundreds of kilograms was revealed first when Saddam's brother-in-law deserted to Jordan in 1995 and revealed all the secrets to the CIA. He carried with him large trunks of documents which UNSCOM got. Evidently, it became so difficult for Iraq to hide the secret stocks that Saddam decided to expel UNSCOM altogether.

Biological Warfare and Bioweapon-Terrorism Compared

Biological warfare means contamination of enemy troops, population, domestic animals or victuals with living microbes (bacteria, viruses or rickets) or toxins. In military use, point contamination (wells, food, troops in closed fortification, reinforce positions, and so on) is usually attempted, epidemics avoided, because they are uncontrollable and may infect the attacker's own troops. In this century, four states (Japan, the Soviet Union, the USA and Iraq) have acquired major bioweapon arsenals but never effectively used them. Moral inhibitions may also have worked, because from the beginning of history use of poisons and diseases in warfare has been considered to be forbidden. "*Armis bella, non venenis geri!*"

In war, a biological agent is usually delivered to the target with a carrier – missile, grenade, aeroplane (bombs or sprays). Large amounts of agent are needed because enemy troops are often sparsely distributed in the terrain and may carry protective masks and garments.

A terrorist needs only a small amount of microbes or spores which must be protected against the environment, usually microencapsulated. The agent is easy to carry through customs in an innocent-looking container – a black champagne or perfume bottle, for instance. Aerosol can be produced by a small spray, which resembles artist's spray or that of a perfume bottle. For increased efficiency, a small bottle filled with pressurized freon and a ventil controlled with a watch may be used. Half a litre of agent is likely to be enough for a terrorist's purpose. Terrorists probably want to have a large effect to get much publicity. They might spray the aerosol in a closed space containing thousands of people, like a railway station, department store, sports hall, airport, or underground station.

In 1993 the Japanese Aum Shinrikyo terrorist organization sprayed from a high building anthrax spores and from a vehicle botulinum toxin over Tokyo, but did not get any noticeable results. It is not easy to produce the right size aerosol, the optimum particle size being 0.002 to 0.005mm. Preparation of the right aerosol requires professional experience.

The Agents

The Australia group, founded by the Conference on Disarmament for control of exports related to WMD, has published lists of biological agents containing 13 bacteria, 4 rickets, 19 viruses and 10 toxins suitable for terrorist use. The US Army Handbook on Medical Management of Biological Casualties[10] has a more concise list containing 7 bacteria, 3 viruses and 4 biological toxins (see Table 1).

Table 1
Most Probably Agents for Biological Weapons

Bacteria	**Viruses**	**Toxins**
Anthrax Brucellosis Cholera Glanders Plague Tularaemia Q-fever	Smallpox Venezuelan equine encephalitis Viral hemorragique fevers	Botulinum Staphylococcus enterotoxin B Ricine T-2 mycotoxin

Taking all aspects into consideration, I consider the following four agents the most probable choices for biological weapons: anthrax, plague, smallpox, botulinum. When infection is obtained as aerosol, lethality with smallpox is 30% and anthrax over 80%. These two are easy to cultivate and very durable. Plague and botulinum toxin are less likely; their large-scale dispersion is cumbersome.

The Relative Advantages of Bioweapons for Terrorism

Although the number of international terrorism attacks per year has been decreasing, their lethality has increased during the 1990s. The new types of terrorist groups seem to aim towards causing larger and larger numbers of innocent deaths to get greater coercive effect and to undermine the morale of the society. Biological weapons have a number of properties which make them at least theoretically very suitable for terrorist use.

Bioweapons are cheap to prepare, easy to move, easy to hide while crossing borders. Theoretically, they can be extremely effective. One gramme of botulinum toxin as aerosol could *theoretically* kill 1.5 million people. However, practice is far from theory.

Bioweapon aerosols can be distributed secretly. They are not observable except quite near the nose of the spray. There is a latent period of a few days (except for toxins which act in a few hours) during which time the terrorist can disappear, even leave the country. The infected persons may spread widely, particularly if the infection was released at a station, harbour or airport. Much valuable time may be lost before it is realized that a pandemic is developing. In terrorist use this may be an advantage, but in military use it is a severe disadvantage, one of the main reasons that military commanders have not been enthusiastic about using bioweapons.

Virulent agents multiply exponentially. Lethality may be very high, almost 100% with inhaled anthrax spores or Marburg or Ebola virus.

Difficulties in the Use of Bioweapons

The use of microbe weapons is not so easy as their production. Spraying an anthrax culture would produce a mist in which the droplet size would be about 0.5mm, that is to say, about 100 times too big. To get the right particle size the bacteria must be carefully washed, lyophilisized, microencapsulated and provided with an electrical potential to repel each other.

Living agents (except spores) are sensitive to sunlight, dry air, too low or high temperature, pollution, rain, mist and other environmental factors which make their virility impossible to estimate. Some of these factors evidently ruined Aum Shinrikyo's attempts, described above, and also decrease the military value of all but thoroughly tested bioweapons, such as the Soviet ones. Therefore, the risk of terrorist use may be smaller than the US Government estimates.

Much has been written about the possibility that a terrorist group would develop, by gene technology, a pathogen, a variant of which would be resistant to antibiotics and vaccines and in some property better than the original. This is theoretically possible but would require professional expertise in gene technology and a lot of very dangerous laboratory and field tests. It is much more probable that for the time being terrorists would use natural pathogens such as anthrax or plague which are extremely efficient and durable in the environment. Every time a microbe is changed by gene technology it loses another property which was developed in natural conditions. But gene technology is a rapidly developing field and almost anything may happen in the future.

Protection Against Bioweapons

Protection entails several successive steps: prevention, observation, identification, sheltering, vaccination, medical treatment of casualties and decontamination.

Disarmament treaties, weapons regimes, and antiproliferation measures are used for prevention.

The task of intelligence is to find out the threat. Efficient customs and police control prohibit imports. Satellite surveillance gives information about bioweapon production. Large ventilation, fermentation, lyophilization, and so on. capacity is characteristic of biomass production and can give a hint. Personnel are experts in pathogens and may participate in international conferences. Phone, radio and e-mail communication can be followed.

Because of the stigma that bioweapons carry, the production plants are today built deep underground (Libya, North Korea). The USA is developing penetrating missile warheads specifically for the purpose of attacking such facilities.

International real time exchange of information between governments about the movement of terrorists would greatly help with protection against their activities.

Detection and Identification

The military have available instruments that can detect microbe aerosols in the atmosphere. The US Army has a laser-based Long Range Biological Standoff Detector System which can spot a particle cloud from 50km distance. It cannot identify the pathogen but gives time to put on protection. For identification of the pathogens, the US Army has the Biological Integrated Detection System (BIDS) which can identify 18 pathogens.

Smaller and more specific instruments based on gene technology are under development in several countries. From the human genome oligonucleotides corresponding to most common pathogens are put in order into thin grooves on a glass or silicon wafer. From the air current passing over it the pathogens present are each fixed to the specific antibody. This is verified by microscope or other counting instrument and the sequence is analysed by a computer which prints the list of pathogens present.

When such a detector has been sufficiently miniaturized and automated it can be used by individual soldiers and civilians. It will serve in case of natural epidemics, too, and therefore is well worth financing. But this will require some time.

Medical Protection

Vaccination and antibiotics can give protection in many cases, but must be given before the sickness breaks out. In the USA the whole military, 2.4 million, is presently vaccinated against anthrax. Vaccination would give good protection against smallpox, too, but smallpox has been eradicated from the globe and the vaccine for it is not produced any more. A hot debate regarding destruction of the two remaining type cultures (in the USA and Russia) is ongoing.

There exist good protective masks and garments against chemical and biological agents and in many countries the military and civil defence personnel are provided with them. They are very expensive and it would be impractical to provide the whole population with them. Children, for instance, would require many sizes and could not don the masks and set the garments properly. For the population civil defence shelters provided with adequate filters are necessary.

The decontamination of everything, particularly the equipment and environment after the bioattack, would be a complicated operation for which no community is prepared today.

The use of a bioweapon may infect thousands or even tens of thousands with a contagious disease. All these people must be collected into quarantine. The Spanish Influenza which killed 21 million people in Europe in 1918-1919 gives an idea of what a pandemic may mean.

The American Antiterrorism Campaign

In April 1995 President Clinton gave the Presidential Decision Directive 39 (PDD-39) which opened a great crusade against terrorism. It focuses on chemical and biological weapons but also covers the security of electronic networks. It is the President's answer to the savage terrorist attacks against US Government properties in the 1990s. The nerve gas strike of the Aum Shinrikyo organization in the Tokyo underground in 1995 may also have influenced the decision.

Clinton has succeeded in sweeping along the usually reluctant Congress so that, in 1999, $10 billion will be devoted to antiterrorism projects. There is no shortage of funds – the 1999 budget surplus will be $99bn, $20bn more than anticipated in January 1999. The 2000 surplus will be $142bn.

The antiterror campaign will comprise, among other things:

- analysis of intelligence information on terrorism by the CIA;
- planning of crisis management, co-ordinated by the Department of Justice and the FBI;
- co-ordinating repairs of eventual damages by the Federal Emergency Management Agency (FEMA).

The programme includes, among other projects:

- training emergency personnel in the 120 largest cities in the USA for chemical and bioweapon emergencies;
- providing airports with special detectors;
- vaccination of all US military personnel, 2.4 million persons, against anthrax;
- producing large emergency stocks of vaccines and antibiotics against biological agents.

Clinton has designated a National Co-ordinator for Security, Infrastructure Protection and Counter-terrorism, who acts through the National Security Council. Such an official is necessary when tremendous funds are channelled through 40 Government agencies to many hundreds of projects.

Devoting huge sums of money as a precaution against bioterrorism which has never been effectively carried out so far may seem an exaggerated response. But the fact that the new terrorism strives for high lethalities and biological agents seem theoretically the cheapest and easiest means to achieve this, may warrant such a precaution. The good thing is that all these efforts are useful in containing natural epidemics, too.

It seems odd, however, that so little attention is paid to the causes of terrorism. A terrorist who is willing to risk his life to his cause must consider it important. Many terrorists fight for political goals, in the Clausewitzian manner, for a better society. They do not accept the US view that the American democracy, free trade and market economy would lead the world to wealth and peace. They see the truth to be different: there are crises around the world – in Africa, south-west Asia, Indonesia, the near East, the Balkans and Russia, for instance. The division between the rich and poor is deepening. While President Clinton presents a vision of historical development towards a "happy global village" (dominated by the USA), peoples with a different historical and cultural legacy (Chinese and Arabic ones) consider the American hegemony threatening. It would not be threatening if it were cultural only, but if there is embedded an economic element, like domination of markets, it raises resistance.

The present development of the Third World seems to be towards poverty under the raw market economy. There still exists much to be improved in the world and terrorists are trying to do it in their way, unfortunately.

References

[1] R J Smith, "Clinton Directive Changes Strategy on Nuclear Arms," *The Washington Post*, 7 December 1997. J. Ehrlich, "New US Nuclear Policy Maintains Ambiguity," *Defense News*, 5-11 January 1998.

[2] Joint Chiefs of Staff, "Doctrine for Joint Theatre Nuclear Operations, February 1996.

[3] "Soil Sample is Key to US Missile Strike to Sudan," *Chemical and Engineering News*, 31 August 1998.

[4] Hans Kristensen, *Nuclear Futures: Proliferation of Weapons of Mass Destruction and US Nuclear Strategy,* BASIC Research Report, 98: 2, March 1998.

[5] V. Loeb, "A Dirty Business," *The Washington Post*, 25 July 1999.

[6] Presidential Decision Directive (PDD) 60, Department of Defense, November 1997. Classified, but partly reported in public.

[7] For a description see, for example, Biological and Toxin Weapons Verification Program, Federation of American Scientists, http://www.fas.org/bwc/papers.htm

[8] C. Piller, "Ultimate Vaccines: DNA – Key to Biological Warfare," *The Nation*, 10 December 1983, pp. 597-601.

[9] R. Preston, "The Bioweaponeers," *The New Yorker* (hardover ed.), 9 March 1989, pp. 52-65, http://www.jya.com/bioweapon.htm

[10] Medical Management of Biological Casualties. Handbook. US Army Medical Research Institute of Infectious Diseases, Fort Detrick, Frederick, Maryland, July 1998.

APPROACHING THE THRESHOLD: CONFLICT, INGENUITY AND ENVIRONMENTAL DEGRADATION IN MALAWI

Diana Cammack

This paper addresses the impact which resource constraints and environmental degradation is having on stability in Malawi and the capacity of Malawians to address these problems.

Conflict and Crime

Violent conflict, crime, and corruption are becoming more regular and more evident in Malawi, especially since the transition to democracy that culminated in a multi-party elections in 1994. A second election was held in June 1999 - widely believed by local NGOs and opposition parties to have been unfair - which saw Bakili Muluzi and the UDF win but by a very narrow margin. The President and parliamentarians were elected not for their stand on issues - for there were none in this campaign - but according to their regional or tribal origins, their religion, and their relationship with the late Hastings Banda (and his colleague, John Tembo).

The pre- and post-election period were marked by episodes of partisan, tribal, religious and regional-based violence, which saw party-affiliated young males attack leaders of the opposition and their rallies, their personal property, such as businesses and cars, and one another. Contributing to the tension was the UDF's claim during the campaign that the MCP (Malawi Congress Party) planned to use the MYP (Malawi Young Pioneers), supposedly still hiding in Mozambique, and arms imported from Angola and Vietnam, to overthrow the UDF after Muluzi won the election. No proof or arrests followed, though Muluzi continues to make these accusations publicly. His mood has continued to be non-conciliatory, unwilling to acknowledge that bias and ineptitude by the Electoral Commission led to a flawed election and that the opposition has some right to complain. At the same time a purge of persons who are not known UDF-supporters is underway in the top ranks of the Ministries of Home Affairs (for example, police), Education (Chancellor College), Information (MBC), Foreign Affairs (embassy staff), Justice (DPP), and so on, and on the boards of statutory corporations.

Meanwhile the opposition parties have taken the Electoral Commission to court, demanding that it overturn the election (on constitutional grounds and because of UDF rigging). The case is dragging on, while new evidence of UDF cheating emerges. The UDF Minister of Justice/Attorney General (and now acting DPP) has stated that regardless of what the court rules, Muluzi will remain President. To press the court to take up the case, the MCP has organized marches in Blantyre, marches that have been broken up by the police using tear-gas. These follow spontaneous riots in the north and in Blantyre in the days right after the election, during which time Blantyre opposition party supporters urged (chanted) the army "to take over" and rioters in the north burnt mosques and forced Yao Muslims[1] to flee for their lives. A roadblock where locals demanded passports from people driving north was purportedly established by opposition supporters at Nkhotakota shortly after the election, which symbolized northern frustrations and aspirations.

All of this politically inspired conflict serves as a backdrop to tension emanating from other sources. Crime has become markedly worse since the transition for a variety of reasons: the MYP, who served as the rural police force, were disbanded by the Army in 1993, and the small regular police force (5,500 people to patrol 10 million) is under-resourced and poorly skilled. The courts are blocked up, partly because there are too few government lawyers (20 when 80 are needed), and the prisons are overcrowded; torture and abuse were still being reported in these by the US Embassy in 1998. Arms (especially AK-47s and hand guns) are flooding into the country, mostly from Mozambique, as criminal gangs (some with links cross-border) hijack cars, smuggle drugs, and carry out burglaries and robberies. A number of shop owners, farmers, and expatriates have been injured and killed – which along with other problems (decaying infrastructure, corruption, and so on) discourages foreign investment.

Meanwhile people in the rural areas, where 85 per cent of Malawians live, complain that crime is so prevalent that women can no longer collect firewood or go to the grain mill in safety. Robberies are common, as small animals, crops, clothes, household goods, and even communally owned property like borehole handles and trees are stolen.

This is having a serious impact on the livelihood strategies of the already largely impoverished rural smallholders: for instance, people will no longer grow certain crops (for example, cassava) or keep animals (for example, goats) that can easily be stolen. People have started keeping their small animals in their huts with them at night. Also, they will no longer store their maize in household silos (*kokwe*) but put it under the bed in sacks. This is dangerous because of the fungus that grows on maize, especially hybrid, which the government is trying to get people to grow, if it is stored without fresh air.

Alternative policing is the result, which ranges from sophisticated security services (with vehicles and radio networks, mostly for large estate owners and the expatriate elite) to spontaneous vigilante gangs, which beat and sometimes kill criminals. There were 105 incidents of "mob justice" in the country between September 1998 and May 1999. These arise in part because the normal rural police service is failing to catch, arrest and keep criminals in jail. Instead, they are sometimes turned in to or caught by the police, but bribe their way out of jail. People also complain that the chiefs who sometimes hear cases are not honest and show preferences for people who bribe them.

Democracy is also blamed: the police can not longer detain criminals without charging them as before, so if evidence is not found and charges are not pressed, villagers complain, the suspects are allowed to go free, to retaliate against those villagers who have turned them in. Secondly, people blame democracy for allowing people to be "free to do whatever they want" without any of the social responsibilities that normally pertain to democratic life. A decline in the "moral authority" and powers of the chiefs continues, partly as the traditional authorities are now paid by the government, instead of being dependent on the patronage relationship with the people; because their powers over resources, especially land (as it now is in scarce supply and control of it is retained by families) have diminished; and because they are increasingly charged with being corrupt (for example, allocating land to outsiders for a fee, giving preference to people who pay them during litigation). Finally, democracy is blamed because the MYP no longer exist to absorb and train the large number of otherwise unemployed young men who live in the rural areas. Some villagers actually call for a return of the MYP to police the area.

Also in parts of the country, including in the high-density suburbs (townships) there have been formed what are called "inkatha" groups, which are comprised of young men who with or without police support, stalk and arrest local criminals. At times they have gone too far and have ended up in jail themselves. A wide range of institutions have thus been formed, some spontaneous and violent, some organized and efficient, to take the place of the police.

Corruption has also been democratized. Where before Dr Banda, Cecilia Kadzamira and John Tembo and a few cronies were allowed to mix government and personal funds, now "thousands have their hands in the till." Moreover, corruption reaches to the very top, which has set an example to the rest of the nation - the civil service especially but also businessmen - that theft is permissible and unlikely to be punished. The existence of the relatively independent press means reports of corrupt acts spread quickly, where before no mention was made of such events as Banda controlled the media. Some observers point to the fact that now, with a five year term politicians feel compelled to steal money early, as they do not know whether they will be re-elected. That, plus the absence of the MYP and of the fear that Banda inspired – even senior people were thrown into jail for petty theft – means that Malawi is becoming as corrupt as most other African states. Corruption takes the form of petty bribes by the public to get services performed by civil servants, to major theft and abuse of power: loans from banks, which are never repaid, to politicians to allow them to set up business in Malawi; cash to politicians to allow people to smuggle drugs and import goods without paying customs duties; kick-backs on road repair contracts, petrol deals, cell phone contracts, school supplies, satellite TV licenses, and so on.

Ingenuity, HIV/AIDS and Malnutrition

Malawi is one of the countries worst hit by HIV/AIDS. This is having an impact on the nation as a whole (15 per cent nationwide), but the most productive people in society – young and middle-aged adults, most of whom live in the cities and are educated, with disposable income and therefore opportunity to catch the disease – are dying (with a 35 per cent HIV positive rate in the cities). The boss of Press Corporation explained to me that no sooner does he get senior staff trained up and ready to take over key positions in the company than they become ill. He half jokingly said he has to hire three people to get two. The extended family structure can no longer cope with the number of orphans in the country, dependency ratios especially among already-impoverished rural female-headed households climb, which means farming is less productive and families get poorer, and the elderly become responsible for the very young. Family savings are wiped out by the illness and funerals. In other words, AIDS is making Malawi poorer at the micro- and macro-economic level. Worse still, AIDS is eliminating the most productive, educated people from society, the cleverest people, those who are needed to solve Malawi's severe economic, environmental and political problems.

AIDS is having an impact on the nation in other more subtle ways. For instance, what one epidemiologist called a "terminal mentality" is evident among many people, who have not been tested but assume they are HIV positive, who live for the present, try to maximize their wealth for their families – often through corrupt means – and pass on the virus with little regard for others or for Malawi's future. (One psychologist noted that the whole country is "clinically depressed", which could well be true as people are burying friends and relatives weekly.) Such attitudes do nothing to help Malawi address or solve its long-term problems.

Evidence also suggests that long-term malnutrition and recurring bouts of diarrhoea in infancy, both of which characterize Malawian society, are detrimental to innovation, creativity and problem solving. Empirical studies are needed to test this hypothesis in Malawi, but anecdotal evidence suggests that 60% stunting points towards the reason why, for instance, villagers "have no strategic approach" to solving their problems, as noted by the government's National Economic Council. [2]

Ingenuity, Social Psychology and Culture

Additionally debilitating are three sets of attitudes and behaviour that characterize Malawian society. Drawing on the work of Geert Hofstede[3], the social traits that once helped promote peace, conformity and social welfare in traditional society do not generate the type of ideas and behaviour needed to create technical and society ingenuity. These are: firstly, hierarchy and "large power distance" where inequality is accepted and where, for instance, decisions cannot be taken in the absence of authority and where initiative and volunteerism are stifled; secondly, conformity and collectivism, where exclusivity can exclude people who are capable and promote those who are incompetent and where jealous (*nsanje*) behaviour can ensure no-one aspires to rise above the mediocre crowd; and thirdly, conflict and uncertainty avoidance, where deviant ideas are not tolerated and where rules and order are deemed essential.

Malawi's neopatrimonial heritage [4] also takes a toll on the creativity of Malawians. In particular, still present to one degree or another are "presidentialism" (the concentration of power in the hands of a single person and weak civic institutions), "clientelism" (patronage network to reward loyal followers), and the use of state coffers to make these payments (through a bloated civil service and revenue overspends). Taken together, these undermine the institutions that are needed to generate or adapt ingenuity.

Resource Scarcity, Underdevelopment and Environmental Degradation

When Dr Banda came to power the population of the country was probably about a third of what it is today. With a growth rate of 3.2 per cent (fertility rate of 6.7) it is one of the fastest growing populations in Africa. (HIV/AIDS is taking its toll, though how significant is its impact is highly debatable [5] and will not be discussed here). Most telling: Malawi has the highest population density of any sub-Saharan African country with a single short rainy season, ranging from 74 persons/sq km in the north to more than 200 in the south. The average nationwide landholding is 0.86 ha (0.7 ha in the crowded southern region). Under such conditions, using primitive technologies, families are unable to feed themselves.

Land is a contentious issue because Malawians feel everyone has a right to farm or be buried on a piece of it. Some was allocated to white estate owners early in the century, but most estates are controlled by Malawians who acquired leasehold land and a license to cultivate and sell barley tobacco when that privilege could only be granted by Dr Banda. Estates cover 13 per cent of the land, while public lands (parks and reserves) account for 21 per cent, and customary land, 65 per cent – the later of which is home to 85 per cent of the population. Pressed for at least a generation some of this group has been farming unsuitable (sloped and fragile) land. Encroachment by the peasantry on both public and estate land is now common, and has the detrimental effect of degrading fragile watersheds and soils in the first instance, and forcing estate owners to keep land planted (not allowing it to rest for fear it will be encroached or expropriated) in the second. Soil erosion and siltation of the rivers and lakes – with serious consequences for bird life and fish – result. Soil infertility has become critical, and is a problem experienced by the peasantry in terms of reduced productivity and ultimately hunger. Unaware of how to address this problem organically, few people have taken up exotic techniques to improve fertility. Meanwhile, for a variety of macro-economic reasons having to do with credit, fiscal policy, transport, and so on, chemical fertiliser is now too expensive for most local farmers to buy.

Naturally, the land issue is politicized, as politicians promise the peasants reapportionment, but know that the well-managed larger estates (many owned by expatriates and foreign companies) earn the bulk of the nation's foreign exchange and use their land well, as do some of the small owner-occupied estates. Most of the medium-sized estates are less productive and have significant amounts of under-utilized land.

Over-exploitation of other natural resources is evident as well. Deforestation is a serious problem as over 90 per cent of the country's energy needs are met by charcoal and wood. Particularly hard hit are peri-urban areas, though deforestation in the north, where new land is still being opened, is fastest. Elsewhere people have run out of new land, and new forests, to exploit. Here they use agricultural waste, which should instead be ploughed back into the fields to improve fertility. Over-fishing is evident in the south-eastern part of Lake Malawi, and in some of the smaller lakes, as siltation, and the use of small-mesh nets and unsustainable techniques - resulting in part from de-mechanisation of the fishing fleet, which forces boats to stay nearer to the shore - have ruined fishing beds. Poaching in the national parks has impacted on wildlife populations (and therefore potential tourism revenues) as well.

Conflict over land is evident at all levels, but has not yet become a major political issue. For instance, chiefs and village headmen report that most of the cases they hear have to do with plots of lands – boundary disputes, inheritance cases, and so on. Their role is also to chastise people for cutting down trees, but this is difficult for them as they have few coercive powers and little support from the police. At planting season they must deal with fights over farm boundaries that arise, especially in areas (near estates or the lake shore, for instance) where land is particularly scarce.

There is, as noted before, pressure on estates and public reserves, as peasants in some places have moved onto lands and pressed (generally unsuccessfully) to have them formally opened up. Areas along the lake shore where chiefs and party officials have sold land to expatriates for hotels and holiday cottages are tense, as fishing villages want access to the beaches and lake and need land to farm. In a few cases MPs have taken requests for land reapportionment to parliament, but these are generally denied by the Ministry of Lands because the areas under consideration are fragile watersheds. Disturbances between different groups of fishermen have broken out, too, as a result of men from one lake encroaching on another lake. Land pressure has built up near the cattle ranch in the Shire river valley, an issue that became politicized as pro-Malawi Congress Party chiefs said they gave the land to Dr Banda while pro-UDF chiefs claimed it was forcibly taken. Fights along the Zambia-Malawi border have also increased recently, as villagers fight for resources and Malawians cross into the relatively underpopulated Zambia to cut trees.

The President's Land Commission's report, due out in 1998, is delayed, partly because it (reportedly) suggests that the land leased by mid-rank civil servants and under-utilized, should be turned back to the communal sector for the use of the peasantry, where public and large estate lands will be maintained. Such a decision is politically unpalatable. In a few areas conflict with estates has broken out as "squatting" has taken place – mostly in the south where land pressure is highest – and government has had to acquire some nearby land for resettlement. These conflicts - near the tea estates - inevitably take on racial overtones.

That the land pressure has not increased tribal tension has to do with the fact that ethnic groups are confined to particular areas and are not living side by side, competing for the same resources. Potentially, land resettlement could cause conflict if handled poorly. For instance, the Lomwe in the Thyolo-Mulanje area are hardest pressed for land, and they want the government to "give them back" the tea estates,[6] a move that would be nationally economically unwise. Moving them instead to, say, the relatively underpopulated north is out of the question for the Tumbuka and Tonga would not permit it. If they were moved to, say, the central region's Kasungu game park the villagers who were forcibly moved to create that park, and who are eager to get the land themselves, would not accept them. (Instead, tribal conflicts in recent months have been where two groups live side-by-side - as mentioned before, in the north with the Yao, and in Lilongwe, where the Chewa and Yao are fighting over burial grounds and political power.

Summary of Conclusions

This project began in 1997 when I asked why Malawi - in spite of its democratic transition, poverty, population growth, and natural resource constraints - has had relatively little conflict. I now believe that conflict and tension are increasing, but that they started from a relatively low base for the following reasons:

- Repression by (and fear of) the MYP and party youth during the Banda years crushed most dissent (and crime) and kept whatever arose secret. In other words, we have little true understanding of how much crime, tension and conflict (from whatever source) there was in that period and more research on this issue is needed.
- Probably resource/economic scarcity was not a serious problem until the early 1980s, when Malawi ran out of good land [7] and when access to land in, and food from, Mozambique was cut off by the war there. The arrival of 1+ million Mozambican refugees put an extraordinary stress on Malawi's land and resources.
- Most people were uneducated, illiterate, rural-based, and isolated from the outside world. They were told by Banda that they were relatively well-off and their expectations were dampened by censorship of news from the outside world. The arrival of the UN and NGOs to care for the refugees brought in new ideas and generated debates and changes previously unknown.
- The Malawian elite was bought off by land, licenses to grow barley, credit, agricultural advice, preferential pricing, civil service posts, and so on. Maize prices were subsidized, which kept staple food prices low and the countryside and urban poor quiet. As the economy declined in the 1980s frustration with Banda's policies grew, especially among the urban, commercial elite (and those in the MCP who were sidelined by John Tembo), who led the opposition revolt. It was at this time that

social unrest began.

- The desire for democracy united two-thirds of Malawi across tribal boundaries [8], though different tribal and regional coalitions voted for their own representatives in 1994.
-

The situation on the ground began to change in the 1980s and the 1993-94 transition resulted. It in turn contributed to further changes, which are now promoting conflict in the society.

- The democratic transition raised expectations, particularly among people who equated it with improved livelihoods.
- Democratic policies, including the elimination of the MYP, the introduction of universal primary school education, the decline of censorship and an increase in the spread of news, the end of detention without trial, and the introduction of other "rule of law" measures combined to "lift the lid" on repression. It is now easier to express and spread discontent.
- Divisions along tribal or regional groupings took on meaning during the 1998-99 electoral process, when people were no longer united by being pro-multi-party, but were divided by ethnically/regionally based leaders along regional and ethnic lines. No issues or cross-cutting identities (for example, class interests) unite people to the extent multi-partyism did. Leaders seem not to understand how the situation is different and more dangerous than in 1994 and continue to mobilize youth on the same regional/ethnic/party lines.
- All three avenues for releasing tension generated by environmental constraints (and long-term poverty) – addressing root causes through the application of technical and social ingenuity, opening escape valves through geographical mobility, and outright repression – have been evident since 1994.
- The second – urbanization and cross-border migration to Mozambique – is important but not vital. In fact urbanization (without urban job creation) stores up future problems for the present regime.
- The first – generating, adopting and applying ingenuity – comes in various forms, including Starter Packs (free distribution of fertilizer and seed), community-based resource management (in fishing, forestry, wildlife management, and so on), the creation of six different effective organic methods of raising soil fertility [9], and so on.
- The third - an abusive use of power - is just becoming evident, in the political field first, where the UDF is purging opposition party members in the civil service and ministries, threatening to ignore court rulings, attacking constitutional human rights guarantees (by undermining the Ombudsman) and so on. This effort will undermine Malawians' generation and use of ingenuity in that it will exclude groups and individuals who are needed to tackle the nation's problems.
- Meanwhile many of the root causes and symptoms of the nation's massive environmental and economic problems are poorly addressed, and conflict, crime, corruption, and social tension are likely to increase.

References

[1] Muluzi is a Yao and a Muslim. Generally they form the trading class outside their own area in the south. Most people in the north are Tonga and Tumbuka, and Christians.

[2] *Qualitative Impact Monitoring of Poverty Alleviation Policies and Programmes in Malawi*, 1999.

[3] *Culture and Organisation: Intercultural Cooperation and Importance for Survival*, 1991.

[4] Michael Bratton and Nicolas van de Walle, *Democratic Experiments in Africa*, 1997.

[5] The government's statistics office says that growth has dropped to 1.9% but is figures are questionable.

[6] History is being slightly re-written here because there were very few Lomwe in that region until the estates were opened at the turn of the century and began recruiting them as labourers from Mozambique.

[7] Elizabeth Cromwell and James Winpenny, 'Does Economic Reform Harm the Environment? A Review of Structural Adjustment in Malawi', *Journal of International Development*, 5, 6, 1993, citing Government of Malawi, *National Atlas of Malawi*, 1983.

[8] While the Chewa/central region formed the base of single-party/MCP/Banda support, not all Chewa voted for single party rule.
[9] Interview with Stephen Carr, Agricultural economist and WB consultant, 2 September 1999.

GLOBALIZATION - MIGRATION - INTERNATIONAL PARTNERSHIP

Annemarie Dupré

The term globalization appears more and more in the recent debate and analysis of world affairs. Whether the concept gets a negative or positive meaning depends on who is doing the exercise of analysing the so-called globalization process. Actually, the process is neither one nor the other. Globalization in itself is a neutral instrument of inter-linking at a global level, which can be used for different aims and objectives, and whether one can speak of a negative or positive development will depend on these aims and objectives. The evaluation also depends on the perspective from which one is looking at the phenomena. Often globalization is only considered an economic process, although it could also become a useful instrument to reach other aims at the global level. The whole debate has become a very hot issue, and often the discussion becomes emotional, with ethical or moral arguments being introduced too quickly, although these certainly cannot be overlooked.

One issue that is often discussed in the context of globalization is the increase of migratory movements. Also migration is neither a negative nor a positive phenomenon by itself. It depends on the causes of migration, on its quantity and its quality.

Finally, the third concept mentioned in the title, "international partnership", needs to be defined in the context of globalization. It could become a tool to avoid unbalanced developments within the globalization process, in that such a partnership could promote a regional and eventually worldwide collaboration and controlling functions.

In the following we shall look at some of these above-mentioned issues:

- globalization: aims; instruments of implementation; some facts and risks;
- increased migration, a consequence of unbalanced globalization;
- international partnership: solidarity, a specific form of partnership;
- some proposals.

Globalization

Globalization is a fact that cannot be ignored completely: international exchange, movements and communication exist. There are economic, political, scientific and cultural inter-linkages that cannot be denied. The process of globalization began with an internationalization of trade and commerce, but nowadays the dominating - and accelerating - factors are transport and communications, reducing distances and saving time. Borders lose more and more importance, but - and this is a major problem - this holds true only for those who have access to modern technology, especially in the field of communications and biotechnology.

Aims and Objectives

Various issues could become objectives of a globalization process. Three major groups of aims seem to be in conflict with each other (see below). On the one hand, there exists the wish to guarantee to everybody acceptable living conditions and a high quality of life, and on the other, a continuous growth of profit seems to be a must. Globalization is considered an instrument to achieve at least parts of these aims. It is a tool to achieve objectives at a global level. Whether globalization will become a positive or negative process for mankind depends on the objectives and on the possibility of promoting the different aims in a balanced way at the global level. Some of these aims are:

- Progress and growth of the economy and of profit: these seem to be a must for a vital economy. This demands a continuous expansion of markets and consumption. National borders are hindrances for such developments. Goods and currency must be able to circulate around the world without difficulties. Financial operations should not be hindered by national law or any other limitations.
- Equal development for all human beings is another aim. First of all this objective wishes to guarantee dignity and adequate living conditions to everybody. This aim is pursued by humanitarian organizations, governments, faith communities, and so on. Economists are looking at the issue in a contradictory way:
  - On one side the objective seems also to be in the interest of the market-orientated thinking. Only a high living standard will guarantee a growing number of consumers, which would mean an expansion of the markets.
  - On the other side there exists a tendency to accept, as an unavoidable fact, that part of the world's population will never have a chance to live in dignity. There are many who are convinced that the number of marginalized people will grow, and that, for economic and scientific reasons, nothing can be done to stop such a development. Some speak of a two-thirds world (2/3 will have access, 1/3 will be excluded). During the so-called Gorbachev Round in September 1995 the 20 to 80 society was discussed (this scenario foresees that in the next century only 20 per cent of the labour will be needed. 80 per cent of the world labour force be jobless).
- The conservation of the geo-biological system of the earth is another aim of many people in all parts of the world, and these people are starting to collaborate worldwide, getting organized in NGOs and lobby groups. Their concept of sustainable development seems to be in contrast to the objective of continuous growth, the latter mainly promoted by economists.

Access to Power

In order to reach these aims the main instrument seems to be power. Some time ago the struggle was for political and military power. Nowadays the struggle has been transferred to economical and financial competition for access to basic resources, and this struggle has become a highly destabilizing factor in world affairs. Over the years the priorities in this struggle have changed:

- Raw material: for a long period it seemed that power mainly depended on access to raw materials. The colonial period was dominated by this rush to ensure one's access to the big reserves of raw materials in various parts of the world.
- Later, access to energy became a major need.
- Another important instrument of power is the ability to control the production and distribution of food.
- Today, success seems to be linked primarily to leadership in the field of communication and information technology.
- The most recent trend is to collect as many bio-technological patents as possible. Knowledge and exclusiveness of such patents has become an instrument of power.

To know how to write computer programmes and the property of genetic codes has replaced the rush for gold or other precious raw materials. The one who has access to these instruments has power and will more easily reach his or her aim. At this moment access is mainly limited to those who have economic power, and this may produce a dangerous imbalance in the globalization process.

Actual Situation: Some Data and Facts

The report on the state of the nations of the United Nations Development Programme (UNDP) for 1999 concentrates on the issue of globalization. The report gives some important data and expresses strong apprehension about unbalanced development.

- 20 per cent of the world population controls 86 per cent of the "Gross World Product" and 89 per cent of all exports;
- today 80 countries in the world have a lower income *per capita* than 10 years ago;
- mainly well-educated and rich people have access to Internet; 88 per cent of all Internet users live in the industrialized countries;
- 80 per cent of all Web pages are in English;
- 74 per cent of all telephone lines belong to 20 per cent of the world population, mainly in the industrialized world;
- 97 per cent of all patents in the world belong to the industrialized countries;

The economic pressure and competition in the globalization process has resulted in policies which reduce jobs and worsen the working conditions. Child labour and irregular labour are continuously increasing. The unemployment rate is going up, and this not only in the so-called Third World. Also in the industrialized countries the gap between those who have access and those who are more and more marginalized is growing.

Risks of Unbalanced Development

The UNDP report warns about the risk of unbalanced development that could lead to high risks for mankind and the whole geo-biological system of the earth.

- The concentration of the financial power in the hands of very few multinational companies results in dangerous developments:
  - the holders of the financial power can easily evade national laws, and are able to dictate rules to national governments, for instance on "Standort" (location) conditions, on taxes, on environmental limits, and so on;
  - these forces are more and more backing out of their social and ecological responsibility.
- The national governments lose power and tax income, meanwhile more and more tasks are allocated to them:
  - as the multinational companies back out of social and ecological responsibility, these tasks are turned over to the governments;
  - the multinationals ask for improved "Standort" conditions; otherwise they threaten to move to other countries, which means loss of employment;
  - the multinational system makes it more difficult for the national governments to receive taxes from these bodies.
- The economic competition is expelling more and more of the labour force. The unemployment rate is growing continuously everywhere.
- International criminality is making good use of modern communications, better transport and the disappearance of borders.
- Growing fear and insecurity are allowing the advance of religious fundamentalism, ethnic and national conflicts. Civil war and violence are growing.

- The UNHCR is denouncing in a contribution of Yusuf Hassan Abdi ("Refugees" no.115/1999) a serious breakdown of human values. Similar apprehension can be found in the UNDP report:
  - the exaggerated role of profit and competition induces individuals to sacrifice human values, such as justice or human rights;
  - the millions of individuals who are marginalized by the economic pressure, without hope of finding work or at least having a minimum for survival, risk to lose, together with hope and dignity, all forms of human values;
  - many children grow up in a "desolate moral vacuum devoid of the most basic human values" in "which nothing is spared, held sacred or protected"; thousands of children serve as child soldiers, they are street children or grow up during long periods of civil unrest and violence; others are abused, used for criminal activities or for prostitution; millions have no access to any kind of educational services.

Awareness of these problems is growing, and the problems are discussed in various situations and at many levels, in particular the role of the international economic and financial institutions. The debate on the Millennium Round or the foreign debt crisis are priorities in the debate of the World Bank or the International Monetary Fund. The civil society is getting more and more involved in these economic debates (e.g., the Movement of Jubilee 2000)

These various risks of a purely economically-oriented globalization lead to various problems such as ecological disasters, growing poverty for large parts of the world population and in particular to a noticeable increase of migratory movements. The actual labour market needs a mobile labour force and large numbers of workers are moving from one country to the other, because the enterpreneurship is moving its production from one "Standort" to the other because of more convenient conditions. Due to the global economic competition, rationalization means reduction of jobs. The unemployment rate goes up continuously. At the same time the demographic development shows - despite big efforts for birth control – a still increasing world population, especially in the developing countries. This continuous increase is adding to the unemployment rate. This surplus of labour is producing an additional flow of migrant workers. The growing weakness of national governments, the advancing of political and religious extremism, growing ethnic divisions are producing civil unrest, violence and, more and more often, civil war. Large numbers of refugees flee from their homes.

The widespread loss of ethical values and the technical possibilities in the sector of communications and transport have increased the involvement of the international criminality in the movement of refugees.

Increased Migratory Movements: A Consequence Of Globalization

One of the most visible results of an unbalanced globalization process is the increase in migratory movements. In all parts of the world people are moving from one country to another. The decision to migrate may have different motivations. Each person has their own migration plan, which can be based on positive or negative reasons: some may move because of their specific skill, being highly qualified and getting interesting working contracts. Others move in order to survive.

Causes of Migration

Labour migration. The movement of people in order to find employment has become one of the major reasons for emigration. On the one hand, the mobility of labour is of great value to the economy. The global inter-linkage of all economic processes makes it necessary to dispose of highly mobile workers in order to remain competitive. The ageing and the low birth rates of the population of the industrialized countries may create "bottlenecks" in the labour market of these countries. Foreign labour will be needed (pull factors).

On the other hand, the continuous reduction of jobs and the global demographic development produce a dangerous increase in unemployment. In order to get acceptable living conditions or only to survive, large numbers of potential workers are moving in search of employment or to get a chance to earn enough to survive (push factors).

Family reunion. In most cases one member of the family - normally the most vital one - will emigrate alone, trying to find a sufficient income in order to support the whole family. The support of the family can then happen in two different ways: the migrant sends part of his/her income home, the so-called remittances that are an important factor in the economy of many countries. If the migrant succeeds in stabilizing himself/herself in the host country he/she will ask for family re-conjunction. The spouse, children and sometimes parents will join him in the host country. Family reunion is an important part of the total of migratory movements.

Refugees. The growing number of violent conflicts and ecological disasters, the loss of any chance to survive, are some of the main reasons for migration today, so-called forced migration. As indicated above the economic competition, the rush of multinational bodies and governments to secure for themselves strong economic positions, the growing gap between rich and poor are resulting in violence, civil unrest, persecution of scapegoats and religious and ethnic radicalism. The victims of these events who are trying to escape can be divided into various categories, all in need of protection.

- Conventional refugees: these are individuals who flee from their country because of personal persecution for reasons of race, political opinion, religion, and so on. These refugees are protected by the UN Geneva Convention.
- Refugees for humanitarian reasons: these are people who are not individually persecuted but flee from violence, war, catastrophes, very often in bigger groups or entire parts of the population of a certain area. The Geneva Convention, as it is interpreted today by most countries, does not protect these people, although they are in need of protection. For these groups better national and international legal instruments are needed. Good examples are the refugee definition of the Organization of African Unity and of the Catagena Convention.
- Internally displaced: these are refugees moving within their country of origin. For these, no international rules for protection exist. They are often particularly at risk without any real support.

Other categories of migrants. The first three categories are the biggest groups, but people move also for other reasons, for example, for tourism, study, health reasons, and so on.

Some Data

- According to estimates of the UN in 1990, ca. 120 million people were migrants. They were distributed on the various continents as follows: Europe 25.8 million (21.1 per cent), Africa 14.2 million (14.7 per cent), Asia 43.9 million (37 per cent), America 30.2 million (25.5 per cent), Oceania 4.1 million (4 per cent).
- In 1997 the estimated total number of migrants went up to ca. 150 million.
- The European Commission gives figures on demographic developments in the EU countries. Expectations until 2015: there will be a 16 per cent decrease of young people entering the labour market. There will be a 26 per cent increase in the age class of 50 to 64 year old persons. Seniors above 65 years will increase from 30 per cent to 40 per cent.
- In 1997 there was a total of 22,729,000 refugees, convention refugees and humanitarian refugees. These are only those who have got some kind of recognition.

- The number of those who ask for asylum or temporary protection is far bigger. In Europe an average 75 per cent of the requests for protection get a negative answer. That means that the real flow of refugees in search of protection is far higher than the official numbers of the UNHCR indicate.
- OCSE-Sopemi gives for 1996 the following percentages of migrants present in the industrialized areas, in proportion to the total population of the respective area: USA 9 per cent, Canada 17 per cent, Australia 25 per cent, European Union 4.8 per cent.
- There is no precise data for the other countries, but it is estimated that the numbers are higher than in the industrialized countries.

Because of restrictive migration and asylum policies all over the world the number of irregular migrants is growing rapidly. These restrictive policies have pushed many migrants into the hands of criminal traffickers. The international criminality is making enormous profits by smuggling refugees and irregular migrants.

Management of Migratory and Refugee Movements

In order to govern migration movements, the causes - economic, political, social and demographic developments - should be considered. Up to now most migration policies are only defensive, trying to stop the flows at the national borders. During the post-World War II reconstruction period migrant workers were welcomed. Since the 1970s the reverse tendency has taken hold: because of economic difficulties (for example, the oil crisis) and the globalization of economy and finances, unemployment rates went up in all parts of the world. Immigration no longer seemed to be desirable, meanwhile the emigration pressure increased notably for economic and political reasons. The various national governments of the European Union, for instance, reacted with purely restrictive border policies. Their main interest is to limit the influx of migrants and refugees and to control irregular immigration. Some concessions are made for seasonal workers. The various governments are reluctant to give up their national competence for this issue, although only a homogeneous policy on a regional level will be effective.

The dimensions of the phenomena will make it difficult to find a definite solution within a short period. In order to achieve first results some regulating instruments should be applied:

- Root causes must be tackled, especially the negative results of an unbalanced globalization process.
- New concepts for development policies should be elaborated and applied. In this context the economic importance of the remittances of migrants to their countries of origin should be considered. The currency flow coming from remittances stands second to oil, looking at the global flow of currency.
- A total closure of borders by the industrialized countries is illusionary. Such measures never have been successful. If the migration pressure is high, it is impossible to avoid irregular migration. Experiences at the Mexican border to the USA or movements in the Mediterranean Sea to Greece, Italy and Spain, and also at the eastern border of Germany, are a demonstration of the impossibility of closing borders. Such policies have in addition a very negative effect, as international criminal organizations will profit from the emergency situation of potential migrants, making considerable earnings from trafficking of human beings.
- Some positive measures could be:
  - an effective policy for all those in need of protection, eventually introducing rules on temporary protection.
  - a quota system for labour migration.

Such measures can only be effective if they are applied homogeneously in an entire region, and this requires new forms of collaboration and partnership.

International Partnership: Solidarity, a Specific Form of Partnership

The Need for Re-Balancing the Globalization Process

Currently, globalization seems mainly to be promoted in the economic and financial sector, and it seems that it is based on competition and a rush for leadership and exclusion. The numerous risks of the actual globalization process, indicated above, make it clear that the process is proceeding in a rather unbalanced way and is dominated by the economic and financial actors.

The report of the UNDF 1999 points to exactly this problem. The rush for increasing worldwide integration is dominated by a philosophy of profit and efficiency. The aspect of the human development and social protection has to be integrated into this development. "Globalization needs a human face."

The call for a balanced globalization process is not only a call in defence of humanity and humanitarian values. The distortions of an unbalanced development put the economic actors themselves and their programmes at risk. The growing instability of society in all parts of the world, the advancing poverty and the continuous ecological devastation will put a limit to economic growth. There will be fewer markets, fewer consumers, less stable "Standort" conditions, more violence and less capacity of governments to face the growing problems.

There is a need to accompany the economic model with a model of social behaviour, which would be able to defend human dignity and human rights, but also conservation of the geo-biological system. Such a model should become a tool to re-establish a correct equilibrium among the various aims of globalization. We need strategies and models capable of overcoming distortions and imbalances, where the dictate of the market and the financial powers are putting at risk human dignity, the survival of large numbers of people and together with it the economic system itself.

The UNDF report expresses this as follows: "It is necessary to put the human dimension again at the centre of the debate, in order to act for humanity and not only for projects". Such a debate and research for strategies working for these aims at an international or even global level need global forms of collaboration. These cannot only be based on competition and exclusive power but should adapt forms of partnership, which must be re-defined in the light of equal participation of all actors, and a consensus on a basic stock of human values and instruments to enforce these values.

A Special Model of Partnership: Solidarity

A special model of partnership, which in Western culture normally is called solidarity, should be considered. It might be a useful instrument if the model could be adapted to the needs of, and the differences between, the various cultures, and if one could develop a consensus building system, which would allow equal possibilities of participation and respect of all actors. To start with, such a system would have to be limited to certain larger regional areas (for example, the Organization of African Unity, the European Union, American co-operation systems, and so on). However, the final objective should be a global system, for example, the re-thinking and re-organization of UN bodies or global economic institutions.

1. In the Western thinking the concept of solidarity has been developed in order to face social problems, such as inequality of social partners, problems of access to resources, and support in specific emergency situations. Two ideological streams contributed to the defining of the concept:

- Christian teaching
- the culture of the working class and the left wing ideology.

In recent European history the idea of solidarity has not been a dominant factor. There have been important episodes of solidarity, but they have been rather few and far between and limited in time and number of people involved. In situations of competition of economic and political powers the idea of solidarity had very little chance. A philosophy of radical liberalism, the various forms of totalitarian regimes, or the dogma of the free market system do not favour models of human cohabitation in solidarity. These models are based on uncontrolled competition, on the dominance of the stronger and the marginalization of the looser.

Even if the industrialized states took the concept of solidarity into consideration, this happened in a Europe-centric way, adapted to the needs and the thinking of the so-called "First World". This resulted in a reduction of the idea, losing a great part of its potential possibilities:

- solidarity has been reduced to a paternalistic form of assistance;
- a division of roles has been introduced: the role of the donor and the role of the assisted, an operation which has practically abolished the basic criteria of solidarity, that is to say, the equal status of all actors within a social system.

The donors decide on the aims and strategies of assistance. The assisted become mere receivers of measures of assistance. The donor decides on priorities and needs. The assisted has to adapt to these decisions in a process of assimilation.

2. A new concept of solidarity, to be acceptable at the global level.

Particularly in the sector governing migration processes and their root causes, that means migration and development policy, this interpretation of solidarity is still dominant.

In order to re-balance the globalization process, and in particular to find a more effective approach to handling the growing gap between rich and poor and the increasing migratory movements, ethical values will have to be re-thought. A redefined solidarity concept may become a useful instrument in this situation. The human dimension has to become part of the debate on globalization, but avoiding "Euro-centrism", keeping in mind that this issue has to be looked at from a global point of view. Such a rethinking of human values and in particular the concept of globalization needs to respect certain criteria:

- Exaggerated forms of nationalism, ethnic pride or religious fundamentalism must be set aside. It will be necessary that the various states renounce a part of their sovereignty, in order to allow international institutions to control the correct application of human rights.
- Instruments and mechanisms to impose the respect of human rights must be introduced in a system of global governance. The numerous violent conflicts in all parts of the world, recently in the Balkan area, have made it clear that military actions by single states or a group of states hardly can stop armed conflicts and cannot protect the population against the violation of their basic rights. The functioning of the international institutions will have to be revised, probably first regionally and later possibly at a global level.
- New forms of correct partnership are needed and must be activated long before a conflict may degenerate into violent confrontation. Instruments of conflict management need to be introduced effectively.
- A consensus on some basic values must be found at the global level, respecting the cultural values and traditions of all partners. There are very different approaches to the theme of human rights, but cultural differences and needs should not become an excuse to allow discrimination or mistreatment of any group of human beings.
- Access to the decision-making process must be guaranteed to all partners. In order to reach some forms of global governance for the globalization process all partners must participate actively and responsibly in the decision-making process. Such a process must allow an analysis of the various interests and ways of consensus-building must be found. All forms of "Euro-centrism" or dominating attitudes of the so-called "First World" need to be reduced or avoided.

Only if governments, multinational economic forces and civil society are capable of finding such new means of real international solidarity, is there a chance to reduce the damage which has been caused up to now by an unbalanced globalization, and to avert the future damage that will otherwise ensue.

Some Proposals

Proposals of the UNDP report 1999:

- a global forum which brings together economic forces, trade unions, NGOs and religious institutions in order to allow a dialogue on possibilities to enlarge global governance, and to give access to rich and poor at the global decision-making processes;
- a mandate to the World Trade Organization (WTO) to introduce a code of conduct for multinationals and new anti-trust rules;
- the empowerment of poorer countries in the decision-making process of the WTO, introducing a kind of independent "Ombudsman", to defend the interests of these countries; the resolving of conflicts by this Institution can only be correct and successful if rich and poor countries have the same access to services and experts who will debate the case;
- service units for globalization issues with high quality expertise should be provided to all developing countries in order to improve the national policy on this problem;
- minimum standards on labour conditions and ecology at the regional level need to be introduced in order to give the developing countries collective power for negotiating at the international level and to support them in the global competition.
- some practical examples: the various Euro-Mediterranean Agreements should consider these proposals; the same criteria should be applied for the renewal of the Lomé Treaties or other agreements between industrialized and developing countries; an important issue is the problem of foreign debts of the developing countries.

Specific proposals on migratory policies:

- a regional homogenous migration policy, which accepts the concept of international solidarity and burden sharing; before reaching a global policy regional efforts should be made; from a European prospective the competence of the Institutions of the European Union need to be enlarged; Europe needs to respond to its international responsibilities in the globalization process;
- improved international rules for temporary protection and the internally displaced;
- minimum standards for the protection of "boat people"; many refugees and migrants try to reach safe havens using unsafe boats.
- a homogenous quota system for labour migration, which foresees a certain influx of migrant workers to the industrialized countries;
- the ratification of the UN Convention for the rights of migrant workers and their families, 1990.

ILLICIT ARMS TRADE IN THE GREATER HORN OF AFRICA AND ITS IMPACT ON DEVELOPMENT

Jacob Kamenju

Africa is awash with small arms. Most of these arms are available as a result of the demise of the Cold War in the early 1990s. During the Cold War, the superpowers competed to make friends and allies with the "non-aligned" nations, notably the Third World in general and Africa in particular. They armed governments, insurgents and all to further their interests. After the Cold War, conflicts continued in many African countries; for instance in Angola, Ethiopia, Somalia, and so on. Thus, the "First" World and mainly countries of Western Europe who manufacture these weapons had a ready market for their arms, ammunition and mines. For these manufacturers it was business as usual. Arms industries continue to thrive.

At the end of the Cold War, the USSR broke up, bringing forth new republics which were ill-prepared to exist on their own and some of whom had sophisticated arms industries. These industries then became the major source of government revenue and employment. Arms manufacturing was strengthened with total disregard to the effects on security and development of the recipient countries. Added to this category of arms manufacturers were the two African countries of Egypt and South Africa whose arms are in circulation in the Sudan and the Great Lakes region. The proliferation of small arms continues to date.

The dangers of large numbers of small arms in societies cannot be overemphasized. The United Nations estimates that over 95 per cent of world deaths in armed conflict since 1992 have been caused by small arms. The scourge of Aids and other diseases, accidents and other kinds of death form a small percentage of the toll. Thus, the proliferation of small arms must be brought to the fore and concerted efforts made to control and manage their use.

The study of the proliferation of small arms is still in a nascent stage in our region and needs the cooperation of all governmental and non-governmental actors/users in research and other endeavours to find out the dynamics, sources, quantities and so on of this problem. Currently, no-one can give the figures of small arms doing the rounds in this region. Estimates available from a 1996 survey indicate that there were over two million small arms in circulation within the seven IGADD countries. This figure is considered conservative and the actual figure could easily be double or even triple this. There is no data available for the Great Lakes region.

Proliferation of small arms is an international problem and cannot be confined to individual states. Our borders are too porous and any attempt to confine the problem within states will come to nothing. The international community has recognized this and has taken action to initiate, facilitate (to a small extent) and co-ordinate small arms issues in furtherance of the international struggle for peace and justice.

Sources of Small Arms

In the Greater Horn region most of the small arms can be traced back to the Cold War when the Russians armed the Somali government but later abandoned it for Ethiopia. All the weapons left behind were unaccounted for and perhaps form the bulk of the current number of firearms in the region. The failed American intervention in Somalia saw an increase in the number of weapons in the wrong hands through stolen and captured arsenals. The fall of Ethiopia's Mengistu Haile Mariam in

1991 and the escape of Somalia's Siad Barre in 1992 left behind armies in disarray whose weapons (380,000 and 75,000 respectively) are still unaccounted for to date. Earlier in 1986, the defeat of Tito Okello's government in Uganda and take over by Museveni's National Resistance Army saw an increase of an estimated 120,000 weapons into the wild arsenal of illicit arms.

In the Great Lakes region the picture is very much the same. The instability in Rwanda, Burundi and the Democratic Republic of Congo with their complex ethnic rivalries has resulted in the formation of rebel alliances thus creating demand for both licit and illicit arms. This scenario has resulted in large numbers of refugees, many of them defeated ex-combatants who may still possess or trade in small arms and use the same for their survival in hostile countries. Most countries in the region see refugees as parasites and destabilizing. The Tanzania/Burundi border is an example of such conflicts. The locations of refugee camps in Kenyan remote areas like Kakuma is another.

The ongoing 17-year-old civil war in the Sudan, the 15-month Ethiopia/Eritrea conflict, and the anarchy and inter-clan rivalry in Somalia are situations undoubtedly conducive to increases in firearms. Uganda is said to aid the South Sudanese Peoples Liberation Army (SPLA) in its fight against the Government of Sudan (GOS) while the GOS hosts and arms the Lord's Resistance Army in their campaign to oust Museveni. This means more arms in the region. A similar situation exists on the Kenya/Somali border (closed since 22 August 1999) where the American intervention of 1992 left many arms with the warlords. These have found their way into the regional conflicts and caused instability in neighbouring countries; for example, the American Army's colt pistol is perhaps the most used weapon in robberies and car jackings in Nairobi.

Another major source of weapons is found in the governments in the region that arm specific ethnic groups to protect their clandestine interests in the name of self protection and the protection of livestock (also known as home guards). These same arms are used for cattle rustling, banditry and outright harassment of neighbouring communities. For example, the Pokot and the Turkana of Kenya, the Karamanjong of Uganda, the Toposa of Sudan and the Dongilo of Ethiopia. Though these weapons can be termed licit by the issuing authorities, their use is outright illicit. These communities are pastoralists and cattle rustling is their age-old activity. But, the traditional spear, bow and arrows have been replaced by modern automatic weapons thus increasing the loss of life manyfold. For example, in raid in October 1992, more than 200 Turkana men, women and children were massacred when a Toposa raiding party attacked them in Narus in Southern Sudan where they had migrated to escape a dry spell.

Legitimately held weapons have found their way into illicit use; for example, in Zaire (now DRC) Mobutu's soldiers went for long periods without pay and had to resort to the use or sale of personal weapons for survival. Security personnel in the region are also known to hire out their weapons for criminal activities due to poor remuneration. (This situation is made worse due to the increasing numbers of retired security personnel who are ill-prepared for resettlement.) Recently, Somali refugees from southern Somalia sought entry into Kenya with the help of the United Nations High Commission for Refugees (UNHCR). Unknown to the UNHCR, were reports of the male fork closing the border under the cover of darkness, hiding their weapons and thereafter posing as family members seeking entry into Kenya as genuine refugees.

It is estimated that the arms trade in the region runs into billions of dollars a year. This money could be used for more needy areas in the social sector such as health, education, and infrastructure development.

Effects of Illicit Arms on Development

The presence of firearms in society causes insecurity to the members and their property. It influences the methodology of conflict resolution of having to use force as a more viable option instead of a negotiated one. The proliferation of firearms, like violence, grows in an upward spiral. A firearm is countered by a firearm. This causes a vicious circle where individuals and communities continue to acquire more arms for their own "protection." This, in turn, causes more refugees and internally displaced persons (IDP). These persons will be competing for resources for survival and may engage in illegal activities to make ends meet. A case in point is the availability of firearms in some parts of

Nairobi which are sold mainly with Somali connections. A survey conducted recently by one of the leading daily newspapers established that you could buy an AK-47, in Nairobi, within 30 minutes at Kshs10,000 (approximately $130). In the North Eastern Province of Kenya you could purchase the same for the price of a goat – $10.

Economic activities are greatly reduced due to insecurity caused by the illegal use of firearms. For example, in Nairobi, businesses in the city centre and its environs close by dusk for fear of armed robberies; a time which is most convenient for workers to do their shopping on their way home. Business is thus lost and most people have to find time, in daytime, to do their shopping; time that would otherwise be used to make more wealth and/or in other productive activities. In Bujumbura, the capital of Burundi, the main market which is within the city centre closes its gates at 3.00 p.m. to ensure people get home before it gets dark due to insecurity.

Private security takes a good percentage of an average family's income. Most homes in urban areas have either privately hired guards, communal ones, vigilante groups or residents take turns to look after their security, especially at night. Homes have become like prisons with sirens, heavily grilled windows and doors, high stone or metal walls, electric fencing, sensors and alarm systems. These items cost a lot of money thus reducing a family's standard of living, cutting on domestic savings and in effect make a long-term contribution to poverty.

The security services sector has become a major industry in the region. In Nairobi alone, there are over 200 registered security companies. These range from international ones like Securicor to some little ones comprising a handful of guards. The top 20 of these have an estimated turnover of Kshs20bn ($27m) annually. This is a colossal amount by any standards and a big share of this would be saved for better use through the removal or better management of firearms and accompanying improved state security.

Some companies are no longer paying salaries from a joint payroll anymore due to armed robberies during pay-days. The result is reduced working hours because such employees have to queue for long periods in banks. The companies reduce the risk of losing the money to robbers but suffer reduced production due to the time taken to go to the banks.

Armed robberies have become the order of the day in the region. Investment, both local and foreign is greatly reduced due to increased incidence of armed robberies. Many investors in the region are averse to high cash turnover businesses like distribution because they have become targets for robberies.

Armed car jackings have replaced what used to be "stolen cars" business that extended from East Africa through Central Africa to South Africa. This change has occurred due to the availability of cheap firearms. This menace is increasing fast. For example, in Nairobi in August 1999 the reported car jackings hit a new peak of ten in one day; the highest ever recorded. This means reduced interest and business in luxury cars especially the four-wheel-drives which seem to be the most coveted by the robbers.

In the rural areas, armed cattle rustling is hampering livestock breeding as the pastoral communities, especially the small ones which cannot afford to protect themselves, are giving up the rearing of cattle, goats, sheep and donkeys as the larger ones will definitely rob them. Some of these communities are favoured and armed by their home governments under the guise of home guards. In some rural areas of Kenya, unless you belong to the major local communities, you may not keep animals for they will be taken away anyway. This is a guarantee of less investment by the communities due to the uncertainty of the security of ownership of such animals.

Many schools in the region have closed down due to insecurity. This has a long term effect as it increases unemployment among teachers, puts more uneducated and unprepared people in the streets and creates tension in society. The result is manifested in the number of street families. The street children of yester-years have come of age and have families whose shelter is what the street can offer. This is a time bomb and a fertile resource which can be easily converted into armed robbers or insurgents.

The insecurity in the region also means less food production activities and reduced humanitarian access to those running away from conflicts. The result is malnutrition, starvation and desperation.

Conclusion

The Greater Horn of Africa is a region full of conflicts ranging from outright wars (Ethiopia/Eritrea), fully-fledged civil wars (Sudan and DR Congo) to anarchy (Somalia) and intervention by Uganda, Rwanda and Burundi in DRC. These create demand for both licit and illicit firearms. Refugees, many of them ex-combatants, add their personal weapons into the cycle. The region is awash with all kinds of small arms from all over the world. The dynamics of these weapons flows have not been clearly established. There is, therefore, a definite need for research in the dynamics of this trade, to find out the source and number of these arms and the routes used by the arms merchants into and out of the region.

The region has been left behind by Southern and Western African countries which have established ways and means of reducing the arms in circulation and managing those in the field. There is need for Eastern and Central Africa region to catch up with the rest of Africa and indeed the rest of the world since the problem of small arms is a global one. There should be urgent concerted efforts to manage, limit and control the number of illicit small arms in the Greater Horn Region.

The Pugwash Conferences should come up with a plan to support the study of the small arms problem in the region with a view to raising the level of research work and the availability of information to be on a par with that in Southern and Western Africa. This will mean involving research groups in the region in Pugwash activities and assisting these groups in capacity building, training and modelling a reliable and efficient system of an international standard.

SECURITY AS AN ISSUE WHEN ADDRESSING THE REFUGEE CRISIS IN AFRICA

Lucy Mulli

The African continent is currently playing host to 7,000,000 refugees.[1] The root cause of this huge number of refugees has been the rising conflicts in the region, which have seen the movement of people from their homes to neighbouring countries. Provision of humanitarian assistance is normally regarded as the key factor when aiding refugees. This perception is a logical one, in that, when refugees are camped in an area this is regarded as a temporary situation, hence the immediate problem is one of providing food, water and medication, the understanding being that once the conflict de-escalates they will return home. This has not happened. Conflicts are taking longer than perceived to be resolved. And it is in this prolonged gap of time that refugee camps pose a threat for the host country, the mother country, or even both. Hence, focusing primarily on humanitarian needs is no longer adequate; the security threat that refugees pose is fundamental.

Security Threats

The first aspect of insecurity is that posed to the host country. In most countries in Africa, it is quite common to have groups of people who share the same language and culture but are divided by a borderline. In the event of war, when people flee from one country to the other, instead of settling in the refugee camps, and in view of the abject conditions of these, people have become illegal immigrants in the host countries. Since they are the same people as the inhabitants of the new region it is hard to detect them. These immigrants, having no means of support, may resort to crime and in so doing create insecurity in the new country. Further tension is created between them and the local population when the latter have felt that refugees are getting better treatment then they, the residents.

In Northern Kenya for instance, Somalis from the former republic of Somalia have left the camps and passed themselves off as Kenyan Somalis. To survive, they have sold the weapons they carried out of Somalia, thus creating an influx of arms into Kenya. Tension has also arisen between the refugees and the local population, especially during periods of drought where refugees have been better catered for than the local population who have had to struggle for sustenance. While these are issues of concern there have been other cases where the tension the refugee presence creates has affected state stability on a higher level then the micro-level referred to in the case of Northern Kenya.

In 1994, fleeing from Tutsi attacks, 1.2 million Hutu refugees fled Rwanda to Eastern Zaire. Some of the refugees were not ordinary civilians but former militias. For two years they lived in refugee camps fearing reprisal from the new government in Rwanda. During the two years the militias were the two controllers of the camps and not aid workers. Indeed, humanitarian assistance provided was used to buy weapons and other military supplies to carry out attacks on the governments in Rwanda and Burundi. And with the help of the then Zairian forces they attacked the local Tutsi population. By October in that same year the Zairian town of Uvira in Kivu was in a state of chaos. The Banyamulenge (local ethnic group) were fighting both the government forces and attacking the Rwandan refugee camps. An already bad situation got worse when the Rwandan army engaged Zairian forces and Hutu militias[2]. Through this illustration we see the various threats posed by the entry of refugees into a country.

The first threat is that posed by refugee camps as a shelter for the perpetrators of criminal activities in their home. The Rwandese warriors who entered Zaire were said to be about 200,000 belonging to the Forces Armees de Rwanda (FAR)[3]; within a mass of 1.2 million people they would have been hard to detect. But even assuming that there had been fewer people arriving, the refugee camp norm of accepting all people on a *prima facie* basis, makes such detection very difficult. And once militias have entered the camp the probability of politicization inside refugee camps is only a matter of time. And soon the refugee camp becomes a point from where they can regroup and carry out counter-attacks. The refugee camp in this case is transformed into a source of instability for both the host and native country.

An alarming trend has also emerged where refugee camps have been used as recruitment grounds for guerrilla movements in the home country. An inter-press report reveals how Farouk Saleh Mohammed Abdallah, who defected from the Sudanese People liberation Army (SPLA), described how he trained child soldiers in Dima refugee camp in Ethiopia around 1990. He says "we collected children and we trained them on how to fight, read and write". After the change in government in Ethiopia in 1991 they moved to Uganda and Kenya and Southern Sudan he continues, "...these refugee camps are not what the world is made to understand, but are in fact a reserve for SPLA recruitment..."[4].

Clearly, when addressing the refugee crises we have moved beyond the problem of addressing humanitarian needs. A new problem – security or lack thereof has arisen.

Security concerns along with humanitarian concerns need to be taken into account in camps. The first suggestion is that refugees be moved from the border of the country. When refugees enter a neighbouring country, they will normally settle along the border. Of course, this is only natural since it is closer to home, but moving them further in-land should be considered, both for their own safety and that of the host country.

Apart from relocation, protection and monitoring of the said camps should be undertaken. An impartial international security force should provide for this operation. In reference to the usefulness of these, a UNHCR spokesman in the great lakes region during the above-mentioned conflict said "... we asked for an international force to intervene to separate the refugees from the criminals, what we got instead was a security contingent made up of ex-FARs (Zairian armed Forces) which was a poor substitute"[5].

The added economic burden of such measures is, of course, a real one. Indeed, as the United Nations High Commissioner for refugees told the Security Council in July 1999, "...only 60 per cent of her office's $137 million budget for Africa had been funded so far this year. In contrast, the United Nations refugee programme for Kosovo had received commitments for $265 million, including $50 million from the United States alone"[6]. It appears that there is a fatigue when it comes to helping solve Africa's conflicts, perhaps because they do not seem to de-escalate and at times they even increase. By containing the refugee problem within a particular geographic area and overseeing the movement of the refugees we will go a long way towards ensuring that the conflict does not escalate, and thus solve the problem in the long term. Providing refugees with occupations will also detract from the temptation of engaging in dubious activities.

Conclusion

The premise of this paper is that the refugee crisis in Africa is an added dynamic to conflicts. Thus there is a need to address the problem not simply as a humanitarian problem but also as a potential security threat. As such, security measures such as sending security troops along with the humanitarian workers and moving refugees away from the region of conflict will have to be considered.

References

[1] President Mkapa "visiting Sweden urges assistance for refugees" Reuters report

//c:\windows\profiles\marna\Application data\RBB\RB18793.htm

[2] Account drawn from ; Thompson Stephen "the Crises in Eastern Zaire" World challenges groups Peace and Justice http//:www.angelfire.com/ab/worldchalleges/cz.htm

[3]Adelman Howard "ethics of Peacekeeping and Humanitarian relief; A pragmatic versus a principled approach- the Case of Zaire" http//:www/stanford.edu/group/CISAC/test/reachung/summerinstiute99/reading/adelman2.txt

[4] Children of war –media abstracts, Sudan http//:www.rb.se/chilwar/ett_07/media.htm

[5] Kaigamba Jean Baptiste "Hiding Killers behind a UN shield" http//:www.mg.co.za/mg/news/97june-refugee.html

[6] United Nations: Politics-Africa- "UN voices optimism after Algiers summit" Reuters report ://C:\WINDOWS\profiles\marna\ApplicationData\RBB\html\RB18793.ht

FIREARMS PROLIFERATION AND VIOLENT CRIME IN MALAWI

Undule Mwakasungura

Levels of crime in Malawi, especially violent crime, have increased dramatically in recent years. A high proportion of these violent crimes involves the use of firearms, circulation of which has increased. The majority of them are armed robberies and the targets are not only business premises but also individual private homes. In many instances, these violent crimes have led not only to loss of property but also to loss of life. The most commonly used gun is the AK47 assault rifle. Other types of guns reported to have been used in the criminal violence are pistols and other types of firearms. Malawi has never been at war and therefore the AK47 rifles find their way into Malawi mainly through Mozambique. There is also evidence that more and more of the light weapons are entering the country via Zambia, from the massive stocks build-up in the Congo and Great Lakes area.

It is important to note that before the democratic change Malawi experienced less of the violent crimes than the country is going through now. This was because during the previous dictatorial regime there were institutions/organizations apart from the police and army that were also involved in providing security. The obvious one is the disbanded Malawi Young Pioneers, a former paramilitary wing of the Malawi Congress Party. The disbanding of the Malawi Young Pioneers, which was a good thing because of the atrocities they committed, brought about a break down in the artificial sense of security that was being sustained through terror. The disbanding of such institutions created a vacuum that requires filling with well thought-through security measures. The problem of security breakdown is compounded by people having no confidence in the police who have taken change of the security issues, because they equally participated in committing the atrocities. That is why it is important that the police win the confidence of ordinary people if they are to participate in bringing about security in this country.

The commitment the government has shown to restoring security is a welcome move. This commitment has been shown by the government taking some initiatives in enabling the police to win back the lost confidence of the members of the public. One such initiative is police reform, which started in 1995 with support from the British Government. The others that are relatively recent, such as the Malawi Community Policing Forum, are working at the Community level to stop the use and flow of firearms.

Proliferation of Firearms

The demand for light weapons is on the increase in the country. Whereas in the beginning the demand might have been basically a criminal one, for criminals to carry out violent crimes that have inevitably brought insecurity to communities, there is also an increase in demand for, and use of, light weapons by ordinary citizens for their personal security. If the situation is not checked, we may experience a further increase in the demand for and use of firearms, fuelled by further intensification of insecurity. Another problem that is likely to come about is that even the firearms intended for personal security might end up in wrong hands and be used for violent crimes. Already there have been incidences where security officers have either provided firearms for criminal use or have themselves participated in violent crimes. Unfortunately, light weapons tend to empower criminals of all types, which jeopardizes the safety of individuals and communities.

It is a known fact that proliferation of light weapons takes place in an easy manner once the demand has been created, be it political, criminal or military, because of some of their unique characteristics. They are by nature light, which facilitates cheap and easy transportation; they can be easily smuggled and concealed, across borders and within countries, using rudimentary forms of transportation. In the case of Malawi, there is strong evidence that the firearms are being smuggled from the neighbouring countries, most especially from Mozambique. The fact that Mozambique is awash with excess light weapons following the end of the civil war is well documented. The majority of the demobilized soldiers were only surrendering weapons of poor quality, keeping back those in better condition. For instance, one FRELIMO soldier gave the following account:

> We knew that guns make good business. So we kept the best for ourselves. I have sold some to dealers from Joni and I keep others for the future. The secret is to keep them in good condition. FRELIMO was never going to pay us for the years we were made to fight. We have to look after ourselves.

The above case confirms the idea that many of the weapons that find their way into Malawi are from Mozambique. Some sources estimate that some 1.5 million AK47s were distributed to the civilian population over the course of the civil war. These weapons are now in the hands of arms traffickers and criminals. Concealing light weapons is not difficult once they have been acquired; caches can be broken up and individual weapons hidden in either urban or rural settings.

Another thing that make proliferation of light weapons easy is that they can be very cheap, which opens up demand from a much larger cross section of groups and individuals. For instance, in Mozambique an AK47 complete with a couple of clips of ammunition can be bought for less than $15, or for a bag of maize. At these prices it is most likely that the majority of people own an illegal firearm. Furthermore, light weapons are durable and are easy to maintain and rarely break down.

The proliferation of light weapons in Malawi became more serious in 1995 when refugees from Mozambique sold or bartered AK47s to Malawians in order to have resources to resettle at home. It is believed that many of these weapons are hidden in Malawi and the police also confirm that cross border trafficking is still continuing.

Control of Firearms

The crucial fact about light weapons is that once control is lost and these weapons enter other networks, the bulk of them are irretrievable. Many gun collection schemes (amnesties, gun buy-back or exchange for services or equipment) in a number of countries, including Malawi, have been ineffective because they failed to gain the active support of the local communities, or because they were not associated with other measures to improve confidence in the police or security services. For instance, in Malawi there are no weapons collection or effective public reporting schemes. There was a Presidential amnesty for those handing in a weapon, but no figures for the final number of weapons surrendered under this scheme, nor information as to what happened to them. There is also a reward scheme (MK5,000, approximately $100) for information leading to the arrest of someone in illegal possession of a weapon – again, the reward appears not to be monitored and there is no way of measuring what impact, if any, either of these efforts have had.

Voluntary weapon collection schemes have their limits. For instance, in countries that have come out of violent armed conflicts such as Mozambique, ex-combatants avoided handing in their weapons. In the case of Malawi, with the deteriorating security situation, ordinary citizens may possess arms illegally because they have lost confidence in the capacity of the police to protect them and they are unlikely to hand in weapons voluntarily. Criminals and arms traffickers are equally unlikely to voluntarily hand in their arms.

These problems will be reduced if there are effective peace-building and development programmes and security sector reforms to improve confidence and reduce the incentive to hold arms for personal security or to make a living. Nevertheless, coercive operations by police and security services to confiscate illicit arms still have an important role to play.

Control of cross border arms trafficking is equally a problem. The police, customs and immigration officers lack basic resources. There are insufficient trained personnel with access to transport and communication. The police cannot adequately patrol the borders because there are numerous border-crossing points with Mozambique and Zambia.

With all these problems, it would appear that a new approach should be explored to combating proliferation of weapons and gun violence. The best long-term approach to combating gun-related crime requires:

- Local communities and police to work together in an ethical framework, building up public awareness and accurate, timely information of the dangers of unlawful small arms transfer and abuse; and
- Special regional cross border initiatives to detect firearms and other serious illicit trafficking.

ENDLESS NIGHT: CONFLICT AND CONFLICT RESOLUTION IN ANGOLA

Hussein Solomon

Since the 1960s, Angola's traumatized population has known nothing but the anguish, agony and anger of war. After an analysis of the current state of play in the conflict-ridden country, this paper notes that prospects for conflict resolution are slim, suggesting that Angola's citizens' nightmarish existence is set to continue.

Current State of Play

Between February 1998 and April 1999, the Union for the Total Independence of Angola (UNITA) has transformed itself from a guerrilla movement into a conventional armed force[1]. Two reasons account for this.

The first reason relates to UNITA's acquisition of sophisticated conventional weapons such as 6 Russian MI-25 combat helicopters, 6 MiG-23 aircraft, 50 tanks, 75 armoured troop carriers, major mobile artillery and hand-carried satellite telephones. Most of this military hardware has been purchased from the Ukraine and North Korea. These weapons are flown directly to UNITA-held air bases in Angola from the Ukraine aboard large former Soviet Ilyushin-76 and Antonov-124 cargo planes.

The second reason relates to UNITA forces also receiving help in "training and logistics" from its old ally, Morocco, as well as South African and Ukrainian mercenaries operating the new equipment for UNITA.

The money for the weapons purchased emanates from the illicit diamond trade. Since 1992, UNITA has earned $2.5–3 billion from sales of rough diamonds from mines it controls. Despite the Popular Movement for the Liberation of Angola's (MPLA) recapture last year of the main diamond-producing centres in Angola's north-west Lunda provinces, UNITA still held mines in the north of the country close to the Democratic Republic of Congo (DRC) border and around Mavinga in the south-eastern province of Kuando Kubango. UNITA has been able to circumvent a UN embargo imposed in 1998 banning the sale of diamonds without a government certificate. This UNITA has done by developing a cross-border network that smuggles diamonds to traders in Zambia, Namibia and the DRC, for onward sale to diamond cutting centres in Israel and Antwerp. In early August 1999, the UN Security Council finalized the formation of two panels of experts to investigate how UNITA managed to evade UN-imposed sanctions and to recommend appropriate action. If recent reports which allude to the fact that UNITA has been building up an arms arsenal for some time are correct, however, these UN panels may well be a case of too little, too late.

The acquisition of these new weapon systems has given UNITA a tremendous degree of strategic and tactical manoeuvrability. The acquisition of combat helicopters and fighter aircraft allows UNITA to challenge for the first time the air superiority of the Angolan Armed Forces (FAA). The range of these aircraft reportedly places oil-producing areas, such as Soyo, well within reach. This would mean that UNITA technically has the capability to destroy the oil revenue the MPLA regime depends upon for its political survival. Second, UNITA's newly-acquired airpower allows it to support its infantry and provide air cover for its artillery. Thirdly, UNITA also allows the rebel movement to use its aircraft for air reconnaissance and therefore to plan its next offensive.

UNITA's tanks, armoured troop carriers and major mobile artillery allows it greater flexibility and mobility which adversely affects FAA's ability to predict the next UNITA offensive. In the DRC, we have already witnessed how the rebel, Congolese Rally for Democracy (RCD) have used Iridium satellite telephones with devastating effect. Similarly, these telephones, in addition to contributing to mobility and flexibility, also enhance UNITA's leader Jonas Savimbi's command and control of different theatres of conflict simultaneously.

In response to UNITA's weapons purchases, the MPLA government has purchased several T-72 tanks. But this is unlikely to shift the balance of forces in Angola in FAA's favour. Other factors also militate against FAA. First, FAA has been too used to fighting UNITA as a guerrilla force and therefore specialized in counter-insurgency warfare. Now that they have to confront UNITA as a conventional force, they are unable to do so. Second, government troops are demoralized, poorly trained and often have to fend for themselves. For instance, FAA soldiers are not supplied with food rations and have to live off the countryside. In this context, desertions are commonplace. Third, FAA has poor intelligence on UNITA. Finally, there are severe policy disagreements between the MPLA government in Luanda and senior FAA officers on the ground. This comprises FAA's military operations. MPLA politicians in Luanda want the FAA to launch an offensive deep into UNITA territory to capture Andulo, Savimbi's birthplace, and Bailundo, Savimbi's headquarters and also the home of King Ekuikui – the traditional and spiritual leader of the Ovimbundu[2]. The politicians believe that the capture of the two towns would strengthen their hand at any future negotiations. FAA commanders, on the other hand, prefer to stabilize the perimeter of Huambo and Kuito rather than trying to take Andulo and Bailundo. This stems from the fact that the FAA's awareness at how difficult it is to destroy UNITA militarily. It is an awareness borne from bitter recent experience. In December 1998 FAA's offensive was decisively stopped at the Wama junction, north of Huambo.

The military superiority of UNITA is seen at various levels. In late March and early April 1999, UNITA besieged and shelled the towns of Malanje, 400 kilometres east of Luanda, and Kuito, 600 kilometres to the southeast. Casualty figures could not be obtained because humanitarian staff had been withdrawn. What we do know is that since the resumption of the fighting in late 1998, almost a million Angolans have been internally displaced by the conflict. Hundreds of thousands more are refugees in neighbouring countries.

In addition to the shelling and siege of Malanje and Kuito, UNITA, in an eclectic display of its conventional and unconventional warfare capabilities, has conducted hit-and-run attacks on roads in most areas of the country. Meanwhile there have been three attacks within 100 kilometres of Luanda itself. In one attack, an electricity pylon had been sabotaged near Catate to the southeast. Two other attacks were reported on the road outside Caxito north of the city.

Moreover, UNITA has also resumed mine laying and the destruction of bridges on the main roads of the northern Uige Province. The bridge over the Kuilo River, near the town of Sanzo Pombo, 110 kilometres west of the border of the DRC, has been destroyed.

UNITA has also captured the northern Uige town of Maquela do Zombo, which is situated on the DRC border. The capture of Maquel do Zombo holds great strategic significance for the rebel movement since it gives UNITA total control of all the frontier regions of Uige Province. This control has facilitated UNITA's diamond-smuggling operations as elucidated above and also the entry of recently trained UNITA recruits from secret training bases in the DRC.

A Regional Conflagration?

The spill-over effects of the Angolan civil war seriously threatens to engulf several central and southern African states in a regional conflagration. Already there are allegations that Luanda is responsible for the recent spate of bombings in Lusaka in retaliation for the fact that UNITA is also being supplied with war materiel from western Zambia. Although Angola and Zambia signed an agreement, facilitated by Swazi King Mswati III, which ended the war of words between Luanda and Lusaka, tensions continue to exist.

Meanwhile, FAA has intervened militarily in support of Presidents Denis Sassou-Nguesso and Laurent Kabila in Congo-Brazzaville and the DRC respectively in the hope of shutting down UNITA's supply routes and training camps. But this action on the part of FAA may be counterproductive since there are allegations that UNITA has forged strategic alliances with the Congolese Rally for Democracy (RCD) rebels in the DRC and their backers in the form of Kigali and Kampala. In addition, there are also reports that UNITA has forged an alliance with the "ninja" militia – supporters of former Prime Minister Bernard Kolelas – that is fighting Sassou-Nguesso's troops in the Pool region of Congo-Brazzaville.

From a conflict resolution viewpoint, as these conflicts in the various countries drag on, the conflict dynamics in each country become increasingly interlocked forming a single conflict system. Such a conflict system needs to be tackled holistically and in an integrated manner as opposed to piecemeal efforts at resolving any "national" conflict.

Towards a Resolution of the Angolan Conflict

Resolving the conflict in Angola is fraught with difficulties at three levels. First, who to negotiate with? The Luanda regime has refused to negotiate with Savimbi's UNITA, preferring instead to negotiate with a tiny breakaway group called UNITA Renovada which consists of some UNITA members in the Angolan National Assembly. This was a position reiterated by President Dos Santos at the Maputo SADC Summit in August 1999 when he stated that there would be no further negotiations with Savimbi's UNITA. In my opinion, the MPLA government has painted itself into a corner. By refusing to engage in dialogue with Savimbi's UNITA, the MPLA has no option but a military confrontation with Savimbi that it is losing.

Second, there is the problem of who would be able to act as an impartial mediator to the conflict? The United Nations has disqualified itself to act as a mediator by the UN Security Council declaring in March 1999 that Jonas Savimbi was the cause of the current conflict in Angola. This followed its decision a month earlier to close the UN Observer Mission in Angola (MONUA) and withdraw UN military observers from Angola. In August 1999, however, the UN announced its plans to boost its presence in Angola and open a new office to "explore effective measures" aimed at restoring peace. It is difficult to see how this new initiative on the part of the UN would work out given the fact that the modalities and mandate for the UN mission is still unclear. Moreover, if this current UN initiative is aimed to bring about peace between the belligerents it is difficult to see how the UN can act as a mediator since it has not recanted from its previous position – holding Savimbi responsible for the resumption of the hostilities.

Similarly, SADC had disqualified itself to act as mediator by declaring Savimbi a war criminal at its June 1998 Summit in Mauritius. The SADC position on Savimbi was reinforced in August 1999 at the organization's Maputo Summit when it agreed to supply FAA soldiers' with uniforms and human and other material needs.

There is also the possibility of track-two diplomacy on the part of the Angolan non-governmental organizations (NGOs). Whilst this is a possibility, I do not believe this to be a viable one. Three reasons account for this. First, there has always been suspicion regarding the neutrality of NGOs on the part of UNITA in particular. Second, Angolan NGOs are weak and do not have the necessary material and human resources to sustain a long-term conflict resolution intervention. Finally, there is the fact that NGOs often compete with one another engaging in useless (and expensive) duplication and working at cross-purposes with each other. In July 1999, for instance, unionists and intellectuals launched a "peace manifesto". The Catholic Church, meanwhile, did not back the manifesto and set up their own independent "movement for peace".

This, however, leads us to the third problem of resolving the Angolan conflict – is there a willingness on the part of both sides to settle this conflict? Here one could answer that no such willingness exists on account of two factors. The first is that the conflict has gone on for so long, it has become personalized. As such, there can be no rapprochement between the parties as long as Dos Santos and Jonas Savimbi remain the main protagonists in this unfolding African tragedy.

Second, the lack of willingness on the part of UNITA can also be explained by its military superiority on the ground. There is currently nothing that UNITA can derive from sitting at the negotiating table that it cannot get by means of the battlefield.

As the war intensifies across the country, Angolans are dying at the rate of 200 a day. For Angolans the dawn of peace has not occurred and is not likely to materialize in the foreseeable future. For Angolans, it seems night has no end.

Notes

[1] Shortly after the establishment of UNITA in 1966, Savimbi and his key officers went to China for guerrilla warfare training.

[2] The Ovimbundu is the majority ethnic group in Angola and UNITA portrays itself as representing this group. UNITA also portrays the MPLA as representing the interests of the Mbundu people and mesticoes (those of mixed European and African descent). This assertion was proven false in the 1992 elections when large numbers of Ovimbundu voted for the ruling MPLA.

ELIMINATION OF ANTI-PERSONNEL LANDMINES: THE AGENDA FOR THE 21ST CENTURY

Lynne Hopkins and Martin M. Kaplan

The past 12 months have resulted in both positive and negative developments in the international effort to eliminate anti-personnel landmines (APLs). In the policy arena the victory of an early entry into force of the Ottawa Convention in March 1999 was somewhat mitigated by the outbreak of hostilities in the Balkans between Yugoslavia and the NATO Alliance, resulting in substantial new emplacements of landmines in the region. The States Parties to the agreement are now faced with the challenge of implementing the crucial elements of their Treaty. In the domain of research and development, international research groups continue to search for more efficient and cost-effective methods for detection and neutralization of APLs in the ground. Some disputes have emerged between NGOs involved in advocacy of the APL issue and those people engaged in actual demining operations in the field. This paper primarily examines the current status of American and European R&D in demining technology, but also presents a brief overview of the political and legal developments over the past year, and discusses some of the political controversies that have arisen in that they relate to research and funding issues.

Pugwash Position

Pugwash has supported the international effort to eliminate the use, production and distribution of APLs through its strong endorsement of the Ottawa Treaty as a legal instrument of international accord, and through its continuing interest in and advocacy of the research and development of new and more efficient technologies for humanitarian demining operations. Pugwash has provided a forum for discussion over the past few years for humanitarian organizations such as the ICRC and for scientists, engineers and policymakers engaged in the global struggle to eliminate the scourge of landmines. Pugwashite Kosta Tsipis of MIT has given his expertise during the past year to a European project to develop new detection technology. Pugwash will continue to provide support to strengthen and implement the elements of the Ottawa Treaty, and to further the work of scientists and NGOs in their goal to have a world free of landmines in the 21st century.

Political and Legal Status of APL Elimination

The Ottawa Treaty - A Unique and Non-traditional International Agreement

In September 1998 Burkina Faso became the 40th country to ratify the Ottawa Convention[1], triggering the six-month period needed before the Treaty entered into force, which took place on March 1 1999. The Treaty represents an unprecedented event in the history of arms control due to its initiation by like-minded States and NGOs acting outside the framework of the U.N. and without the involvement of such major powers as the USA, China and Russia. (For an in-depth discussion of the evolution of the Ottawa Treaty, refer to report by the authors presented at the 48th Pugwash Conference.[2]) Critics point out that without the participation of these major powers, which represent some of the major producers and users of APLS, the Treaty also lacks the strong verification and compliance components integral to other global treaties. However, a major goal of the Treaty has been achieved - the creation of a global norm for landmines elimination which even the United States aspires to, and has promised to join by signing the Treaty by the year 2006, albeit with the proviso that "...suitable mine alternatives can be found."[3]

The Treaty also represents a new process that in a sense bestows a partnership role on a variety of non-governmental organizations and other "stake-holders" involved in the landmine issue, such as the ICRC and advocacy groups like the International Campaign to Ban Landmines and the Mines Advisory Group (MAG), a British NGO which recruits and trains people worldwide for demining operations in the field.[4] The adherents to the Treaty consider this historic relationship to be an essential foundation to the Treaty's continued effectiveness and success and are committed to continuing the strong interaction between governments, international agencies in the UN system and non-governmental organizations in future mine action operations.

The States Parties to the Treaty and the NGOs must now confront the challenge of implementing their Treaty, which they view as a "comprehensive framework for Mine Action."[5] A stated major goal among the signatories of the Treaty is to return people in landmine-infested regions to their homes and livelihoods with as much speed, and efficiency and safety as possible. To achieve this goal mine-clearance operations have to be improved. This might not necessarily mean more sophisticated technology which is expensive and takes years to develop, but simply training more people to do what people are already doing in the field, with the most effective equipment available at the present time.

In the first meeting of the States Parties in Maputo in May 1999, an important step towards implementing the Treaty was the establishment of the "Intersessional Work Programme" with five "Standing Committees of Experts" who are to address the key thematic issues of the Treaty:

1. The general status and operation of the Treaty
2. Mine clearance
3. Victim assistance and mine awareness
4. Stockpile destruction
5. Technologies for mine action[6]

These Standing Committees are to meet in Geneva periodically and the recently established Geneva International Centre for Humanitarian Demining (GIC) will facilitate their work.

The Geneva International Centre for Humanitarian Demining (GIC)

The GIC was established in 1997 with Swiss Ministry of Defense funding as a much needed entity to strengthen the co-ordinating role of the UN Mine Action Service (UN MAS), to exploit the valuable experience accrued by the NGOs, to represent the needs of people working in demining operations in the field and to try to help Mine Action groups avoid competition and duplication of efforts, whether technical or political.[7] It promises to be an effective and valuable instrument for establishing cooperative links with other nations involved in projects for landmine elimination. The recently appointed Technical Director of the Centre, Brigadier General Patrick Blagden, was formerly the Clearance chief for the United Nations, and has many years experience in both the political and technical arena of landmine issues.

Since the Geneva office opened in the autumn of 1998, its representatives have attended many international meetings to establish contact with mine action groups, with research teams and NGOs. One of the assignments of the GIC is to develop what they call an Information Management System for Mine Action (IMSMA). This is a database that can be used as a management tool by governments and donors to prioritize activities and allocate resources accordingly. The system consists of two modules - a Field Module to aid acquisition and collation of information at the field level for later distribution, and an HQ Module to collate, process and disseminate information at UN HQ. An important component of the Field Module has to be a Geographic Information Systems (GIS) for generating good quality maps and other geographic documents. The concept of GIS will be discussed later in this paper.

Besides information management, the Centre will organize training courses for national mine action managers and information technology specialists with the goal of increasing effectiveness of national mine action planning and execution. These courses will be carried out in conjunction with the UN and other organizations. For example, the UNDP has initiated a study on mine action training and will work closely with the GIC on this project.

The GIC will also cooperate with the European Commission's Joint Research Centre (JRC) at Ispra, which in turn is associated with EUDEM (European Union in Humanitarian Demining) and the ESPRIT program (ESPRIT is the information technologies programme of the European Union, which co-ordinates and funds commercial and academic research in various fields, including humanitarian demining). The work of these groups will be discussed in more detail later in this paper.

The United States - Landmine Policy for the 21st Century

The USA, along with Russia and China, did not sign the Ottawa Treaty. These three nations have signed, and the USA has ratified, the Amended Landmines Protocol (Protocol II) to the Convention on Conventional Weapons (CCW). Although the signatories to the Ottawa Treaty view the CCW as an inadequate agreement, major Ottawa Convention nations such as Canada, Britain, France and South Africa have also signed the Protocol. The amended Protocol, which the USA is already in compliance with, strengthens the original Convention by restricting the use, production and transfer of APLs. For example, the new protocol prohibits the use of non-detectable APLs and severely restricts remotely delivered mines. Use of non-self destructing and non-self-deactivating mines are only allowed in clearly marked and recorded areas. Export of mines is forbidden to non-States Parties.[8]

At a time when other major western powers are destroying their stockpiles and committing to end production, use and transfer of APLs, the United States is still manufacturing and using them (as of 1999, there is still a US moratorium on exports). They have retained the right to use non-self destructing mines in the Korean demilitarized zone. President Clinton has reiterated the United States' willingness to sign the Ottawa Treaty by 2006, and to end the use of all APLs outside of Korea by 2003 "...if Washington successfully identifies and fields suitable alternatives to its APLs and mixed anti-tank systems." [9] Current US landmine stockpiles consist of one million mixed anti-tank systems, nine million self-destructing APLs and one million non-self-destructing APLs retained for Korea.

The United States is spending money on mine clearance, however. The White House has appointed Donald Steinberg as US Special Representative to the President and Secretary of State for Global Humanitarian Demining. $25 million has been allocated to DARPA over three years for research on promising sensor technology, and on 19 August 1999 the Department of State announced that they had awarded a Performance-Based Service Contract to RONCO Consulting Corporation, "an experienced mine detection and clearance firm, in an amount not to exceed $250 million over five years. The contract is comprised of one base year plus four one-year options. It will enable the United States Government to task RONCO, on a worldwide basis, to provide highly professional mine clearance, mine detection dog support, logistics and supply services and other programme management support to countries receiving US humanitarian demining assistance."[10] President Clinton's speech to the Maputo gathering of the Ottawa States-Parties further indicated that Washington planned to dedicate $100 million to humanitarian demining activities in the year 2000.[11]

Landmines in Kosovo

"I think you could argue that Kosovo today is among the world's most dangerous real-estate" was the comment from Ambassador Donald Steinberg, the US Special Representative for Global Humanitarian Demining. Although the full extent of landmine contamination in the region is still unknown, the UN has so far identified 800 mined areas.[12] With refugees pouring back into the region, KFOR and relief workers are worried about the danger to civilians. Most of the mined areas established by Serbs appear to be along the borders with Albania and Macedonia. Unexploded bombs also represent a threat, but many of these are of NATO origin and pretty good records exist of their location.

The war in the Balkans earlier in the decade offers some insight of the problems facing demining teams in Kosovo. According to the International Committee of the Red Cross, the Bosnian and Herzegovinan conflict produced some 30,000 minefields containing around 750,000 mines, covering some 300 square km in area.[13] Records are available for 17,854 of those minefields. By mid 1997 only an estimated one per cent of mine-contaminated land had been cleared to humanitarian standards. However, as landmines were generally deployed defensively and their locations recorded, identifying many dangerous areas should not be impossible. The problem has been resources. The UN Mine Action Center believes that with 2,000 de-miners in the field the problem in Bosnia and Herzegovina could be brought under control within three to five years.[14]

If, like Bosnia and Herzegovina, Kosovo minefields were predominantly laid for defensive purposes, and if sites are well recorded and marked, then the task of mine clearance will be laborious and risky, but ultimately feasible. As part of the cease-fire agreement, Serbia was to provide records of landmine deployment, but as of June 1999, NATO had only received two maps from the Serbs. Officials are also worried that the area around Pristina may be heavily mined. Both the Serbs and the KLA used landmines not just as a defensive weapon but also to disrupt civilian life.[15] It is feared that many booby traps were laid by retreating Serbians to make it difficult for Kosovars returning to the region.

Issues of Numbers and Funding

It was admitted in the spring of 1999 that inflated statistics were generated by the United Nations in the early 1990s on the numbers of landmines in the ground.[16] These figures were used by Nation States, NGOs and international agencies such as the ICRC to win over a shocked public to support and promote the global campaign to ban landmines, initiating the Ottawa Process leading to the Treaty. The publicity generated over these statistics was perceived in certain circles as a "double-edged sword." Some mine clearers apparently believe that resources for mine clearance have been diverted to advocacy groups such as the ICBL.[17] They also believe that the inflated numbers have scared off potential donors. Numbers such as "110 million landmines worldwide" and statements that it costs $1,000 to clear each mine, and that it will take thousands of years to clear minefields make demining operations look hopelessly inefficient and costly. The example cited is that the billionaire financier George Soros announced that he would not fund mine clearance until it became more cost-effective.[18]

The US State Department in its 1998 revised report on landmines *Hidden Killers: The Global Landmine Crisis*, addresses the issue of numbers:

> In 1998, a growing consensus is emerging to suggest that the previous estimates of the number of landmines worldwide may have been high.... New estimates of the number of uncleared landmines are in the range of 60-70 million. Official and unofficial reports on the landmine problem continue to cite a range of 80-110 million, notwithstanding that some countries experiencing a mine problem in 1994 have since declared that they are no longer affected. The new number has been calculated on the basis of survey material and the experience of demining operations in several key countries. In-depth surveys have been conducted in a number of key regions, and the presence of international organizations, commercial entities and NGOs has helped to refine the data on suspected minefields. In addition, as demining teams have carried out their work, they have been able to verify - or refute - the presence of quantities of landmines in any given area. Finally, the relatively low number of armed conflicts and vigorous mine-clearing programs in several countries, notably in Central America, have reversed the trend of previous years when more landmines were laid then removed. In some areas, the combination of peace and intensive demining programs has made so much progress that they offer the hope of being able to declare some mine-affected countries, such as those in Central America, mine-free in the near future.[19]

Another concern is whether such numbers prompt governments to fund research for improved methods of detection and elimination instead of financing more and better training and more equipment for the existing methods of demining operations. Some technology proposed by researchers is too impractical to work in the field, is too many years from implementation, and too costly to produce and maintain.

In consequence of this controversy, the remainder of this paper will examine the scientific work being carried out in Europe and the USA that promises the most effective and realistic results in improved technology for humanitarian demining.

Scientific and Technological Efforts to Detect and Neutralize Anti-Personnel Landmines

> The ultimate goal of mine action is to alleviate suffering and remove obstructions to the rehabilitation, development and reconstruction of the economy and environment. The success of demining projects must not be calculated on the basis of the number of mines removed, but rather by factors such as the rate of return of land to economic use, the reduction of casualties, etc.[20]

Official policy on landmines from those who support the Ottawa Treaty and from the United States is to spend resources on speeding up demining efforts and to return people to areas that have been abandoned out of fear of hidden landmines. Economic hardship resulting from the vast areas of mine-infested land standing empty worldwide only serves to perpetuate instability in post-war regions.

Essential to the success of this policy is global cooperation and co-ordination of demining activities, including research and development of improved technologies. The scientific community involved in humanitarian demining is being urged to listen to the needs of operators in the field and concentrate their efforts on practical and speedy solutions for detection and clearance problems while continuing to pursue longer-term research on more high-tech systems.

Priorities in the European Community

The most common mine clearance tool is a metal detector and a pointed stick and mine clearers, "...like archaeologists on a dig, extract the mines with small digging implements."[21] In many locations, like Cambodia, getting to mined areas is a major undertaking with mine clearers having to hack their way through dense vegetation. Clearance workers continue to ask for rapid, safe and low-cost technology for use immediately or as quickly as possible. However, while field operators should receive adequate technical support, scientists should not be discouraged from pursuing research for solutions to improve the primitive, laborious and dangerous techniques currently employed.

With these considerations in mind, the European Union's Joint Research Centre (JRC) at Ispra in Italy held an organizational "End User" meeting in January 1998, including field operators and demining experts, as well as scientists, EC and UN officials. "The aim was to ensure that the field experience of those engaged in demining operations was fully taken into account in framing R&D programmes."[22]

Some major priorities were:

- Improved survey capability
- Development of fast and reliable whole-area clearance system, and of multi-sensor detector systems; improvement of existing tools, especially hand-held detectors; demining equipment must be reliable in the field and sustainable.
- Information management capacity - this means primarily the introduction of a standard Geographical Information System (GIS). Several GIS were already available but lack standardization and interface
- A database comprising physical and chemical characteristics as well as relevant signatures of different mines.
- Development of an effective exchange of information between stakeholders, including NGOs and involvement of NGOs in R&D programmes.
- Provision of enhanced tools for vegetation cutting/removal
- Research into sensory capabilities (artificial dogs' noses)

Work Sponsored by the European Union

The information technologies programme (ESPRIT) is an integrated programme of industrial R&D projects managed by DG III, the Directorate General for Industry of the European Commission. It comprises a network of private industry, research institutions and universities. The Humanitarian Demining programme of ESPRIT has called for projects that are working to fulfill the criteria of the priorities set out above. In particular, R&D is requested on

- improved hand-held APL detectors - projects should aim to improve existing detectors and development of multi-sensor systems. A significant drawback with current detectors is their high false alarm rates and limited detecting capacity non- or low-metal APLs. A promising solution to these problems is to apply fusion of data or information on multi-sensor outputs.[23]

In response to this request, several projects have been proposed and some are already underway to develop multi-sensor systems, with time-lines of R&D of 24 months and less. One of these projects (referred to as "29870 Hope - a Hand-held Operational Demining System") combines a metal detector, a ground penetrating radar and a microwave radiometer. The resulting product is a hand-held unit that will be able to detect plastic and minimum metal, will be reliable by achieving nearly 100 per cent clearance, have a maximum weight of 7.5 kg, speed up demining operations by more than 30 per cent, reduce the false alarm rate by more than 50 per cent and cost less than 12,000 ECUs. The development project involves collaboration with 14 institutions, with a start date of January 1999, and a duration period of 24 months.[24]

Other desirable areas for research are air-borne or vehicle-based platforms for remote surface and ground penetration radar or sensing operations. More reliable vapor sensing detectors are being researched. Dogs are currently the best detectors of explosive vapors, but are not always reliable; many groups are conducting research to develop "electronic dog-noses" as discussed later in the paper.

Geographic Information Systems (GIS)

GIS is used as an increasingly powerful tool in many professions, and its application to detection and removal of landmines is entirely appropriate, providing the data used is reliable. GIS is a visual relational database that can link information to physical objects in maps. It is a way to produce computer-generated maps of where objects are located, and link a table of information to that object or geographic region. It is necessary to have good quality satellite imagery or aerial maps that can be digitized with geographic coordinates for the system to be useful. When landmines are deployed during a military operation it is hoped that their geographic location is recorded accurately. With this sort of record, together with field use of GPS (handheld global positioning system using satellites to record ones location on the planet) applied to a good GIS, locating and removing mines would be much more rapid and efficient.

The Geneva International Centre for Humanitarian Demining, and James Madison University (JMU) in the United States are both working on developing a standard GIS for landmine detection. Better quality visual data is needed of mine-infested regions. New aerial photographs using high-resolution visible-spectrum photography with a resolution of 1 metre or less would help update the maps currently in use. False colour imagery, for example, might show where digging has taken place.[25] Groups such as the GIC who are working to develop a standardized Geographic Information Systems would find such aerial photographs invaluable; satellite imagery would also greatly enhance work on developing the GIS database. The limitation is cost. Some existing mapping systems that be obtained free and could be used to superimpose mine information are: the Tropical Forest Information Systems (TFIS), which is the global land cover map being developed by the International Geosphere-Biosphere Program (IGBP) which conducts research on global climate change;[26] and maps from the Space Applications Institute used in the Tropical Ecosystem and Environmental Observation by Satellites (TREES) project.[27] The GIC has also asked Ambassador Donald Steinberg, the US Special Representative on Humanitarian Demining to help them get access to US Government material.[28]

A project called TELSAT4, initiated in the spring of 1999 by scientists at the University of Brussels, is a study to see if satellite image sequences can be changed to enhance detection of minefield indicators.[29] This research group collaborates with other institutions on ESPRIT sponsored projects, including ground penetration radar[30] and airborne remote sensing imaging.

Research in the United States

> Dogs are one of the best kept secrets of the mine action world....The use of sniffer dogs in Afghanistan and Mozambique shows that in certain locations and in particular types of minefields, dogs can make a spectacular contribution to operational speed. In certain circumstances, dogs and their handlers are 140 times faster than manual clearance teams.[31]

Some promising US research is in the area of "electronic dog-noses."

An interdisciplinary team under the direction of DARPA has embarked on an aggressive three-year $25 million programme to try to electronically replicate the ability of canines to detect landmine explosives. According to Regina Dugan, who runs DARPA's canine nose research, dogs can find landmines with around 95 per cent detection rate success and with a very low false-alarm rate.[32] Some of the institutions involved in this programme are:[33]

- California Institution of Technology
- EIC Laboratories
- Johns Hopkins University
- Massachusetts Institute of Technology
- Rockwell Science Center
- Texas A & M
- Texas Instruments
- Tufts University
- Auburn
- Pennsylvania State University

Conclusions

In the wake of the political progress of the landmine elimination movement, the agenda has shifted from mobilizing resources and public opinion to support the passage of legal agreements to the serious work of turning the words of the two major Treaties on landmines into deeds. Regardless of whether there are 110 million or 60 landmines lying in wait for victims and preventing land from being returned to useful purposes, this is still an enormous and challenging task for those involved in the work of tracking down and removing or destroying the mines on a daily basis.

Over the past year and a half, landmine policy-makers both in Europe and the USA have adopted a realistic approach to mine clearance. The huge amount of money recently awarded to RONCO International, the private demining enterprise, demonstrates that although the United States continues to support research for high-tech solutions, it also acknowledges that the current methods of manual detection and removal are the most effective methods available right now. Deminers should receive the funding necessary for adequate equipment and well trained personnel for using that equipment. For the future, research should continue to find a more effective technology for demining, and international co-operation on the landmine problem must be consolidated. Ensuring the efficient exchange of information at all levels of mine action is very crucial to the success of the global effort to ensure a mine-free world for the 21st century.

References

[1] Known formally as the Convention of the Prohibition of the Use, Stockpiling, Production and Transfer of Anti-Personnel Mines and their Destruction.

[2] Lynne Hopkins and Martin M. Kaplan, "Report on the Global Effort to Eliminate Anti-personnel Landmines: The Political and Technological Challenges for the 21st Century," proffered paper for Agenda 4, Item c: Demilitarization, Demobilization and Reconstruction, 48th Pugwash Conference on Science and World Affairs: "The Long Roads to Peace," Oaxtepec, Mexico, 29 September - 4 October 1998.

[3] *USIS Geneva Daily Bulletin*, March 4 1999, p. 3

[4] ICRC Overview, 1999, "Landmines Must be Stopped", p.34

[5] The term "Mine Action" is now used widely by groups involved in landmine issues to describe any activity related to eliminating APLs, whether through political advocacy, fieldwork operations or R&D.

[6] *The Geneva Reporter*, Quaker United Nations Office, Volume 18, #2, April-June 1999, p.3

[7] Geneva International Centre for Humanitarian Report on Activities 1998 and Perspectives 1999, Geneva, 20 November 1998, p2-3. Http://www.gichd.ch/chron2.html#1

[8] "US Ratifies Amended CCW Landmines Protocol" in *Arms Control Today*, April/May 1999. Http://www.armscontrol.org/ACT/aprmay99/ccwam99.htm

[9] "Ottawa Convention States-Parties Hold First Conference; US attends," in *Arms Control Today*, April/May 1999. Http://www.armscontrol.org/ACT/aprmay99/aplam.htm

[10] US Department of State, Office of the Spokesman, Press Statement by James P. Rubin. Http://secretary.state.gov/www/briefings/statements/1999/ps990820a.html

[11] "Ottawa Convention States-Parties Hold First Conference; USA attends," in *Arms Control Today*, April/May 1999. Http://www.armscontrol.org/ACT/aprmay99/aplam.htm

[12] USIS Geneva Daily Bulletin, June 22 1999, p.2

[13] The Silent Menace: Landmines in Bosnia and Herzegovina, ICRC, 1997, p.12

[14] *Ibid.*, p37

[15] *Ibid*

[16] Michael Flynn, "Political Minefield," in The Bulletin of the Atomic Scientists, March/April, pp. 49-53

[17] *Ibid.*, p.52

[18] *Ibid.*, p. 52

[19] *Hidden Killers: The Global Landmine Crisis* - Report released by the US Department of State, Bureau of Political-Military Affairs, Office of Humanitarian Demining Programs, Washington, DC, September 1998 Http://www.state.gov/www/global/arms/rpt_9809_demine_ch3a.html

[20] *Demining "End User" Forum, EC's JRC, ISPRA, January 1998. Summary of the Discussions and Recommendations.* Ed. Alois J. Sieber, Geoffrey Van Orden, Thierry Van der Pyl

[21] "Land Mines: Horrors Begging for Solutions", by A. Maureen Rouhi, in *Chemical & Engineering News*, March 10 1997. Internet document: http//pubs.acs.org/hotartcl/cenear/970310/land.html

[22] *Demining "End User" Forum, EC's JRC, ISPRA, January 1998. Summary of the Discussions and Recommendations.* Ed. Alois J. Sieber, Geoffrey Van Orden, Thierry Van der Pyl

[23] ESPRIT HPCN Humanitarian Demining 1998 Workprogramme. Http://www.cordis.lu/esprit /src/hpwphude.htm

[24] For more information on this project contact Mr. Michael Rothe at VallonGmbH, +49 7121 9855-31 e-mail: Vallon-GmbH@t-online.de

[25] "Humanitarian Demining and the need for Technology," a speech by the Technical Director, Patrick M. Blagden, of the Geneva International Centre for Humanitarian Demining. No date

[26] http://www.cnrm.meteo.fr:8000/igbp/

[27] *Demining "End User" Forum, EC's JRC, ISPRA, January 1998. Summary of the Discussions and Recommendations.* Ed. Alois J. Sieber, Geoffrey Van Orden, Thierry Van der Pyl. Http://www.cordis.lu/esprit/src/hphduser.htm Also, see website for description of the TREES project: http://www.jrc.org/sai/mtv/activities/trees.asp

[28] Geneva International Centre for Humanitarian Report on Activities 1998 and Perspectives 1999, Geneva, 20 November 1998, p5. Http://www.gichd.ch/chron2.html#1

[29] TELSAT4 project (T4/03/55): Change detection in Satellite Image Sequences for minefield delineation. http://etro.vub.ac.be/minedet/.

[30] "The principle of the radar lies in the emission of an electromagnetic wave towards the soil, and the registration of the reflected waves. These waves will then carry information about encountered objects. Our work will lie in extracting this information." See web-site: http://etro.vub.ac.be/minedet/

[31] United Nations Study Report: The Development of Indigenous Mine Action Capacities. www.un.org/Depts/Landmine/Reports/study/study.htm

[32]"Unexploded Ordnance Detection/Neutralization", report by Dr. Regina Dugan of the Defense Advanced Research Projects Agency (DARPA). Internet document http://www.darpa.mil/ARPATech-96/transcripts/dugan.html

[33] For a detailed description of the work of this team, consult the DARPA Dog's Nose/UXO Detection web-site: http://www.darpa.mil/dso/rd/applied/uxo/programs/index.html, or contact Dr. Regina Dugan, Phone: 703-696-2296, Fax: 703-696-3999. E-mail: rdugan@darpa.mil.

DEMINING: A CHALLENGE TO SCIENCE AND TECHNOLOGY AND TO INTERNATIONAL SOLIDARITY

Vladimir Knapp

The world total of some 100 million mines, mostly antipersonnel mines, according to UN statistics, abandoned on the former battlefields of Asia, Africa and Europe, present a frightening threat to hundreds of millions of civilians, preventing their return to normal life. However, one aspect that adds to the size of the problem, but is missing from the summary statistics, is their distribution.

During the era of the Cold War between the nuclear superpowers, only wars by proxy in remote backward areas of Africa and Asia could be waged without too much risk to operators. Now that the Cold War is over, those areas are of little interest to their war sponsors, but the effects of war persist all the same, especially the mines. Those, as a rule, underdeveloped regions destroyed by wars, are now heavily handicapped in recovery because of mine infestation. Out of those 100 million mines, the largest portions are in such countries and regions.

UN statistics show that the average cost for removal of a single mine is about $300. Probably, a better measure is the cost of demining per unit area. It ranges from $0.5 to 2 per square metre, depending on the method and the country. In small countries of Europe, mine-infested areas cover many thousands of square kilometres, and many tens of thousands in large countries of Asia and Africa. So, these countries have to spend many billions of dollars before they could start their economic development seriously.

Clearly, with their financial constraints, only those demining methods which rely on cheap labour are viable, that is finding the mines with prodding sticks in human hands. Unfortunately, this is not only dangerous but also a very slow method. A trained deminer can demine 20-30 m^2 a day. A force of one thousand deminers (Afghanistan, for example, has about 3,000 trained deminers) will clear between 6 and 10 km^2 in a year. When you add to a figure of 100 million mines also the information on the area of mine infestation, then it becomes quite clear that these unfortunate countries and regions have no prospect of normal economic development for many, many decades.

Even neglecting the fact that the extent of war damage and destruction, including mine infestation, in these poor regions was possible only because warring factions were supported and armed by their respective sponsors, who ought now to feel some obligation to help their unfortunate former allies, such regions of persistent poverty, and consequently instability, are against the interests of the wider world. If these regions are not given hope of a better future, their people will try to find it elsewhere, adding their numbers to migrating millions from Asia, Africa, and possibly the Balkans. Quicker demining may not be a sufficient condition for development, but it is certainly a necessary condition.

The challenge to achieve quicker and more efficient demining is not only financial, it is also scientific and technological, as non-metallic mines, especially small antipersonnel mines, cannot be detected by conventional metal detectors. Many developed countries are responding to the challenge, and a number of new methods and devices are being developed and tested. In the long run these devices will reduce the cost of demining; in the short run their cost is likely to be an obstacle to their use in poor countries. When considering the ways of applying new technologies for mine detection, one very important observation must be made. It is that not only positions of individual mines are unknown, but that positions and borders of minefields are also uncertain. As a result of that *suspected mined areas* are much larger than the *actually mined areas*. An enormous step forward in demining would be the reduction of suspected mined areas to the level of actually mined areas. This is the opinion and advice of most experienced people in demining.[1]

Through the efforts of many research centres and groups, the technologies and equipment for identifying the minefields, as distinct from larger suspected mined areas, are in sight. They are not simple ones, but their use could, in the course of a few years, significantly decrease the areas to be demined. Croatian demining experience indicates a reduction by a factor of about two, at least. Even larger factors can be expected in other regions. In this paper we plead for their introduction and propose ways to do it. Their application could, in a relatively short time, open large areas to civilian use and allow the concentration of demining efforts on actual minefields. The phase of area reduction is expected to be cost effective, and a few years after completion of this phase several new and better methods for individual mine detection and removal will come into operation.

Now that the campaign to ban antipersonnel mines has achieved great success through the Ottawa ban, and with justified expectation of its essential universality in the near future, the problem of removal of those 100 million mines already in the ground can and should be focused upon. As this is not only a political and economic concern, it is also a scientific challenge, a Pugwash initiative is expected and logical.

Technologies for Mine Detection and Area Reduction

In humanitarian demining, as distinct from military operation, an area is considered clear for civilian use if at least 99.6 per cent of the mines have been removed. To achieve this goal is in any circumstances a very difficult task, much complicated by the low or very low metal content in many antipersonnel mines. By making metal detectors very sensitive, and checking every signal, sometimes many hundreds of false alarms per genuine one, with prodding stick in human hands, this target can be attained, albeit with an unacceptably slow pace of about 20 square metres per day per deminer, and with great risk to operators.

Several new detection principles, relying on detection of bulk explosive, rather than the minute metal content, are in various stages of development. Methods may be atomic or nuclear; nuclear quadruple resonance, x-rays backscattering, neutron interaction. Electromagnetic methods cover a wide range of wavelengths; they may be active, with irradiation of the ground with EM waves of several metres down to decimetre wavelengths, or passive, in the infra-red and optical region. Promising work is in progress in a number of chemical laboratories, where methods rely on detection of explosive vapours. With some of them extraordinary sensitivity is reported (the so-called microcalorimetric method, developed in Oak Ridge National Laboratory, University of Tennessee and Environmental Engineering Inc., Knoxville, reported at Monterey Symposium in 1998). Another promising result may come from non-linear acoustics. Surprises from biology should not be ruled out either. There is certainly extensive research in progress with many new ideas, so that a measure of optimism is justified.[2] However, a note of caution is appropriate. The road from laboratory device to a reliable operational device is full of obstacles, which scientists without cooperation with experienced deminers sometimes overlook. No device can be successful in all terrain and weather conditions. At present none of the devices in development is sure to offer the needed 99.6 per cent detection efficiency. With all reserves, research is so wide and intense that it would be very surprising, if by the years 2003-2005 a few devices were not operational and capable of satisfying the criterion, either as a single sensor or in a combined multi-sensor system.

In the meantime, pending development of such operational systems, a rational and cost effective approach is to give priority to the less demanding task of the capability to locate individual mines with great precision. In most of the mined region one could differentiate between areas suspected to be mined and the actual minefields within them. The more fighting was done by irregular army units, the larger is likely to be the suspected mined area relatively to the real minefields. A great advance in returning the territory to civilian use would be the finding of actual minefields, which would reduce suspected mined and closed areas by a factor of two, at least.

The systems capable to cover large areas quickly should be placed on aerial platforms. This, of course, limits their operation over wooded regions. Fortunately this limitation can be removed by complementary use of ground area survey methods. Use of dogs in area reduction mode is a viable solution. The so-called MEDDS [3]method developed in South Africa has proved itself of capable of performing this task in the regions of Angola and Mozambique. The MEDDS method uses dogs in the laboratory, which has advantages and limitations. More conventional field dogs in combination with suitable mechanical devices can also be a successful method for area reduction[4].

Multisensor Airborne Systems for Area Reduction

An operational aerial survey system requires development of remote observation sensors and their integration on a suitable airborne platform. Sensors, capable of operation at distances of tens of metres or more, are electromagnetic, working in the optical and infra-red spectral regions, or in the radar region. In order to improve spatial resolution of relatively small airborne antennas, synthetic aperture radar (SAR) is applied. With respect to thermal imaging, mine detection is much improved by the use of hyperspectral scanners which record images in several infra-red wavelength regions.

A number of major research centres, both civilian[5] and military,[6] are developing sensors or complete systems for aerial surveys, some of which have reached the field testing stage.

The European Consortium on Airborne Detection of Mine Fields has been testing their system in Belgium and Mozambique, with results to be published very soon.

The Humanitarian Demining R&D Program underway at the US Army Night Vision and Electronic Sensors Directorate has aggressively sought technologies to improve landmine detection. In the FY 1997 efforts focused on the development and evaluation of the Camcopter, an airborne platform designed to carry various sensor payloads for mine detection and mission planning. A Camcoder system containing thermal and monochrome CCD camera is being tested in South Africa.

Testing of the Camcorder system is in negotiation between the US Humanitarian Demining Assistance Program of the DoD and the Croatian Mine Action Centre (CROMAC). It is proposed to equip Camcorder with a hyperspectral imaging scanner and synthetic aperture radar. The Croatian side would provide all necessary activities regarding tests and evaluations, for example, preparation of the basis for the ground truth planning and mapping, interpretation, quality assurance (QA) and the quality control (QC) of the testing procedures, evaluation of detection efficiency based on the manually removed mines, and precise determination of their locations with the help of DGPS on the ground. The project has several tasks, amongst them; to test the ability to detect non-metal, minimum metal and metal landmines, to detect mines below the surface and on the surface, to test the ability to detect borders of the areas polluted by mines and the areas without landmines. Hopefully, the testing should be underway in the first half of the year 2000, with the results available by June 2000.

Concluding Remarks

Science and high technology needed to develop better and faster demining equipment is available in the West. Whilst it is a challenge, it is certainly not like that met in the Manhattan project when Western science and technology produced a weapon to defeat Japanese militarism. In the present mine threat, lives of Western citizens are not in danger. The challenge is not to produce new weapons, but to save lives of, perhaps distant, but all the same innocent, people who are killed daily, long after the wars are over, when happier people in Europe or the USA have difficulties in remembering what they were all about. Surely, there are differences, but should they act against Western action and help? There is much talk nowadays of global unity, but we are only slowly starting to understand what it should mean. We understand that the environment is one and global. So, for example, combustion gases, emitted largely from the developed countries, affect the global climate. Restraint on emissions is required and many in developed countries will argue that successful emission control is impossible if restraints are not applied on developing countries as well. But globality cannot and should not be a one way street. Development everywhere is a global interest, as poverty combined with political systems, which thrive on poverty, can be a global threat in the world in which weapons of mass destruction are increasingly accessible. Thus, in the name of true global solidarity, timely and efficient help to remove the mine threat from developing countries and regions should be a good investment in a better and more stable future world. Furthermore, introduction of better demining technology should not be left only to commercial enterprises, as it may lead to a situation where those who profited from mine export could make even greater profits from removal of their products. With its international, humanitarian and scientific aspects, the problem of demining deserves full Pugwash attention. We believe that it can be expressed in the following statement and proposals.

Proposal for Pugwash Conference Statement on Demining

a) In view of the fact that insufficient knowledge of actual minefield positions and borders results in very large loss of efforts and time, priority should be given to area reduction from suspected to actually mined areas. From past experience, area reduction could be very considerable, with corresponding large economic benefits both from savings in demining work and because of earlier return of land to normal use.

b) Intense efforts in a number of research centres are promising more efficient methods of mine detection. Also, new developments offer a technology capable of approaching the problem of area reduction. At least one airborne multisensor system, using radar (SAR) and optical (hyperspectral) sensors, is likely to be available in the years 2001/2. By giving priority to area reduction in the early years of the new century, several new mine detection methods are likely to be available for remaining and reduced actually mined areas.

c) Airborne multisensor systems represent a high technology and correspondingly high demand on the most mine infested countries such as Afghanistan, Angola, Mozambique and the Balkan area. Should these regions be left to manual demining of very large suspected areas, there is no chance that the "Demining 2010 Initiative" by the USA could be successful. The 2010 target can be achieved, but only with concerted international effort and the application of more efficient methods.

d) We, the participants of the 49th Pugwash Conference, believe that every effort should be made to speed up the use of new technologies in the demining process. To this end we call for technical development, especially testing and bringing to operational level equipment for multisensor aerial surveys. Developments, particularly in the USA and in Europe are such that the required equipment could be available in the years 2001/2, provided that development support and corresponding political decisions are made in the USA and Europe, at the appropriate levels. Similar attention should be paid to complementary ground systems for rapid survey of vegetation-covered areas and also to general rapid survey ground systems.

e) The countries most heavily in need of help, and which would greatly benefit from use of high technology methods, are at the same time those not financially and technically well prepared to apply them. In view of this, it is our opinion that the feasible and efficient way to resolve this problem would be by establishing *international expert groups* which would visit mine-infested areas and perform aerial surveys with their own equipment. The task of local logistics and ground preparation should be performed by local deminers and organizations.

f) We propose, further, that the work of international expert groups be organized and financed by the UN, USA and EU authorities. Notwithstanding the fact that many of the wars fought in these unfortunate regions were wars by proxy between great powers, which should now feel an obligation to remedy the consequences, we are convinced that such help would have global political and economic values. Politically it would mean a gesture of true solidarity between rich and poor countries, between developed and developing world. Its great economic importance would be in removing a major obstacle to recovery and development of these countries. In view of the scientific and technological developments in progress, in our opinion the time is right for such an initiative.

We appeal to the US Government and the European Union to support actions and bring decisions required for implementation of the above proposals.

References

[1] For example Patrick Blagden and Colin King, at the Second International Conference on The Detection of Abandoned Land Mines, 12-14 October 1998, Edinburgh.
[2] Proceedings of several recent conferences deal with the development of new mine detection methods:
SusDem 97,International Workshop on Sustainable Humanitarian Demining, Zagreb, 29 September-1 October 1997.
Third Symposium: Technology and the Mine Problem, Monterey, 6-9 April 1998.
Second International Conference on the Detection of Abandoned Landmines, Edinburg, 12-14 October 1998.
[3] Mechem Explosive and Drug Detection System, developed by Mechem company.
[4] Demonstrated by Croatian Mine Action Center, in region Sunja, Spring 1999.
[5] Joint Research Centre, Ispra, is one of the most important centers, many research groups associated with Universities and Academies.
[6] US Army CECOM RDEC Night Vision and Electronic Sensors Directorate, Humanitarian Demining Program
DERA, Defence Evaluations and Research Agency, UK.
FOA, National Defense Research Establishment, Sweden, are some of important military research and development centres.

WORKING GROUP 3

DEVELOPMENT

BEYOND AID: CO-OPERATIVE APPROACHES TO ACHIEVE EFFECTIVE INTERNATIONAL PROGRAMMES FOR POVERTY ALLEVIATION

Morris Miller

The Need to Re-think Aid Policy

> "The poor always ye have with you", said Jesus and 2,000 years of history have not proven him wrong.... Over the past 50 years rich nations have given $1 trillion in aid to poor ones. This stupendous sum has failed spectacularly to improve the lot of its intended beneficiaries...Countless studies have failed to find a link between aid and faster economic growth. Poor countries that receive lots of aid do no better, on average, than those that receive very little... (However) aid can work – but only if it is limited to countries with sound economic management."
>
> "Helping the Third World: How to make aid work", *The Economist*, 26 June 1999

Has aid policy come to this: triage? One can safely assume that those who write for *The Economist* - and those who share the current "politically correct" view of aid policy[1] - are aware of the worrying dynamic and the troubling ethics of the *denouement* of this "tough-love" approach. After all, with this set of policies there would almost certainly be a further widening of the income gap between the rich industrialized donor countries and the poorer – and economically weaker – developing ones. It would not seem to be much of a stretch of imagination to envisage the yawning income gap between the rich and the poor, already as wide as a figurative Grand Canyon, leading to an even wider gap than the current ratio of 1:50, that is, less than $500 annual income per person as compared to over $25,000 with all that this implies.[2] And with the onset of a new type of global economy, the so-called "knowledge economy" and "the winner-take-all society",[3] the economic income/power balance would tilt even more steeply downward for the poor, a phenomenon that applies with special force to the developing countries and, to a lesser extent, to industrialized countries - and especially the USA.[4]

There would, therefore, appear to be a hidden agenda in this analysis with its draconian recommendation. The message of that agenda is not openly avowed but neither is it well hidden: the essential characteristic of "soundness" is "openness", that is, the integration of the economies of the recipient developing countries into the global economy. This would have obvious implications with regard to their policies and practices as they relate to trade and investment, to the role of government and privatization, and so forth. As an experienced UN official, Richard Jolly, has observed, "not surprisingly, much of the leverage with aid today is directed towards requiring the country to adopt free market policies, including openness to trade, foreign investment and free market institutions."[5] The advice, in short, is self-serving on the part of the donor community.

This policy stance would ease the entry for foreign investors that are principally the large transnational corporations that are economically and financially powerful with operations that straddle continents and account for an exceptionally high percentage of trade and investment flows between the industrialized and the developing countries.[6] This is assumed to be the path to economic salvation in as much as it brings with it capital and technical/managerial know-how and the capability to compete in international markets. In effect, it is advice that would place increased reliance by the developing countries on private sector capital flows and thereby justifying the diminution of the Overseas Development Assistance (ODA) flows and, at the same time, ODA's greater concentration on a few select countries, and, so as to help assure "soundness", the imposition of severe conditionality in the ODA that is provided.

This advice on aid policy needs to be clarified and questioned with regard to the issue of the nature and degree of the dependency of the recipient countries on the inflow of ODA and of foreign direct investment (FDI) and, since the public and private sources play different roles, how these two types of sources of capital are related. Several questions come to mind:

- • can one generalize about the impact of either ODA and private capital inflows on the poverty of those in most of the developing countries when the poor, and especially the rural poor, are marginalized, that is, essentially living and working outside the money-centred modern economy?
- • how much of an impact can be achieved by ODA flows especially in light of the fact that its allocation is extremely skewed with 10 developing countries receiving about 80 per cent of ODA leaving 164 out of the loop?
- • how much of an impact can be achieved by FDI flows (that currently are about five times greater than ODA:$250-$300 billion per year as compared to about $55-$60 billion) when 10 countries receive about three-fourths of these flows, the 47 poorest developing countries receiving less than 1 per cent?
- • how much of an impact can be achieved by both ODA and FDI flows when together over the last decade they accounted for less than 15 per cent of the capital invested in the developing countries?

The facts are that, with the exception of the very poorest developing countries, the overwhelming proportion of the investment to finance the growth of the developing countries has come from their own internal savings and that the net foreign capital inflows have been very unreliable from one year to the next.[7] To put the contribution of ODA into perspective one need only look at the lending of the major contributor to ODA, the World Bank: its total lending over the last five decades has represented only about two per cent of the GDP of the recipient developing countries and, as Professor Anne Krueger notes in a recent article, "even though World Bank projects returned an impressive 10 per cent real return, they contributed only 0.2 per cent to growth rates."[8]

In any case, it should be recognized that it is impossible to assess in measurable terms the contribution that is made by foreign capital inflows both because one cannot know what would have happened to the economy without this inflow and because that contribution is bound to be partly or wholly intangible and diffused. It may well be that the key common feature of the "successful" cases that *The Economist* cites may not be that they received ODA and that they used the ODA well within the framework of "sound" economic management, but rather that they were lucky in their special resource endowments, for example, Botswana and diamonds, Brunei and oil - or in their particular geographic location – for example, the countries of southeast Asia that played a role in providing a proximate locale for enormous US investment and technical and managerial know-how in pursuit of the Korean and Vietnamese wars that was then followed by Japanese investment and know-how.

To take this locational factor further, there is a significant correlation of all countries in the tropics being poor and those in the temperate zones being rich and those that are coastal are generally wealthier than those that are landlocked.[9] One should, therefore, be very cautious in assuming a causal link between the way ODA has been used and its rate of growth, except, of course, where, on the recipient side, there is rampant corruption and sheer incompetence in the use of ODA funds and, on the donor side, there is laxity in the way funds are doled out. One thing, however, should be clear, namely that ODA would have to be increased by several orders of magnitude and distributed very differently to make a real difference, especially for the 164 countries that receive only a fifth of total ODA flows.

Even if it were to be suggested that the overwhelming percentage of the ODA flows be directed to the countries that have the greatest need for concessional aid and, at the same time, these funds were to be supervised more diligently to minimize corruption and mismanagement, the advice would likely fall on deaf ears. It should be recognized that the political leaders of the donor countries would likely regard such diversion as difficult to achieve, but, more importantly, would not likely have the will to do so. After all, the present list of ODA claimants are not chosen on the basis of their need for ODA funds and on the basis of how well ODA funds are likely to be used but for geo-political and/or business reasons that clearly benefit the donor community in the short-term. Accordingly, it should come as no surprise to find that much of the bilateral ODA is "tied" to oblige the recipients to accept donor supplies and talent and that the 10 developing countries that receive about 80 per cent of the present ODA flows happen, by and large, to be the same 10 that receive about the same percentage of the private foreign direct investment flows. The rationale for the World Bank's contribution to this biased allocation of ODA funds has been articulated by some officials of the World Bank: their loans can thus have a *leveraged impact*.. This may be good public relations but it runs counter to the role of public institutions that have been established to assist those that are most in need of assistance both for growth and for stability when, as so often happens, FDI flows are volatile and, therefore, cannot be relied upon.[10]

Then there is ODA as debt relief for the group of developing countries that have to contend with a situation of a negative *net* flow due to their obligations to service their foreign debts. The inflow of ODA increases a debtor's capacity to service their debts and enables the creditors to secure swifter repayment of the outstanding debt. This obscene situation of capital flowing from the poor to the rich has prompted gestures from the donor community that principally take the form of ODA as "debt relief" for those designated as "highly indebted (desperately) poor countries" (HIPC). Thirty-three have been designated as HIPC and thus are candidates for the publicized debt relief initiatives, the latest of which was recently formulated at the 1999 summit meeting in Cologne and labeled "the Cologne Initiative". It reeks of false generosity since these countries are never likely to repay their outstanding foreign debts that, in fact, have grown three-fold since the early 1980s as the unpaid interest has been added year-by-year to the principal outstanding.[11] What adds insult to injury is the conditionality associated with this debt reduction initiative that makes it unlikely that even that source of relief will amount to much.[12]

In this connection it is worth noting that many of those developing countries that are heavily burdened with debt have been the beneficiaries of ODA in the form of "structural adjustment loans" (SALs) with its quick disbursement as contrasted with the conventional project and sector lending project cycle of several years from identification to preparation to appraisal and, finally to disbursement, often in tranches spread over months and years.[13] As *The Economist* noted in a 1993 article focused on sub-Saharan Africa, "structural adjustment has neither restored growth not eased poverty.[14] Its record elsewhere has hardly been better from the point of view of the recipient developing country. Even the IMF, whose lending always has conditionality clauses that calls for "structural adjustment" of the recipient's economy as an integral characteristic of its macro-economic "advice", has acknowledged that:

> on the basis of existing studies, one certainly cannot say whether the adoption of (structural adjustment) programs led to an improvement in inflation and growth performance; in fact it is often found that these programs are associated with a rise in inflation and a fall in the growth rate.[15]

Yet, donors, and in particular the World Bank, have adopted an aid policy that makes a virtue of the shift to SALs from lending for projects that has been the traditional form of its lending program and still comprises about three-quarters of its lending program. Again we find that the policy shift towards this form of ODA was not undertaken for the poor of the recipient countries (as was advertised) but adopted for the benefit of the banks of the donor countries that were in "crisis" condition with their very dubious accounts receivable exceeding the value of their capital assets.[16] The most telling demonstration of this bias in the motivation underlying the SALs lies in the fact that there is, *prima facie*, a high social cost in the "belt-tightening" conditionality of SALs that is mainly imposed on the poor; subsequent evidence confirmed that outcome.[17]

Clearly there is a need to re-think aid policy. In this connection, at the outset, it should be recognized that it would be a mistake if the persistence of poverty in recipient countries were to be attributed to the weaknesses and failures of aid policy and practices for this line of reasoning would, by implication, assume that correcting aid policies and practices could be successful in closing the income gap and reducing widespread and deep poverty. *This approach would be too limited in as much as it would not be addressing the need to make significant changes in the prevailing global system and its power relationships.*

Openness and the panoply of related policies are not likely to constitute a feasible and successful strategy for the majority of the developing countries - unless and until special conditions prevail in the global economy. To take but one example, the necessary counterpart to a successful export policy is the willingness of the industrialized countries to have an accommodating import policy, that is, the willingness to import a flood of exports from the developing countries. In short, while this openness strategy advice may have contributed to the rapid growth of a few developing countries, it suffers from "the fallacy of composition": what may be a feasible and successful strategy for a few developing countries may not be so for the many unless and until special congenial conditions prevail in the global economic/financial system.

The challenge and the cure, therefore, goes beyond considering aid policy in isolation from the broader context of the global economy. Bearing this in mind, it is clearly necessary to re-think aid policy and practices with aid policy conceived in the broadest way to include the range of other policies that impact on the development process. It would, therefore, seem appropriate to focus on three facets of the aid policy issue:

i) *on the traditional forms of ODA* where the relationship of donor to recipients is on a country-by-country basis and delivered by the plethora of bilateral and multilateral donor agencies that rarely coordinate their efforts;

ii) *on the innovative initiatives where donors collaborate* in an attempt to address the problems impacting on development that are beyond the range of possible action by individual countries whether on the recipient and the donor sides of the aid process; and

iii) *on the policies that are not labeled "aid policy" as such but which have a direct and a significant indirect impact on developing countries*, such as those that pertain to trade and investment rules, debt relief initiatives, and such.

Re-Thinking Aid Policy and Practices

> Creativity is needed to bridge the huge gulfs between human needs, scientific effort and market returns....We urgently need new creativity and a new partnership between rich and poor.. if the extremely poor are to enjoy a chance for human betterment.... An important challenge, as yet mainly unrecognised, is that of mobilising global science and technology to address the crises ..
>
> Jeffrey Sachs, "Helping the World's Poorest", *The Economist,* 14 August 1999

ODA as the Conventional Mode of Transferring Capital and Know-how

In rethinking aid policy and practices a realistic perspective about the past and present record should be the starting point. First, the record of ODA over the last five decades has not been the utter failure that is implied by the critics since it can be claimed that in some measure – that happens to be immeasurable – it has contributed to progress on many fronts in the many of the developing countries. For example, over the past two decades alone average *per capita* income levels in the developing countries has increased by about 25 per cent or 1.3 per cent per year;[18] their infant mortality rate per 1,000 live births has been cut dramatically, in the worst case "low income countries" from 139 to 88 and for all developing countries from 104 to 59; their average life span has risen by about 9 years in the last decade alone; their percentage of adult literacy has risen from 46 to 70 per cent, and so on. In effect, if one leaves aside the trends in sixty of the developing countries, as the *World Development Indicator 1998* report states, "the developing world today is healthier, wealthier, better fed and better educated."[19] But, given the rapid rate of population growth in these countries, the number of those still existing in absolute poverty has remained steady at roughly one-third of humanity. The least that can be said on the positive side is that *without* ODA the indicators of well-being would almost certainly have been worse.

Looking to the future the relevant issue is how ODA could be made more effective in achieving its declared objective. Determining how ODA's impact can be improved calls first for a consideration of past and current process, that is, the ways in which ODA has been delivered, that is,

the "traditional" approach of development agencies. That approach has the following elements in common:
i) establishing a macroeconomic framework that is conducive to raising the incomes of the marginalized rural and urban poor,
ii) providing the poor access to productive resources, especially productive land and, through micro-credit programs, capital,
iii) strengthening the quality and delivery of services of education and health,
iv) strengthening extension services backed by strengthened research,
v) strengthening the managerial capacity of those in governmental institutions and of those engaged in the private sector, and, in so doing, encouraging and facilitating the participation of the poor,
(vi) encouraging and facilitating regional cooperation and a coordinated focused approach by donors.

All of these have no doubt been helpful to some degree. But clearly they have fallen short of the requisite level of achievement with regard to the objective of reducing the widespread poverty in developing countries to a significant degree. Nor are these measures likely to be effective enough in improving the living and working conditions of the world's rural and the urban poor if changes are not made that go beyond aid of the conventional variety that takes a country-by-country approach in aid programmes that lends funds and provide technical assistance for projects and sectoral programs.[20] Understandably, the country-by-country mode of lending has been favored since it assures that some governmental entity has undertaken an obligation to use the funds in a specific way for a specific purpose and, most importantly, to repay. Lending for projects and programs of international scope is not possible and making outright grants poses problems of a financial nature for international developmental institutions that depend on maintaining their credit standing in the international markets from which they must borrow. Thus, the commitment of funds for international programs by these institutions has been minimal, worthy as these programmes have proven to be.

Aid as International Programs Focused on Specific Deprivations and Opportunities

Problems of global scope have prompted the United Nations system to convene large-scale conferences on themes related to these problems most of which focus on aspects of the development *problematique*. Global targets have been set by politicians in these highly publicized conferences to meet concerns of large segments of the world's population about such issues as persistent poverty and the growing gap between the rich and the poor, environmental degradation, population growth and resource depletion fears, gender and other forms of bias, and so on. These conferences and their ensuing "programs of action" reveal good intentions as is evident when one peruses the list of resolutions that constitute the 21st century strategy aptly titled *Strategy 21* . The *1998 World Development Indicator* lists some of the main targets as follows:
i) reducing by half the proportion of people in extreme poverty by 2015,
ii) achieving universal primary education for all countries by 2015,
iii) demonstrating progress towards gender equality,
iv) eliminating gender disparities in primary and secondary education by 2005,
v) reducing the mortality rates for infants and children by two-thirds and the mortality rates of mothers by three-fourths by 2015,
vi) providing access to reproductive health services for all individuals no later than 2015,
ii) implementing national strategies for sustainable development by 2005 to ensure the current environmental losses are reversed by 2015.

We might take the first item of this list as illustrative of the obstacles in the path of achieving the hoped-for targets. The act of setting a target of reducing the number in extreme poverty by half by 2015 may seem to advance the cause of poverty reduction by placing the issue on the international agenda in a publicized manner. But it is an act that will undoubtedly remain a vain hope without some backup in terms of money and, more importantly, in terms of organization and will. It is pertinent – or perhaps impertinent – to ask in the light of past experience: how strong is the global collective will to achieve that target? What steps are envisaged and with what reinforcement beyond what has been provided to date?

It should be apparent by now that, with rare exceptions, targets set in the context of large-scale international conferences are exercises in futility since they are usually couched in general terms and have no teeth for non-compliance and no process established to achieve the hoped-for targets.[21] The best case that can be made for them is that they raise the public's consciousness about issues of importance and highlights some possible approaches to resolving or attenuating them. But, as almost all such conferences - especially the highly publicized one - are quickly forgotten, the *net* effect is more justifiably deemed to be negative in as much as an illusion of progress in the public's collective mind has been mistaken for reality. The illusion in the public's mind that there is political concern is the pay-off that keeps this ritual alive. What is wanted may be clear but how to get there is vague.

It should be recalled that poverty reduction has been given lip service continuously for over five decades by almost all of the leaders of the donor countries and of the international development agencies, that is, since the Roosevelt-Churchill declaration in the Atlantic Charter of 1944. Of the "four freedoms" that were specified, "freedom from want" had a featured place and the conference at Bretton Woods, New Hampshire, in that same year gave it validation by establishing the International Monetary Fund (IMF) and the International Bank for Reconstruction and Development (IBRD or better known as the World Bank). Almost two decades later, in a 1973 speech in Nairobi the President of the World Bank, Robert McNamara, highlighted the fact that 40 per cent of humanity had been left behind in the global income surge of the 1950s and 60s and declared that the World Bank would launch a poverty eradication initiative on behalf of "the bottom 40 per cent".[22] Almost two and a half decades later the same sentiment is being voiced with much the same rhetoric by the present president of the World Bank, James Wolfensohn, who declared in the foreword to the World Bank publication, *World Development Indicators 1998* , that:

> the World Bank is committed to helping countries fight the scourges of poverty, illiteracy, disease, hunger and environmental degradation...We believe that by reporting regularly and systematically on progress toward the targets the international community has set for itself, we will focus attention on the task ahead and make those responsible for advancing the development agenda accountable for results.

One is tempted to ask: accountable to whom? and what is new?[23]

Apart from the global-scale health-related initiatives of the World Health Organization (WHO) in collaboration with other UN agencies and in particular, UNICEF, and the global-scale agriculture-related program of the Consultative Group on International Agricultural Research (CGIAR), there has been no global-scale initiative put forward and launched at international fora *and activated* by one or more international agencies acting in concert that is commensurate with the challenge of substantially reducing global poverty or, at least, some of its key deprivations. There has been a great deal of rhetoric, target-setting and reliance on moral suasion to achieve the targets. What has changed are the content of declarations, slogans and publications advertising a shift towards greater involvement with the private sector and with non-governmental organizations and towards programs under the rubric of inclusion, participation, transparency and accountability. "*Plus ça change...*"

The organizational part of the necessary change calls for greater reliance being placed on international programs that focus on programs designed to alleviate specific deprivations associated with poverty such as the lack of access of the poor to cheap and reliable energy sources with low life-cycle costs, and to adequate education and health services and related equipment and supplies. For the first time in history scientific and technological advances are bridging distance at low cost and discovering new vaccines, but the poor lack access to the fruits of those scientific and technological advances, especially with regard to energy, health and informational/educational technologies for those hundreds of millions in the rural villages where most of the desperately poor still happen to live. The poor do not provide a market of interest to the private sector whose decision-makers, understandably, are not motivated to devote any funds to undertake research that would be of direct benefit for them. And public sector financing of such targeted research is grossly under-funded and poorly organized with two exceptions: the international programs of the WHO and UNICEF that have focussed on research and follow-up implementation programs to address specific health problems (such as eliminating or substantially reducing such diseases as "river blindness", malaria, diphtheria, etc.) and the international programmes of the Consultative Group on International Agricultural Research (CGIAR), founded and funded by the World Bank, FAO and private foundations, that are designed to enhance agricultural productivity.

The two programmes are a form of ODA that has proven to be exceptionally helpful with regard to eliminating or easing two of the worst deprivations of the poor, namely, ill-health/premature death and hunger. They are the type of initiative that can serve as suggestive models of organizational arrangements that involve international cooperation. The WHO programmes of immunization and disease-carrying pest control have prevented and/or eliminated widespread debilitating and fatal diseases and the Consultative Group on International Agricultural Research (CGIAR) which, since it became operational in 1971 in response to recurrent large-scale famines of the 1950s and 1960s, has played an important role in tripling world food production while global population doubled over the years since it became operational.

If the will is there, the *first* step is the recognition that a coordinated international-scale effort that goes beyond ODA of the conventional variety can be very helpful though the focus may be only on some selected facets of the poverty condition as there has been on health and hunger. There would likely be a consensus as to the priorities of providing improved access, i) to health services and inoculation programs that have already achieve considerable success but still need to be accelerated and broadened in scope, ii) to primary and secondary education, and iii) to reliable low-cost *in situ* energy (solar, wind, biomass, etc.) for rural villagers living beyond the electric grid network.

- • if research were to bring down the costs of the equipment, related software and connections and appropriately adapt them, *modern communication technology* that bridges great distances would be capable of ending the isolation of rural villagers that has been one of the principal factors that have kept them so impoverished in terms of disposable income, education, and health. Modern communication technology has the potential to enable these rural villagers to benefit from contact with qualified specialists living in the urban centers so as to know about and benefit from more adequate health services both curative and preventive; it would also enable them to benefit from contact with qualified teachers living in the urban centers so as to meet the need for quality primary education in rural communities that could yield the well-recognized benefit of educating young women that is said to yield the highest rate of return in the broadest social sense.
- • if research were to bring down the initial capital and life-cycle operating costs, *modern energy technologies* that utilize local sources of energy in biomass, solar, wind, shallow gas deposits, micro-hydro, etc. would enable rural villagers to drill deeply enough for fresh water, to overcome back-breaking tasks borne by women and children, and to have light and power for the relevant health and educational equipment and services.
- Such initiatives, that through research would enable the mobilization of science and technology, could enable the rural poor to gain non-income benefits, that is, rural and urban life of the desperately poor could be improved in terms of access to the modern amenities of sanitation, lighting, communication and such well before their *per capita* incomes were to increase. But such research would also undoubtedly provide the pathway to increased incomes and the improved well-being that comes with it. [24]

How such programmes might be designed and how much it might cost and what sources might be tapped for the needed financing are issues best left to feasibility studies that should most appropriately be sponsored and managed under the auspices of one or more international agencies. But – as discussed later – the inclusion of the private sector is essential given the large size and scope of their research programmes and such issues as patent rights and a host of other factors that are very complex and would need to be resolved. With a combination of incentives and regulations the focus of such research programs might be amenable to a change of emphasis that would be more helpful for the poor in the developing countries, but the public interest would necessitate a leading role for one or more international organizations.

Aid as "Global Structural Adjustment": A Cooperative Approach that Goes Beyond Aid in its Traditional Forms

There is a further step that would go beyond the traditional forms of ODA and the international-scale cooperative approach a là the WHO and CGIAR inoculation programmes, but which are not generally recognized as ODA, namely, cooperative policies of the larger industrialized countries that are designed to create the key conditions in the global economic/financial system that are conducive to a faster growth of the developing countries. The elements of the needed congenial conditions are interest and exchange rate levels and their volatility, trade and investment flows and their terms and volatility, debt relief across the board to ensure a higher net ODA and other factors that pertain to the nature of the global market context. A realistic appraisal of the global economic/financial situation would suggest the metaphor of developing countries trying to swim against a strong current or trying to run up a down escalator.

The underlying assumption is that even if each developing nation were to be able and willing to "correct" their institutional, policy and cultural failings by undertaking the prescription of a program of "structural adjustment", only a few of them could hope to sustain the effort required to reverse these deplorable trends *so long as the international milieu remains essentially the same.* This thus puts the spotlight on the prevailing global economy *as a system* which has the following characteristics that are impediments to the growth of the developing countries:

i) declining and skewed ODA flows from levels that were already far short of the needs of developing countries,

ii) historically and unsustainably high and volatile real interest rates on capital borrowed from abroad,

iii) secularly deteriorating terms of trade for those developing country economies greatly dependent on the earning from the export of primary commodities the prices of which have been severely depressed since the mid-1970s,

iv) the persistently higher protectionist barriers of the industrialized economies in relation to imports from the developing countries as compared to imports from other industrialized countries,

v) exchange rates that have fluctuated widely and wildly ever since 1970 when President Nixon closed the "gold-window" and, thereby, effectively terminated the 1944 Bretton Woods agreement,

vi) the ineffectiveness of global economic/financial leadership on the part of the USA and of its G-8 partners who convene annual summits of political leader and finance ministers, ostensibly to achieve coordination of policies to foster global growth with equity and stability.

If the economies of developing countries are to get back onto a growth path that is desirable in terms of both pace and pattern – and necessary if the targets set at UN conferences are to be met - "correcting" all six of these unfavorable factors is a *necessary* condition (though, as noted earlier, given the limitations of ecological, climatic, cultural and other conditions, correcting them may not be a *sufficient* condition). Creating a congenial international economic/financial environment within which the developing countries must trade, borrow, service their debts, attract capital investments, and such is tantamount to advocating *systemic* change.[25]

Knowing or sensing this, there is, understandably, a sense of despair on the part of the poor in the developing countries, an acute awareness of their impotence in contending with the outside world with its global condition of a grossly unequal distribution of power. How under these conditions, they ask, can the developing countries expect to achieve a sufficiently rapid rate of growth of their economies to enable *per capita* incomes to rise at the target rate of three per cent per year and, at the same time, change the nature of that economic growth to ensure that it is more equitable and environmentally benign? They are, after all, at a stage of development that is capital and energy intensive and all that that implies in terms of dependency on the outside world for its money and its technology. Therefore, attaining the ambitious objective would seem to call for the use of resources *in a different way* and *for different purposes*, changes that are, in a word, *systemic* in their nature and scope.

One of the first orders of business for the reformed institutions of global governance operating under a new Bretton Woods mandate would be creating a congenial environment for the developing countries by striving to establish the six necessary conditions of lower real interest rates, greater transfer of capital, etc. It may suffice to focus the discussion on three of those factors: the issues pertaining to interest rates, to ODA and to global leadership.

Lowering Real Interest Rates to Their Historic Levels of About One Per Cent

Lowering real interest rates calls for an international cooperative effort with the USA in the linch-pin role by virtue of the size of its economy and its importance in world trade and capital flows and, therefore, of its impact on the supply and demand for capital that, in the final analysis, determines the price of capital or the real rate of interest. Japan, the European Economic Community and, to a lesser extent, others have a role to play in this process since any actions by one to set interest rates could be offset or nullified by the actions of others. But there is little hope of lowering interest rates without action taken to arrest the decline in world savings. These fell in the industrialized countries, in aggregate terms from 25 per cent of gross national disposable income in 1965 to less than 20 per cent in 1999 and this decline has continued in all the major industrialized countries, with the exception of Japan and Germany.[26] The chief culprit has been the US economy that has had a low and even negative rate of savings and is, therefore, very heavily dependent on the import of foreign-generated savings which has its counterpart in a historically high trade deficit with imports far exceeding export earnings. In effect, the USA has played the role of taker rather than giver, self-centered rather than benign leader.

Increasing Financial Flows to Developing Countries

Without taking the governments of the recipient countries off the hook for corruption, misguided policies, mismanagement and inefficiency and such, the attribution of failure or disappointment with both the size and the geographic and sectoral allocation of ODA must rest in great measure on the shoulders of the politicians in most of the donor countries who, whatever their declared intentions with regard to poverty alleviation, have, with few exceptions,[27] been miserly regarding the amount of their commitments in financial terms and very conservative and unimaginative with regard to the modalities to achieve this objectives. Political leaders have rationalized their lack of financial commitment to ODA (as contrasted with their rhetoric laced with crocodile tears) by pointing to the contribution of private capital flows but the recent resurgence of private capital flows does not answer the need for capital of most developing countries so as to ensure a sufficiently high rates of *per capita* growth for *all* the developing countries, defining "sufficiently high" as three per cent per year and stressing "all" to place the emphasis on growth *with geographic equity*.

The basis for this pessimistic judgment is evident when one looks at the gap between estimates of the minimum amount of capital required and - if one judges by the past record - the likelihood of such sums being forthcoming on commercial and on concessional terms. If there is to be any hope of achieving the alleviation of poverty on a significant scale the span of time as a framework for policy and related institutional change should be no longer than one or one and a half generations, a span short enough to enable the widespread perception of improvement that engenders hope and long enough to be feasible since systemic changes take time. This, rather than an estimate of the availability of the requisite amount of capital, provides the basis for the target set in The *Human Development Report 1999* for wiping out the worst forms of poverty in the developing world. The challenge is formidable: achieving the target of an annual rate of increase in *per capita* income of three per cent for 30 years is more than double the average rate of *per capita* growth achieved over the past three decades, a rate that averaged 1.3 per cent per year since 1970. And it has not been improving of late: in 1998, for example, the rate of growth of their economies averaged less than two per cent, lower than the rate of increase of their population.

Nonetheless, just looking at one key factor, capital, the questions arises: how much capital is required to achieve this ambitious target and where are the funds to be found?

Based on various "reasonable" assumptions several modelers have arrived at median estimates of capital requirements for the low and middle-income developing countries to achieve this three per cent rate of growth. They range from an *additional* annual *net* capital inflow of about $60 billion to about $125 billion if the developing countries were to finance 80 per cent of their own investment requirements from internal sources.[28] That would mean a doubling or tripling of the present levels of ODA. But this would not be all the claims that are likely to be made on the donors, the suppliers of ODA, and on private investors. There is the estimated capital requirements of the former Soviet republics and the Eastern European countries that should amount to another $60 to $90 billion.[29] Then there is the capital required to meet the minimal targets of UNCED's *Agenda 21* that is estimated to range from $100 to $300 billion annually over the next 20 years just to meet minimal global environmental quality standards.[30] In addition, there is the capital needed as reserves to be set aside to meet emergencies, a recurrent event given the volatility of private capital flows and of the size and volatility of the movement of currency transactions that is estimated to now be running at over $1.8 billion a day, most of it motivated by speculation rather than by productive investing and trading.[31]

There has been a plethora of proposals as to where to find the needed new sources of funds. Some of the more frequent ones are the following:
i) some variant of a "Global Marshall Plan";[32]
ii) issuing a substantial amount of Special Drawing Rights (SDRs) by the IMF (say, $50 billion over three years) to needy developing countries rather than in accordance with quota entitlements which favour the industrialized nations with the larger quota entitlements;[33]
iii)diverting a stipulated percentage of disarmament "savings" to development and environmental programmes (the "peace dividend");
iv) imposing a global carbon tax on the users of fossil fuels (which, it is estimated by the Worldwatch Institute, at $50 per ton of carbon emission would yield annually about $28 billion in the USA alone), and
v) taxing the $1.8 trillion per day volume of foreign exchange transactions handled by forex centres at a miniscule rate (say at 0.2 per cent) that, nonetheless, would yield large sums (the Tobin tax proposal that was initially put forward by him as a means of dampening the speculative nature of the global currency flows).[34]
vi) taxing, fining and extracting royalties on commercial operations utilizing and polluting the "global commons" (principally the upper atmosphere, the high seas, Antarctica and areas not under any one nation's jurisdiction).or the payment of "rent" by the developed industrialized countries for their *disproportionately* heavy use of the "global commons" as a huge and seemingly limitless garbage disposal bin.[35]

These measures and the many more that are not listed can only be undertaken through both *ad hoc* agreements between nations on a regional or global basis and through permanent institutional arrangements to ensure their implementation and enforcement. This is tantamount to saying that the appropriate action would be to *launch negotiations to establish a "global structural adjustment program"* akin to the process that established the Bretton Woods agreement of 1944 that put in place a global system of rewards and penalties designed to secure adherence to agreed-upon rules governing trade and capital movements.

Providing Benign Leadership in Forging a very Different Relationship Between "The Haves" and "The Have-Nots"

To get some perspective on the issue of leadership it is noteworthy that the industrialized donor countries have devoted less than five per cent of their own savings to the capital transfer process that is so vital for the developing countries. This average conceals the fact that the USA has been, as noted before, a taker rather than a giver of capital so that the USA is absorbing rather than contributing to the global pool of savings that are being transferred. This should be contrasted with the historic record which the Nobel Prize economist, Professor James Tobin, alluded to in a recent talk when he pointed out that:

> before 1914 the United Kingdom lent overseas as much as half its national saving, financing the economic development of the Americas, Australia, India and other realms of the British Empire in Asia

> and Africa...and the Bank of England served as a sort of world central bank and lender of last resort... Relative to the size of the world economy, the transfers of saving from (all the) developed to developing economies are now much smaller than at the beginning of the century.[36]

The challenge is how to achieve this large transference without the imperial trappings that characterized the essential nature of the UK's policies that were designed in large part to establish, expand and maintain a parasitic empire.[37]

While this state of affairs prevails, the leaders of the most economically powerful nations should be judged as having failed the test of global leadership. That test is not only to be judged on the basis of whether cataclysmic breakdown has been avoided but, as importantly, whether the global "public good" has been promoted towards achieving a state of affairs where there are rules of behavior in commerce and every-day activities that are characterized by the attributes of a desirable world in terms of stability, fairness and openness.[38] Attaining this objective calls for institutional changes at the level of global governance and benign leadership by the most economically/financially powerful nation, the USA, and supportive collaboration by the other members of the G-7 to undertake or support the necessary changes. If "benign leadership" is one characterized by a concern for the well-being of the global community over the long-term, it is quite clearly lacking if judged by the current polarizing trends. Instead what we find is the world's most economically and financially powerful nation responding in its policies to crass self-interest that is as short-sighted as it is selfish. As one American columnist put it, "greed is in the saddle."

These attributes are difficult to change when the USA is the home base of most of the very large multinational corporations who exercise enormous lobbying power and has a political system of check-and balances that is a prescription for policy paralysis rather than active promotion of a more stable and equitable world order. To take one aspect, the US system of government is one in which the legislators in the House of Representatives have two year mandates that force them – if they wish to be re-elected - to have short-term and parochial horizons, short term because their mandates are so short and parochial because over half of its members do not even own a passport and are either lacking in knowledge of the outside world or lacking in compassion about the fate of that world beyond their borders. Under the circumstances it should occasion little surprise that the stance taken by these legislators is not likely to be a global one as befits a nation that has pretensions to be a benign leader.[39] Even a very popular President like Franklin Delano Roosevelt found it politically difficult to accept a boatload of refugees from Nazi Germany and to enter the war on the side of the Allies.

It is difficult enough to steer the process of change through the political reefs in the face of powerful cross-currents of conflicting ideas and headwinds of vested interests in the *status quo*, but when the corporate vested interests are exceptionally influential it could be asserted with justification that we already have a great deal of global governance and that those who wield the power are not the elected leaders of governments but corporate executives with funds enough to influence public opinion and to gain access to legislators and their support. "There are," a former Canadian Ambassador to the USA, Allan Gotlieb, has asserted on the basis of his experience, "a thousand points of entry into the American political system." When more than half the total international trade is between branches of the *Fortune 500* multinational companies, the value of whose output accounts for about a quarter of the global total and whose power is reflected in the fact that a handful of companies control over half the global market in a wide range of key industries, it is essential that this power-behind-the-throne be recognized. And that their objectives also be recognized, namely, an obsessive focus on the financial bottom line of the next year-end statement. This is not meant to condemn the corporate leaders in terms of personal traits, but rather to appreciate that their conduct is inherent in the nature of their function within the prevailing system. But this perspective and conduct is clearly not compatible with benign global leadership that strives for the collective "public good."

As a practical matter for the short and medium-term, the challenge is to find a means of gaining the cooperation of the powerful transnational corporations without succumbing to their power in setting priorities with respect to objectives and modalities to achieve them. In the process, it is essential to re-think the role of the international public institutions, principally the U.N and its system of specialized agencies so as to assure leadership that can assure that the "public good" is being well served by this collaborative approach. To achieve this it is essential that boldness and imagination be key ingredients in the strategy and tactics of the international community: boldness given the formidable nature of the obstacles that must be surmounted, imagination given the *mega* nature of the institutional and policy changes required to achieve the desired objectives.

Strategy and Tactics

> The world may be moving inexorably towards one of those tragic moments that will lead historians to ask: why was nothing done in time?
>
> Ethan Kapstein, "Workers & the World Economy", *Foreign Affairs*, May-June, 1996

Realists urge that we recognize the necessity of providing some incentives for the private sector firms, especially the transnational corporations to shift their research priorities from its narrow financial criteria to a broader social one with global dimensions. One could appeal to these corporations to make such a shift to meet the desired goals. This would require that governments launch programs of moral suasion in asking the corporate executives to look to the longer term when their market would be expanded by virtue of the developing world being made healthier, better educated and more prosperous.[40] It is a "soft" approach that is not likely to be persuasive.

Professor Sachs has put forward a concrete suggestion with regard to promoting research for the direct benefit of the poor in the developing countries in the sector of health or life-sciences but which is applicable as well to research pertaining to the development of energy and informational technologies that can be adapted in design and cost to be of direct benefit to the poor in the developing countries. He makes a realistic assumption that with respect to applied research "there is no escape from...public-private collaboration" given the dominance of private sector research that dwarfs the public funding for such research. In this connection he points to the fact that the funding for the worldwide network of public-sector research programs in one field of life-sciences is dwarfed by the budget of a single multinational firm, Monsanto, that is attempting to patent the genetic code for a variety of seeds which, if successful, would oblige farmers to buy new seed for every planting with profound implications. This comparison between public and private research programs in terms of expenditure and patent rights is applicable as well with regard to research programs in the fields of energy and of informational technology, including its application in the field of education. He has ventured to assert that his proposal "could galvanize the world of the private-sector": have rich countries put up contingency funds that guarantee a market to the corporations that are prepared to engage in research and development of relevant products that are of direct benefit to the poor.[41]

The issue is not only one of the degree and the manner of that collaboration which is essential if realism is to hold sway over preference. But, as importantly, there is the issue of control of policy with regard to the application and the financial beneficiaries of the research results. Public financing of the desired research programs would assure that the inevitable conflict between property rights and people rights would be won by the people. A public-private collaboration requires careful design to guarantee the same outcome. That outcome is especially important with regard to the rights - and royalties – ensuing from any private-public arrangement since private ownership of key patents is a prescription for a further polarization of incomes and all that that implies. Ex-post regulation would not suffice; public leadership is essential to assure that the public interest is being served.

This role of protector of the "public good" would have to be assigned to one or more of the specialized U.N. agencies and, in doing so, would add *new* responsibilities, responsibilities that these institutions have, in any case, begun to assume – as the World Bank has done in its role with regard to CGIAR and other programs of international scope to which it devotes as yet only a very tiny fraction of its budgetary outlays. The CGIAR model that involves such collaboration between the World Bank, bilateral aid agencies and privately-endowed foundations (such as Rockefeller, Ford and Kellogg) is one suggestive model of an organizational arrangement that commends itself for applied research in environmentally-benign energy, especially applicable for rural areas beyond the electric grid and for applied research in informational technology, especially applicable for educational purposes.

The "newness" could be understood in its institutional aspect as:
i) the strengthening/expanding/rationalizing the operations of the existing international agencies and/or ii) the establishment of additional institutions on their own or as adjunct to those already in operation. The first approach is clearly preferable since starting new institutions is a difficult and lengthy process and, in any case, the existing ones may well be amenable to more rapid change in their structure, in their polices and in their mode of operations to adapt to new challenges and opportunities as conditions change. Most of the necessary changes could be made without any modification of the prevailing constitutional frameworks of the international organizations.[42] In any case, incremental changes in the system of UN agencies have taken place over time through policy decisions or modes of operating. In some instances the changes have involved establishing new entities as, for example, the World Trade Organization, that had its genesis as an idea in 1944 when the Bretton Woods twins, the IMF and the World Bank, were born. Its establishment now, fifty years later, is eloquent testimony to how institutional changes in the realm of global governance follow underlying changes in the global economic system and shifting power relationships.[43]

Cooperation of international and bilateral aid agencies and private foundations that are active in the field of development and related research would make the efforts of these international institutions more effective with regard major policy and institutional issues of global scope. Achieving this cooperation and with it, coordination, must be regarded as a key aspect of any strategy for launching a concerted program to alleviate the many deprivations of global poverty. In this connection, there is a critical aspect that needs to be addressed: the frequent stone-walling stance of the USA when other countries share a common but contrary view of a significant problem and the approach to resolving it. These countries are rarely prepared to confront the USA and this is especially so with respect to aid policy. A case in point is a process related to an important component of ODA, namely, the International Development Association (IDA) replenishment ritual that the World Bank negotiates every three years.[44] In these negotiations, all the other countries have usually waited for the USA to take the lead in the pledging process and, thereby, they have lent force to the US withholding tactic that, to put it bluntly, is a way of holding up the World Bank (or any other financially dependent international agencies involved) for ransom: we will not pay our fair share that is expected of us – and to which we agreed - unless and until such and such is agreed upon as policy and practice by the World Bank.

The same US opposition to an idea others consider worth discussing is evident in the case of the so-called "Tobin tax", a suggestion put forward in the early 1970s by the Nobel Laureate, Professor James Tobin, and which in recent years has gained widespread support as an idea worth serious examination. The US response has been stone-walling in a very public manner that is virtually unprecedented: Senator Dole and three other of his colleagues introduced a bill in the Senate in 1996 to prohibit UN officials from considering the Tobin tax or any other international taxation proposal.[45] Senator Jesse Helm went so far in his opposition to the Tobin proposal as to state that the US Congress would consider quitting the UN if its officials. were to ignore that message of censorship. As Künibert Raffer observes in his review of the Tobin tax, "Senator Dole's bill is remarkable since it was designed to limit freedom of thought and research by government action, a practice unusual in democracies."[46] The reason for this strong reaction is not difficult to fathom: it is that same as the logic applied to IDA replenishment. It is effective simply because other nations wait for the USA to make its pledges and its payments in accordance with the understanding about each nation's quota that is based on some agreed-upon formula incorporating such factors as the economic size and the population of the contributing countries.[47] To counter this withholding strategy of the USA on the occasions when it is used to kill an initiative that has the support of a group of "like-minded" countries, might it not be appropriate to consider how to proceed *without* US involvement - at least at the initial stage? Their participation would soon follow.[48]

Devising and implementing effective strategy and tactics that would achieve the desired objectives is a formidable challenge at the best of times, but especially so in this time of transformation or *mega*-change. The appreciation of the dire consequences of not rising to the challenge can, perhaps, be a forcing mechanism. The fear of mutual peril might have more persuasive power than the appeals to team work and the appeal of a humane and prosperous world for *all* its current and future inhabitants that, for the first time in history, human ingenuity has made a realizable dream. Morality is necessary but, it seems, hardly sufficient. When, however, the prevailing condition is seen to be unsustainable *in its dynamic*, the issue becomes more troubling even for the cynic since it raises the issue of global stability and security. After all, it takes little imagination to envisage what might happen if these trends persist: at some point in time, say, the decades when the next generation exercises power, there is a high probability that the globe could well be embroiled in turmoil beyond our imagining. If we ignore this danger we would seem to be sleep-walking to the edge of a figurative precipice.

The necessary policy and institutional changes will be made by volitional decisions in a controlled and orderly way or they will be made by *force majeure*, that is, in an uncontrolled and precipitous manner. Pending that outcome that may be a long time in coming, international initiatives should be taken, hopefully, well before that denouement. We know what could be done in a shorter time frame within the basic framework of current institutions that could alleviate some of the most distressing attributes of poverty. An international dialogue leading to feasibility studies is an essential first step on the road to action to launch these international programs and, at the same time, to make the global economic/financial system more congenial for the developing countries and, thereby, more equitable. In a phrase, what this process involves is a frontal attack on the poverty issue by focused programs regarding health, energy and informational/educational technologies and, at the same time, mapping out a strategy for *global-scale structural adjustment* with appropriate institutional arrangements, in a phrase, a 21st century version of a Bretton Woods type of agreement with new objectives and capabilities that are appropriate to the times. Since, for the first time in history, something can be done that is commensurate with the challenge, there is no longer a need to accept, as Jesus did, the prevalence of the scourge of global-scale poverty.

References

[1] See, for example, David Dollar, *Dividends and Disappointment: Learning from the Success and Failure of Foreign Aid.*, World Bank Policy Research Working Paper, Washington, D.C.,1998; Paul Collier & David Dollar, *Aid Allocation and Poverty Reduction*, World Bank Working Paper #2041, 1999, and David Dollar, Assessing Aid: What Works, What Doesn't and Why, Oxford University Press, N.Y., 1998 (web site:worldbank.org/research/aid/aidtoc.htm). This point of view has been clearly, if crudely, articulated by the secretary-general of OECD who in a recent (July 1999) speech in New York placed all hope for development of the Third World countries on the private sector and advises the developing countries to pay heed to the issue of corporate governance through accommodating policies and practices to entice foreign investors: *"strong corporate governance rules will be the basis on which capital will eventually flow to the poorest countries and it is private sector investment – not government or World Bank handouts – that provides the best way of narrowing the huge gap between the world's richest and poorest nations. Bringing sound corporate governance practices to the world is not going to be an easy or quick task but someone has to do it."*

[2] This income gap is reflected in some revealing statistics from World Bank and UN sources: in 1996 the *per capita* income gap was a ratio of about 1:50+ between those in "low income" countries (with average *per capita* incomes of $490 that contain more than 56 per cent of the world's population of 6 billion) and those in "high income" countries (with average *per capita* incomes of $25,870 or more that is home to about 16 per cent of humanity); a quarter century earlier that ratio was approximately 1:30. Lest it be thought that this is a poor country-rich country phenomenon, it should be noted that when the comparison is made between the poorest 20 per cent of humanity and the richest 20 per cent, that is, abstracting from country categories, the ratio is roughly estimated to be 1:140+.

[3] For an interesting approach on the issue of income polarization, see three articles by three American professors: Robert H. Frank, "Talent and the Winner-Take-All Economy", *The American Prospect*, Spring 1994, Paul R. Krugman, "The Right, the Rich and the Facts: Deconstructing the Income Distribution Debate", *The American Prospect*, Fall 1992, and Barry Bluestone: "The Inequality Express", *The American Prospect*, Winter, 1994.. This last article could be subtitled "Murder on the Orient Express: Who Killed Equality ?"

[4] The fact that it is a global phenomenon is evident in considering the US economy: Professor Paul Krugman has stated that in the USA during the 1980s "the rise in incomes at the top is an major economic issue" and offers "the now infamous 'Krugman calculation' that 70 per cent of the rise in average family income has gone to the top 1 per cent of families…and many measures of real wages for typical workers show a decline." "The Rich, the Right, and the Facts: Deconstructing the Income Distribution Debate", *The American Prospect*, Fall

1992, page 22. Since 1992 all these trends have greatly accelerated. For a broad-based summary of US trends see the book by Mickey Kraus, *The End of Equality*, Basic Books, N.Y., 1992. For a commentary on that book see the article in *The New York Times* of May 8,1994, "'Winner-Take-All' Economics: Life's Hard? Blame the Market."

[5] "The Myth of Declining Aid", ed., Roy Culpeper, Albert Berry & Frances Stewart, *Global Development Fifty Years after Bretton Woods*, St. Martin's Press, N.Y., 1997, page 132.

[6] Over a third of international trade is intra-firm trade (about $1.5 trillion of $4.5 trillion in 1996). The power of these corporations is reflected in a few simple statistics: as few as five of them account for half the global sales in the automotive, aerospace, consumer durables, and steel, and account for about 40 per cent of the global market for oil, personal computers and media; in the period from 1980 to 1993 their sales increased by 140 per cent and their assets increased in value by 230 per cent while 4.4 million of their employees were "downsized". These statistics are elaborated more fully in *The Economist*, March 1996 and in Sanjaya Lall's, "TNCs: The New Custodians of Development?", ed., Roy Culpeper, Albert Berry & Frances Stewart, *Global Development Fifty Years after Bretton Woods*, St. Martin's Press, N.Y., 1997, page 171.

[7] As, for example, in 1993 when the total ODA flows dropped significantly and in 1998 when FDI in the developing countries fell by 4 per cent while globally the total FDI flow rose 40 per cent.

[8] "Wither the World Bank and the IMF?", *Journal of Economic Literature*, December 1998. She adds, "it is difficult to understand how the (contributions of the) international financial institutions …might have been expected to result in a quantum leap in economic performance."

[9] For an interesting paper on this theme, see *Geography and Economic Development*, John Gallup, Jeffrey Sachs and Andrew Mellinger, NBER Working Paper,#W6849, Cambridge, 1999.
Professor Sachs makes the same point in his invited article in *The Economist* of August 24th, "Helping the World's Poor": *"As it happens, the poor live in different ecological zones, face different health conditions and must overcome agronomic limitations that are very different from those of rich countries. Those differences, indeed, are often fundamental cause of persisting poverty." He notes that 93 per cent of the combined population of the 30 highest-income countries lives in temperate and snow zones. Of the 47 highly indebted poor countries . by contrast, 39 are tropical or desert societies. There are only three in a substantially temperate climate, and those three Laos, Malawi and Zambia) are landlocked and, therefore, geographically isolated."*

[10] Recently published statistics reported by UNCTAD reveal the volatility: in 1998 foreign direct investment increased globally by almost 40 per cent over the previous year to reach $644 billion while, at the same time, FDI flows to developing countries fell by 4 per cent. See *The Economist*, June 26, 1999. As for ODA flows, they peaked in the early 1990s and dropped dramatically in 1993 only to recover some ground slowly.

[11] Almost two decades after "the international debt crisis" erupted in 1982, the outstanding debt of the developing countries still remains high and a heavy burden on the poorest of the developing countries. Only about $20 billion is still owing to private banks; the World Bank and the IMF and a few other international agencies now hold most of the outstanding debt. Their approach has prompted Professor Jeffrey Sachs to write an op-ed article in *The International Herald Tribune* of June 12-13, 1999, titled, "Release the Poorest Countries from Debt Bondage", in support of the cause of the organization, *Jubilee 2000*, that is dedicated to achieving outright elimination of the foreign debt of the poorest countries: *"The financial bankruptcy of the poorest countries has been evident for at least 15 years, but the IMF, the World Bank and the rich countries have delayed real solutions to this chronic problem. The World Bank and the IMF take great pride in claiming that their loans never go bad, so instead of recognizing reality, they lend the poorest countries new money to repay the old debts, claiming that the loans are still sound...the only winners (being) the staff of the IMF and the World Bank which have invented a perpetual motion machine for missions to these countries."*

[12] In 1996 a "bold" initiative to reduce the debt of the poorest countries was announced but after three years only two countries of 44, Bolivia and Uganda, had qualified for any relief. The proposed measures of 1999k the Cologne Initiative for debt relief to the "Heavily Indebted Poor Countries (HIPCs) is another attempt. Its promise is a reduction of the outstanding debt of the 33 countries that are eligible countries from $130 billion to $60 billion or, in present value terms, to $27.5 billion, and make $20 billion of concessional loans available. But the relief is contingent on the indebted countries being not only desperately poor but also agreeable to adopting "good economic policies on a sustainable level". The eligibility criteria with respect to "the sustainable level" is that the present value of the debt be no more than 150 per cent of annual export revenues or 250 per cent of annual budgetary revenues. *The Economist* (June 26, 1999. Page 23).observed that "the promised relief may prove elusive or come at the expense of other aid and other recipients." In any case, since these are the very countries that could not service their outstanding debts and have been judged to be unlikely to do so now - or ever, the Cologne Initiative is a hollow gesture. The irony is that the elimination of the entire debt is feasible without much strain on the global community's richer members and on the international institutions that hold most of the remaining debt. For example, the IMF could write off the debt of $7.8 billion owing to it by selling

off about one-third of its gold reserve: it has $22 billion unrealized capital gains given that gold is valued on their books at $47 per ounce while the market price is over $200 per ounce. The World Bank has special reserves set aside for loan losses that could cushion the largest part of the financial blow that would be incurred in writing off those debts.

[13] On this theme, see my paper, *The Folly of Demonology: How Critics View the World Bank and How Critics Should be Viewed*, University of Ottawa Faculty of Administration Working Paper, 1995. Available also on my home page: www.gutenberg.com/~miller.

[14] "Nothing to lose but your chains". *The Economist*, May1, 1993. For a detailed assessment of SALs, see Tony Killick, "Low-income Countries and the IMF- a Case of Structural Incompatibility", *Global Development Fifty Years after Bretton Woods,* op.cit., and also Killich's contribution to a conference in 1995, "Adjustment and Economic Growth", *Fifty Years after Bretton Woods: The Future of the IMF and the World Bank*, op.cit., pages 139-159. His assessment is that "the evidence on the economic consequences of structural adjustment programs does not point to strong results (but) the assessment is fraught with difficulty, particularly because we can only guess at what would have happened without such a program...(However) in light of the evidence it seems justified to describe the effectiveness of structural adjustment programs as patchy, at best." (pp.141 & 142).
See also, Paul Mosley, Jane Harrigan & John Toye, *Aid and Power: The World Bank and Policy-based Lending*, Routledge, London, 1991. The World Bank assessments are set out in *World Bank Structural and Sectoral Adjustment Operations: The Second OED Review*, June 1992, *The Third Report on Adjustment Lending: Private and Public Resources for Growth*, 1992, and *Adjustment in Africa: Reforms, Results and the Road Ahead*, 1994.

[15] Mohsin Khan, "The Macroeconomic Effects of Fund Supported Adjustment Programs", *IMF Staff Papers*, Vol.37, No.2, 1990, p.222. Cited in Michel Chossudovsky, *The Globalisation of Poverty: Impacts of IMF and World Bank Reforms*, Zed Books, London, 1998, p.69.

[16] For a full treatment of this issue, see M. Miller, *Coping is not Enough! The International Debt Crisis and the Roles of the World Bank and the IMF*, Dow Jones-Irwin, Urbana, Illinois,1986, and M. Miller, *Debt & the Environment: Converging Global Crises*, U.N. Publications, N.Y.,1995.

[17] Indeed, one critic, Professor Michel Chossudovsky, whose writings are extremely critical of the World Bank and the IMF, in discussing the issue of SALS, in one of his latest books goes as far as accusing these institutions of imposing conditions that have the effect of "brutally dismantling the social sectors of (the recipient) developing countries." (*The Globalisation of Poverty: Impacts of IMF and World Bank Reforms*, Zed Books, London, 1998, p.68.) He charges the IMF and the World Bank with imposing through the SALS "a consistent and coherent...programme of economic and social collapse." (p.69). I was on the Board of the World Bank in the early 1980s when the SALs were initiated and expressed my opposition to this shift in lending at the expense of project lending on the grounds of their likely harmful impact on the poor of the recipient countries through such measures as reducing subsidies and welfare programs, etc. In any case, this was the type of lending that the IMF does with questionable results. Since SALS accelerated World Bank disbursements and, thereby, enhanced the ability of debtor countries to service the outstanding debts that were owed to the creditor banks located in the richer industrialized countries, the interest of the creditor banks was apparently the paramount motivation for pressing the World Bank to engage in SALs.

[18] The period of the two decades after World War II was one of even more rapid growth: *per capita* growth of 2.4 per cent annually in the 1950s and from 1965-70, 3.5 per cent with the slowest rate in sub-Saharan Africa averaging 2.3 per cent.

[19] The progress that has been made is sketched in detail in *World Development Indicators 1998*, pages 3-27 and in the *Human Development Report 1999* where, on page 22 there is "a balance sheet of human development, 1990-97" setting out some other indicators of global progress side by side with indicators of global deprivation that define the challenge..

[20] These measures are discussed in some detail in M.Miller, *The Challenge and the Opportunity: Approaches to Alleviating Rural Poverty in Southern Africa*, University of Ottawa Working Paper, Ottawa, Canada, 2/96. This is available on my home page: www.gutenberg.com/~miller.

[21] There are exceptions, as for example, the Montreal conference on ozone depletion that resulted in the Montreal Protocol that has been quite effective in addressing the problem through international cooperation. In this case three of four preconditions for success were present: first, a reasonably solid scientific base existed, second, there were substitutes available or could be produced to phase out the harmful CFC and a secretariat to facilitate this process, third, there was a way of monitoring compliance and a secretariat to undertake the evaluations of compliance. The fourth relates to enforcement through fines or other means but as world government does not exist to undertake this task, moral suasion has to suffice. For an elaboration on this theme, see my paper "The Futility of Large-scale UN Conferencing: Energy as a case in point", (prepared for the Rockefeller LEAD conference paper, Bangalore, India), University of Ottawa Working Paper, 1995. Available on my home page: www.gutenberg.com/~miller.

[22] At that time I was involved as a member of the World Bank's Policy Planning Division working with Mahbub ul Haq on the issue of how the World Bank could operationalize the rhetoric, namely, doing something significant about the absolute poverty of "the bottom 40 per cent".

[23] It is pertinent to recall that poverty alleviation has always been the declared objective of the Bank but since 1973 it was made very explicit. However, there were only a few modest changes made in operational terms: there was a slight shift in the lending program towards "integrated rural development", education, health, environment and other such "soft" projects and programs. In 1990 the World Bank published its *World Development Report* with poverty as its main theme. In the case of the World Bank – as with many other development agencies - one of the major causes of failure to live up to the promise of helping the poor lies with the prevailing culture of the organization that rewards managers by the amount of lending their departments can achieve, an incentive that introduces a bias favoring capital-intensive programs and projects and against those that are designed to alleviating poverty in the rural sectors where projects and programs are characterized by much greater social and political complexity for each unit of "throughput". Managers are not fools: they know to which signals they should respond if promotion is important. At the same time, the poor countries left out of the lending loop have little power to change the priorities of the lending institutions and of those with the power to set the priorities of these institutions.

[24] This, in effect, merely recognizes that the concept of the gross national product (GNP) does not measure "well-being" - as the annual *UN Human Development Reports* emphasize in their ranking of countries by misery indices or human development indices rather than by the average GNP *per capita* numbers alone.

[25] For a full treatment of this issue, see M. Miller, *Debt and the Environment: Converging Crises*, U.N. Publications, N.Y., 1991 and M.Miller, "Where is Globalization Taking Us?: The Need for a New (Improved) Bretton Woods", *Futures: Special Issue for the U.N. Summit on Social Development*, Oxford, April, 1995.

[26]See K. Schmidt-Hebbel, *Savings across the World*, World Bank Discussion Paper #3354, Washington, D.C.,1994. Also N. Loayza, K. Schmidt-Hubbel, & L. Servén. *What Drives Saving across the World?*, World Bank, 1999; B.B. Aghevil and J.M. Boughton, "National Saving and the World Economy: Why have savings rates declined since the early 1970s?", *Finance & Development*, IMF & World Bank, Washington, D.C., June 1990, pp. 2-5; A.L. Bovenberg, "Why has US Personal Saving Declined?", *Finance & Development*, June 1990, pp. 10-11; and J.J. Polak, "The Decline of World Savings", *Economic Insights*, Institute for International Economics, Washington, D.C., September/October 1990, pp.18-19

[27] The exception are a few donor countries – Denmark, Norway, Sweden and The Netherlands - that contribute much more than double the percentage of their GNP than do the members of the donor community as a whole and who contribute as a percentage of the GNP about 8 times that of the USA The present level of ODA remains far short of the target of 0.7 per cent of their GNP of the donor community that was promised many decades ago at the onset of the so-called "Development Decade."

[28] In an "overview paper" titled *A Popular Coalition for Action* that was prepared for the 1995 Conference on Hunger and Poverty, the authors set out an arithmetic exercise to calculate the "global poverty gap" that is defined as the extent of the shortfall of the income of the poor in relation to a "poverty threshold" below which they would not have enough purchasing power to provide adequate food, shelter, clothing and access to a "reasonable" quality of basic services of education and health. In sub-Sahara Africa that deficiency gap was estimated to be about 20 per cent, in East Asia about 3 per cent, and globally about 10 per cent of the poverty-line income.

[29] The Institute of International Economics has estimated $1,500 billion per year would be needed over ten years to catch up to US *per capita* income levels. A Munich-based economic forecasting institute, IFO, has estimated that the annual capital needs of eastern Europe would be between $75 and $100 billion and of the former Soviet republics and Russia, between $185 and $235 billion. Assuming these countries manage to provide about 4/5 of their capital requirements from internal sources, the estimated gap would range from about $60 billion to $90 billion per year.

[30] For a breakdown of the estimates see, M. Miller, *The Energy/Environment Connection: Overcoming the Institutional Obstacles to 'Doing the Right Thing"* , World Bank/EDI Working Paper, Washington, D.C.,1990, and "Getting Grounded at Rio", *Ecodecision*, Montreal, April 1992. The carbon emission reduction program alone runs into hundreds of billions of dollars with one estimate for a 40% reduction estimated at $300 billion per year and $400 billion to achieve a 60 percentage reduction target.

[31] The estimated flow of financial transactions is estimated to have a value of about $1.8 trillion a day of which it is estimated about 5 per cent is to finance trade and about 10 per cent to finance investments, the remainder being speculative in nature. The Bank for International Settlements (BIS) estimates that about 40 per cent of these transactions involve round-trips of 2 days or less and 80 per cent of 7 days or less. (See Inge Kaul et al, *The Tobin Tax: Coping with Financial Volatility*, Oxford University Press, 1996, p. 3) The speculative component does have a positive function in making currency markets more liquid, but hardly in proportion to its disruptive

role as a contributor to volatility with which developing countries find great difficulty. For a summary of the issue, see the article by Kunibert Raffer, "The Tobin Tax: Reviving a Discussion", *World Development*, Vol.26, #3, 1998, pp.529-538

[32] See, for example, Philip Shabecoff, in an article entitled "A 'Marshall Plan' for the Environment", (*The New York Times*, May 3, 1990). The same idea has also been put forward under various labels with the necessary funding being raised through such means as the issuance of environmental mutual funds, tax-free environmental bonds, environmental flow-through shares, environmental capital venture funds, an emergency environmental fund, and such. A "Global Environment Facility" has been set up and is operational, but with funding far far short of the scale of the Marshall Plan.

[33] R. Dobell and T. Parson ("A World Development Fund", *Policy Options Politiques*, 12/1988) have suggested the establishment of a World Atmosphere Fund to be financed in part from the auctioning of pollution-emission permits, a proposal that is already incorporated in the seabed mining provisions of the U.N. Convention on the Law of the Sea. *The Brandt Commission reports of 1980 and 1983 put forward a similar proposal to establish a World Development Fund that would collect and administer the incomes generated by such royalties and taxes. There is treatment of this theme in the Brundtland Commission report, *Our Common Future*, (Oxford University Press, 1987), chapter entitled, "Managing the Commons".

[34] Professor Tobin has observed that a rate of 0.2 would not interfere with currency transactions that are related to trade and productive investment: "*it means nothing for a round trip of a year or more from one currency to another and back, but for one week round trips it would be equivalent to a difference between interest rates in the two markets of 10 or 210 per cent per year, a palpable protection of monetary authority.*" ("Financial Re-globalizaton"., *Policy Options/Options Politiques*, op.cit. p.22). In a recent paper (the prologue to Kaul et al, *op.cit*., page .ix), he has calculated that a tax of 0.2% on a round trip to another currency would cost 48% a year if transacted every business day, 10% if every week, 2.4% if every month and a trivial charge on commodity trade and even less on long-term foreign investment.

[35] The Tobin Tax can be conceived as a variant in as much as it would involve charges (as tax, rent or royalties) on the users of the global commons, the upper atmosphere when satellites are parked there in orbit and are used to send financial bits of information between forex centers or the deep oceans when cables are used for the same purposes. Professor Ignacy Sachs has suggested that toll gates at Gibraltar, Panama and Suez could levy charges for those who want to cross the oceans. On this issue see Arthur Cordell et al, *The New Wealth of Nations: Taxing Cyberspace", Between the Lines*, Toronto, 1997.

With regard to charging for the past presumptive right to pollute the environment that was exercised by the industrialized nations that treated the garbage-can capacity of the environment of air and water to absorb pollutants as a free good, see D. Helm and D. Pearce who have noted that the practice "can easily turn the morality of the polluter-pays principle upside down (to become) the victim-pays principle". They had predicted, in this connection, that "the issue of 'side payments' would likely dominate the 1992 United Nations Environment Conference in Brazil." ("Assessment: Economic Policy Towards the Environment", *Oxford Review of Economic Policy*, Vol.6,No.1,1990). Also in the same issue see Professor D. Dasgupta's interesting article, "The Environment as a Commodity".

[36] Reprinted as an article, "Financial Re-organization" in *Policy Options /Options Politiques,* Institute of Research in Public Policy (IRPP), Montreal, July-August 1999, p.22.

[37] Angus Maddison's writings on the historic global-wide growth record and the factors related to this growth include an analysis of the capital flows related to empire-building and maintaining that empire in an exploitative manner. He found that the UK sent out more of its savings than it received in return through the exploitative process, but that while this was a negative factor for the people of the UK as a whole, a small percentage of the UK's population gained enormously from the imperial domination.

[38]"Public goods" or "collective goods" are defined succinctly by Professor C.P. Kindleberger, ("International Public Goods without international Government," *The American Economic Review*, March 1986), as "that class of goods (or services) like public works where exclusion of consumers may be impossible, but in any event consumption of the good by one consuming unit - short of some level approaching congestion - does not exhaust the availability for others." In the international arena he lists "an open trading system, well-defined property rights, standards of weights and measures that may include international money, or fixed exchange rates, and the like." Dr. Sylvia Ostry (*Governments and Corporations in a Shrinking World: Trade and Innovation Policies in the United States, Europe and Japan,* (Council of Foreign Relations Press, N.Y., N.Y., 1990) states that "the primary 'international public good' of the multilateral trading system is stability. Greater co-ordination of trade policies among the major economies would increase global stability and, thereby, global welfare."

[39] That parochialism is reinforced by the fact that members of congress must reside in the constituency where they live, unlike the arrangements in a parliamentary system. If they wish to be re-elected this residency requirement ties them tightly to the interests of their locality or region and this often conflicts with the national

or the global interest. Checks and balances are fine to avoid tyranny but hardly fitting for dynamic leadership that requires enough flexibility of governmental leaders to override national interests for the sake of the global ones that, in the long term, are bound to be in the self-interest of the leading nations. It is relevant to contrast this congressional system with its three poles of power with the parliamentary system in its basic principle. Edmund Burke expressed it well in the letter he wrote to the constituents of Bristol, England: he asked his constituents to elect him on the understanding that if there was to be a conflict between their local interests and the national interests he would be duty bound to support the latter.

[40] An excellent illustration of this approach is the suggestion of Professor Jeffrey Sachs in his recent article in *The Economist* (op.cit.) that a Millennium Vaccine Fund be established. "We need," he writes, " a new and creative institutional alliances (such as) a Millennium Vaccine Fund which (would) guarantee future markets for malaria, tuberculosis and AIDS vaccines. (This is an approach that "is administratively straightforward, desperately needed and within our technological reach. Similar efforts to merge public and private science activities will be needed in agricultural biotechnology." (p.20). On the latter aspect of his proposal, the omission of reference to CGIAR is surprising given its focus on research to increase agricultural productivity and, in that connection, includes bio-technological research. Yhe private sector participation in CGIAR is limited to foundations such as the Rockefeller, Ford and Kellogg who initiated the idea of CGIAR.

[41] *The Economist*, August 14th, p.19. This suggestion would, of course, stipulate a minimum purchase price for a product that meets minimum conditions of efficacy. He points out that in the case of medicinal drugs there is a threat to the drug firms of the poor countries resorting to copies or generic drugs that would be cheaper by not paying for the research required for their development. The same recourse could be expected with regard to informational and energy technologies. However, this would not stimulate the research needed to bring down their costs and adapt their design for the needs of the poor in the developing countries. That stimulation is essential and incentives of a financial nature are apparently required to gain the necessary cooperation of the multinational firms already engaged in relevant lines of research. The proposed approach raises issues of ownership of patent rights and the role of the relevant specialized UN agencies.

[42] Edward Bernstein of the Brookings Institution is the most articulate advocate for this role to be left to the IMF. His position is well summarized in the article he wrote for the 40th birthday of the Bretton Woods twins that was published in the September 1984 issue of *Finance & Development* entitled "Do we need a new Bretton Woods?". For a critique of Bernstein's views see, M. Miller's, *Coping is not Enough!, Ibid,* pp.155-171.

[43] The USA scuttled the proposal for an International Trade Organization that was to be the third institution emerging out of the Bretton Woods conference of 1944. The republichan whip, Newt Gingrich threatened to follow that precedent on the grounds that he was against organizational arrangement that would, as he saw it - in the words of a *New York Times* editorial of May 10, 1994 - "subject Americans to rules imposed by Third World countries". The NYT editorial that is captioned "Mr.Gingrich's False Alarm on Trade" puts forward a revealing line of reasoning in arguing for the WTO ratification by noting that Gingrich did not have much to fear about "subjecting Americans to rules imposed by Third World countries" since "the WTO would be more pussycat than tiger - and would protect U.S, interests better than the existing GATT." In this connection it noted that the WTO's primary function would be to resolve trade disputes under procedures that the US delegation fought hard to achieve: panel decisions would prevail unless all parties to WTO decided to set them aside and since the USA brings more complaints than any other country, this would favour the USA; furthermore. should the dispute panel decide against the USA there was little to fear since *"the WTO bares no fangs: it cannot force Congress to change (its) trade laws (and) few countries would (dare to) engage in masochistic tit-for-tat retaliation with the USA...The WTO (only) sounds scary..."* So much for the spirit of international cooperation as a give-and-take process.

[44] These negotiations involve the pledging of funds to IDA, a subsidiary of the World Bank, that then lends the funds at very concessional rates (near zero for long periods including a long grace period before repayment begins) to countries that are too poor to borrow at the prevailing rates of interest that the World Bank secures in the capital markets that it taps for funds. These loans are called "soft" loans and are not treated differently in Bank operations. IDA is resorted to in order to gain funds outside the capital market since the Bank's charges are well below what the Bank would have to pay for these borrowed funds. It is a way of avoiding a deterioration in the Bank's credit standing which is the best there is and is an advantage passed on to Bank borrowers.

[45] This act was introduced in the Second Session of the 104th Congress and was titled, "Prohibition on United Nations Taxation Act of 1996".

[46] "The Tobin Tax: Reviving a Discussion", *op.cit., page 529.* He observes that no such exceptionally strong political resistance was provoked by similar suggestions, as in John Maynard Keynes' *A Treatise on Money – The Applied Theory of Money*: ("a transaction tax could strengthen the weight of long-range fundamentals", "risk and cost would reduce short-term international flows without harming long-term investment", "a substantial government transfer tax would mitigate the predominance of speculation over enterprise' (and) might prove the

most serviceable reform available."), in the *Brandt Commission Report,* and in Joseph Stiglitz's article in 1989, "Using tax policy to curb speculative short-term trading", *Journal of Financial Services Research (3(2-3), pp.101-115)*

[47] When one of my predecessors on the World Bank board, Claude Isbister, pledged the full three year commitment to IDA in one payment, the US Treasury officials were furious and called their counterpart officials in Ottawa to make their displeasure known. When I was in the same seat years later I was warned "don't dare do an Isbister."

[48] It would be interesting to observe the reaction of the multinational firms that compete for a large share of contracts related to World Bank projects if their participation in the bidding process were to be conditional on the USA contributing its obligatory share to IDA. It seems safe to assume that the USA could not afford to play the spoiler in this game if the stakes were high for its own corporations.

NATIONALITY AND CITIZENSHIP IN A GLOBALIZED WORLD

Hugo Estrella Tampieri

The Present System

In the historical development of nation states, the very concept of nationality and citizenship has varied many times. There has been a growth of those considered citizens in direct relation to the success of the struggle for political and civil rights: former slaves, the illiterate, women, minorities, have been slowly accepted among those who can decide on the common destiny of a country.

In the present world relation characterized as "globalization", where the role of nation states is in question, I think that is appropriate to review also those concepts. Traditionally, the acquisition of nationality came from two sources: "the right of blood" and "the right of the land". Each of these has a traditional Latin name: *Jus Sanguinis*, and *Jus Soli*.

Jus Sanguinis implies the right to acquire the nationality of the parents, the forebears, and is an attempt to keep the unity of the group, the link to the original community. No matter where the person is born, what language is spoken in the place, what culture is surrounding them, what environmental conditions are going to determine their temperament; they belong to the national group of their parents or grandparents. It is the prevailing system in Europe.

Jus Soli, instead, means the link of individuals with the land where they are born. It recognizes the ability of the person to be fully integrated in the place, and where they are going to live and interact with other people, the original culture and the environment. This idea tries to build up a new nationality, enriched by what people from various ethnic and cultural backgrounds bring to the common heritage and thus, is intended to create a new relation, more natural, with the land. It is the prevailing system in American countries.

To sum up: *Jus Sanguinis* perpetuates the past, no matter how far away in time and space; *Jus Soli* builds for the future, no matter from how far the family comes, and how ancient it is.

What Is Left of Our Origins?

To complete the idea of the worldwide rule of the "right of the land", as this paper is intended to propose as a matter of concern, we can take a look at the result of keeping a system that has, in my view, a primary responsibility as a root of conflict, as *Jus Sanguinis* has. When I first learnt about what happened during the holocaust, one of the most touching facts was the discrimination not just as a social matter, but as a legal one. People bearing a yellow star on their clothes, forced to live in ghettos and with passports labeled "JUDEA", was something that deeply influenced my abhorrence of ethnically-based systems. Another case study: in Yugoslavia during the half century of communist rule, many fundamental aspects of public life were changed. Not the original nationality. So, in the Federal Republic, each citizen was a member of the Federal Republic, but always retained - legally and clearly printed in official documents - the data of their origin, whether Serb, Croatian, Bosnian, and so on, and when conflicts arose, it was easier to establish racial-ethnic laws and divisions, able to force a border between people that had attended the same schools, shared the same working spaces and lived in the same neighborhoods for decades. In South Africa, tribalism was one of the most evil tools for the oppression of the different groups that live together in the country, and was the mined land that the apartheid system left for the future. They had treated in different ways the diverse groups, in order to have a weaker – divided - opposition.

In any case, the reasons have been similar: the fear of disappearing as a nation, the end of the group. In every case, the fighting factions have gone backwards, trying to dive in to the past in order to tie themselves to the most primitive differences, especially religious ones. Behaving like that meant to fall in the usual "The self accomplished prophecy": passing racial or ethnic laws, terrifying the "enemy", showing each other their mutual strength and ability to destroy themselves.

On the other hand, countries with a long tradition of an equal nationality system, have had their own conflicts, based mainly on racial terms, like in the USA with native people or the black population, and in many Latin American countries with their indigenous groups. In those cases, the problems were more in terms of the exercise of citizenship, civil rights, and income distribution, rather than the nationality issue. Anyway, they pose another important ethical question to be fostered in any legal system: the right to keep the differences, the level of tolerance required to build up a pluralistic society. Such society requires deep integration in the public aspects and an absolute respect of the differences. A society with that level of inclusion can only enrich a culture, and help us learn from one another. This reality has made countries rich in economic and human terms. I come from one of the most successful examples of the "melting pot" experience, and the very basis of that success has to be sought in the basic laws of the last decades of nineteenth century. There was a call for immigrants, giving them the guarantee of never being diminished for their particularities. Religious liberty, an integrative public school system, secular civil institutions and full participation in civic life, were the basis for building a new nation. Fifty years later, that nation was one of the five richest countries on earth according to income. When institutions were broken, civil liberties surpassed by authoritarianism and the preeminence of Catholic, European and anti-communist groups was forced into that society, everything was lost: freedom, wealth, future.

What About Globalization?

Today, for the first time, we are moving towards one of the dreams of the French Revolution: an individual can be a citizen of the world. Science and technology have played a decisive role in this reality, and the transnational economic integration seems to be doing the rest. Economy, in the case of Europe, seems to have planted the seed for the whole European nationality. The European Parliament is a good example of civic participation on a multinational level. In this case we can consider that Europe is moving, then, to the *Jus Soli*, allowing a Greek to work in England or someone from France to represent a political view together with a German. But still there are many countries that by rejecting their recent authoritarian past are facing the danger of other nightmares: former communist republics have established common citizenship but legally keeping nationalities. In Slovakia, for instance, a citizen has their nationality written in his or her passport: Czech, Romany, Hungarian, even though they have been born in the same town and hospital with a Slovak. This situation results in the formation of political parties based on those national origins, instead of basing them in a political programme; the mere existence of such a distinction between citizens is, in my view, a seed of conflict for the future. How can someone feel legally bound and a full citizen of a country that establishes such divisions? When will economic or social problems result in a discrimination of those "second class" citizens for their jobs and public services? We should be aware that the enemies of pluralism, like Mr. Le Pen in France or Mr. Haider in Austria, are strong and not just an invention of Third World countries.

We still have time to prevent the terrible ideas and events of the 1930s happening again. Considering and legally enforcing people to be equal is a good starting point for that.

SCIENCE IN THE PROCESS OF GLOBALIZATION

Stefan Luby

The engines of economic growth are increasingly globalized. Globalization is propelled by information technology. The problem that should be resolved now is how science and technology will be organized and supported in the new environment at the interface between national and global policies. It is assumed that the process of globalization and political and economical integration will not change the structure and supporting schemes in science and technology too much. However, common sets of rules for financing, evaluation and application of results have to be accepted by all countries wishing to be active in science and technology. This process will proceed or even finish within the next decade.

In the post-Cold War era economic growth became the top priority of the strategy of world nations. Both science and technology play a pivotal role in addressing this challenge: they are the keys to modernization and rising international competitiveness. Governments of industrialized countries take a more active role in promoting development in the sci-tech area. As an example of a national strategy, the White Paper on science and technology by the National Science Council in Taipei can be quoted.[1] On the other hand, the engines of economic growth – capital, labour and technology are increasingly globalized. Globalization is propelled by information technology. Today, business mail routinely reaches its destination in a few seconds instead of a few days as before. National information structures are tied together like national power grids or telephone circuits. The resulting global information infrastructure has created not only a possibility but also a demand for industry and trade to operate all over the globe.

The problem that should be resolved now is how science and technology will be organized and supported in this new environment at the interface between national and global policies. Will science also become globalized? Will it separate from technology like it was in the 19th century? Or will it split into realms of national and supranational interests? For the sake of this discussion we will distinguish:

1. pure science – fundamental research looking for new phenomena and principles governing nature and society;
2. applied research - transforming new knowledge into applications in physical, biological and social sciences;
3. industrial development, although with high level of instrumentation and computerization, sometimes better than in fundamental or applied research.

Industrial development is characterized by self-reproduction of technology. Most typical examples are: the innovations and introduction of new generations of semi-conductor devices and circuits following the 1965 Moore's law: the number of components on a chip quadruples every three years without involving any new physical discovery. It is expected that we will come to the end of the present silicon era only in the year 2010.[2] Microelectronics like many other advanced technologies - computers, telecommunications, new materials, biotechnology, aeronautics, belong in the area of global activities. Their further development will be financed by supranational companies. The commercialization of technology expands even into the space, as can be documented by the project of the International Space Station, built by 16 countries.[3]

For a long time science contributed little to economic growth and industrial development. However, in the 20th century the close link between the growth of knowledge and the rise of technology has permitted market economies to achieve unprecedented prosperity.[4] The "western miracle" was born due to well-organized and generously supported applied research in this part of world. It was applied research which created a broad gap between rich and poor nations. Its other role was the defence and conservation of these differences as the story of the Cold War shows. At present the role of applied research is very similar. One of the leading ideas of the World Conference on Science (Budapest 1999) - that science will help to suppress the huge differences in the living standards of countries - does not seem very realistic. We can expect that in future applied research will follow the strategic interests of various poles of the multipolar world created after 1989. It will be supported by corresponding powers and pacts. The NATO scientific programme and the recently launched 5th Framework Programme of the European Union are good examples. Especially the 5th Framework Programme should help to solve the European paradox: excellent fundamental research is not outweighed by the ability to apply its results effectively. We are speaking here about high-tech applied research. The participation in the 5th Framework Programme is a key to European integration for Central and Eastern European countries. (The countries which are already integrated may partly resist the "import" of European priorities formulated by the Programme.)

Fundamental research will also remain in the future part of national scientific policy and legislation. It is a part of national culture, especially if scientists obtain prestigious awards. As a rule they are unknown to the public before. The scientific programmes are oriented inwards following the national priorities of the respective country and outwards, usually in areas where the quality of research corresponds to the top international level. In big science, like particle physics, the integration of resources in unavoidable. Also in fundamental research some features of globalization appear mainly due to brain drain and the different price of scientific manpower. The consequence of this type of globalization is paradoxically the concentration of scientific elite in rich countries or in countries with active immigration policies (Israel).

In fundamental research two myths should be eliminated. One of them is the belief that the results available in scientific journals or over the Internet could generate economic benefit for any country capable of using them.[5] This myopic vision is typical for politicians wishing to squeeze national support of science. The second is that the distinction between fundamental and applied research is to a large extent arbitrary, or even destructive. It is often used to mask overpopulation in fundamental research or to move it to the more applied topics.

Concluding this discussion, I assume that the process of globalization and political and economic integration will not change the structure and supporting schemes in science and technology too much. However, it will increase the level of standardization in this area. A common set of rules for financing, evaluation and application of results has to be accepted by all countries wishing to be active in science and technology.

References

1 White Paper on science and technology, NSC Taipei 1997

2 M. Schulz, *Nature* 399 (1999) 729.

3 Declaration of UNISPACE III Conference, Vienna 1999.

4 N Rosenberg, L. E. Birdzell, *Scientific American* 263 (1990) 18.

5 M.L. Good, *Physics Today*, August 1996, 23.

INFORMATION TECHNOLOGY IN DEVELOPMENT SITUATIONS

Paul Guinnessy

This paper briefly looks at the problems associated with increasing information technology (IT) use in developing countries, and looks at some solutions that will soon be available.

The numbers of computer and Internet users in the developing world is gradually increasing. However, their use outside professional organizations, such as governments and universities, is severely limited by a number of factors. They are:

- Erratic or non-existent telecommunications structures.
- Computers are still expensive despite dramatic price drops in America and Europe. A cheap personal computer is at least twice as expensive in South Africa as it is in Europe.
- The high cost of using the Internet. On average, between 4-10 times more expensive than in Europe (data from the OECD).
- Lack of broad technical skills among the general population to use these systems.
- Erratic or expensive energy supplies to run these systems.

However, there are a number of projects that may indirectly reduce some of these problems, and solve the information drought in places such as Africa. They may also help its residents to improve their technical skills, increase knowledge of their environment, and improve their health and living conditions.

Currently, most IBM compatible computers are fragile, complex, too powerful and unreliable for most computer needs. Their operating systems are also expensive and difficult to master. At the end of 1999, a new range of computer systems will start to come onto the market which will provide a more simplified, toughened machine for approximately $150, half the cost of the cheapest current computer system. These new machines will also run alternatively to the standard Windows operating system, which will either come free or cost in the region of $1-2. The manufacturers, who include groups such as SUN/America Online, Microsoft, Sega, Sony, Apple, Psion and 3Com, believe that within two years this price may drop even further. The low cost of these systems, and their ability to browse the internet, word process and send email, makes them ideal for developing nations. Moreover, some countries such as Mexico are starting to switch their computers over to the Linux operating system which is powerful and free. So far they have saved their schools over $200,000 in licensing fees.

Secondly, both Africa and the Americas will soon have their telecommunications infrastructure dramatically improved by new dedicated fibre optic networks surrounding the two continents. A satellite communication system called Teledisc, which becomes operational in 2002, will also provide relatively cheap and high speed Internet access from and to anywhere in the world. The mobile phone manufacturers have also developed new protocols which, for the first time, make high speed Internet access accessible from the existing mobile phone infrastructure. This infrastructure is being rapidly expanded worldwide. Mobile phone networks are significantly cheaper for most countries to install for local, regional phone systems than traditional copper or fibre optic networks.

Finally, online learning is becoming more accepted and used in the developing world. Namibia for example, has over 1,600 schools. Currently only 28 per cent of them have a telephone, and even fewer have computers. As of last month, only 16 schools in Namibia had Internet access. But these 16 schools have linked to their main Natural History museum. Over 100 students are helping the museum to catalogue the biodiversity in their country. By involving the students, the Namibian government will get a fully computerized database of some of its wildlife, while the students will hopefully become interested in science, and improve their technical skills. By expanding programs such of this nature, students could interact with leading experts on a wide range of issues at a very low cost to government or non-profit resources.

Although this paper has indicated some areas of optimism, the equipment is still expensive compared to the average salary of most individuals working in the developing world and the resources for their governments. The access to reliable energy sources will remain a drawback to using this equipment for some time. But the outlook for increasing the literacy and knowledge of all individuals worldwide has never looked brighter.

SOUTH AFRICA AS LEADER OF THE NON-ALIGNED MOVEMENT - SOME ISSUES FROM SRI LANKA

D.L.O. Mendis

South Africa has been the leader of the Non-Aligned Movement (NAM) since taking over the chair from Columbia in 1998. At the 48th Pugwash conference in Mexico, in 1998, a question was raised at the final plenary session, concerning the void created by the collapse of the former Soviet Union, leaving a single superpower in post-Cold War global politics. Superpower politics in the Cold War era was based on countervailing power that springs from the barrel of a gun, as stated by Chairman Mao long ago. But, the unique experience of Nelson Mandela in South Africa seemed to bring back a long absent element of humanity in global politics. The question was whether this could create a resurgence of moral strength from the NAM under Mandela, as a force of countervailing power in global politics. The answer from the panel in Mexico was an unequivocal negative: in the real world, military power is real and irresistible; moral power in comparison is illusory. Nevertheless, moral power has been effective in such situations as when governments have been pressurized to release prisoners of conscience, for example. The very example of Nelson Mandela's lifelong struggle against the oppression of white supremacy adds to that of Gandhi's in another day and age. For these reasons two examples from Sri Lanka are presented in this paper where some aspects of moral power are invoked: one to check the destructive plans of a multinational corporation in Sri Lanka, aided and abetted by the local comprador class; and the other simply in the cause of local engineering research and development, seen as the cornerstone of non-dependent economic development.

At the Student Pugwash USA (SPUSA) meeting in San Diego in 1999, several presentations in Working Group 8, on Science, Technology and Culture, touched on this approach to development.

Rian Leith from South Africa presented the concept of Ubuntu, and the "Collective Fingers Theory", as the philosophy underlying the new South Africa, also expressed in its motto "Unity is Strength". Leith said that the Xhosa proverb *umuntu ngumuntu ngabantu* means "I am because we are", and added: "Ubuntu will continue to be entrenched in South African society mainly because of Mandela's successor Thabo Mbeki, and his visionary policy of an *African Rennaissance*".

Deborah Hussey from the USA, writing on "Democratizing Science and Letting Ethics into the Lab" described the unethical activities of some multinational corporations that impact adversely especially on Third World countries that are hard pressed to defend themselves. She said: "...science can be brought into balance if we value collaboration as well as competition, the rest of nature as well as man, and the emotional as well as the rational; after all like the wave-particle, nature-nurture, and mind-body dualities, these distinctions are artificial and by no means absolute".

Akifumu Fujita from Japan, discussed the development of nuclear technology in Japan, after World War II ended with the needless atomic bombings of Hiroshima and Nagasaki. He described how the Japan government had: "... proposed a budget of 235 million yen (derived from U-235) for the building of nuclear piles. In protest at this undemocratic way of making decisions the Science Council of Japan made an appeal to the public stating three principles for peaceful use of atomic energy. These three principles, *democracy, autonomy, and openness* were enacted in the Atomic Energy Act of December 1995". He went on to say that: ".... the development of nuclear energy in Japan has been an undermining process of the Three Principles under the domination of the USA in the global framework of the Cold War. Can the spirit of the Three Principles survive today? *It can survive and it must survive*".

These examples from the Student Pugwash conference encouraged me to hope that the younger generation cares, and will try to correct some of the wrong trends of their elders. It is mainly for this reason that the following examples are presented in this paper.

Eppawala Phosphate in Sri Lanka and Baja California Saltworks in Mexico

A proposal by a US multinational corporation to mine to exhaustion, in 30 years, a rare non-renewable igneous phosphate rock deposit in Eppawala, Sri Lanka, is before the Sri Lanka authorities. There has been public agitation against the proposal ever since it became public knowledge less than two years ago, long after negotiations had started, but apparently before an agreement had been signed. This was discussed at the 48th Pugwash conference in Mexico last year and mentioned in the Working Group 2 report. The Pugwash Newsletter, April 1999, announced the publication of my book entitled *Eppawala - Destruction of Cultural Heritage in the name of Development.*

In the case of the Baja California Saltworks in Mexico, a US multinational corporation has entered into an agreement with the Mexico government to build a saltworks that will result in destruction of the last nursery of the grey whale. Thirty-four renowned scientists, including nine Nobel laureates, have addressed an Appeal against this proposal that has been given publicity in the media. After the SPUSA meeting, attempts were begun in the USA to bring the Eppawala proposal to the individual attention of these 34 scientists and seek their support against the proposal, on account of the adverse impact on our cultural heritage in Sri Lanka, comparable to the adverse impact of the Baja California project on the grey whale nursery, and the environment.

This was discussed in Sri Lanka, 12 August 1999, at a presentation on "Social Responsibility of Scientists - Eppawala phosphate and Baja California Saltworks", at which the Chairman of the Board of Investment, Sri Lanka, which is responsible for the Eppawala project, was present. He stated that the proposal to exhaust in just 30 years the non-renewable resource, which will last for over a millennium if used to supply local needs alone, was taken up because of the need for foreign exchange. He however disclaimed responsibility for the decision, passing it on to the politicians. It is known that hard currency is needed not only for normal development, but for purchase of ordnance for the on-going ethnic conflict. Thus the issue is tied up with the most vital problem confronting Sri Lanka today, the separatist war.

The situation then is that officials disclaim responsibility for such a far-reaching action. Politicians are blamed, but they do not take the people into their confidence at all. There is a need for outside pressures to be brought to bear on the Sri Lanka government by concerned scientists and citizens from around the world in a manner similar to the pressures brought on governments for release of Prisoners of Conscience, for example. Communications should be addressed to the President of Sri Lanka urging that the proposal for mining the Eppawala phosphate deposit to exhaustion in 30 years be abandoned.

The National Engineering Research and Development Centre of Sri Lanka

The need for an engineering research and development centre in Sri Lanka has been recognized for a long time. In the late 1960s attempts to start a Technology Development Centre in Kandy floundered, but the idea was not abandoned. In the early 1970s, outstanding engineering R&D work in the State Engineering Corporation (SEC) in the decade of the 1960s was identified as a possible nucleus around which to build a full-fledged National Engineering Research and Development (NERD) centre. A Cabinet paper was prepared by me in the Ministry of Planning and Economic Affairs for formal approval by the government. After a three year period of internal manoeuvering, lobbying, buck-passing, and dilly-dallying, a full-time Chairman was finally appointed in 1974. Meanwhile, however, the inspiration for setting up the Centre, the research and development work in the State Engineering Corporation, had taken a nose-dive. The reason was the enforced departure of the Chairman around whom the R&D work had been done, on account of a typical politically motivated witch-hunt, after a change of government in 1970, followed by the exit of many talented persons in his wake. Thereafter the NERD centre, under an inexperienced academic, functioned for three years from an office in Colombo, on the basis of research grants given for such activities as the manufacture of a dental chair, making a mockery of the whole concept of engineering research and development for national economic non-dependence. A project for the local manufacture of a low cost personal transport vehicle, and another for local development and manufacture of a two wheeled tiller, were done outside the NERD Centre at this time, in the Ministry of Industries, and the Ministry of Planning and Economic Affairs, respectively. For diverse reasons both these projects were later abandoned.

With the next change of government in 1977, the former Chairman of the SEC was recalled to take over the NERD Centre, and activate it to equal or surpass the achievements of the former R&D section of the SEC for which he had been responsible. An extensive plot was obtained in a suburban industrial estate and the NERD Centre was soon in business as intended years before. With the retirement of the Chairman, his successor continued the good work and the NERD Centre recently celebrated its 25th anniversary with the opening of a Technology Park. This is the first of its kind in Sri Lanka and has tremendous potential for sharing and propagating local, indigenous and transferred science-based technology with practical applications in the economy. As a strategy for development, it is considered necessary to bring this to the attention of Pugwash and other foreign scientists, so that there could be mutual benefits from cross-cultural collaborations.

Already such cross-cultural exchanges have started. One example is a proposal from an organization in California for research and development for manufacture of a bladeless turbine. In this day and age, information is freely available on the Internet web pages. The web page for the Tesla Engine Builders Association (TEBA) has been very useful. Hopefully, this will be a joint venture based in the NERD Centre, and other such projects will follow.

Finally, two other endeavours merit mention here. First is a project to set up a Science and Technology Centre at the Faculty of Engineering in the University of Peradeniya. This is an initiative of faculty alumni, some of whom are in good employment in foreign countries. It will be developed around a small applied mechanics laboratory set up over the years as a labour of love by a distinguished faculty member. The second, the singular endeavour of a Sri Lanka scientist, is the recently inaugurated Pemberley Study Centre, a unique international study centre on the lines of the Rockefeller Centre in Bellagio, Italy. My hope is that the Centre will be the venue for many Pugwash Workshops in the future. The attention of Pugwash scientists is drawn to these endeavours.

THE INTEGRAL PLAN OF ECONOMIC DEVELOPMENT FOR THE TEHUANTEPEC ISTHMUS, MEXICO: SOME FACTS BEHIND THE MASK OF PROGRESS

Tannia Falconer

The Tehuantepec Isthmus is located in the Southeast of Mexico. It embraces in the first place, the states of Oaxaca and Veracruz, and part of the states of Chiapas and Tabasco. Its length is approximately 140 km, and it connects the Pacific Ocean to the Atlantic Ocean, one of the main reasons why, through history, the region has been attractive and has awakened the ambition of many people, even before the conquest.

Along with its geostrategic position, the Isthmus holds an enormous importance at both national and international levels, in terms of biodiversity, oil and mining resources, social and cultural diversity.

Biodiversity

According to Estudillo,[1] Mexico holds fourth place in the world's biodiversity. The Chimalapa's region (Oaxaca, Veracruz and Chiapas), is the greatest of the rain forests in Mexico, and as for biodiversity, it is the most important region of the Northern Hemisphere. This richness is due to its position at the confluence of the Neotropical and the Nearctic regions.

The environmental impacts and the alterations in the Isthmus area are notorious, and they all have come from "development projects" of many kinds. Examples of this are great extensions of rain forests cleared to relocate indigenous people affected by dams; extensive cattle ranching and overgrazing; deforestation of 53,000 hectares to build a dam and irrigation district, both useless by now; indiscriminate use of chemical fertilizers and pesticides; running water contamination by industrial and urban wastes; oil and chemical spills; serious damage of soil and its productive capacity, and so forth.[2]

Oil and Mining Resources

The Isthmus region is also known for its hydrocarbon sources, mainly petroleum. Mexico is a very important trade gate for the USA, the EU and Japan. Mexico is the third largest producer of crude oil (Saudi Arabia being the first, and Iran the second). Some experts say that the Tehuantepec region in its southern portion is also potentially exploitable, but at least in Mexico there is not enough information available that shows how large these oil reserves are.[3] This is a very delicate issue considering the prevailing international economic order, and its dependence on hydrocarbon energy.

Social and Cultural Diversity

The Tehuantepec region is very complex, not only in environmental or resources issues, but in the social, political and economic aspects too. In Oaxaca and Veracruz there are twelve different indigenous groups that account for 482,860 people, who represent 24 per cent of the total regional population. There are Zapotecos, Nahuatl, Mixes, Popolucas, Huaves, Chinantecos, Zoques, Mixtecos, Chontales, Mazatecos, Tzotzil and Chochos.

From the total regional population (almost two million), a quarter is indigenous population, 62 per cent of these indigenous people live in Oaxaca (330 communities), and 38 per cent in Veracruz (209 communities).

In Oaxaca, 32 per cent of the indigenous people of 15 years and older are illiterate; and 49 per cent of them have incomplete primary school education. In Veracruz, the percentages are 35.5 per cent and 73 per cent respectively.

In a general context of both states, 38.7 per cent of the indigenous population do not have electricity; 64.6 per cent do not have running water; 81.5 per cent do not have drainage infrastructure, and 56 per cent have earthen floors. These numbers just give a brief idea of the prevailing conditions for the indigenous population.

The Integral Plan of Economic Development for Tehuantepec, better known as the Tehuantepec Isthmus Megaproject

There have been many "development projects" dedicated to the Tehuantepec region, due to all the aspects of resources, environment and society that were mentioned above. There is one project that has become more and more known over time. Its name has been changed as well as some of its characteristics, from project "Alfa-Omega", to "Programa de Desarrollo del Istmo de Tehuantepec" (Development Plan for the Tehuantepec Isthmus) to "Programa Integral de Desarrollo Económico para el Istmo de Tehuantepec" (Integral Plan of Economic Development for Tehuantepec). At the present this project is better known – especially among researchers and people interested in this area – as "Megaprojecto del Istmo de Tehuantepec", (that is, Megaproject of Tehuantepec Isthmus).

The reason why it is known as "Megaproject" is because it includes many projects that shape it. The Megaproject includes about 125 different projects, with both private and public investment, from which almost 60 per cent are planned for Oaxaca, and 44 per cent for Veracruz. The kind of projects are: 20 per cent urban; 18 per cent oil and petrochemical industries; 14 per cent fishing ; 9 per cent industrial; 7 per cent railroad infrastructure; 6 per cent harbour infrastructure; 6 per cent forest plantations; 6 per cent highway infrastructure; 5 per cent land and cattle industry; 5 per cent mining; 3 per cent tourism.

All these projects are supposed to bring many employment opportunities to the local population, but considering the present conditions of the indigenous people, they are at a great disadvantage, mainly due to the high illiteracy that prevails in the vast majority of the communities. This situation disqualifies them from access to employment promised by the Megaproject.

Tourism projects have proved to generate uneven conditions among the population. A clear example is Huatulco, in the state of Oaxaca, a harbour which is supposed to be a development pole, and that has been receiving great investment, has actually between 80 and 100 per cent of indigenous peoples' dwellings with earthen floors, no running water and no drainage, a situation that leads to unhealthy conditions in those areas. This proves that investment is not addressing local people's needs, and it is obviously excluding indigenous people from the projects. Half of the active population of this place is employed in the services sector, while in the rest of the region the majority of the economically active people are employed in the primary sector.

The Megaproject is already happening. Considering that this region is a very rich one in many aspects, international interests are looking at it in terms of big profits, and not caring about the impacts that they will have on the environment and on the culture of local people. If the working labour cannot be integrated into the system imposed by multinational corporations, they can be easily forced out of their lands to be relocated, as has happened to many indigenous groups.

Conclusions

The Megaproject is not really promoting the concept of national development, and therefore, uneven conditions, and the uneven distribution of wealth and resources, are the prevailing patterns in this region. The implementation of this kind of project can be a real threat to societies and the environment, if there are no specific studies about the whole context in which they are to be established.

It is necessary to conduct more research in the Tehuantepec region in order to avoid the negative impacts of those projects on society and the environment. There is a need to use laws to include local people in the decision-making process, and to generate conditions for people to use their own land; local people need autonomy over their own resources.

Many people are just opposed to the implementation of the projects that the Megaproject includes, without even analysing the whole context; this is one extreme. At the other extreme are the people interested only in getting profits out of this, no matter the consequences or the impacts. Neither of these extremes are reasonable. The ideal position would be a convergence of all the people involved, in order to analyse the positive and the negative points, and to work for the benefit of the local people, and try to avoid damage to the environment.

References

[1] Jesus Estudillo Lopez, "Consideraciones de la biodiversidad en la zona de Istmo: aves y plantas" in seminar *Reflexiones sobre el Megaproyecto del Istmo de Tehuantepec,* Mayo-Julio, 1999, UAM: Mexico.

[2] Nemesio J Rodríguez, *Un Sueno Cortesiano: el Istmo de Tehuantepec*, in *El Estado del Desarrollo Economico y Social de los Pueblos Indigenas de Mexico*, INI/PNUD, Mexico.

[3] Miguel Garcia Reyes, "Recursos Geologicos del Istmo" in seminar *Reflexiones sobre el Megaproyecto del Istmo de Tehuantepec,* Mayo-Julio, 1999, UAM: Mexico.

THE MICROTRON MT-25 REGIONAL LABORATORY PROJECT: AN EFFORT IN COOPERATION FOR DEVELOPMENT

María Elena Montero-Cabrera

One of the main difficulties in developing applied nuclear research in Central American and Caribbean countries, as well as in Latin America in general, is the lack of nuclear facilities. It is necessary to strengthen regional cooperation in nuclear applications, in particular in the common use of accelerators and other nuclear facilities, based on a cooperative scheme in which each country contributes with its own facilities, and researchers from other countries that have none could participate.

There is a microtron in Cuba now. Its main application is gamma-ray activation analysis, but it is possible to use it as a source for fundamental nuclear physics studies, as well as for condensed matter research. The Microtron MT-25 is an electron cycle-accelerator which allows researchers to obtain three kinds of radiation: electrons, gamma rays and neutrons. It was designed and assembled in the Nuclear Reaction Laboratory of the Joint Institute for Nuclear Research at Dubna, as a result of a Cuban-Russian joint project. The construction of this accelerator was completed in 1990 and, after testing, it was dismantled, packed and sent to Cuba, where it arrived in 1992. The accelerator has been kept in good condition since then, and awaits the construction of the laboratory where it will be installed.

Historically, the Microtron Project was part of a planned nuclear development in Cuba. Besides a large nuclear research centre with a research reactor, a cyclotron and a microtron, the original plan included the building and implementing of nuclear power plants. Top high school graduates were directed to nuclear careers, such as nuclear physics, radiochemistry, nuclear engineering, nuclear and conventional electronics, condensed matter physics, and so on. The Instituto Superior de Ciencias y Tecnologia Nucleares (ISCTN) was created in Havana, in an effort to produce part of these university majors in Cuba. But after 1990, the economic crisis brought nearly the whole nuclear programme to a halt. From the Nuclear Research Centre planned, only the Isotope Products Processing Centre and the Radiation Protection Centre were finished. One of the results was that many nuclear professionals were no longer needed and had, at least, to change their profile; furthermore, the microtron accelerator is now in storage. The Cuban Physical Society (CPS), worried about this situation, decided to request financial and technical support from an international agency.

The CPS, in a joint meeting of the Latin American Federation of Physical Societies (FELASOFI) with the UNESCO Physics Action Council (UNESCO-PAC) during the first Canadian-American-Mexican Physics Meeting, held in Cancun in 1994, orally applied to UNESCO-PAC for the financing of a project for laboratory building and auxiliary systems, as well as for installation, starting, spectroscopic and nuclear equipment.

In 1995, the CPS, along with the ISCTN, presented to FELASOFI and CLAF (Centro Latinoamericano de Fisica, which represents UNESCO in this field) a written project for the laboratory. In mid-1996, UNESCO-PAC expressed its willingness to support the project, and since November 1996 three User's Club Meetings have been held with UNESCO support, with the participation of physicists from Cuba, Mexico, Brazil, Guatemala, Russia and Italy. During the first meeting, UNESCO-PAC, FELASOFI and CLAF confirmed their support for the project. During the second meeting, Mexican and Brazilian specialists, together with Cubans, participated in the detailed writing up of the project. During the third meeting, the Russian chief-of-staff from Dubna confirmed their willingness to participate in the project. Since 1998, the Cuban Nuclear Agency has supported a project for the Assembly, Checking and Characterization of the Ultra-High Frequency Generator of the Microtron MT-25. But as yet, no-one has committed to covering the basic budget for the laboratory project.

The Project

The ISCTN has both the accelerator and the space to build the laboratory, with personnel to assemble, operate and maintain the accelerator, providing support to the outside users of the laboratory. This Microtron is the building block for creating a CLAF/UNESCO Regional Laboratory, for fundamental and applied nuclear physics research. The Regional Laboratory will be a non-profit organization devoted to strengthening cooperation and collaboration among Latin American and Caribbean countries.

Besides the formation of personnel to work on nuclear-related activities, the ISCTN is also devoted to the development and management of facilities used in nuclear research. The research performed at ISCTN facilities is divided into two main branches: basic and applied nuclear physics. Work on basic nuclear physics, besides the obvious objective of generating new knowledge, serves as a perfect setting for the formation and training of personnel, as well as the development of new techniques and procedures. The applied nuclear physics programme is intended as a means to provide a state-of-the-art nuclear physics laboratory, with electron, gamma and neutron beams, gamma and charged particle spectroscopy systems, and a highly-trained personnel to assist the industrial, medical and academic communities of the Caribbean and Central American regions. Operation of the Regional Laboratory will also permit training and research in areas like electronics, control, vacuum and radio-frequency, which are related to the accelerator and its instrumentation and that will be maintained and upgraded.

UNESCO, IAEA and International Agency Input. Budget in hard currency for the laboratory building, auxiliary systems, and installation and starting equipment are requested from UNESCO. Expert service in accelerator installation and implemention, spectroscopic and nuclear equipment, as well as training are requested from the IAEA.

Objectives of the Project. The goal of this project is, with the support of UNESCO, the IAEA and other national and international agencies, to create a Regional Laboratory on the basis of a Microtron existing in Cuba. It is mainly oriented to the development of fundamental and applied research through the fulfillment of projects approved by a Programme Advisory Committee.

Other Research Centres and Universities, Beneficiaries of the Project.
Cuba: Centro de Estudios Aplicados al Desarrollo Nuclear (CEADEN) and Centro de Isótopos (CENTIS) of the Nuclear Agency (CITMA); Instituto de Materiales y Reactivos para la Electrónica (IMRE) and Faculty of Physics, of Havana University; Centro de Investigaciones del Petróleo (CEINPET); Instituto de Ecología y Sistemática; Instituto de Geofísica y Astronomía. Instituto de Investigación de Medicamentos; Instituto de Investigaciones de la Agricultura Tropical; Instituto de Investigaciones del Arroz.
Mexico: Instituto de Física de la Universidad Nacional Autónoma de México.
Brazil: Instituto de Física, Universidad de Sao Paulo.
Guatemala: Facultad de Ingeniería, Universidad de San Carlos.

Impact. The Microtron MT-25 Regional Laboratory should promote scientific development and cooperation in the Caribbean Basin and both Central American and South American countries. Its exploitation should improve the regional analytical and irradiation capacities, prepare personnel for teaching, conduct fundamental and applied research on industry, mining, agriculture and the protection of the environment.

WORKING GROUP 4

THE ENVIRONMENT

CONFLICT AND COOPERATION OVER FRESH WATER

Peter H. Gleick

In the past decade there has been a major shift and rethinking of the field of international security *analysis*.[1] With the end of the Cold War and the breakup of the Soviet Union, the world has moved away from a focus on political and military conflict between two major superpowers. That focus has been replaced by a growing concern over regional conflicts, civil, religious, and ethnic wars within regions, and the connections between environmental degradation, scarcity of resources, and regional and international politics and disputes. Traditional political and ideological questions that have long dominated international discourse are now becoming more tightly woven with other variables that loomed less large in the past, including population growth, transnational pollution, resource scarcity, and inequitable access to resources and their use. These issues have loosely been termed "environmental security" and interest in them has engendered a wide range of innovative and controversial analyses and studies.

This change results from a growing recognition that new and large-scale environment threats with political ramifications are beginning to dominate international discourse about environmental concerns. During the 1970s and early 1980s, environmental concerns focused primarily on local or regional air or water pollution problems. In the mid-1980s, however, new environmental problems with global ramifications became more evident, including global climate change, the depletion of atmospheric ozone, population pressures on limited resources, and inequities in resource use between developed and developing countries. Many of these new problems affect more than one region or nation and involve problems that cross political borders.

While many of the past, present, and future causes of conflict and war may seem to have little or no direct connection with the environment or with resources, a strong argument can be made for linking certain resource and environmental problems with the prospects for political frictions and tensions, or even war and peace. History shows that access to resources has been a proximate cause of war, resources have been both tools and targets of war, and environmental degradation and the disparity in the distribution of resources can cause major political controversy, tensions, and violence.

At the centre of the ongoing debate is the assertion that resource scarcity and certain forms of environmental degradation are important factors contributing to political instability or violent conflict at local, regional, and interstate levels. A related argument is that, as population growth continues and certain large-scale environmental problems worsen, cases of environmentally related instability are likely to proliferate in the future.[2] [3] [4] [5] [6] [7] [8] [9]

In short, there is a growing perception that local, regional, and global environmental deficiencies or resource scarcities will increasingly produce conditions that may lead to conflict. Despite academic disputes over definitions and terms, a wide variety of recent comments by senior diplomats and policymakers, some of which are presented below, leave no doubt that issues related to environmental security have moved to the highest levels.

In 1987, the General Secretary of the Soviet Union, Mikhail Gorbachev, stated:

> [The world] is not secure in the direct meaning of the word when currents of poison flow along river channels, when poisonous rains pour down from the sky, when an atmosphere polluted with industrial and transport waste chokes cities and whole regions, when the development of atomic engineering is

justified by unacceptable risks. The relationship between man and the environment has become menacing. Problems of ecological security affect all - the rich and the poor. What is required is a global strategy for environmental protection and the rational use of resources.[10]

In 1989 then United States Secretary of State James Baker said:

The strategic, economic, political, and environmental aspects of national security and global well-being are, today, indivisible.[11]

In November 1989, Prime Minister Margaret Thatcher gave a speech to the United Nations General Assembly saying:

While the conventional, political dangers - the threat of global annihilation, the fact of regional war - appear to be receding, we have all recently become aware of another insidious danger. It is as menacing in its way as those more accustomed perils with which international diplomacy has concerned itself for centuries. It is the prospect of irretrievable damage to the atmosphere, to the oceans, to earth itself.

In 1994, United Nations Ambassador Madeleine K. Albright (now US Secretary of State) stated:

We believe that environmental degradation is not simply an irritation, but a real threat to our national security. . . Left unaddressed, it could become a kind of creeping Armageddon. . . it could, in time, threaten our very survival.

In April 1997, as United States Secretary of State, Ms. Albright went on to say:

Not so long ago, many believed that the pursuit of clean air, clean water, and healthy forests was a worthy goal, but not part of our national security. Today environmental issues are part of the mainstream of American foreign policy.

The argument and focus of debate has now shifted from "whether" there is a connection to "when," "where," and "how" environmental and resource problems will affect regional and international security. There is also a long way to go before the United States and the rest of the world produce a common policy agenda or set of initiatives that truly incorporate environmental and resources issues into approaches to reduce the risk of regional and national conflicts. Nevertheless, there is a new framework under construction that is permitting scholars and policymakers to apply new tools, to set new priorities, and to organize responses to environmental threats to peace and security.

This framework is particularly well developed in the area of water resources. In the past several years there has been considerable progress in both understanding the nature of the connections between water resources and conflict and in evaluating regional cases where such connections may be particularly strong.[5 12 13 14 15] There has also been progress in trying to identify policies and principles for reducing the risks that water disputes will lead to conflict and to better understand how to promote cooperation and collaboration over shared freshwater resources.

Progress has been more than academic. In October 1994, for example, Israel and Jordan signed a peace treaty that explicitly addressed water allocations, sharing water information, and joint management policies for the Jordan River Basin. In 1996, India and Bangladesh signed a formal treaty that moves toward resolving their long-standing dispute over the Farraka Barrage and flows in the Ganges/Brahmaputra system. In 1997 the International Law Commission, after nearly three decades of negotiations, drafting, and discussion, finalized the Convention on the Non-Navigational Uses of Shared International Watercourses, which was approved by the General Assembly in April.

And countries like Brazil, South Africa, and Zimbabwe are incorporating mechanisms and principles for resolving conflicts over shared waters in their new water laws.

Conflicts over Shared Water Resources

Fresh water is integral to all ecological and societal activities, including the production of food and energy, transportation, waste disposal, industrial development, and human health. Yet, freshwater resources are unevenly and irregularly distributed, and some regions of the world are extremely short of water. As we move into the twenty-first century, water and water-supply systems are increasingly likely to be both instruments of political conflict and the objectives of military action as human populations grow, standards of living improve, and global climatic changes make water supply and demand more problematic and uncertain.

There are four major links between water and conflict. Water has been used as a military and political goal. Water has been used as a weapon of war. Water-resources systems have been targets of war. And inequities in the distribution, use, and consequences of water resources management and use can be a source of tension and dispute.[5] These links can occur at local, subnational, and international levels.

Water as a Military and Political Goal

Where water is scarce, competition for limited supplies can lead groups, communities, and even nations to see access to water as a matter of highest concern. This aspect falls into the most traditional Cold War/realpolitik framework where water can be a defining factor in the wealth and power – and in the economic and political strength – of a nation. Access to water resources may then provide a justification for going to war or it may serve as a focus of dispute. While it may never be the sole justification for conflict, history suggests that it has already proven to be an important factor.

Four important characteristics must be considered in order to determine the likelihood that water will be the object of military or political action: (1) the degree of water scarcity; (2) the extent to which the supply is shared by two or more groups; (3) the relative power of those groups; and (4) the ease of access to alternative sources.

Water is unevenly distributed throughout the world. Some regions receive enormous amounts of rainfall or river flow, while others are extremely dry and suffer from "absolute" scarcity. Human factors, such as high population densities or intensive industrial development may cause conditions of "relative" scarcity. In the past, absolute scarcity was a more important factor in regional water issues. Increasingly, however, relative scarcity is the problem leading to water disputes, as population pressures come up against resource limits.[5,16,17] Of course if water is not scarce, conflict over access to water is less likely.

The problem of shared resources complicates the problem of scarcity. When water is extensively shared, misunderstandings or lack of agreement about allocations are more likely. Fresh water is very widely shared because political borders rarely coincide with watershed boundaries. At the international level, over 220 river basins are shared by two or more nations.[18] But even countries with few or no internationally shared rivers or aquifers often have internal water disputes among states, ethnic groups, or economic classes trying to gain access to additional water supplies.

If there are great disparities in the economic or military strength of the parties involved, unilateral and inequitable decisions are more likely. While a weaker party will rarely provoke or initiate military action – and even more rarely prevail – against a stronger adversary, the reverse often occurs. If a weaker nation either controls a water source or is dependent on water from an outside source, disputes and conflicts may occur. When adversaries are equally matched economically or militarily, negotiation and cooperation are more common outcomes.[5]

Finally, if there are few technologically or economically attractive alternative sources of supply, the potential for conflict is higher. If water is scarce and shared, but there exist alternative sources such as other rivers, groundwater or even expensive desalination, conflicts are less likely to occur. There is a high economic, social, and political cost to conflicts; they are likely to be avoided if substitutes can be found.

Over the course of human history these different factors have come together many times to produce a wide range of disputes over access to shared freshwater resources. Forty-five hundred years ago, the control of irrigation canals vital to survival was the source of conflict between the states of Umma and Lagash in the ancient Middle East. Twenty-seven hundred years ago, Assurbanipal, King of Assyria from 669 to 626 B.C., seized control of wells as part of his strategic warfare against Arabia. In the modern era, the Jordan River Basin has been the scene of a wide variety of water disputes. In the 1960s, Syria tried to divert the headwaters of the Jordan away from Israel, leading to air strikes against the diversion facilities.[16] The 1967 war in the Middle East resulted in Israel winning control of all of the headwaters of the Jordan as well as the groundwater of the West Bank. In these cases, water was certainly not the sole issue precipitating conflict, but it was an important factor.

Water remains an important factor in the politics of the region. The multilateral and bilateral peace talks conducted in the 1990s, which led to the interim agreement between the Israelis and the Palestinians and to the peace treaty between Israel and Jordan, explicitly included negotiations and agreements on the shared water resources of the Jordan River. Israeli and Syrian concerns over the Banias, which originates in the Golan Heights, remain an important unresolved issue. Jordanian concerns about Syrian dams on the Yarmouk, the major tributary to the Jordan, are still unanswered.

Disputes over the allocation of water occur at the subnational level as well, and have the potential to turn violent. In California in the mid-1920s, farmers repeatedly destroyed an aqueduct taking water from their region to the urban centres of southern California. The governor of Arizona called out the local militia in the 1930s to protest the construction of water diversion facilities on the Colorado River between Arizona and California. This dispute was eventually resolved in court.

Court decisions do not always successfully end disputes. An interim court decision in India to allocate additional waters from the Cauvery River – which originates in the state of Karnataka – to Tamil Nadu actually precipitated violent conflicts resulting in the deaths of over 50 people.[5] In 1997, the World Court issued a decision in the dispute between Hungary and Slovakia over the Gabcikovo-Nagymaros project on the Danube, effectively refusing to decide and returning the dispute to the two parties for more negotiation.[13,19] As populations continue to grow, regional water scarcity may lead to more frequent examples of this kind of water dispute and conflict.

Water as an Instrument or Tool of Conflict

The usual tools and instruments of war are military weapons. But the use of water as both offensive and defensive weapons has a long history. One of the earliest accounts is an ancient Sumerian myth from 5,000 years ago, which parallels the biblical account of the great flood. In this myth, the Sumerian deity Ea punishes humanity's sins by causing a great flood. In 695 BC, Sennacherib completed the destruction of Babylon by diverting irrigation canals to wash over the ruins of the city.

In the last several years, two examples suggest that the use of water as a weapon continues to be considered. North Korea announced plans in 1986 to build a major hydroelectric dam on the Han River upstream of South Korea's capital, Seoul. The project would provide much-needed electricity to the North, but South Korea sees only its potential as a military weapon. South Korean hydrologists calculated that the destruction of the dam by the North and the sudden release of the reservoir's contents would destroy most of Seoul. While the project currently remains on hold, due to serious political and economic difficulties in North Korea, South Korea has built a series of levees and check dams above Seoul to defend against any such threat.

Another use of a large dam and reservoir as a weapon of war was proposed during the Persian Gulf war. The allied coalition arrayed against Iraq discussed the possibility of using the Ataturk dam in Turkey on the Euphrates River to shut off the flow of water to Iraq, which is highly dependent on flows in the Euphrates for water supply. No formal request to Turkey was ever made, and Turkey subsequently stated that it would never use water as a means of political pressure, but the possibility remains a concern in the region.[5] Both Syria and Iraq continue to have disputes with Turkey over the Ataturk and the level of flows in the Euphrates reaching both downstream countries.

In 1997, a fresh dispute arose between Singapore and Malaysia. Singapore has never been self sufficient in water because of its high population density and small size and it depends on piped water from Malaysia for nearly half of all its needs.[20] In addition, Singapore imports water from Malaysia that it then treats and sells back under an agreement signed in 1965.[21] Relations between the two countries have long been clouded by economic competition, and religious, political, and ethnic differences that have flared periodically since their separation in 1965. In early 1997, these relations soured again after comments were exchanged by senior politicians about mutual concerns, leading to Ahmed Zamid Hamidi, the head of the youth wing of Malaysia's ruling party, urging the government to review the basis of water agreements with Singapore.[22] Chief Minister of Johor state in Malaysia went further, suggesting that they appropriate two of the three water-purification plants operated by Singapore in Johor.[23] Singapore is clearly worried that Malaysia might use water as a political and strategic weapon against Singapore – a point made to Malaysian Prime Minister Mahathir Mohamad by Singapore's Prime Minister Goh Chok Tong: "an agreement by Malaysia to meet Singapore's long-term water needs beyond the life of the present water agreements would remove the perception in Singapore that water may be used as a leverage against Singapore".[21 24] Singapore has also launched a campaign to increase water supplies and to reduce consumption through an aggressive conservation programme. Among their supply plans are new desalination plants that would produce water at about eight times the cost of current supplies.[20] Whether this high price moves Singapore toward more diplomatic negotiations with Malaysia remains to be seen.

Water and Water Systems as Targets of Conflict

Where water resources and water-supply systems are a strategic resource with economic, political, or military importance, they become targets during wars or conflicts. In modern times, dams and hydroelectric facilities were regularly bombed as strategic targets during World War II and the Korean War. The United States targeted irrigation levees in North Vietnam. Syria tried to destroy Israel's National Water Carrier while it was under construction in the 1950s. The Persian Gulf war saw several examples of this problem: the Iraqis intentionally destroyed the water desalination plants of Kuwait and in turn suffered from the destruction of their water-supply system by the allied forces assembled to liberate Kuwait.

Inequities in Water Distribution, Use, and Development

Tensions and conflicts may also result from such indirect factors as the inequitable distribution, use, and development of water resources. Many rivers, lakes, and groundwater aquifers are shared by two or more nations, and many more cross other kinds of political borders. More than half the land area of the world, and perhaps 70 per cent of usable land area, is in an international watershed, where river flows or lakes are shared. These include the Nile, Jordan, Tigris, Euphrates, and Orontes rivers in the Middle East, the Indus, Mekong, Ganges, and Brahmaputra rivers in Asia, the Great Lakes, and the Colorado, Rio Grande, Amazon, and Paraná rivers in the Americas, and Lake Chad and the Congo, Zambesi, Lake Chad, the Niger, Senegal, Okavango, and Orange rivers in Africa, to name only a few. This geographical fact has led to the geopolitical reality of disputes over the uneven distribution of shared waters. Table 1 lists internally available renewable water resource and total water withdrawals for countries in four major shared river basins, revealing the enormous disparities in resource supply within specific regions. In particular, note the countries whose water use exceeds internally available water resources. These countries, including Egypt, Iraq, and Syria, have a particularly strong interest in continuing to receive water from river flows that originate outside of their borders.

Equally uneven is the level of water use. Many industrialized nations and nations with extensive irrigated agriculture withdraw more than 1,500 cubic metres of water per person annually for all uses. At the other extreme, nations with limited supplies or low levels of economic development may use fewer than 100 cubic metres per person per year. Table 2 lists the 15 countries that use the most water per capita and the 15 countries that use the least. Such a low level of water use has direct and undesirable human consequences, including adverse impacts on health, the inability to grow sufficient food for local populations, and constraints on industrial and commercial activities.

Table 1
Water Supply and Demand by Country in Four Major River Basins

River Basin	Countries	Renewable Internal Supply (*) (km^3/year)	Water Withdrawals (km^3/year)
Nile	Sudan	35	17.8
	Ethiopia	110	2.2
	Egypt	1.8	55.1
	Uganda	39	0.2
	Tanzania	80	1.2
	Kenya	20.2	2.1
	Congo (Dem. Republic)	935	0.36
	Rwanda	6.3	0.8
	Burundi	3.6	0.1
Zambesi	Zambia	80.2	1.7
	Angola	184	0.5
	Zimbabwe	14.1	1.2
	Mozambique	100	0.6
	Malawi	17.5	0.9
	Botswana	2.9	0.1
	Tanzania	80	1.2
	Namibia	6.2	0.25
Tigris and Euphrates	Iraq	35.2	42.8
	Iran	128.5	70
	Turkey	196	31.6
	Syria	7	14.4

* These data include only internally available or generated water resources, not water flows into a country from outside its borders.

One of the most important water constraints facing many regions is insufficient water to grow food. Sandra Postel[25] suggests that as annual water availability drops below 1,700 cubic metres per person, domestic food self-sufficiency becomes almost impossible and countries must begin to import water in the form of grain. The number of countries in this category will continue to rise with population growth and overall dependence on grain imports will deepen and spread. Food insecurity is a political concern and can lead to economic weakness and other regional problems.

Table 2
Countries with the Largest and Smallest Reported Per-capita Water Withdrawals (*)

Country	Withdrawals (m3/yr)	Country	Withdrawals (m3/yr)
Congo, DR (formerly Zaire)	9	Suriname	1,181
Bhutan	15	Pakistan	1,277
Guinea-Bissau	17	Australia	1,306
Solomon Islands	18	Bulgaria	1,600
Comoros	18	Chile	1,625
Burundi	20	Madagascar	1,638
Congo	20	Korea DPR	1,649
Uganda	20	Afghanistan	1,702
Papua New Guinea	25	Canada	1,752
Central African Rep.	26	Tajikistan	2,065
Benin	28	United States of America	2,162
Togo	28	Iraq	2,367
Gambia	29	Kyrgyzstan	2,527
Cameroon	31	Turkmenistan	6,346 **
Equatorial Guinea	31	Guyana	7,616 **
Lesotho	31		

* These data include only reported water use. Data on rainfall used in agriculture and unreported water use are not available.

** These numbers are reported by FAO but may be in error.

Finally, there are often adverse consequences of water development and use, and people who do not receive the benefits from water projects may feel these consequences. Examples include contamination of downstream water supplies or groundwater aquifers, dislocation of people because of dam construction, and the destruction of fishery resources that support local populations.

What is the connection between these issues and conflict? For the most part, inequities will lead to poverty, shortened lives, and misery, but perhaps not to direct conflict. But in some cases, they will increase local, regional, or international disputes, create refugees that cross, or try to cross, borders, and decrease the ability of a nation or society to resist economic and military aggression.[57]

Reducing the Risk of Water-Related Conflict

The ultimate goal of any assessment of the risks of conflict over water resources must include efforts to reduce those risks. Various regional and international approaches exist for reducing water-related tensions. Among the approaches are legal agreements, the application of proper technology, institutions for dispute resolution, and innovative water management. Unfortunately, these mechanisms have never received the international support or attention necessary to resolve many conflicts over water. Efforts by the United Nations, international aid agencies, and local communities to ensure access to clean drinking water and adequate sanitation can reduce the competition for limited water supplies and the economic and social impacts of widespread waterborne diseases. Improving the efficiency of water use in agriculture can extend limited resources, increase water supplies for other users, strengthen food self-sufficiency, reduce hunger, and lower expenditures for imported food. In regions with shared water supplies, third-party participation in resolving water disputes can also help end conflicts.

The Role of Law

International law has an important role to play. There are typically two forms of international codes of conduct for resolving disputes over water: general principles of international behaviour and law, and specific bilateral or multilateral treaties. Several legal instruments, for example, prohibit environmental warfare or the use of the environment as a weapon of war, including the Environmental Modification Convention of 1977, the 1977 Bern Geneva Convention, and the 1982 World Charter for Nature. In this area, the problem is not lack of law but lack of enforcement and acceptance by the international community.[5]

International law in the area of shared water resources is both well advanced and, in what may appear to be a contradiction, largely ineffective. More than 30 years of negotiations and discussions have occurred since the original statement of the 1966 Helsinki Rules governing international waters, and the new 1997 Convention on the Non-Navigational Uses of Shared International Watercourses took 27 years to formalize.[26 27] The negotiations and the final product reflect the difficulty of integrating legal principles with the complexities of the hydrologic cycle.

Among the most relevant sections are Article 7 of the Convention, which obliges States to take all appropriate measures to prevent harm to other States from their use of water, Article 8, which obliges watercourse States to cooperate on the basis of equality, integrity, mutual benefit, and good faith in order optimally to use and protect shared watercourses; and Article 33, which offers provisions for the peaceful settlement of disputes by negotiation, mediation, arbitration, or appeal to the International Court of Justice. Among the weaknesses of the Convention are the inherent conflict between equitable uses and the obligation not to cause appreciable harm. These principles are not binding, and offer little concrete guidance to countries trying to allocate scarce water resources, but they may prove effective at encouraging states to negotiate water disputes and to implement joint management activities.

In addition to basic principles of international law, as represented by the Convention, there are hundreds of different bilateral and multilateral treaties signed by parties to international rivers that allocate water, regulate navigation and power, monitor and control water quality, and affect all other aspects of joint management. Each of these treaties is individually negotiated. Some have been highly effective at reducing water-related conflicts. Historical evidence suggests that these forms of agreements tend to be more effective, though they also have their limitations. Since 1814, approximately 300 water-related treaties have been negotiated dealing with management, water allocation, or flood control and hydropower. In the twentieth century alone, Wolf [15] has identified 145 treaties dealing only with water.

Despite the effectiveness of some legal tools, there are concerns that existing international water law may be unable to handle the strains of ongoing and future problems. It is also possible that international water law may simply be an inappropriate mechanism for addressing some of these problems, such as sub-national and local disputes. Many political entities claiming water rights or allocations will not be served by the International Court of Justice, or the 1997 Convention, or other international bodies.

In spite of these concerns, good progress has been made recently to resolve long-standing water disputes, including a peace treaty between Jordan and Israel that includes water provisions and a new treaty between India and Bangladesh over the Farakka Barrage and the Ganges River. These positive developments offer the hope that cooperation will continue to be the preferred mode of resolving water-related disputes. At the same time, new unresolved problems are arising in new regions, such as southern Africa. While diplomatic and institutional solutions are being explored here as well, the appearance of new international and intra-regional water disputes due to inequities driven by population growth or economic competition raises concerns for the future.

Summary

Conflicts are caused by many factors, including religious animosities, ideological disputes, arguments over borders, and economic competition. After more than a decade of academic arguments, political rethinking, and increased participation in security discussions by environmental scientists and analysts, it is widely accepted that resource and environmental factors can play a tangled but definite role in local, regional, and even international disputes. Not all water-resources disputes will lead to violent conflict; indeed most lead to negotiations, discussions, and non-violent resolutions. But in certain regions of the world water is a scarce resource that is becoming increasingly important for economic and agricultural production, and increasingly scarce due to growing population and resource degradation. In these regions, shared water is evolving into an issue of high politics and the probability of water-related disputes is increasing. Policymakers must be alert to the likelihood of disagreements over water resources and to possible changes in international water law, regional political arrangements, and patterns of use that could minimize the risk of conflict.

References

[1] Portions of this paper appeared in P H Gleick, 1998, *The World's Water 1998-1999*, Island Press, Washington, DC.

[2] Myers, N. 1989. "Environment and security." *Foreign Policy*, Vol. 74, pp. 23-41

[3] Gleick, P.H. 1990. "Environment, resources, and international security and politics." In E.H. Arnett (ed.) *Science and International Security: Responding to a Changing World.* American Association for the Advancement of Science (AAAS), Washington, D.C. pp. 501-523.

[4] Gleick, P.H. 1991. "Environment and Security: The Clear Connection." *Bulletin of the Atomic Scientists* (April), pp. 17-21.

[5] Gleick, P.H. 1993. "Water and conflict." *International Security* 18, No. 1, pp. 79-112 (Summer 1993).

[6] Homer-Dixon, T. 1991. "On the threshold: Environmental changes as causes of acute conflict." *International Security* 16 (2), pp. 76-116.

[7] Homer-Dixon, T. 1994 "Environmental scarcities and violent conflict: Evidence from cases." *International Security* 19 (1), pp. 5-40.

[8] Homer-Dixon, T. And V. Percival. 1996. "Environmental Scarcity and Violent Conflict: Briefing book." The Project on Environment, Population, and Secuirty. American Association for the Advancement of Science and the University of Toronto.

[9] Carnegie Commission. 1997. Final report of the Carnegie Commission on Preventing Deadly Conflict. Carnegie Corporation of New York, New York.

[10] Gorbachev, M. 1987. "Reality and Guarantees for a Secure World." *Moscow News*, No. 39, p. 3287.

[11] Baker, J. 1989. Speech by Secretary of State James Baker, August 7, 1989, Mexico City, Mexico. Seventh US - Mexico Binational Commission meeting.

[12] Gleick, P.H. 1994. "Water, war, and peace in the Middle East." *Environment 36*, No. 3, pp.6-on. Heldref Publishers, Washington.

[13] Klötzli, S. 1993. "Der slowakisch-ungarische Konflikt um das Staustufenprojekt Gabcíkovo." Environment and Conflicts Project Occasional Paper No. 7. Center for Security Studies and Conflict Research/Swiss Peace Foundation Berne, Switzerland.

[14] Klötzli, S. 1994. "The water and soil crisis in Central Asia - A source for future conflicts?" Environment and Conflicts Project Occasional Paper No. 11. Center for Security Studies and Conflict Research/Swiss Peace Foundation Berne, Switzerland.

[15] Wolf, A. T. 1997. "'Water wars' and water reality: Conflict and cooperation along international waterways." NATO Advanced Research Workshop on Environmental Change, Adaptation, and Human Security. Budapest, Hungary. 9-12 October.

[16] Falkenmark, M. 1986. "Fresh waters as a factor in strategic policy and action." In A.H. Westing (ed.) *Global_Resources and International Conflict: Environmental Factors in Strategic Policy and Action.* Oxford University Press, New York. Pp. 85-113.

[17] United Nations. 1997. *Comprehensive Assessment of the Freshwater Resources of the World.* Prepared for the Commission for Sustainable Development and the General Assembly. Stockholm Environment Institute and the United Nations, New York.
[18] United Nations. 1978. *Registry of International Rivers.* Centre for Natural Resources, United Nations. Pergamon Press, New York.
[19] Lipschutz, R. D. 1997. "Damming troubled waters: Conflict over the Danube, 1950-2000." Environment and Security Conference, Institute of War and Peace Studies, Columbia University, October 24.
[20] Zachary, G.P. 1997. "Water pressure: Nations scramble to defuse fights over supplies." *The Wall Street Journal*, December 4, p. A17.
[21] Dupont, A. 1998. *The Environment and Security in East Asia.* Adelphi Paper, forthcoming. Research School of Pacific and Asian Studies, Strategic and Defence Studies Centre, Australian National University, Canberra, Australia.
[22] Lee, J. 1997. "KL uses water as political tool." *The Jakarta Post*, July 7.
[23] Jayasankaran, S. and M. Hiebert. 1997. "Snipe, snipe: Malaysia-Singapore spat reflects growing economic rivalry." FEER, June 5, p. 24.
[24] *The Straits Times.* 1997. "PM Goh's speech in Parliament: I shall work for a new era of cooperation." *The Straits Times*, June 6.
[25] Postel, S. 1997. "Water for food production: Will there be enough in 2025?" Presentation at the International Water Resources Association IXth Congress, Montreal, Canada, 3-7 September.
[26] McCaffrey, S. 1993. "International water law." In P.H. Gleick (ed.) *Water in Crisis: A Guide to the World's Fresh Water Resources*, Oxford University Press, New York.
[27] United Nations. 1997. "Convention on the Law of the Non-Navigational Uses of International Watercourses." UN General Assembly A/51/869 (April 11). United Nations Publications, New York.

WATER SUPPLY AND DEMAND: A TENTATIVE INVENTORY

André L. Mechelynck

After air, water is probably the single most important commodity sustaining life on Earth. Man is currently using - and polluting - a sizeable portion of this natural resource, and a reasonable estimate of the current situation and its future evolution appears to be in order.

Much scattered information is available. There are, however, notable gaps in these data, and the construction of a complete picture is very difficult and fraught with large approximations.

I am trying to present a first draft of such a picture; there are, no doubt, many errors in the following, which, I hope, will be corrected by further studies and contributions from more competent colleagues. This work should not, under any circumstances, be considered other than a preliminary excursion into a rich and only partially explored territory.

Water Uses

There is very little actual *consumption* of water, but rather *use*; the water thus used being returned to the environment, either as more or less polluted liquid water or as water vapour (transpiration), which is returned to earth as rain.

The consideration of water use must thus be further divided, according to the characteristics of the water needed and those of the effluent water, essentially their degree of purity or pollution. This is extremely variable (see Table 1).

The reclamation of this polluted water depends, of course, on the nature of the pollution; until the middle, or maybe the end, of the 19th century (and until now in many developing countries), the policy was to let nature take its course: the abundance of water in lakes and rivers, the small degree of pollution (related to the low level of population and the absence of industry) and its nature (mainly organic: garbage, excrements, etc.), meant that the natural powers of recuperation of surface waters (through oxidation by naturally present dissolved oxygen) were sufficient, in most cases, to regenerate the effluent waters to some kind of purity. (There were, of course, notable exceptions, mainly around large conurbations with little sanitation; this probably contributed in a large measure to the Black Death (14th century, 25 million deaths); the Great Plague of London (1664-65, 70 000 deaths); or the 1894 epidemic (20 million deaths around the world in the course of the next 20 years).) Protection measures were progressively developed (use of underground waters, drainage and sewer systems, separate supplies of drinking water, etc.).

Around the end of the 19th century, the growth of cities and the development of industry called for stronger, positive measures towards the reclamation of used water (see Table 2). This has now, since the mid-twentieth-century, become a fully-fledged industry in the developed countries.

The major user of water is the *biosphere* (Man, wild and domestic animals, plants, trees, crops, etc.). Living matter is made mostly of water; on top of this, breathing and perspiration liberate water vapour to the atmosphere. The remainder is returned to the environment as excrements, garbage, refuse or dead bodies.

The water thus consumed must be reasonably clear and germ-free. The amount used (metabolized) is strictly tied to the world quantity of living matter, which varies little with the passage of time.

On top of this, man uses *household water* for various uses (washing, cooking, cleaning, flushing, etc.). This water could be of a lower quality than the previous fraction. According to living habits, the amount used can vary from a fraction to a rather high multiple of a person's biological consumption.

Then comes *industrial water*, where the uses are numerous and multiform, either as a cooling or heat transfer medium, a primary material, or a practical conveyor for miscellaneous materials. The quality needed may be very variable, from highly demineralized water (boiler feed water) to "ordinary" surface water.

Tables 3, 4 and 5, as well as Figure 1, give estimates of the various transfers. We estimate the *total current use of water* to be in the region of 32×10^{12} m^3 of water per year.

This is likely to grow in the future, although not at a very great pace; what will be needed, however, is an increase in water quality in the developing nations and an increase in reclamation of polluted water in the developed nations.

Water Supply

Water is probably the most abundant mineral on our planet (Table 6 gives the quantities in various reservoirs). The most abundant reserves (over 97 per cent), however, are the oceans, which contain about 35 kg of salts per m^3, making the water unfit for human or agricultural use.

Man (and the whole biosphere) has thus essentially to depend on *precipitations* (rain and snow), which may be used directly (plant growth) or be collected in various reservoirs, either on the surface (glaciers, lakes, rivers) or underground.

But precipitations are rather unevenly distributed. Eighteen per cent of the earth is desert, with very little precipitation, and does not support life. Humanity is in fact dependent on the flow of water, either from heaven or from the streams which carry it to the sea. (This should not be underestimated; the Netherlands, for instance, use roughly *twice* as much water as they collect as rain.) Assuming, on the whole, that 50 per cent of the precipitations are used, and that 50 per cent of the water used is recycled, we would obtain a *potential supply* of some 100×10^{12} m^3 of water per year, which should be amply sufficient to cover our future needs.

Provisional Conclusions

There does not seem to be a lack of water on Earth, now or for the foreseeable future. Several conditions, however, will have to be met, and definite objectives reached. We must, here, make geographical and economic distinctions.

Regions enjoying *abundant fresh water* are, in general, heavily populated, sometimes up to saturation point. The emphasis, there, will be put on water conservation, transportation and recuperation (Tennessee Valley Authority, Aswan High Dam) calling, in some cases for important international cooperation (the Danube basin, for instance, 817×10^9 km^2, crosses or borders some 10 different countries).

In *developing countries* the continued growth of industry will call for increased emphasis on pollution limitation and reclamation of effluent waters. Water sanitation - supplying clean water for domestic purposes - is and will remain a constant preoccupation.

Desert reclamation is getting more attention; Egypt has started a project to irrigate their share of the Libyan desert;

There is thus a call for:

- improved technical developments for water purification;
- important civil engineering constructions (dams, canals) to control and transport water;
- legal penalties on sources of pollution;
- international cooperation.

It is thus up to engineers, lawyers and politicians to get together and do the job.

Table 1
Water Quality

a) Water supply

The characteristics to be measured are:

- Physical characteristics
  - temperature
  - colour
  - turbidity
- Chemical characteristics
  - pH
  - dissolved oxygen
  - hardness
  - presence of various ions or chemical substances
- Biological characteristics
  - taste
  - smell
  - toxicity
  - micro-organisms

The admissible values will differ according to the considered use; there is, however, a wide difference from country to country, from continent to continent, as to what is indispensable, useful, or simply desired. We do not hesitate, in the West, to wash ourselves, our dishes and our clothes, flush our toilets, sprinkle our gardens and clean our cars with city (drink) water; many people over the world use river or lake water for drinking purposes. The needs vary also from industry to industry

b) Polluting agents

- Domestic effluents
  - Excrements
  - Food residues
  - Cleaning agents residues (soap, detergents, etc.)
  - Microbes and viruses
- Agricultural effluents
  - "Natural" fertilisers (manure, etc.)
  - Mineral fertilisers (nitrates, phosphates, etc.)
  - Pesticides and weed control agents
- Industrial effluents
  - Chemicals, organic and inorganic
  - Heat
- Natural production
  - Vegetal (algae, etc.) (eutrophication)
  - Animal (molluscs, arthropods, diatoms, micro-organisms, etc.)

Notes

10^{12} m^3 is equal to 1,000 cubic kilometres (or 10^{12} tonnes)
10^9 m^2 is equal to 1,000 square kilometres

Table 2
Water Reconditioning

Used or polluted water may be reconditioned, according to the nature and intensity of the pollution, by natural, biological, mechanical, chemical or electro-chemical means, in increasing levels of cost.

a) Classical means
- Filtering (slow)
- Sterilizing (Cl, HOCl, O_3, etc.)
- Flocculating ($Al_2(SO_4)_{2,\ etc.}$)
- Filtering (fast)
- Demineralization

b) Special
- Desalinization (sea or brackish water)
    - Distillation
    - Ion exchange
    - Electrodialysis
    - Reverse osmosis
    - Freezing

Table 3
Water Supply Methods

Classical
Total potentially "renewable" fresh supply: 61×10^{12} m³/y
- surface water 51×10^{12} m³/y .
- underground water 10×10^{12} m³/y
:

Other suggested schemes
- Towing icebergs to temperate zones
- Capturing sweet water at river mouths in large "bags" towed to consuming areas
- Spraying coal dust on glaciers to increase melting
- Cutting "corridors" in the Arabic (eastern) range of Egypt to increase rainfall
- Letting seawater into Sahara
- Spraying clouds with AgI or frozen CO_2

Table 4: Water Use (Developed countries, agriculture excluded)

a) Uses and major users	**per cent**	
Drinking water	16	Population
Cooling water	53	Metals and oil industry
Boiler feedwater	12	Electricity and oil industry
Rinsing or washing water	7	Metallurgy
Process water	12	Paper industry
b) Effluents and means of reclamation		
(in order of approximated increasing costs)		
Hot water	3	Cooling
Solids in suspension	62	Filtering
Organic matter	28	Oxidation
Liquids in emulsion	4	Flocculation, flotation, filtration, ...
Inorganic matter in solution	3	Chemical treatment

Note: desalinization of sea water is probably the most expensive way to obtain pure water; its use is limited to regions where fresh water is scarce, sea water abundant, and energy cheap, Kuwait being the classical example.

Table 5
Flows (see Figure 1)

Notes on Table 5. Please refer to Figure 1 for a graphical presentation of these figures.
Total flows and turnover periods to and from:

	Total flows (x 10^{12} m³/y)	Turnover periods
- Oceans	423	3,239 years
- Icecaps	2	14,500 years
- Cultivated lands	70	
- Uncultivated lands	41	
- Surface water	71	22 months
- Underground water	10	956 years
- Users	32	
- Atmosphere	496	10 days

(The figures are highly approximate and meant to give orders of magnitude)

From	**To**	**Nature**	**Flow (10^{12} m³/y)**
Oceans	Atmosphere	Evaporation	423
Atmosphere	Oceans	Rainfall	386
Atmosphere	Icecaps	Snowfall	2
Atmosphere	Cultivated lands	Rainfall	67
Atmosphere	Uncultivated lands	Rainfall	41
Icecaps	Oceans	Melting	2
Cultivated lands	Atmosphere	Evaporation	35
Cultivated lands	Surface water	Runoff	15
Cultivated lands	Users (biosphere)	Consumption	20
Uncultivated lands	Surface water	Runoff	36
Uncultivated lands	Atmosphere	Evaporation	5
Surface water	Cultivated lands	Irrigation	3
Surface water	Users	Miscellaneous uses	10
Surface water	Underground water	Seepage	10
Surface water	Atmosphere	Evaporation	13
Surface water	Oceans	Runoff	35
Underground water	Surface water	Sources, wells	8
Underground water	Users	Miscellaneous uses	2
Users (biosphere)	Atmosphere	Transpiration	20
Users	Surface water	Used water	12

Table 6
Reservoirs (see Figure 1)

Oceans:	**1,370,000 x 10^{12} m³**	
		399 x 10^{12} m² (79% of the earth's surface)
		35 kg of salt per m³
Ice caps and glaciers	29,000 x 10^{12} m³	
		16 x 10^{12} m² (3% of the earth's surface)
Deep groundwater (> 750 m)	5,300 x 10^{12} m³	
Shallow groundwater	4,200 x 10^{12} m³	
Lakes	125 x 10^{12} m³	
Soil moisture	65 x 10^{12} m³	
Rivers	1 x 10^{12} m³	
Biosphere	0.6 x 10^{12} m³	
Total land:	**38,700 x 10^{12} m³**	
		106 x 10^{12} m² (21% of the earth's surface)
Atmosphere	**13 x 10^{12} m³**	

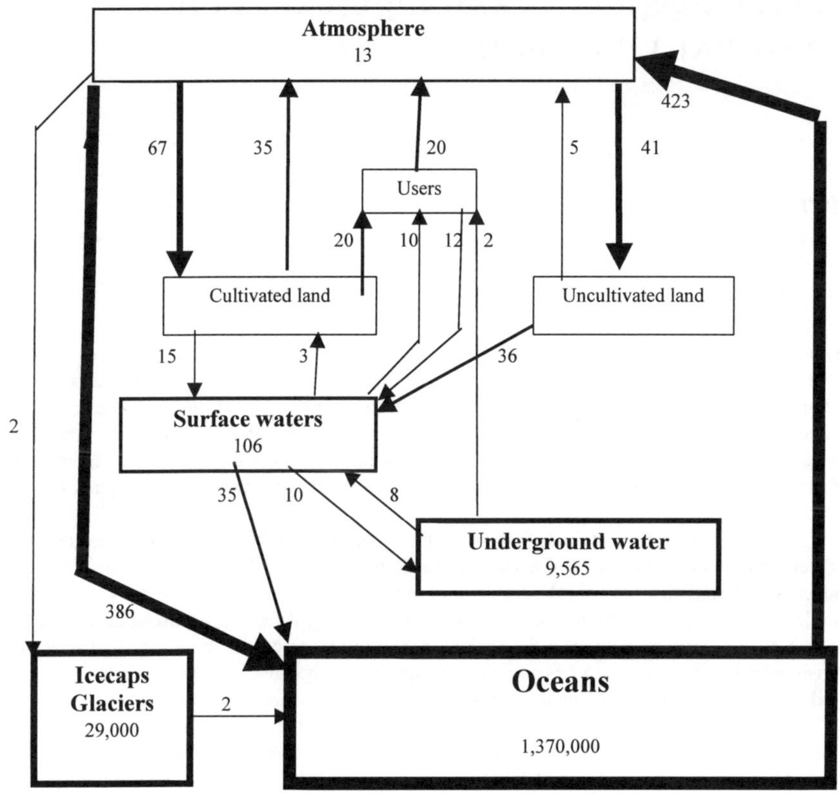

Figure 1. Water reservoirs and flows (many figures are approximations or guesses; see text)

Reservoirs in 10^{12} m^3
Flows in 10^{12} m^3/y

WATER RESOURCES AND SURVIVAL: A LOOK AT PASTORAL LIVELIHOOD

Peter Otim

Human ecology as a discipline emphasizes that Man, like other living organisms, requires energy to live and be able to reproduce. The main source of this energy is the sun, which, through the process of photosynthesis, is stored in plants. It is from the various forms of plant life that we extract this energy either directly as primary consumers, or as secondary consumers (from other primary consumers). This forms the basis of our survival, and all the strategies for adaptation are geared towards this basic requirement for our species to exist. However, one of the basic requirements for plant life to exist is water.

Indeed, literature abounds that shows the importance of fresh water, sometimes giving annual water demand in quantities for drinking, domestic purposes, and for food production. Whereas these macro level statistics are useful because they indicate global trends – especially at the turn of the millennium when most nation states are involved in making plans to bring about desired development - it is equally important to specifically look at the micro processes in society that make the water crisis relevant. This is what I am going to attempt. This is because it is the micro-level processes that individuals have to deal with on a daily basis that impact on larger scale relations.

In this paper, I try to highlight the plight of one of the pastoral groups to whom the arid-to-semi-arid Sahel region of Africa is home. I try to show how scarcity of water negatively affects the life of the Karimojong, and how this in turn acts upon the lives of their neighbours. This is because as they strive for survival, their activities affect not only the physical environment they occupy, but also the social environment through the other societies they come in contact with. It is argued that East Africa probably has the largest concentrations of cattle with an estimated 15 to 20 million pastoralists, and yet pastoralism continues to be discouraged by virtually all governments as one that is irrational and encourages the destruction of the environment.

The Karimojong are one of the pastoral groups occupying the semi-arid northeastern region of Uganda. The Karimojong are known to maintain two types of settlements: the permanent settlements (*ere, ngireria pl.*), where the women, children and the elderly stay; and the temporary settlements (*nawii, ngauiyoi pl.*), that are set up in the grazing areas, which is basically the dwelling place of the warriors.

One significant factor affecting the survival opportunities and options in Karamoja (the region) is its rainfall, which is not only small (annual average ranging from 350 mm in the drier eastern parts, to 750 mm in the wetter western areas) but also unreliable with regard to when, where, how much, for how long it fall, and what area it will cover. Table 1 shows the mean annual rainfall for Moroto district for the years 1987-1994.

The rains are scattered, varying in amount from year to year and from one area to another. While one place receives a sprinkle, another receives a heavy storm – and yet another may only experience a gush of rainwater running down a river.[1]. These rains are also seasonal, coming between April and August. The result of this pattern of rainfall is a low resource base characterized by seasonal variations in productivity and patchy conditions presented by the vegetation even within the same locality.

Table 1
Mean annual rainfall for Moroto district, 1987-1994

Year	Mean annual rainfall (mm)
1987	877
1988	1,135
1989	835
1990	711
1991	761
1992	569
1993	572
1994	764

Karamoja: A Brief Retrospect

Karamoja as a region has had a history of neglect by the central government in Uganda right from the colonial times. It did not offer any attraction to the colonialists, who at the time were interested in agricultural societies to produce the required raw material for the agro-based industries at home, and therefore remained unadministered for a long time. There exists much literature that looks at the impact of colonialism on the livelihood of the Karimojong people, and it all seems to have a consensus on one common point: the results of this intervention were largely disastrous to the Karimojong - mainly because of the attitudes that governed the policies that were employed at the time. This is mainly because national development plans are made by policymakers who are mainly from sedentary groups for sedentary populations.

Indeed, the stand taken by post-independence governments in Africa in general against pastoralists has not been any different from the one that was taken by the colonial governments; instead it is more of a continuation of the colonial policies. This has led to poor livelihood standards not only in Uganda but also in Africa in general, which has contributed to the lumpenization of the pastoralists. The Karimojong have benefited little by the attainment of independence because of the colonial policies that continue to be upheld. Because the Karimojong remain largely insignificant both socially and politically, most governments have had little concern with their fate. While addressing a delegation of European Union ambassadors who were on a fact finding tour of Karamoja, the then Minister of State for Karamoja lamented that "... Karamoja had all along been neglected and was always considered last in government development programmes".[2]

Even in a country like Somalia, where the majority of the population are pastoralists, they are increasingly getting marginalized. The obsession by governments and policymakers with the "cultivation culture", and therefore settlement, as the best mode of land use, has resulted in the view that good land should only be put to crop production and not used for livestock[3].

The Adaptive Regimes of the Karimojong

The adaptation of the Karimojong to this limiting environment that they occupy is characterized by opportunistic management in the widely scattered resource base. The practice of mobility is central to their systems - where animals are moved to natural forage in intricate but clearly designed cycles. Mobility is crucial to the survival of the Karimojong herds and this, therefore, means constant periodic movement. These movements are dictated mainly by the availability of water for the herds, such that when the rains return, herds are driven to the eastern part of the region in order to take advantage of the new grasses in this area, which quickly sprout with precipitation but which will shrivel away soon after the rains stop because of the poor rocky soils. When the distance to watering points increases, and as the forage diminishes, the herds are moved to the central zone of permanent settlement where the herds are fed on the available pastureland and the stalks in the then harvested fields. It is also at this time that the grasses in the western part of the region are set on fire so that by the time the herds are moved there fresh and tender grass will have sprouted[4].

So, as the rains stop and as the dry season intensifies, the herds are moved to the western region where the most reliable watering points in the region are to be found. The cycle is repeated seasonally as rainfall dictates. However, it should be noted that the duration spent in any particular area depends on the availability of water for the herds and can be from just a few weeks to months. It is the ability to access the dispersed forage and watering points through mobility that offers the best protection against the unreliability of rainfall.

My recent findings show that shortage of water for livestock presents the most crucial factor for the survival of the Karimojong. From the period just after the massive raids by the Karimojong against their agro-pastoral neighbours, there have been repeated demands by those affected that the government stop the Karimojong from crossing into their districts. The argument is that since the watering areas are shared, the Karimojong have ample time to survey which kraals are most stocked and normally raid those stocks as they return to Karamoja when the rains return.

In response, the Karimojong herders insisted that they were only following their water that ran down into the large swamps that form the borders of the region. It is search for water that prompts the majority of the movements of the herders, and they sometimes move away from certain locations even when some pasture is still available because of lack of water. Since during the dry season most of the grasses fed on by stocks are dry, there is increased demand for water for the stocks, yet ironically this is the season when water is least available. The result is poor productivity that increases the demand for cattle. To satisfy these demands, they raid.

Alternatives for Survival

It is in relation to such an uncertain environment that the Karimojong have adopted herding as the best means of survival. Whereas subsistence agriculture continues to be practiced in Karamoja, it has been the main victim of the unpredictable weather. Sorghum is the main crop planted by the Karimojong and is used for both food (mainly for making porridge) and for making beer. The decision of whether or not to plant seed is normally a gamble because of the unreliability of rain. So seeds are normally sown with the first rains, but if there is a dry spell even as short as three to four weeks during the growing season, the young crop normally withers away and new seed has to be planted with the next precipitation. Sometimes drought strikes during or about flowering – in which case it would be too late to plant another crop because the rains would soon be stopping. This results in crop failure.

It has been argued that crop failure in Karamoja occurs in cycles of every four to five years. Although agriculture remains an important source of food for the Karimojong, the unreliability of harvest means that the people cannot depend on rain-fed production, and its role as a source of food remains as a supplement to animal products. However in the homesteads much of the food consumed includes some agricultural products as compared to the grazing areas where virtually all food is animal products.

Violence in Karamoja

Cattle raiding is a common phenomenon among pastoralists in general and it all centres around struggle for resources and survival, in most cases at the expense of others. Among the Karimojong, raiding of cattle was for a long time carried out as a method of traditionally replenishing stocks lost in times of hardship like drought and epidemics. It was a conscious community effort to ensure its sustainability and was sanctioned by the elders. Dyson-Hudson gives an account of the various raids that occurred in the 1940s and 1950s and what clearly comes out is that these conflicts were over the the most important resources for the survival of the community - water and pasture.

Over time, as the Karimojong continued to move into neighbouring districts in search of water for their stocks, incidents of raiding continued to escalate until it climaxed between 1987-1989 when seven districts had their stocks virtually completely rustled. As a result of the proliferation of small arms in the region, which has been estimated as 400,000 in a population of just over 400,000, the situation has continued to deteriorate with massive raids carried out within the region. At the end of July 1999, over 100 people belonging to the Bokora group were killed by a rival Karimojong group as a result of conflict over a watering place.

Lessons to be Drawn

Water scarcity is increasingly becoming an issue of conflict in the delicate region of Karamoja. As the Karimojong population continues to grow without any alternatives for survival, issues of water scarcity, and therefore alternatives survival strategies, need to be addressed sooner rather than later. As described briefly above, survival on rain-fed agriculture is a near impossibility. This therefore calls for development of water catchments that can provide rainwater over a longer period to irrigate fields. Also there is need for the government to plan for alternative survival for the Karimojong.

Formal education has not been popular in Karamoja because the national timetable is designed for an agricultural population which requires one to be attending classes during the period the Karimojong are furthest from their settlements because of lack of water and pasture. The result is that enrollment and dropout rates are highest in Karamoja.

Otherwise, with the ever increasing militarism in the region, this type of conflict, which has also drawn in neighbouring communities from Kenya, is likely to metamorphose from being local to include neighbouring countries.

Pastoral violence should not be looked at as a local problem but as a problem that has the capacity to become regional due to the sheer numbers of pastoralists in eastern Africa.

References

[1] An article run in the New Vision newspaper of 1 June 1995 reported that 14 families who had temporarily occupied a valley in Nadunget sub-county, Moroto district, were in May 1995 washed away by this kind of floods with a loss of 9 lives, 930 heads of cattle, and 1,286 goats.

[2]. See *New Vision*, 15 November 1995.

[3]. For instance the Resident District Commissioner Mbarara challenged the visiting vigilante commanders from Karamoja to adopt mixed farming and water harvesting for a more settled life.

[4]. This is a contested practice with some scholars blaming it as causing environmental deterioration. However, works, like that of Mamdani, argue that it is consciously done not only to provide tender grass but also to rid these tall grasses, especially, of ticks which cause disease to livestock.

WAR OVER WATER – THE CASE OF THE ILISU DAM PROJECT IN TURKEY

Carsten Rohr

Water Disputes in the next Millennium

Water is considered likely to be a major cause of international conflict in the next millennium. "Wars of the next century will be over water." This warning comes from the World Bank, the largest international investor in water projects. Klaus Töpfer, head of the United Nations Environmental Programme, warns about extreme conflicts caused by already severe water scarcities. NATO's Committee on the Challenges of Modern Society (CCMS) is dealing with water as a security risk. In 1994, the Center of Strategic and International Studies in Washington identified 10 regions where water could become a reason for conflict; almost every dam project increases tension between or within states. The CIA knows that water interests and water problems are nowhere greater than in the Middle East. On 21 May 1997, the United Nations General Assembly adopted the Convention on Law of Non-Navigational Uses of International Watercourses, designed to prevent border disputes and wars between states that share water resources; Turkey was one of only three countries (together with China and Burundi) to vote against it.

The Case of the Ilisu Dam in Turkey

The South-Eastern Anatolia Development Project (GAP) is a huge $32 billion hydropower and irrigation scheme, which was initiated in the 1970s and is due to be completed in 2010. It involves 22 dams, 19 hydroelectric power plants and a network of irrigation canals for the rivers Euphrates and Tigris. It is planned that at full development over 1.7 million ha of land will be irrigated and 27 billion kWh of electricity will be generated annually with an installed capacity of over 7,500MW. Both rivers rise in Turkey, where they gather about 90 per cent and 50 per cent of the water respectively, before flowing through Syria and Iraq.

The Ilisu dam is the latest $1.5 billion project, currently the largest of Turkey, located on the Tigris River, 40 miles upstream of the Syrian-Iraq border, in the heart of the Kurdish populated area. A rockfill dam, 1,820 metres long and 135 metres high, will create a reservoir with a maximum volume of 10.4 billion cubic metres and a surface area of 313 square kilometres. It has a capacity of 1,200MW, and is expected to produce 3,800 GWh of hydroelectric power annually; it will also be used for irrigation. Demand for electricity is increasing by 8% a year in Turkey and frequent power cuts are inhibiting economic growth.

The claims of Turkey, Syria and Iraq on the water of the Euphrates and Tigris exceed the capacities by 55 per cent and 12 per cent respectively.

Consequences

The dams on the Euphrates, used primarily for irrigation, reduce the average water flow by almost 50%; the Tigris projects, primarily used for power production, will reduce the average water flow by 10%. In addition, the water quality will be degraded through salination, pesticides, and a reduction in mud (as natural fertiliser) and fish.

The dams give Turkey control over the water supplies of Syria and Iraq, as spare capacity of the reservoirs – the difference between maximum and normal operational capacity – will be sufficient to block any water flowing to Syria and Iraq for several months.

The Ilisu reservoir will flood 52 villages and 15 towns, including Hasankeyef, a Kurdish town of 5,500 people, which is the only town in Anatolia that has survived since the Middle Ages and is under archaeological protection. It will displace approximately 16,000 people in the troubled Kurdish region.

Turkish Objectives

The main (official) objective is the development in the poor south-east, both economically and socially, including communication, housing, industry, education, health and other services. According to Olecay Ünver, President of the GAP Regional Development Administration, "The GAP Master Plan's basic development scenario is to transform the region into an agro-related export base," and "... prosperity will be evident for everyone except the most prejudiced and opinionated." Displacing thousands of rebel Kurds gives an opportunity for controlling them better.

The development of the region is achieved by increasing the water resources for irrigation and domestic and industrial consumption (at the expenses of the downstream countries), and the large-scale and centralized generation of electricity, which is also easier to control than the small-scale and distributed.

The control over water supplies gives Turkey a strategic advantage, for example, as a bargaining tool against Syria's and Iraq's support for the PKK.

Contractors, Creditors and Foreign Involvement

The World Bank refused to back the project in 1984. The Ilisu project appears to violate five of its guidelines: OD 4.00 (Annex A, Environmental Assessment, and Annex B, Environmental Policy for Dam and Reservoir Projects); OD 4.30 (Involuntary Resettlement); OP 7.50 (Projects on International Waterways); OPN 11.03 (Management of Cultural Property); and BP 17.50 (Disclosure of Operational Information), and it also contradicts the UN Convention on Law of Non-Navigational Uses of International Watercourses. Instead the Turkish government appointed a Swiss hydro-electric company, Sulzer Hydro, and ABB Power Generation, a Swiss-Swedish company, to organize the project. The firm then approached the British company Balfour Beatty to head the construction consortium in nine countries including Impregilo (Italy), Skanska (Sweden), and the three Turkish companies Nurol, Kiska and Tekfen. Balfour Beatty was the company at the centre of the Pergau dam battle in Malaysia. That project became mired in controversy after Britain ignored warnings that the project was unsuitable on environmental grounds and gave a large chunk of its aid budget to build the dam, because Malaysia was buying arms from British manufacturers. The financial package is being arranged by the Union Bank of Switzerland (UBS). The World Bank's refusal to back the project meant that private banks regarded the project as too risky and the consortium had to find governments that would underwrite the project. They have submitted applications for export credit guarantees of about $850 million to nine countries (Austria, Germany, Italy, Japan, Portugal, Sweden, Switzerland, the UK, and the USA).

Most export credit agencies have few or no environmental or ethical standards, including Britain's Export Credit Guarantee Department (ECGD), which is part of the Department of Trade and Industry (DTI). British trade minister Stephen Byers announced only on 2 August 1999, after much pressure that there will be a review of the criteria for export credit guarantees. By contrast the German government still refuses to do so.

Some people warned that the Ilisu project could risk involving Britain in armed conflict between Syria, Iraq and Turkey (a NATO ally) over the right to water from the Tigris. If supported by Britain (and other countries), this project would be in sharp contrast to New Labour's ethical and environmental foreign policy.

Possible Alternatives

In the environmental impact assessment, commissioned by Sulzer and ABB, no supply-side or demand-side alternatives to Ilisu have been considered as part of the feasibility studies. Renewable energies such as small-scale hydro-electric power, passive solar, photovoltaics and wind energy are certainly good alternatives, as they can be distributed (with no or little transmission losses) and tuned to needs and potentials. To generate the same amount of energy as the Ilisu dam project by photovoltaics, for example, only 7 per cent of the total flooded area would be necessary. Political decision-making processes seem to be the biggest obstacle to distributed and decentralized schemes.

The solution to the world's water crises lies less in the supply than in the demand. Improved water management, water recycling, efficiency improvements, avoiding of waste, and new technologies are just a few points.

Conclusion

In the Ilisu case, the project is largely political and predominantly motivated by the strategic interest of the Turkish government in strengthening its position of power in the Middle East, and controlling the unruly Kurdish areas. This puts downstream countries in a strategically and materially disadvantageous position, and therefore threatens their national security.

In November 1998 Turkey showed its military strength in a 50,000 men military exercise close to the Syrian border. As long as one country is clearly militarily superior, the risk of a military conflict over water is reduced, as the smaller country has no chance of winning a war – until the situation is desperate enough to risk it anyway.

Negotiations about fair sharing of water resources are urgently required.

WATER AND SOCIAL STABILITY: THE SOUTHERN AFRICAN DILEMMA

Anthony R. Turton

This paper attempts to highlight some of the fundamental development-related problems that are confronting the Southern African region at present. The range of the problem is wide, but one aspect stands out as significant - the dilemma arising from the environmental sustainability of current economic development initiatives *vis-à-vis* social stability - making this an issue increasingly likely to affect regional stability in the early part of the 21st Century. In this regard water is a key indicator of where we have been and where we are headed. This paper presents some of the latest research that is being done at AWIRU regarding water and social stability. It argues that two potential future scenarios exist. The first results from an end condition that has been defined as "water poverty",[1] which is likely to result in ecological collapse in conjunction with a high level of social instability. The second results from an end condition that has been defined as "structurally-induced water abundance"[1], which is likely to result in ecological sustainability in conjunction with a degree of social stability. Clearly, the latter is the policy objective that ought to be striven for, yet the current data available tentatively suggests that this is not the case.

Statement of the Problem

If one performs an analysis of Southern Africa, one is immediately confronted by two significant factors.[2] These relate to precipitation and people, and it is the interaction of these two fundamental variables that essentially underpins the developmental dilemma of the region. Let us look at these in a little more detail.

Precipitation[3] is the first fundamental variable of which four distinct aspects are important. The first three of these are generally of greater importance than the remaining two for the purposes of this paper, however. Firstly, there is a marked spatial maldistribution of water, with a distinct latitudinal and longitudinal trend. There are generally higher precipitation levels in the north, decreasing progressively to the south. Superimposed onto this, there are also higher levels of precipitation in the east, decreasing dramatically to the west. This is shown in Table 1. Secondly, there is a temporal dimension to this maldistribution of water. This means that rainfall tends to be distinctly seasonal in pattern, but these seasons vary greatly over time. Thirdly, there is the volume aspect. This is shown in Table 1 under the average rainfall column. The amount of precipitation that falls is subject to a high degree of variation, as shown in under the rainfall range column. These three elements combine to cause a reasonably high level of stochasticity in the overall precipitation patterns. This has significant implications for aspects such as crop planning, surface runoff, soil erosion and river flows. These factors in turn impact on water management because it means that large storage reservoirs[4] have to be planned and built because the reliability or predictability of the precipitation patterns are of a low order of magnitude. The less important aspect of precipitation (at least for the purposes of this paper) is water quality, which is generally of a reasonably high standard except in certain specific areas where acid rain falls. Eutrophication and salinization do occur in impoundments due to the return flows of effluent and evaporative losses however.

Table 1. Rainfall and evaporation statistics for selected countries in Southern Africa, showing rainfall range and estimated surface runoff.[5]

Country	Rainfall range (mm)	Average rainfall (mm)	Average rainfall (km^3)	Potential evaporation range (mm)	Total surface runoff (mm)	Total surface runoff (km^3)
Angola	25 - 1 600	800	997	1 300 – 2 600	104	130.0
Botswana	250 - 650	400	233	2 600 – 3 700	0.6	.35
Lesotho	500 - 2 000	700	21	1 800 – 2 100	136	4.13
Malawi	700 - 2 800	1 000	119	1 800 – 2 000	60	7.06
Mozambique	350 - 2 000	1 100	879	1 100 – 2 000	275	220.0
Namibia	10 - 700	250	206	2 600 – 3 700	1.5	1.24
South Africa	50 - 3 000	500	612	1 100 – 3 000	39	47.45
Swaziland	500 - 1 500	800	14	2 000 – 2 200	111	1.94
Tanzania	300 - 1 600	750	709	1 100 – 2 000	78	74.0
Zambia	700 - 1 200	800	602	2 000 – 2 500	133	100.0
Zimbabwe	350 - 1 000	700	273	2 000 – 2 600	34	13.1
TOTAL			4 665			599.27

People are the second fundamental variable of which three aspects are important. Firstly, there is a spatial distribution of people in Southern Africa that is generally at variance with the availability of water. In other words, population distribution tends in general to be concentrated in areas that are far from supplies of water. Secondly, there is a significant temporal aspect to the population base in the sense that the whole of Southern Africa generally has a very high population growth rate. Thirdly, there is the aspect of migration patterns. This is a complex issue in its own right. Suffice it to say, for the purposes of this paper, that there are population migration pull-factors and push-factors. Pull-factors are aspects such as perceptions of better job opportunities in the larger urban areas, which tend to be focussed on places like the Gauteng area of South Africa (which is on a high plateau, far from water supplies and facing a significant water shortage as a result), Gaborone in Botswana, and Harare and Bulawayo in Zimbabwe (both on a watershed and with unique water supply problems of their own). Migration push-factors are aspects such as the loss of livelihoods due to overpopulation, poverty, declining levels of land per capita (which is significant for a subsistence farming economy) and drought.

This leads logically onto the third fundamental variable. If the water is concentrated in areas that are spatially distant from the population centres, then effectively what needs to happen is to bring the water to the people (or vice versa). This is where the heart of the political ecology dilemma lies. In order to move water over long distances *pipelines*[6] are employed. For the purposes of the political ecology discourse it is these pipelines that make the debate lively. Because pipelines bring life-giving and job-creating water[7] to areas where the demand for water is high, they perform a function that can be regarded in its broadest sense as being allocative. In short, pipelines allocate a given volume of water to a specific spatial entity. It is precisely this allocative aspect of pipelines that results in the next variable, *power*[8]. In its crudest form, power[9] is derived from the privilege or the relative advantage[10] that the receiving entity derives from having been favoured over the non-receiving entity. Seen in this way, pipelines become conduits of power because they allocate water from an area of relative abundance, to an area of relative scarcity. The water so allocated therefore has an economic cost, whether this aspect is actually reflected in the final price that is charged for the water or not; and the water so allocated results in a relative advantage of one form or another.

This raises two distinct new aspects of the problem, however. Firstly, the water supplied by these large pipeline schemes comes at a high ecological cost, opening the debate on sustainability. Secondly, in many cases there is simply no more water left for further mobilization in readily accessible river basins. This means that the pipeline projects are becoming increasingly complex, costly, international in nature and therefore prone to political factors beyond the control of any one government.

Theoretical Aspects

Recent work that has been conducted[1] shows that in terms of water and social stability, two possible scenarios are likely to be relevant to developing countries in semi-arid regions. In this regard, by linking different combinations of a first-order natural resource scarcity (water) with a second-order social resource scarcity (adaptive capacity), the authors have developed some key concepts. Of these concepts, the most important for the purposes of this paper are those relating to the definition of "water poverty"[11] and "structurally-induced water abundance"[12]. Taking these as two possible end-conditions for future scenario planning purposes, a set of policy options can be generated. "Water poverty" is the end condition that is likely to result in significant environmental degradation in conjunction with a high level of social instability. This is the condition that rational policymakers would logically choose to avoid. "Structurally-induced water abundance" is the end condition that is likely to result in economic sustainability, and probably even a degree of environmental sustainability, coupled with a higher degree of social stability. This is the condition that rational policymakers would choose as a desirable end goal. It is illustrated graphically in Figure 1.

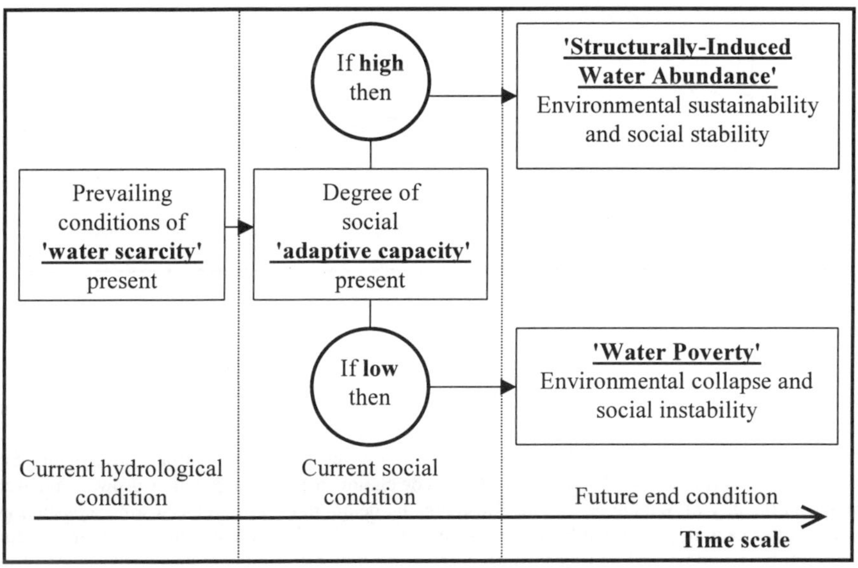

Figure 1. Schematic representation showing the hypothesized two end conditions that are likely to occur when combining both a first-order and second-order resource in the definition of key variables.

Thus in terms of this theoretical conceptualization, the interceding variable between "water scarcity" and either "water poverty" or "structurally-induced water abundance" is the existence of social "adaptive capacity[13]", loosely defined by Ohlsson[14 15] as the ability of a social entity to adjust to the increasing levels of "water scarcity". "Adaptive capacity" is a nebulous concept to quantify, however, prompting Turton & Ohlsson[16] to hypothesize that Water Demand Management (WDM) is an empirically testable manifestation of the "adaptive capacity" of a given social entity. To this end, the existence of reflexivity[17] in the total water consumption curve of a country can be taken to indicate a relatively high level of effective coping strategies. This enables case studies to be interrogated with the specific intention of determining the degree of "adaptive capacity" in existence. In this regard, the existence of a reflexive water demand curve, specifically with regards to agriculture, can be assumed to be a precursor of "natural resource reconstruction" and therefore empirical evidence of environmental sustainability.

Some Empirical Evidence

The first piece of evidence comes from Israel, where reflexivity has been achieved as a result of the judicious use of "inter-sectoral allocative efficiency"[18]. This is a classic example of "structurally-induced water abundance" and illustrates the very real difference between "water scarcity" and "water poverty". Israel is "water scarce", but due to the high levels of "adaptive capacity" present, it has managed to avert the debilitating conditions of "water poverty".[19] This case is illustrated in Figure 2.

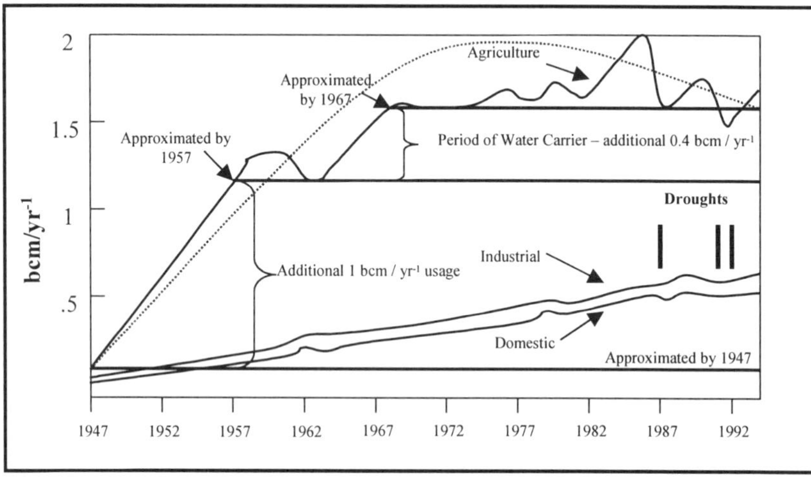

Figure 2: Israeli water consumption from 1947-1993 showing the effect of the National Water Carrier.[20] The dotted line illustrates the reflexive nature of the agricultural demand curve.

The second piece of evidence comes from Namibia. This is a country with a high level of "water scarcity" and no perennial rivers flowing across its soil, with the only exception of a small area known as the Caprivi Strip (where the Okavango and Kwando River are under Namibian sovereignty for a short duration only). Figure 3 shows the direct correlation between population growth and water demand in the city of Windhoek.[21] This illustrates the demographically-induced water consumption curve that has been hypothesized by Turton & Ohlsson.[22] Attention is drawn to the fact that no form of reflexivity is evident, so this trajectory is clearly unsustainable in the long-term, suggesting "water poverty" as a possible end condition if all things remain equal. Major supply sided solutions are being suggested to meet this growing demand. The Eastern National Water Carrier (ENWC) brings water South from Grootfontein. A pipeline is planned from the Okavango River at Rundu[23] to augment the ENWC. A series of pipelines are being planned to take water from the mouth of the Zaire River, through Angola and ultimately into the Kunene, Okavango and Zambezi River Basins.[24 25]

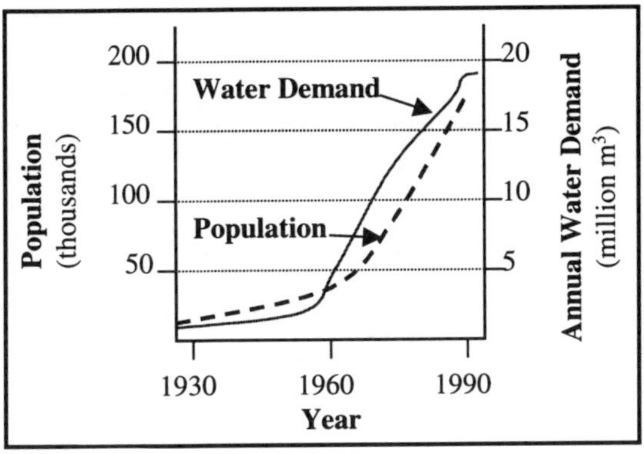

Figure 3. Example of Demographically Induced Water Consumption for the City of Windhoek[26].

Significant insight into the typical Southern African dilemma is given in Figure 4, which shows the distribution of water demand as a function of income category in the city of Windhoek.[26] This shows graphically that informal settlements use less water per capita than more affluent cohort groups.

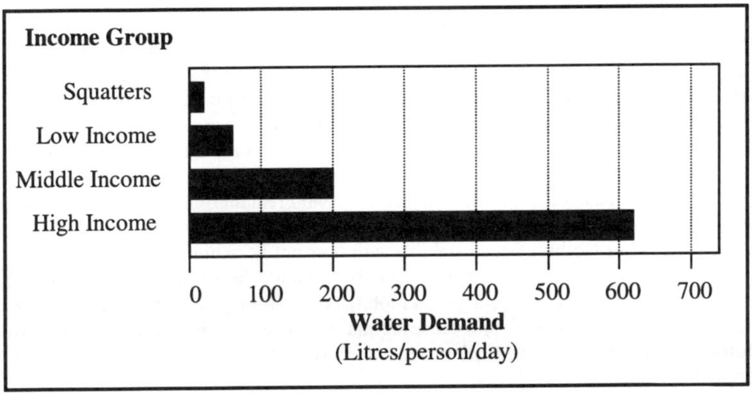

Figure 4. Daily Water Consumption per Person Expressed as a Function of Income for the City of Windhoek.[26]

The third piece of evidence is from South Africa. This illustrates the dilemma that was first shown in Figure 4, but this time on a national scale. These data offer a unique insight into the problems confronting developing countries in arid regions. South Africa has recently undergone a major political transformation. Prior to 1994, the form of government was an oligarchy with a low level of legitimacy. In this case, a high level of "resource capture" occurred,[27] resulting in institutionalized or structural scarcity.[28] It has been shown that this institutionalized form of "resource capture" was one of the subtle forms of power that propped up the otherwise insecure South African regime prior to 1994.[2] After 1994 the first democratically elected government under the presidency of Nelson Mandela, sought to right these wrongs of the past. One of the policy vehicles for this was the Reconstruction and Development Programme (RDP) which actively sought to redistribute scarce water throughout society in a more equitable manner. The effects of this policy are shown dramatically in Figure 5.

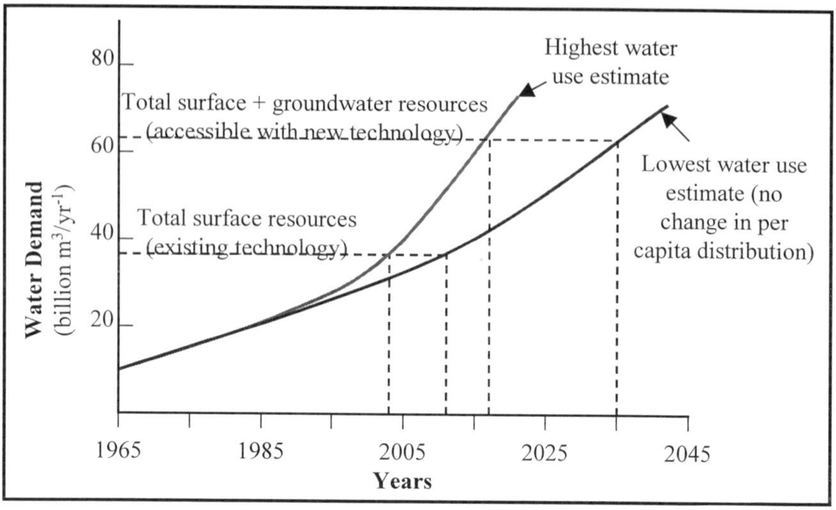

Figure 5. The Water Supply Dilemma in South Africa

Two scenarios are sketched on Ashton's data series.[25] It should be noted that the total surface and groundwater resources have been inflated under the assumption that new technology would make additional resources available. This projection is thus more optimistic than the prevailing reality.

The lowest water estimate assumes that existing per capita consumption remains the same as it was under apartheid, but that consumption is extrapolated in keeping with the population growth projections. Significant in this scenario is the fact that structural scarcities from the apartheid era will not be redressed so the balance of privilege in society would remain roughly unchanged. Two aspects of this scenario are significant. Firstly, it would de-legitimize the existing regime, which would result in long-term political instability. Social stability will thus not be achieved as "resource capture" will not be addressed to any significant extent. Secondly, it would be ecologically unsustainable in the long-term, with all surface water having been mobilized by around the year 2012. In other words, at this time, no river water would flow into estuaries and increasing volumes of groundwater would have to be abstracted. By the year 2035 all groundwater and surface water resources will have been mobilized in this scenario.

The highest water estimate assumes that all of the RDP projects will deliver water as promised by the government. This greater availability of water, coupled with increased security of tenure of the consumers, will result in a dramatic increase in consumption, along the lines of the phenomenon illustrated in Figure 4. Significant in this scenario, the structural scarcities of the apartheid era will be addressed and the balance of privilege will change in society. Two aspects in this scenario are significant. Firstly, the current regime will derive a high level of legitimacy from this policy. It will also result in a greater degree of social stability than was the case under conditions of "resource capture" during the apartheid era. Secondly, it will be ecologically unsustainable, with significant levels of environmental degradation to be expected in the short-term. This is at odds with the new water legislation, which aims at "natural resource reconstruction" by granting ecosystems the legal right to sufficient water for their own use. The effects of this environmental unsustainability may be felt as soon as during the next election cycle, introducing a sense of urgency into the problem.

Thus the South African government is in a serious water supply dilemma. If they supply the water as promised in terms of the RDP, then ecological degradation is almost surely imminent, with the early effects being felt within the next election cycle. If they choose instead to fall short of the RDP promises, then social stability will become an acute issue in the short-term, but with environmental degradation only being a reality in the medium to long-term. This is like being caught between the proverbial rock and a hard place.

What is absent from this data set is any indication of reflexivity at the national level. This is worrying, as it implies unsustainability over time. The window of opportunity is open however, but is likely to remain that way for only a short period of time. The high level of regime legitimacy enables far-reaching water sector reforms to be implemented in the short-term. Major reforms have already taken place, but these seem to have resulted in an increase in demand for water, thereby hastening the ecological degradation that seems to be imminent. Clearly, what now needs to happen is for South Africa to actively develop a series of water-related development policies aimed specifically at averting "water poverty" and encouraging "structurally-induced water abundance".

Is this likely to happen within the desired time frame? Barbara Schreiner, in her capacity as the Chief Director: Water Use and Conservation at the South African Department of Water Affairs and Forestry (DWAF) noted that there is a major capacity problem at present, and she cautioned participants at the IUCN WDM workshop in terms of their expectations of DWAF.[29] This implies that while government intentions are noble, in reality South Africa may be incapable of mobilizing the necessary social resources to cope with the problem of "water scarcity", and thereby avoiding the debilitating effects of "water poverty". This is the reason why it is being hypothesized that social stability will be a function of the magnitude of second-order resources available to a society facing long-term "water scarcity".[1 30] This is the scope of a new research programme being conducted by Ohlsson & Turton with a Southern African regional focus.

"Virtual Water" and Social Stability

Having shown somewhat pessimistically that social instability is a likely outcome of the water deficit within Southern Africa, it becomes necessary to focus on possible coping strategies that can be considered by governments. In this regard, the trade in "virtual water"[31] has been recognized as a politically silent approach to managing water deficits in the MENA[32] region.[20 33 34 35] Indeed, this coping strategy has become strategically significant to developing countries in semi-arid regions as trade in "virtual water" prevents water crises from becoming water wars.[36] This being the case, it becomes instructive to see how "virtual water" has been traded in the Southern African region in the recent past. There is a distinct trend in the reduction of wheat production, accompanied by an increase in wheat imports. This is likely to become a key stabilizing factor in future as countries in the region grapple with the problem of maintaining social stability in the face of water deficit. Additional research is being done on the strategic significance of this factor.

The water management goal therefore becomes one of keeping this social instability within reasonable parameters. In this regard, "virtual water" can play a crucial role as a viable coping strategy in Southern Africa as it has done in the MENA area.

Quo Vadis SADC?

Given the magnitude of the task at hand, coupled with the fact of the interconnectedness of the entire Southern African region, SADC as the regional structure could play a crucial role in fostering "structurally-induced water abundance". To this end, there is a Protocol on Shared Water Systems in existence. It is very weak, but efforts are under way[37] to strengthen this legal instrument. WDM is also being actively spoken about and implemented by various countries in the region. The IUCN have played a significant role in getting stakeholders[38] together, but there is still a long way to go before these policies start to impact on water consumption. What is needed is a concerted research effort, specifically aimed at determining the social implications of inter-sectoral allocative mechanisms. It is only by taking advantage of the gearing effect inherent in allocating water from economic activities or sectors with a low return to water, to activities with a higher efficiency ratio, that reflexivity is likely to be achieved at the national or regional level. This is socially disruptive, however, and opens the politically sensitive debate on food security versus national self-sufficiency amongst other issues.

Trade in "virtual water", both between SADC and the rest of the world, and within SADC itself, is likely to play an increasingly important role in maintaining social stability while these structural adjustments are being made.

Conclusion

Although South Africa is intuitively considered to have a higher level of technological resources available relative to other Southern African states, this may create the wrong impression and misguide researchers. A more fruitful research direction is related to the existence of "second-order scarcities" of the social resources needed to adapt to increasing levels of "water scarcity". This type of approach opens up various research horizons, necessitating the development of new definitions, concepts and models. In terms of this, Turton & Ohlsson[1] have hypothesized two end conditions. The first is "water poverty" with a low level of social stability combined with ecological unsustainability. The second is "structurally-induced water abundance" with a high level of social stability combined with ecological sustainability. Existing data tentatively suggests that various states in Southern Africa are not necessarily headed for the latter. This suggests that the interaction between water scarcity and social stability is likely to become a significant threat to regional security in the early part of the 21st Century. Trade in "virtual water" is likely to become strategically significant as a mitigating factor against social instability in future, buying time for the necessary social transformation to take place. Existing data shows that it has already become relevant. A concerted effort needs to be launched now at the policy-making levels of the region if social stability is to be maintained over time. This effort will of necessity involve cooperation between research partners, stakeholders, foreign donor agencies, SADC and various national governments in a common approach that has been hitherto impossible to achieve.

References

[1] Turton, A.R. & Ohlsson, L. 1999. *Water Scarcity and Social Adaptive Capacity: Towards an Understanding of the Social Dynamics of Managing Water Scarcity in Developing Countries.* Paper presented in Workshop No. 4: "Water and Social Stability" at the 9th Stockholm Water Symposium, "Urban Stability through Integrated Water-Related Management", hosted on 9-12 August by the Stockholm Water Institute (SIWI) in Stockholm, Sweden. Also available as MEWREW Occasional Paper No. 18 from the School of Oriental and African Studies (SOAS) University of London Website
http://www.soas.ac.uk/Geography/WaterIssues/OccasionalPapers/home.html

[2] Turton, A.R. 1999. Precipitation, People, Pipelines and Power: Towards a Political Ecology Discourse of Water in Southern Africa. Paper presented on 2 February 1999 at the School of Oriental and African Studies (SOAS) Geography Department Research Seminar, January – March 1999. Forthcoming in a book provisionally entitled *Political Construction of Environmental Knowledge.* Also available as MEWREW Occasional Paper No. 11 from the School of Oriental and African Studies (SOAS) University of London Website
http://www.soas.ac.uk/Geography/WaterIssues/OccasionalPapers/home.html

[3] For the purposes of this paper, the variable called "precipitation" is taken to mean water in whatever form it is found in the hydrological cycle. This is not strictly in accordance with other definitions of precipitation. Essentially, the water found in the lower Zambezi River that is being allocated to the generation of hydro-electric power in Mozambique, is the same water that fell as precipitation (in the strict hydrological sense of the word) over the highlands of Angola.

[4] This in itself is a problem due to the prevailing levels of evapotranspiration. For example, evaporation from Lake Kariba accounts for 20-25% of the annual flow of the Zambezi River at that point (Mac Donald, M. & Partners. 1990. *Sub Saharan Africa Hydrological Assessment: SADCC Countries: Regional Report.*). Evaporation from the Omatako Canal, a component of the Eastern National Water Carrier (ENWC) in Namibia, accounts for 70% of the water carried by this scheme at that point (Davies, B.R., O'Keefe, J.H. & Snaddon, C.D. 1993. *A Synthesis of the Ecological Functioning, Conservation and Management of South African River Ecosystems.* Water Research Commission Report No. TT 62/93. Pretoria, p. 163).

[5] Pallett, J. (Ed.) 1997. *Sharing Water in Southern Africa.* Desert Research Foundation of Namibia: Windhoek

[6] It must be noted that the term "pipeline" is being used in the widest possible context. Huge schemes such as the Lesotho Highlands Water Project tend to transfer water from one river basin to another. Strictly speaking, it is incorrect to assume that the water stays in pipelines for the entire distance of the transfer. Normally the water is pumped via pipeline either over a watershed, or through a mountain (by tunnel) until the next river basin has been reached, where the water is discharged into the receiving river. For this paper, the variable called "pipeline" will be used to mean any form of water transfer scheme, be it reservoir, canal, tunnel or pipeline.

[7] Conventional wisdom in SADC, the regional development structure for Southern Africa, propagates the idea that "more than anything else, availability of water, whether from the natural environment or an artificial supply, determines how and where development can take place" (Pallett, 1997, *op.cit.* p. 1). This means that competition for the resource is likely to rise in direct relation to its increasing scarcity, elevating it to an issue of strategic significance for the various governments in Southern Africa in the near future.

[8] Bryant offers a useful conceptualization by stating that "power is reflected in the ability of one actor to control the environment of another". This is used as a definition for purposes of this paper. Bryant, R.L. 1993. Power, Knowledge and Political Ecology in the Third World: A Review, in *Progressive Physical Geography*, No.20.

[9] Refer to Bryant & Bailey for a more detailed analysis of power vis-à-vis political ecology. Bryant, R.L. & Bailey, S. 1997. *Third World Political Ecology*. New York: Routledge, pp. 38-47.

[10] In India, for example, water distribution in irrigation systems is regarded as "an inherently political activity" (Mollinga, P.P. & van Straaten, C.J.M. 1996. The Politics of Water Distribution, in Howsam, P. & Carter, R.C. (Eds.) *Water Policy: Allocation and Management in Practice*. London: E & FN Spon.).

[11] "Water poverty" is defined as the existence of both a first-order natural resource scarcity (water) and a second-order social resource scarcity (adaptive capacity) simultaneously (Turton & Ohlsson, 1999, *op.cit.,* p. 4).

[12] "Structurally-induced water abundance" is defined as the condition that exists when a social entity has both a first-order resource scarcity (water) and a second-order social resource abundance (adaptive capacity) simultaneously (Turton & Ohlsson, 1999, *op.cit.*, p. 4). The nearest example of such a condition is Israel which is "water scarce" but not "water poor" (Feitelson, E. 1999. *"Is the Commodification of Water Commensurate with Sustainability Notions?"* Paper presented at the Conference on Sustainability, Risk and Nature: The Political Ecology of Water in Advanced Societies, 15-17 April 1999, Oxford University, hosted by the School of Geography and the Oxford Centre for Water Research.).

[13] For a deeper analysis of the concept of "adaptive capacity" refer to Turton, A.R. 1999(b). *Water Scarcity and Social Adaptive Capacity: Towards an Understanding of the Social Dynamics of Water Demand Management in Developing Countries.* Also available as MEWREW Occasional Paper No. 9 from the School of Oriental and African Studies (SOAS) University of London Website http://www.soas.ac.uk/Geography/WaterIssues/OccasionalPapers/home.html

[14] Ohlsson, L. 1998. *Water and Social Resource Scarcity - An Issue Paper Commissioned by FAO/AGLW.* Presented as a discussion paper for the 2nd FAO E-mail Conference on Managing Water Scarcity. WATSCAR 2.

[15] Ohlsson, L. 1999. *Environment, Scarcity and Conflict: A Study of Malthusian Concerns*. Department of Peace and Development Research. University of Göteborg.

[16] Turton & Ohlsson 1999, *op.cit.*, p. 11.

[17] Beck defines reflexivity as being concerned with the unintended risks, consequences and foundations of society. See Beck, U. 1999. What is a "risk society"? In *Prometheus*, Vol. 1; 74-79. In this regard, environmental sustainability is a key indicator. A reflexive water demand curve therefore means a curve that moves from a condition of "water deficit" to the sustainability level of supply, which coincides with the maximum volume that can be effectively mobilised by pipelines without significant degrees of environmental destruction. This is illustrated in Turton & Ohlsson, 1999, *op.cit.*, p. 14.

[18] This is defined by Turton & Ohlsson as allocating water away from an economic sector or activity that has a low "return to water", usually agriculture, to another economic sector that has a higher "return to water", usually industry. This can be described as the "more jobs per drop" option. Turton & Ohlsson, 1999, *op.cit.,* p. 16.

[19] Feitelson, E. 1999, *op.cit.*

[20] Allan, J.A. 1996. The Political Economy of Water: Reasons for Optimism but Long Term Caution, in Allan, J.A. & Court, J.H. (eds.) 1996. *Water, Peace and the Middle East: Negotiating Resources in the Jordan Basin.* I.B. Taurus Publishers: London

[21] Jacobson, P.J., Jacobson, K.M. & Seely, M.K. 1995. *Ephemeral Rivers and their Catchments: Sustaining People and Development in Western Namibia.* Desert Research Foundation: Windhoek.

[22] Turton & Ohlsson 1999, *op.cit.*, pp. 5, 14.

[23] Ramberg, L. 1997. A Pipeline from the Okavango River? In *Ambio*, Vol. 26, No. 2; 129.

[24] Heyns, P. 1999. Personal communication with the author during the Stockholm Water Symposium, 9-12 August 1999.

[25] Ashton, P. 1999. Personal communication with the author at the CSIR, July 1999.

[26] Jacobsen *et al.*, 1995, *op.cit.*, p. 57.
[27] Homer-Dixon, T.F. & Percival, V. 1996. *Environmental Scarcity and Violent Conflict: Briefing Book.* Washington: American Association for the Advancement of Science
[28] Percival, V. & Homer-Dixon, T. 1995. *Environmental Scarcity and Violent Conflict: The Case of South Africa.* Washington: American Association for the Advancement of Science
[29] IUCN. 1998. *Proceedings of the Workshop on Water Demand Management.* Pretoria 20 - 21 July 1998. Pretoria: CSIR (Environmentek)
[30] Ohlsson, L & Turton, A.R. 1999. *The Turning of a Screw.* Paper presented in the Plenary Session of the 9th Stockholm Water Symposium "Urban Stability through Integrated Water-Related Management", hosted on 9-12 August by the Stockholm Water Institute (SIWI) in Stockholm, Sweden. Also available as MEWREW Occasional Paper No. 19 from Website
http://www.soas.ac.uk/Geography/WaterIssues/OccasionalPapers/home.html
[31] It takes around 1 000 tonnes of water to produce 1 tonne of wheat, so importing a tonne of wheat is tantamount to importing 1 000 tonnes of water.
[32] Middle East and North Africa.
[33] Allan, J.A. 1996. Policy Responses to the Closure of Water Resources: Regional and Global Issues. In Howsam, P. & Carter, R.C. (eds.) *Water Policy: Allocation and Management in Practice.* Proceedings of International Conference on Water Policy, Cranfield University, 23-24 September 1996. London: E & FN Spon
[34] Allan, J.A. 1996. Water Use and Development in Arid Regions: Environment, Economic Development and Water Resource Politics and Policy, in *Water Use and Development*, Vol. 5, No. 2; 107-115.
[35] Allan, J.A. 1998. "Virtual water": An Essential Element in Stabilising the Political Economies of the Middle East. Yale University Forestry & Environmental Studies Bulletin, No. 103; 141-149.
[36] Allan, J.A. 1998. Virtual water: A Strategic Resource. Global Solutions to Regional Deficits. In *Ground Water*, Vol. 36, No. 4; 545-546.
[37] To this end a regional conference is being planned for the year 2000. This will seek to determine, via a high level of stakeholder participation and consultation, what is necessary to empower the SADC Water Protocol.
[38] The IUCN hosted a workshop on Water Demand Management in conjunction with Rand Water, the Department of Water Affairs and Forestry and the Council for Scientific and Industrial Research (CSIR) from 20 - 21 July 1998.

CITY ENVIRONMENTAL PROFILE AS A BASIS FOR PROGRAMMES OF URBAN SUSTAINABLE DEVELOPMENT

Alexander Ginzburg and Veronica Ginzburg

Urban planning combines the necessity of sustainable development and strategic planning. The role of cities in regional, national and global environmental management is increasingly coming to the fore. The sphere of influence of a city extends far beyond its own territory and it could be called the major producer of pollution and waste. Thus, one of the central elements in creating urban environmental sustainability is the adoption of strategy for integration of environmental issues in strategic planning. So, the principles for planning residential environments should reflect a philosophy of sustainable development.

Deterioration of the urban environment today negatively influences public health and life quality and diminishes the ability of big cities to promote economic growth. Anthropogenic influence in big cities has resulted in a specific urban environment that has an adverse effect on human health, both physical and mental.

However, the desire of people to live in a big city is naturally determined. Higher life quality and better communication links can be achieved in cities. Despite the high population density, a man is more independent there and has more opportunities for self-realization. It is the life in cities that determines the level of modern civilization. The share of urban population keeps on growing, though social contrasts and environmental problems are becoming more and more acute.

The growing interest in strategic planning issues was evoked by the appearance of the sustainable development concept which started attempts to break down the barriers between environmental and development policy-making and to elaborate a strategy of long-term management. One of the key elements of the sustainable development concept is the statement that environment and development must be treated in an integrated manner.

Local Agenda 21 stresses the role of local authorities in the integration of economic development with environmental protection and the need for cooperation. Among the main functions of local development plans is providing a strategic and long-term context for decision making with respect to land use change and development.

Urban policy, for the sustainable development of a city, must be oriented on the citizens' needs. It is supposed to cover all aspects of urban management such as economic growth, land planning, architecture and construction, industry, transport, energy sector, housing and communal services. Special attention should be paid to the role of the private sector as well as to the market and economic mechanisms.

United Nations Sustainable Cities Programme

The UN Sustainable Cities Programme (SCP) was established to implement the principles of sustainable development on urban management. The first and one of the main steps in the development of sustainable urban strategy is the realistic evaluation of the current "state-of-the art" in the city. The city environmental profile is aimed to provide such information and to serve as a basis for the programme of the city sustainable development.

In the current ecological situation the Sustainable Cities Programme is a very important tool of the UN for the implementation of Agenda 21 and Agenda Habitat. The programme is aimed to facilitate improvement of the urban planning and management of cities by the support of the city authorities and their partners in the state, private, and non-commercial sectors of the municipal infrastructure.

The Second United Nations World Conference on Human Settlements (Habitat II – the "City Summit") in Istanbul, 1996, underlined the direct and vital contribution that productive and sustainable cities can make to social and economic advancement. This understanding has become the basis for new concepts and approaches to urban environmental-development relationships. Three main points of this basic understanding are listed in the Environmental Planning and Management (EPM) Source Book:

- sustainable cities are fundamental to social and economic development,
- environmental degradation obstructs the development contribution of cities,
- environmental degradation is not inevitable.

The first statement is widely accepted. Cities play a vital role in the social development of all countries and build a diverse and dynamic economy. Thus, sustainable developed cities are essential for national economic growth and welfare. At the local level, strong urban economics generate the resources needed for public and private investments in infrastructure, education, health and improvement of living conditions.

Unfortunately, at the present stage, urban development is usually accompanied by environmental degradation. Aside from its obvious effect on human health and well being, environmental degradation directly influences socioeconomic development. For example, inefficient use and depletion of natural resources raises input prices and operating costs throughout the economy, as well as environmental pollution creating extra costs for all sectors of the economy. Failing to deal with this problem today leads to greater problems in the future. Thus, cities must find a way of balancing urban development needs with environmental protection.

Currently, many cities around the world have serious difficulties on their way to sustainable development that obviously lead to severe environmental and economic damage. However, there are many examples that clearly prove that environmental degradation is not necessarily a result of urbanization. The experience of different cities all over the world shows this fundamental challenge has to do with urban governance: *How to plan better, and more effectively manage, the process of urban development avoiding or alleviating problems while realizing the positive potential of city growth.*

Since market forces independently cannot solve environment problems, the principles of sustainable development must include long-term planning of urban infrastructure and a systems approach to the urban environment problem-solving. New development approaches will help to mobilize and effectively apply local resources.

While elaborating and implementing the principles of sustainable city's development, the UN Sustainable Cities Programme calls for special attention to the following problems:

- a profile of sustainable development (for example, city as a centre of tourism or culture or services) is not obvious and always constitutes a contentious issue for discussions;
- a choice of activities to be considered during the elaboration of the concept of sustainable development (industry, transport, non-commercial organizations, energy, recreation) may have a lot of variations;
- a choice and definition of different indicators (social and economic, space, environment) are connected with a great degree of uncertainty;
- the way from the measurement of indicators and description of the critical state of urban environment to the elaboration and implementation of real projects of sustainable development policy is quite complicated;
- transformations in the planning and land use policies including the change of different activities, cause a lot of discussions and disputes concerning the consequences of these changes in terms of economic, social and environment effect;
- city budget substantially limits the freedom and flexibility of launching new urban environment programmes;
- small local but real improvements in the environment quality are often more appreciated than the strategic improvements that are aimed at long-term improvements and do not bring immediate results;

- the development of the city transport network results in many external problems;
- the realization of urban environment complex management policy (the concept of management from the source to flow) requires great efforts on its implementation.

Based on information from a wide variety of cities, a number of guidelines are suggested in the EPM Source Book, each of which identifies and describes ways which cities have found to be effective in moving towards sustainable development:

- cities improve environmental information and technical expertise,
- cities improve environmental strategies and decision-making,
- cities improve implementation of environmental strategies,
- cities institutionalize environmental planning and management,
- cities make more efficient use of resources for effective change.

For the improvement of city environmental information and technical expertise it is necessary to define, clarify and prioritize urban environmental problems as well as mobilize active participation of various actors and stakeholders involved in different aspects of urban environmental planning and management.

The Istanbul Manifesto recommended as a starting point the collection of city environmental information involving all those whose cooperation is necessary for effective urban environmental planning and management. This will help to define all stakeholders both in the private and public sectors, NGOs and business, simulate their interest and provide participation of all actors. The next step should be identification of prioritized environmental problems through broad agreement among the stakeholders.

The criteria for prioritization have to be worked out through a participatory process and typically derive from consideration both of the impacts associated with each environmental problem and of the local capacities to respond. Criteria for prioritizing issues are numerous but the most important are the following:

a. the magnitude of the health impact associated with the problem;
b. the size of urban productivity loss caused by the problem;
c. the potential for local capacity building;
d. the potential for local resource mobilization;
e. the relative impact of the problem on the urban poor;
f. the degree to which the consequences are short- or long-term;
g. whether or not the problem leads to an irreversible outcome;
h. whether special circumstances offer special opportunities;
i. the degree of social/political consensus on the nature or severity of the problem;
j. whether the problem is significantly affected by local responses and actions.

Thus, the UN Sustainable Cities Programme can roughly be divided into three broad phases: Assessment and start-up, strategy and action planning, follow-up and consolidation. In the first stage, urban environmental issues are identified, key actors are drawn in, political commitment is achieved and priorities are set. This being achieved through an urban environmental assessment and a broad-based consultative process. The objective of the first stage is to identify and clarify environmental issues, involve all stakeholders and to prioritize the issues to be addressed. This phase has three key outputs: a formal project agreement, an environmental profile, and city consultations.

City Environmental Profile

According to the experience of many cities, clarification and systematic identification of the interaction between environment and development issues gives a better understanding of the whole complex of city environmental problems in their inter-linkage and relationship. One of the effective ways to do this is through the development of a City Environmental Profile, which is a synthesis of the current knowledge and data focused on the factors of environmental and development management especially in relation to the roles and interests of different stakeholders.

The environmental profile serves two purposes. First, the environmental profile is a tool to identify, brief and involve all stakeholders in the city's Sustainable Cities Programme process. Second, the environmental profile has a substantive purpose: to give an overview of the interaction between the environment and development and the role of the city's activity sectors *vis-à-vis* this interaction.

The main thing that should be considered when preparing a city environmental profile is: *The environmental profile is not just any environmental report giving environmental information about a city but it is a part of the Sustainable Cities Programme and focuses on the environmental - development interaction, being the effect that development has on the urban environment and vice versa.*

The environmental profile should present information in such a way that the city's stakeholders understand and feel involved in the Sustainable Cities Programme process and the key environment-development issues in the city.

The process of formulation and periodical revision of a City Environmental Profile is concerned with establishing the overall context for urban environmental management, with a particular emphasis on inter-relationships among issues and actors. Successfully done, the process of working out a City Environmental Profile will provide a valuable starting point and common information base as well as promote a wider mutual understanding among the key actors and institutions to be involved.

A deep understanding among different actors involved in the process of environmental management in the city should result from the formulation of, and regular data update in, a City Environmental Profile. Successful environmental planning and management requires understanding, agreement and co-ordinated action by the full range of public, private and popular stakeholders including those which may be outside the formal planning system such as women or marginal and disadvantaged groups.

Clearly, it is not possible to realize successfully all environmental and development issues at once. Thus, experience has shown the importance of being selective and of setting priorities so that attention and action may be focused on a limited area of problems in a strategic sequence.

Moscow Sustainable Cities Project

In March 1998, the Moscow government, the Fund Programme Management Branch of UNEP, the UN Centre for Human Settlements (Habitat), and the Centre for International Projects (CIP) signed an agreement on carrying out the Moscow sustainable cities project in 1998-99 in the framework of the world Sustainable Cities Programme (SCP).

The Moscow sustainable cities project (MSCP) is aimed at conducting municipal, national and international activities in the framework of the SCP, preparing and publishing the local options of Habitat agenda for Moscow as well as for other cities in Russia and the CIS. The Urban Environment Forum "People and Sustainable Cities in the XXI Century" and the annual meeting of the Sustainable Cities Programme (SCP-98) were the first steps in the project implementation. They were held in June 1998, just before the World Environment Day celebrated in Moscow on 5 June 1998.

Execution of the Moscow project should facilitate social and economic development of the city-members. It will allow efficient and accurate use of natural resources, to diminish environmental risks, provide social equity, to improve urban planning and management in order to get a friendlier environment, and encourage the growth of an environmental mentality.

To implement the decision of sustainable development, the Moscow authorities together with federal, municipal and self-government institutions should accurately distribute all the powers and responsibilities.

For further research and development the following issue-specific cross-sectored working groups were established in all areas of the MSCP:

- information flow development;
- improving of urban environmental planning and management capacity;
- sustainable development of the megalopolis and its suburban territory;
- social and economic aspects of urban environment development;
- strategy of dissemination and replication of the MSCP experience.

The main steps of Moscow participation in the SCP are as follows:

1. Organizing the International Congress “People and Sustainable Cities in the XXI Century” and the World Environment Day in Moscow in June 1998.
2. Creating an Internet page of the state of the environment in Moscow.
3. Preparing and publishing the Moscow environment profile.
4. Conducting consultations on the rational management of urban environment.
5. Organizing working groups for the elaboration of the strategy and an executive plan to improve the urban environment.
6. Working through the process of planning and management of urban infrastructure to improve the environment.
7. Replicating Moscow experience in other cities of Russia and the CIS.

Information exchange in the framework of the Moscow project must demonstrate the leading role of Moscow in the project implementation and achievement of the results in other Russian cities.

Preparation and publishing of the Moscow city environmental profile constitutes a key stage in the implementation of the Moscow sustainable cities project.

Moscow, as well as many other big cities, and in particular capitals, is constantly growing and occupying greater territory. Within the concept of sustainable development it is a natural process that requires a scientific basis for environmentally sound urban development, sustainable use of natural resources and normalization of the economic situation. Since market forces cannot independently solve environmental problems, the principles of sustainable development must include long-term planning of urban infrastructure and a systems approach to urban environment problem-solving.

Moscow’s environmental profile was prepared on the basis of documents provided for the Urban Environment Forum “People and sustainable cities in the 21st century” (Moscow, June 1998) by the different departments of the Moscow Government, Moscow scientific and public organizations, scientists and experts. The Moscow City Committee on Environmental Protection (Moscompriroda), Moscow’s Mayor Department and the Master Plan Research Institute made the biggest contributions. The Moscow Mayor Department and Secretariat of the Moscow Sustainable Cities Project conducted organizational arrangements of the profile preparation.

The other important activity of the MSCP is the organization of a national conference “Social, economical and environmental problems of the sustainable development of Russian cities” which took place 24-26 June 1999. Mayors of many Russian cities participated in the conference. Participation in the conference was very important not only for Russian megalopolises but also for small and science-orientated cities, needing to change their urban policy under the pressure of changes in city industrial, economic and social profile.

Participants, guests and organizers of the conference pointed out that cities, centres of social progress, culture, science and education, are still destroyed during local and international conflicts as well as by natural hazards, bad management and other factors.

It is stated in the Call to all people around the world that cities are the centres of human civilization. Their destruction is a violence against human society. It is necessary to stop and prevent the destruction of cities.

The Role of the City Environmental Profile in Moscow’s Urban Development

The Moscow city environmental profile helps to realize the main principles of the environmental policy in Moscow’s urban development.

The development of Moscow and the Moscow region as an integrated regional economic complex plays one of the key roles in the Russian economy. It should be based on a comprehensive resolution to problems connected with the formation of the economic complex with due regard to the capital city functions as well as interrelationships between plants situated in Moscow and the Moscow region.

The main principles of the environmental policy in the field of the Moscow region development should be clearly defined in the General plan of development of Moscow and its entire region. It is centrally important that the development, approval, and implementation of general plans for Moscow and the Moscow region should be co-ordinated.

The state interests both of the Russian Federation and of its two central subjects should be aimed at the provision for:
- conditions for the sustainable development of Moscow as the Russian Federation capital, and of towns and settlements in the Moscow region and in the whole region;
- reliable functioning of the state systems of technological and transportation infrastructures;
- conservation of natural resources.

Public interests of the population of Moscow as well as towns and villages in the Moscow region should be provided by:
- creation of favourable conditions for dwelling, labour, recreation, and tourism;
- restriction of injurious environmental impacts of economic and other activities;
- general improvement of the environmental situation;
- development of engineering, transportation, and social infrastructure in settlements and adjoining areas.

Private interests of citizens and legal entities in relation to the land plots belonging to them should be provided for in course of construction and conservation activities by the observance of legal rights of owners, land users, and tenants of adjoining land plots and other real estates.

State, public, and private interests should focus on providing for preservation and conservation of historical and cultural as well as natural heritage sites being under protection of the federal and regional legislation.

It is possible to provide conditions for the sustainable development of Moscow, towns, villages, and the whole territory of the Moscow region only after resolving the following main problems:
- raising of quality of the domestic water supply in Moscow and in adjoining towns in the central part of the Moscow region;
- establishment of an integrated system of especially protected areas for the whole region;
- establishment of an integrated system for the treatment, disposal, and utilization of domestic and industrial wastes in the whole Moscow region;
- establishment of a special planning and environmental control in the areas near ecologically dangerous plants;
- reducing the proportion of environmentally polluted areas in the total territory of Moscow and its region;
- organization of agricultural production of ecologically sound goods in zones where the agriculture is environmentally safe.

The general improvement of the environmental situation in Moscow and its region depends on many issues. The most important of them is implementation by Moscow and the Moscow regional authorities of co-ordinated and concerted regional policy in the sphere of environmental protection, aimed at an ecological regeneration of the whole region as well as rehabilitation of ecological crisis zones. These should be provided by:
- establishment of ecologically sound zones with the status of specially-protected regional-level natural territories that include sources and basins of rivers serving as water supplies; these territories should be subordinated to the Moscow regional government; their control and funding should be carried out together with the Moscow City Government;
- establishment of regional-level ecological rehabilitation zones in the Moscow region with the status of specially-protected natural territories; their control and funding should be carried out jointly by the Moscow region and Moscow authorities;
- elimination of pollution sources in the regional territory as well as elimination of Moscow's pollution sources;
- ecological regeneration of the Moscow region areas situated within the impact zones of plants; this involves an increased amount of compensatory spending on sanitation and reclamation of disturbed areas;
- maintaining of agricultural zones without any negative influences that can impair the quality of products intended for the supply of Moscow and the Moscow region population.

The integrated environmental policy of Moscow and the Moscow region provides that the conservation of natural resources should be achieved through:

- formation of an integrated system of suburban green areas in Moscow and in the towns of the Moscow region carrying out ecological, sanitary and hygienic, and recreational functions; it should include forested areas, recreational zones, natural reserves, wildlife refuges, national parks, and water protection zones which are regarded as necessary to the population according to regulatory documents;
- establishment of new parks and green areas within built-up territories of the Moscow region in order to restore continuity of the Moscow green "wedges" and natural recreational zones of the Moscow periphery and of adjoining towns in the Moscow vicinity;
- assignment of territories for active recreational use with the development of ecologically-oriented urbanization forms that do not destroy the wholeness of the natural complex;
- establishment of zones of a special city planning control in the areas of underground water extraction and along routes of canals and pipelines used for water supply of the region centre.

The integrated environmental city planning policy should pay special attention to the protection of historical and cultural as well as natural heritage sites that is provided by:
- establishment of an integrated regional system of historical and cultural as well as natural refuges, reserves, national parks, and other specially-protected areas;
- organization of an integrated regional system of ecological tourism using the historical and cultural potential of Moscow and the Moscow region; this can be implemented through establishment of tourist centres on the base of the most valuable historical towns of the Moscow region as well as the Moscow's historical and cultural heritage;
- transformation of tourism and ecological tourism into a priority branch of the Moscow region economy that should create a wide range of new working places for inhabitants of Moscow and the Moscow region; it should be based on active development of the system of tourist and recreational services including construction of basic accommodation facilities for specialized kinds of tourism (business, sport, educational and training, and ecological) as well as creation of tourist centres in accordance with international standards.

The creation of favourable dwelling, labour, and recreation conditions for inhabitants of Moscow and the Moscow region can be provided only by the implementation of an environmentally-oriented city planning policy. Such a policy aimed at restraining and stabilization of the territorial growth of Moscow and the adjoining settlements, involves their internal restructuring and the use of resources of the Moscow region peripheral zone for residential purposes. The development of the settlement system of Moscow and the central parts of the Moscow region under conditions of severe territorial strains can be carried out only with an emphasis on the conservation and development of the natural complex.

The development of the labour market system in Moscow and the settlements of the Moscow region includes:
- primary development of the services sphere, development of high technology production facilities, small businesses, reduction of harmful as well as resources and power intensive production branches;
- establishment of an integrated regional recreational system involving intensification of usage of the existing natural, recreational, historical and cultural potentials of the territory in order to meet recreational and rehabilitative needs of the residents of Moscow and its region;
- improvement of existing settlements, traditional summer-cottage areas, rural settlements systems used for recreational purposes, and organization of areas for a "family recreation" in the Moscow region on their base;
- establishment of the spa system on the basis of health resorts where mineral waters and therapeutic mud-baths are available, construction of health centres, curative and medical facilities on the base of local sources for the residents of Moscow and the Moscow region; these involve the use of the potential of Moscow's medical institutions in terms of diagnostics and specialized treatment.

In order to preserve the environment, the health of the population, and providing for favourable living conditions for future generations, the environmental and city planning policies in Moscow should focus on:
- development, approval, and implementation of an integrated city planning and environmental programme for the Moscow region;
- development, approval, and implementation of the General Plan for Moscow including the Scheme of the city natural complex;

- conservation and development of the Moscow protective green belt; ban on reduction of recreational and green areas in the city and its region;
- economic stimulation for implementation of resource-saving technologies and equipment;
- ceasing of construction of new industrial facilities within the city; development of the industrial potential of the city only on the account of specialization change or reconstruction of existing facilities.

Problems of environmental conservation and management in the Moscow region cannot be solved without a permanent mechanism for prevention, localization, and elimination of negative man-induced impacts from projected and performed activities as well as their consequences.

Conclusion

Moscow is one of the most actively developing European megalopolises. Therefore, the participation of Moscow in the UN Programme on Sustainable Cities is of paramount importance. In July 1999 the Moscow City government approved the Moscow Master Plan for the period up to 2020. Principles of sustainable development in Moscow based on the international achievements in the field of conservation and development of the urban environment, urban planning and development of business, dwelling, industrial and recreation areas in the biggest cities of the world should become the scientific basis of the plan.

Consideration of citizens' requirements for a healthy environment and improvement of life quality is necessary for sustainable development of a big city in the modern world. This should be remembered when planning all fields of city policy, such as demography, social, economic, living, industrial, energy and environment policies. In the capital of the Russian Federation, one of the biggest and fast changing cities in the world, significant changes in the legislation, social, economic and environmental conditions have occurred in recent years. The most important legislative acts of the Russian Federation and Moscow have been passed during this period. They determined the status, regulation, administrative and territorial structure, economy and legal mechanisms of the city management. The requirements of the citizens as well as national, historical and cultural interests and environmental specificity of the territory should be also taken into account.

The Moscow Master Plan up to 2020 called for an increase of life quality, decrease of negative environmental effects from household and industrial activities, preservation of the architectural and historical heritage of Moscow for future generations, and complex and balanced development of all branches of the city infrastructure.

A general improvement of the environment in Moscow and the Moscow region visualizes the implementation of corporate regional policy on environmental protection and rehabilitation of areas of environmental crisis. Creation of favourable living conditions should be based on environmentally-friendly construction, and economic and social policy aimed at the sustainable development of Moscow.

Moscow and the Moscow region constitute an interrelated regional economic and natural complex, which plays a key role in the economic development of the Russian Federation. Sustainable development of this region should be based on the corporate decisions of economic and environmental problems taking into account the functions of Moscow as the capital as well as the interests of the Moscow region.

Meetings and consultations of all stakeholders - elected authorities, executive authorities, business, NGOs, and so on - play an important role in the process of wide replication and practical realization of the ideas of sustainable development. Such meetings and consultations are an important part of the Sustainable Cities Programme. The main goal of the Moscow City Environmental Profile is to provide the participants of municipal consultations with basic materials for discussion of the possibilities and perspectives for the sustainable development of Moscow.

The experience of the realization of the UN Sustainable Cities Programme and the Moscow sustainable cities project shows that the City Environmental Profile is a necessary basis for any programme of urban sustainable development.

A VIEW OF THE ENVIRONMENT FROM A STUDENT PUGWASHITE

Eva-Jane Haden

When I was first introduced to the idea of the Pugwash Movement, I was very excited, and determined to accomplish enormous goals that I had set in my mind to save the world. After having participated in several Pugwash conferences, I realized that so much was to be done, and I felt helpless and hopeless when faced with the enormity of such issues. Sometimes I feel that discussing issues is not sufficient, because I cannot see an immediate result from our conferences, although I suppose it can only be positive towards Humankind, and hopefully will, in time, contribute to the resolution of global problems.

The reason I started my paper on my feelings about Pugwash is because I feel likewise about Man's reaction to the problems of the environment. Once the confrontation between the superpowers became somewhat less important, the media started to create an awareness about such issues as the environment. People seemed to be motivated to solve these problems (especially that of climate change), yet once face to face with such problems, have preferred to remain oblivious, and to continue to use as much electricity, water and petrol without feeling guilty doing so. The main difficulty is to consider long-term as opposed to short-term effects, which means that the general public might rapidly lose interest in such matters. The use of pesticides, for example, brings an immediate gain to farmers who want to produce more, but a monstrous long-term weakness to the future use of what was arable land, as well as being a threat to the wildlife that lives on that land, which can in turn disturb a population's mixed diet. Human beings can prove themselves to be selfish, and not even consider saving any source of energy for the benefit of the survival of Mankind. How many people have planted a tree in exchange for the thousand of sheets of paper that they have used? This is probably a very naive starting point, but I feel the need to remind all of us of the slogan "think globally, act locally."

There seem to be several cycles that are formed when an environmental instability occurs. For instance, if the agriculture of a country is not sufficient, then that population will migrate to another place, which will modify the balance of the production there, and so on. As well as that, such events can cause conflicts, and lack of food or drinking water can be used as weapons. It is obvious that these vicious cycles have to be reversed so as to preserve nature and to ensure the survival of Mankind. Humans are known to be the most able of all species to adapt to different environments. The issue here is that by adapting, humans are ruining nature, and create a whole new environment for the generations to come. It would be inconceivable to reverse Human evolution in the sense of returning to living in primitive huts and ignoring the amazing progress in technology, but an evolution of human mentalities is needed. Humans are responsible for the amount of pollution there is today, thus creating the Antarctic ozone hole discovered in the mid-1980s, and clearly showing the instability of the environment we live in. Recently, it has been discovered that there are more and more cases of people (especially children) with asthma and eczema than ever before, not necessarily resulting directly from air and water pollution, but definitely made worse by it. I feel that it is a responsibility for every human being to be aware of such problems and to limit one's personal use of any source of energy to the minimum. I have lived in Geneva all my life (18 years), yet I am only starting to become aware of the amazing good fortune it is to do so. Geneva, and most of Switzerland as a whole, is a very clean and well looked after region. Hydroelectricity is used frequently, and in Geneva itself, buses, trams and some trains are all electric. It is also evident that Switzerland is a wealthy country and that is why it is possible to have such means.

The previous example reveals the importance of a stable and efficient economy to overcome any disruption of the environmental system. The majority of the developing countries lack in economic and political stability, and thus are victims of environmental problems. The agriculture in developing countries is the most important income for the economy. Any kind of environmental instability, such as deforestation or the greenhouse effect, which leads to the degradation of agricultural land by damaging the soil, puts a stop to the agricultural production and hence to the economy of such countries. A vicious cycle is once again created: the worse economic condition a country is in, the more vulnerable it is to environmental difficulties due to the lack of wealth, health and education, and the worse its economic situation becomes, and so on. There can also be a shortage of water in these countries, which the developed world does not experience thanks to a technology which enables it to desalinate and cleanse the available water. This demands a great budget which is not procurable, and thus the population of the developing world suffer from diseases due to lack of, or polluted, water. I believe that the developed world has a certain responsibility to help the developing countries by bringing the necessary technology there (solar energy, hydroelectricity, and so on), and not to ignore the environmentally-based problems. (The ethical questions of such interventions will not be considered here.) Sometimes contaminated produce is sold to the Third World because the developed countries will not buy them. This also reflects the need for the developed world to test such products before exporting them.

Conclusion

Why, might you ask, do I still then come to these Pugwash conferences? The reason is simple: because I continue to believe that it is possible for humans to change their way of living, and their way of thinking. The media is spreading a minimum amount of awareness and there seem to be an increasing amount of effort of recycling all materials, which leaves me slightly optimistic. My impressions are somewhat biased by the fact that I live in Geneva, but I still think that everyone of us should save energy and be concerned and active about the survival of our environment. An education of individuals, companies and governments is required to solve environmental problems, and to motivate or re-motivate my generation by showing how fundamental this issue really is.

SUSTAINABILITY AND DEVELOPMENT PATTERNS IN THE SOUTH UNDER GLOBAL CHANGE AND MARKET

Luiz Pinguelli-Rosa

The Global Change and Market

One of the most important achievements of the Rio Conference in 1992 was to differentiate between the responsibility of the North and South regarding the Climate Convention of the UN. The goal of reducing gas emissions in 2000 to the same level as 1990 was agreed only for countries listed in Annex 1, which includes all developed countries. Now in 1999, we know that the above goal will not be achieved and it was changed by the Kyoto Conference in 1997, while different mechanisms have been proposed for investments from the North to avoid carbon emissions in the South at low costs. The former Joint Implementation mechanism did not work on the necessary scale and now it is restricted to Annex 1 countries. The new Clean Development Mechanism (CDM) was approved in Kyoto as a consequence of the Brazilian proposal supported by the G-77, the Group of Developing Countries in the Climate Convention. However, CDM needs to be regulated and the Buenos Aires Conference in 1998 did not advance much in this direction.

Using a simple metaphor we can write:

CDM = Avoid Carbon Emission + Sustainable Development

What is sustainable development? Consumption patterns are exported by the industrialized countries in the North to the developing countries in the South, through the market. This is reinforced by the increase in the speed of communications which characterizes the so-called globalization process. In the South, market demand for production of goods and services, however, is mainly derived from only a part of the population which is affluent and able to imitate the consumption patterns of the North, including the high levels of per capita energy consumption. In some developing countries, like Brazil for instance, the higher income classes exhibit levels of energy consumption as high as those in the industrialized world, although never at levels comparable to those prevalent in the USA. The majority of the population in the developing world remains outside of this process, and exhibits very low energy consumption levels, as a consequence of the inequality in income distribution which leads to unequal distribution of goods and services.

It is important to note that the consumption patterns exported to the South by the market create an expectation of high energy consumption in the low income classes of these countries, in spite of the prevalent low overall levels of energy per capita. This expectation generates a latent demand for goods and services which is ready to surface at the smallest increase in income.

In general, the industrialized countries possess high levels of technical efficiency, which imply low energy intensity levels (toe/$US). However, as research by COPPE (Graduate School of Engineering of Federal University of Rio de Janeiro) has shown, there are examples where the higher energy intensity levels in the developing countries do not necessarily imply physical energy intensity (toe/tonne). The reason being that the monetary values ($US/tonne) can be too low in some cases. Also, in the industrialized world, the greater technical efficiency in transportation, for example, has not led to a decrease in the use of fossil fuels, due to the consumer preference for larger cars and the use of high speed motorways. Therefore, with the fall in the price of oil, there has been an increase in the energy per capita spent on transportation, which has led to a low overall level of social efficiency. The upper income part of the middle class of Brazil now has the same tendency of buying a very large van or pick up with about three times the mass of normal cars. The engine efficiency improvement is negated by the excessive size of the vehicle.

Any form development takes in the South will affect climate change since energy consumption per capita in these countries is still very low, and therefore can only increase. The patterns of consumption, however, which are exported by the industrialized countries in the North to the developing countries are the ones which reinforce the use of fossil fuels, since they are based on the technology prevalent in these countries. Sometimes this occurs at the cost of displacement of renewable energy sources and technologies. This is the case with the alcohol programme in Brazil, which was faced with competition from the oil-based automobile industry, and the adoption of thermoelectricity generation in place of hydroelectric sources.

In addition, in the North, society can regulate unwanted market activities through strong democratic institutions. The South, in actual fact, exhibits a defective imitation of this process, in which a weak State is accompanied by greater susceptibility to corruption by large enterprises and to pressures from multilateral institutions that defend the position for an unregulated market.

Consumption patterns dictate demand for economic production of goods and services in the economy. The level of production determines employment and income, which feed back into demand through the market. The formal market in the developing economies is formed by only a part of the population which is affluent and imitates the consumption patterns of the North. Therefore, the part of the population left out of this process generates a demand which is met by the formation of informal production sectors, that produce inferior goods at lower costs, and offer only under employment. As a result, the economy does not generate enough jobs to absorb the population of working age, which itself perpetuates low income and unemployment, and lack of demand for formal market goods. The weakness of the democratic institutions in regulating the economy only adds to the problem.

At the international level there is a differentiation between energy intensity levels and income per capita. In the South, energy intensity varies: Brazil is in a better position than China and Asia; in the North, both income and energy levels are much greater, but also vary between countries: North America (except Mexico) is greater than Japan and New Zealand, for instance. This demonstrates that the North should try to diminish its levels of energy per capita whilst striving to preserve its levels of income. The South should increase income per capita without necessarily increasing its use of energy, which entails attaining a greater level of efficiency, and utilizing renewable sources. The countries from the South have greater emissions intensity, particularly China, and lower levels of energy per capita than those of the North. Therefore, the South should increase energy per capita without the same proportional increase in emissions, for instance, by increasing the use of renewable energy. The North should strive to diminish its emissions per capita.

Brazil has had until now the advantage of a high percentage of renewable primary sources in its energy matrix. In transportation, since the oil shocks, alcohol has been used in cars. But in a deregulated market it cannot compete with gasoline while the oil price is as low as it is now. So the deregulation policy in Brazil, since the beginning of the 1990s, led to the substitution of alcohol by gasoline in new cars, as well as of hydroelectricity by fossil fuel electricity generation in future plants. The market does not work according to clean development in this case.

Sustainable Development and the Role of Science and Technology

The previous section has shown a particular aspect of the problem of achieving a sustainable development in the South, whose solution could receive some contribution from the development of science and technology in the case of Brazil. This country has a national system of research and development but institutional tools are needed to link universities, institutes and private enterprises.

The problems that we face in Brazil, as well as in Latin America in general, arise from the contrast between the modern side of the country and the poverty of most of the population. However, the very existence of the modern part indicates the potential for the changes that are needed. How to achieve these changes, how to use science and technology for sustainable development as one tool among many others: this is out task.

There are many reports and studies by governmental agencies, international organizations and NGOs, with many ideas and suggestions. How to select priorities among them to start? How to choose some short-term measures to be taken? How to fill people with enthusiasm to implement the measures?

With the purpose of presenting a rational approach to developing science and technology we start with a rather obvious method. It consists of the following steps:

a. identification of the problems to be solved and the goals to be achieved;
b. statement of the postulates to be obeyed according to first principles;
c. observations to select relevant information
d. analysis of information in the framework of an approach for understanding the present situation and the possibility of change;
e. proposals to be implemented.

The first step has been considered at the beginning of the present section. Our postulates can be stated as follows:

a. We do not agree with the separation often made between poverty alleviation and modern economic development.
b. We understand that modernization means to expand the modern economy in such a way as to get poor people out from poverty in a degraded subsistence economy, to become a worker and a consumer, as well as a human being with dignity in the society, without the need of charity or alms.
c. To the above purpose, which is the most important and to which technology must contribute, we will avoid two frequent biases: they are (i) it is not true that everything must be planned by government; (ii) it is also not true that nothing must be planned, assuming that pure market forces in an open economy solve every problem.
d. Science and technology are different things. Science must be linked to technology because frequently new scientific results are transferred to technology; in addition, technology changes too fast, while science fundamentals, on which it is based, remain valid. However, this does not mean submission of science to applications, because creativity works better in an environment of academic freedom. And freedom is necessary in an open society for criticism of political acts as well as of the economic power of big companies.
e. Technology cannot be seen as a goal in itself, but it must be useful to solve problems of society and to develop the country. However, it must be clear that technology is a tool and it is not enough to solve all problems, most of them depending on economic rationality, social equity and political and ethical decisions.

There is a quite an impressive amount of studies on Brazil in publications and reports by international organizations. The first question which arises is why this international interest in Brazil? A general answer is the potential economic growth of the country, now under a democratic regime, originated from a recent history of struggle by Trade Unions, political parties of the left and centre, and society against military dictatorship. But the main reason for the choice is probably the prospects for business in a large market. There are special factors which make the country interesting. They are listed below, but adding the limitations of each one according to our diagnosis.

1. The modern economy does exist, but it is too concentrated socially and geographically. There are very modern urban areas in contrast with most of the country.
2. The elite has good education and there are people with enough technical and scientific knowledge, but it is not available to most of the population.

3. The great majority of the population lives in very hard conditions of poverty, with little chance for improving the patterns of life because unemployment is too high and increasing. The majority is not integrated in the economy whose development has not benefited them.
4. However, the country has comparative advantage in many sectors, such as industry in general and manufacturing in particular. The country is rich in minerals, there is significant industry, agriculture and cattle farming with high potential for growth.
5. The population is relatively large and the area of the country is big, which means a low density of population, although a large part of the territory is not developed. A large part is the Amazon forest with very high biodiversity. There are regional economic disparities.
6. There are many kilometres of roads. The telecommunications are some of the best in Latin America. The energy matrix contains a very high component of renewable sources, including hydropower, alcohol used in car engines and charcoal used in the steel industry. Electricity consumption increases and there is considerable hydroelectric potential. The discovery of exploitable oil deposits and the development of offshore technology in the country are important.
7. The potential for industrial development is promising because of modern infrastructure. The private sector is the most important in the economy and there is a reasonable financial sector. Monetary policy has been successful in inflation control but with frozen salaries and high unemployment. Deregulation of the economy, however, produced a rather unstable equilibrium in some aspects, in particular negative export-import balance and exponential growth of government debt.
8. The government has had a role in economic development. The political system is stable and international relationships are very good. Recent privatization of public services, mainly through the sale of State companies to foreign and international corporations, have not come up to expectations until now, especially in electric energy supply.

The simple analysis of the above points allows us to see that only point 3 is purely negative. Points 1 and 2 show the contrasts, as do points 4 and 5. The last ones, from 5 to 8 indicate that there are possible ways to solve the problems. Most of the analysis considers either only the positive points of the economic policy or purely the negative ones, according to who does it: an enthusiast of liberalization of the economy or someone concerned with social inequalities. We try here to put all of them together. The next step is to discuss how to solve them, which is not easy to do.

Using a very naïve metaphor it is possible, based on points 1 to 4, to select the variable to be changed: (1) expansion of modernization; (2) education of all population and diffusion of science and technology; (3) increase of formal employment and creation of jobs; and (4) income distribution to benefit the poor people.

According to the same analogy points 5 to 8 represent something like functions which act on the variables 1 to 4 to change them: (5) natural resources exploitation and development of agriculture; (6) infrastructure and energy supply for industrial development; (7) private dynamics of modern sectors of economy; and (8) role of government.

If we go on with the analogy, we remember that in mathematics the number of variables not exceeding the number of equations is a necessary but not sufficient condition for the existence of solutions. In our case, there is not a theoretical functional dependence expressed in a mathematical form. There are not deterministic equations to solve social and economic problems, and even in recent mathematical descriptions of nature the extreme sensitivity to present conditions of some natural systems does not allow us to make predictions. But the future can be improved depending on present actions. So it is necessary to look for the correct actions which point in the direction of economic as well as social development.

The actors are the private sector, workers, government, civil society and international organizations. They should manage potentialities (5 to 8) for changing variables (1 to 4). Natural resource exploration and agriculture are more related to job creation and income distribution; infrastructure is more related to modernization and income distribution (household electricity supply, water, cross subsidies); private dynamics of modern sectors economy is more related to modernization and jobs; government role is more related to education and income distribution (through tax, social actions). For that purpose actions need to be co-ordinated through a social and democratic consensus, but with efficient institutional tools, involving small and large firms, the state, NGOs, and multilateral organizations. For instance, the government must act on the state role (8) for public education (2). But it must be done to increase the skills of population and to improve the scientific and technological knowledge of professionals for better employment (3). Economy sector actions must be integrated in a network (5,6,7) for growth. But it should be done by creating employment (3), not only through direct formal jobs, but also by contracting small enterprises or individuals. Education together with employment can improve income distribution (4) and integrate the poor people in the modern economy. So, new consumers will expand the market for growing the modern economy (1).

Proposal for Short-Term Demonstration Effect[1]

(a) Natural resource exploration and agriculture are more related to job creation and income distribution;
(b) infrastructure is more related to modernization and income distribution (through household electricity supply, water, cross subsidies);
(c) private economy dynamics of modern sectors is more related to modernization and job creation;
(d) government role is more related to education and income distribution (through tax, social actions).

There are a number of possible ways for the government to start the process with the main goal of taking people out from poverty, by education, job creation, and income distribution, which have the aim of including all the population in the modern economy.

This approach does not consider separately the modern economy and the subsistence economy of poor people, but it covers both.

Expansion of modern economy sectors to include all society, education, reduction of poverty, income distribution, all these changes involve scientific and technological knowledge, training and vocational skills. In this aspect the present work proposes special programmes for short-term demonstration effect:

a. Appropriate technology for low cost housing with local materials, local job creation and local design.
b. Energy for rural area, renewable with preference, as small hydro and, for isolated communities, solar.
c. Labour intensive technologies to make locally some products, selected to be protected temporarily through agreement against importation.
d. Wood industry development through plantation of trees.
e. Technology for increasing the added value of exports on a large scale.

Points (b) and (d) could have support from the Clean Development Mechanism for Abatement of Carbon Emission to the Atmosphere.

All these programmes have been chosen according to the approach developed above to create positive feedback in the economy through starting action by government, on education, job creation, income distribution and on the economy, with the goal of poor people becoming consumers in the modern economy. The programmes are suggestions with examples to be discussed.

The most important is the demonstration effect to fill people with enthusiasm. There are many ideas and proposals from former reports as we have stated at the beginning.

---

[1] This is an abridged text. The main relationships between variables and functions are described in an Appendix to this paper, not reproduced in this volume but available on request from the London Pugwash Office. The full paper also contains a number of Figures also not reproduced above.

AVOIDING AN EPIDEMIC IN SANTIAGO

Catterina Ferreccio

Background

In Santiago Metropolitan Region (SMR), for decades there has been a practice of using raw waste water for agriculture. It is used to irrigate about 130,000 ha - including some 7,000 ha used for growing vegetable crops in the immediate vicinity of Santiago. This has been linked with the unexpected high rates of enteric infections in the SMR. Classical control measures (food handlers control, routine vaccination of school children), had practically no impact on this disease, which persisted with rates above 50 cases per 100,000 population. The control measures that the Chilean government took in response to the cholera epidemic in Latin America had an impact on typhoid and was a natural experiment to prove that the persistence of typhoid was caused by the consumption of contaminated crops.

Besides the increase of incomes, the high levels of education, nutrition and health care, ample coverage with good quality water supply and sanitation, Chile has high rates of some enteric infections that are linked to poverty and lack of hygiene: Typhoid Fever and Infectious Hepatitis.

The classical mechanisms of transmission of enteric diseases are by poor personal hygiene or the "short cycle" (excreta-hand-mouth), and by environmental pollution or the "long cycle" (infected individual-sewage-water pollution-food-people). The short cycle of contagion has been significantly reduced in the SMR, as demonstrated by the diarrhoea mortality and by the relatively lower rates of hepatitis. On the other hand, the long cycle has persisted and even increased as environmental pollution has increased. This last mechanism of contagion is the one that better explains the epidemiological pattern of high typhoid and relatively lower infant diarrhoea and hepatitis, in the SMR.

Some epidemiological facts that also relate typhoid with the environmental pollution are:

- Historically, the incidence of typhoid has been greater in the SMR than the rest of the country, where the practice of irrigating with raw water is less prevalent.
- The causal agent of typhoid fever (salmonella typhi) has been isolated directly from irrigation water in the SMR. The occurrence of typhoid fever increases seasonally in the SMR following the summer pattern of increase in sewage irrigation and the rise in consumption of raw vegetables. Seasonal variations have been much greater in the SMR than in the rest of Chile.
- Viral hepatitis has the highest rates of all diseases of obligatory notification in Chile. Over 80 per cent of the cases are caused by viral Hepatitis A. Rates are higher in the extreme northern region of Chile, with intermediate rates in the SMR. In contrast to typhoid, hepatitis has higher rates in the cold months and affects younger ages more.

The above-mentioned evidence stimulated at least two interventions aimed at breaking the long cycle. The first was in 1984, when in response to the highest outbreak of typhoid in Chilean history, the health authorities prohibited the cultivation of vegetable crops consumed raw, and initiated a public education campaign on the risks of sewage irrigation for health. From 1983 to 1985, in the SMR, typhoid was observed to drop 59 per cent and hepatitis fell 11 per cent, reaching its endemic levels by

1986. The second, in 1991, was in response to the cholera epidemic in Peru, and is the subject of the remainder of this paper[1]

Cholera had disappeared from the Americas by the end of the nineteenth century, until the recent epidemic in Peru, in January 1991. It created an emergency in the continent and particularly in neighbouring Chile. Peruvians had similar rates of typhoid as observed in the SMR, and the mechanisms of transmission of the two diseases are the same, with predominance of the "long-cycle". Therefore, it was to be expected that cholera would travel quickly to the SMR and elsewhere in Chile.

Cholera, rapidly reached different countries and by the end of 1991 fifteen countries had been affected, from Mexico to Chile, and from Atlantic to Pacific coasts. The epidemic peaked in 1991, slowed down in 1992 and dropped in 1993. By the end of 1993, only Uruguay in the Latin American continent and the Caribbean islands remained free of the disease. It should be noticed that in these countries typhoid is much lower than in Chile or Peru. In Chile, in April 1991 there was an outbreak that reached to 41 cases, with 70 per cent of them linked to consumption of vegetables from Santiago; by the end of 1993, only 142 cases and 3 deaths had accumulated, much lower than typhoid or hepatitis rates and much lower than cholera among its neighbours. Table 1 presents cholera in Chile and its neighbouring countries.

Table 1
Cholera in the Americas
1991-1993

Country	Cases	Deaths
Argentina	2,623	33
Bolivia	31,655	625
CHILE	142	3
Peru	606,652	4,211
Rest South America	197,906	2,340
Mexico, Central America	102,827	1,395
TOTAL	941,805	8,607

The Interventions

To avoid the spread of cholera in Chile, the authorities took a series of actions specifically aimed at breaking the long cycle: the prohibition and removal of vegetables grown with sewage irrigation and restrictions on transporting irrigated vegetables and fruits outside of the SMR, intensification of water quality monitoring, the chlorination of irrigation water in canals, the prohibition of serving raw vegetables, and an intensive campaign of social communication informing the population about the risks of vegetables and the need to wash and cook them. New regulations were promulgated extending prohibitions to other vegetables normally consumed raw, and field inspections were intensified.

Results

The programme interrupted the long cycle mechanism of transmission, mainly through changes in the population's vegetable consumption patterns and by changes in irrigation practice.

Public Health outcomes. There was a rapid control of the spread of cholera, and also significant reductions in typhoid and hepatitis throughout the country and especially in Santiago. The typhoid model is used to estimate the degree of association between health outcomes and sewage irrigation, and to extrapolate it to cholera. Through statistical analysis of typhoid and hepatitis time-series data and existing epidemiological studies it was possible to establish the pollution-irrigation-health linkages, as well as to assess the relative effectiveness of interventions taken. The 95 per cent confidence band for 1986 was used to provide a basis of comparison for the other years in the series. For typhoid fever, these comparisons showed that cases in the SMR in 1991 and 1992 dropped below the minimum expected number of cases and revealed a significantly lower reduction of typhoid in the rest of Chile compared with the SMR. Hepatitis, cases in the SMR for 1991 and 1992 also fell significantly below expected minimums, but the fall was less notable than that observed for typhoid. For the rest of Chile, the reductions observed in hepatitis cases were equal to those observed for typhoid (Table 2).

Table 2
Typhoid and Hepatitis Before and After the 1991 Interventions
(average cases per year)

	Typhoid Fever		Hepatitis	
	SMR	Rest of Chile	SMR	Rest of Chile
Pre-intervention, 1985-90	3,558	2,971	4,040	7,474
Post-intervention, 1992	454	1,429	1,430	3,845
Reduction	87%	52%	65%	49%
Impact in SMR in Chile vs Rest of Chile	+35%		+16%	

Health Care Burden in the SMR. The success of the emergency interventions allowed better estimates of the number of cases of typhoid in the SMR that can be attributed to sewage irrigation of vegetables. Taking 1992 as the year representing a controlled situation (that is, without cases transmitted by sewage-contaminated vegetables), a 95 per cent confidence band was constructed for the maximum and minimum expected weekly cases of typhoid for that year. The annual time series of weekly cases for the period 1985 to 1991 were compared to this band and the number of imputed cases for each year was estimated (the sum of the weekly observed cases in each year in excess of the maximum expected weekly cases in 1992). This permitted the calculation of the percentage of cases attributed to sewage irrigation, yielding a range from 49 to 79 per cent as shown in Table 3.

Multivariate analysis of interventions. To measure the relative contribution of the various factors on the variation in incidence of typhoid, a multivariate analysis of the interventions was carried out. The model chosen analysed the impacts of temperature, drinking water supply and "press" (a variable which includes social information by mass media and sanitary education). It was not possible to find adequate variables to reflect other interventions. Using this model, it is possible to conclude that the explanatory variable "press" is significant and explains half of the fall in typhoid fever cases. The remaining half could be attributed to direct controls on the production and sale of vegetables. The model is less successful in explaining the behaviour of hepatitis which appears to be less attributable to vegetable consumption.

Outcome in waste water use. According to Francisco Brzovic, the total area planted with prohibited crops decreased from 6,970 ha. in 1991 to 4,964 ha in 1993. The remaining 2,000 ha, either are irrigated with well water or are growing permitted vegetables or fruits. Also, the corresponding volumes traded in wholesaler markets in Santiago dropped in the first year by 43 per cent. In the second year of restrictions, however, a recovery to previous levels in volumes traded was observed.

Table 3:
Typhoid Cases in the SMR Attributable to
Sewage Irrigation, 1985-1992

Year	Observed Cases (a)	Expected Cases * (b)	Imputed Cases	
			Number (a-b)	Per cent
1985	4,444	935	3,509	79.0
1986	2,907	728	2,179	75.0
1987	1,180	645	535	45.3
1988	1,662	768	894	53.8
1989	2,718	877	1,841	67.7
1990	1,471	674	797	54.2
1991	1,773	517	1,256	70.8
1992	332	30	2	0.6

* The sum of the weekly observed cases falling below the maximum expected weekly cases in 1992, the control year.

Conclusions

From this analysis, several conclusions can be drawn:

(a) The long cycle of transmission of typhoid and cholera, and to a lesser extent hepatitis, have been successfully interrupted in part by changes in the population's vegetable consumption habits and also by changes in irrigation practices for vegetables. Nevertheless, since there was no intervention on the structure of the problem, the stability of these changes is difficult to predict.

(b) Between 500 and 3,500 cases of typhoid could have been avoided yearly, had these measures been applied in 1986.

(c) Critical factors for the success of the measures were: public support, press commitment and active participation, decisive leadership of the health sector, and the acceptance of the measures by farmers. It should also be mentioned that in 1991 the country was experiencing a process of reorganizing the democratic system and there was an atmosphere of collaboration among all parties.

(d) The power of the synergetic actions of population, government and farmers was demonstrated. Nevertheless, to guarantee safe crops for the SMR population, it is necessary to implement structural changes that will not depend so heavily on people's behaviour.

Reference

[1] This study is based on: "Epidemiologic Associations between Water Pollution and Health Impacts in the Metropolitan Region of Chile," 1993, The World Bank and PAHO.

LARGE DAMS, SUSTAINABLE LIVELIHOODS AND THE USE OF ORAL HISTORIES: THE CASE OF CAHORA BASSA

Allen Isaacman

Since the 1960s, when Lisbon announced that it was going to build the Cahora Bassa Dam and harness the mighty Zambesi River, the hydro-electrical project has attracted considerable international attention. Portuguese officials heralded the dam's potential to bring progress and prosperity to the backwater Zambesi valley. The proposed dam was also the object of a highly visible, but unsuccessful, protest campaign. Anti-colonial and anti-apartheid activists contended that it was an integral part of an economic and military alliance between Portugal and South Africa to perpetuate white rule in the region. Adopting a very different stance, engineers and hydrologists hailed the dam's technical complexity and the skill required to build the world's fifth largest hydroelectric installation in a region which lacked the most basic economic infrastructure. For them, the completion of Cahora Bassa in December 1974 was proof that capricious natural forces could be harnessed and a biophysical system transformed to serve humankind. Six months later Mozambique became independent. For almost two decades thereafter South-African-backed guerrillas engaged in a highly publicized campaign to sabotage the dam's powerlines and destabilize the new FRELIMO (Front for the Liberation of Mozambique) government. With the dismantling of the apartheid regime in the early 1990s, policy-makers focused on Cahora Bassa's potential to alleviate the looming energy crises in South Africa and to help stimulate economic development in the broader region.

For all the attention which Cahora Bassa has received, there has been a conspicuous absence of serious research and public debate on the dam's social and ecological consequences. We know little about the impact of the project on the daily lives of the African workers who, often at great personal risk, constructed the dam. We know even less about the lived experiences of the thousands of peasants forced to relocate from their historic homelands. Their story remains hidden in the opaque shadows of the dam. Similarly, the impact of Cahora Bassa on down-river communities, whose river-fed gardens and grazing lands are no longer seasonally irrigated by the Zambesi River and whose fishing lagoons have been greatly reduced, has not received serious attention. Finally, questions regarding the dam's and the reservoir's role in altering seasonal flooding patterns and disrupting ecological communities have been broached only in piecemeal fashion.

It is precisely these issues which this paper begins to explore as part of an alternative history of Cahora Bassa.[1] Informed by the literature on "sustainable livelihoods," it stresses that human security and environmental health and degradation are inextricably intertwined.[2] [3] [4] Specifically, this paper examines how the socio-economic and ecological changes brought about by Cahora Bassa affected people's access to resources and their capacity to utilize these resources effectively. It highlights the critical issue of what exactly is being sustained and for whom.

The construction of Cahora Bassa has much in common with other hydroelectric projects in Africa and other parts of the Third World, but the political and social context and dynamics were quite different. That Cahora Bassa was the last great colonial project in Africa, that it was the world's largest national hydroelectric project designed with the sole purpose of exporting cheap energy, and, above all else, that it was an integral part of an anti-guerrilla military strategy, make it unique.

Security concerns demanded that the colonial state construct a wall of silence around the dam, which helps to explain why so little is known about the social and ecological consequences of

Cahora Bassa. During the colonial period the government silenced all skeptics and critics of the project.[5] This wall of silence continued after independence. Despite nominal oversight by the Energy Ministry, the Portuguese-dominated Hidro-Electrica de Cahora Bassa (HCB) treated the dam as its own private domain and gave out very little information. Moreover, South African backed RENAMO guerrillas transformed the Zambesi Valley into a major battleground and disrupted all social and environmental research in the region.[6] Because fieldwork was simply too dangerous, scholars either abandoned or curtailed the small number of scientific investigations.[7]

The fragmentary written data pose obvious evidentiary and methodological problems. Most of the extant documentation from the pre-impoundment period comes from official reports written by local administrators, employees of the Zambesi River Authority (Missão do Fomento do Zambesi) and the international consortium (ZAMCO) designated to build the dam. These accounts chronicle in great detail the engineering and logistical challenges of constructing the hydroelectric project as well as the anticipated benefits. Taken together, they tell a story of the triumph of technology but simultaneously mask and disguise the social and ecological realities of rural life.

Because this paper is concerned, above all else, with the sustainable livelihoods of communities affected by the dam, their accounts figure prominently. These oral testimonies not only challenge the prevailing colonial formulation but offer an alternative narrative - a detailed interior view of life before and after the construction of Cahora Bassa. They also provide important insights about how peasants perceived, explained and coped with ecological changes in the river basin over time and open up new areas of inquiry about the environmental impact of the dam. One of the central propositions of this paper is that despite difficulties of interpretation, a careful and critical exploration of the lived experiences and the ecological memories of basin residents revealed through oral testimonies, is essential if scholars are to begin to comprehend the extent to which Cahora Bassa has transformed the social and cultural ecology of the Zambesi Valley.

The Hydrology of the Zambesi and Indigenous Agronomic Practices: An Overview

Three hydrological factors are critical in understanding the natural history and agronomic practices of the region. First, although the Cahora Bassa dam and reservoir are contained entirely within Mozambique territorial boundaries, the vast bulk of the Zambesi drainage basin, the third largest river system in Africa, lies outside of the country. Since Mozambique is at the end of the stream, it is dependent on its neighbours for access to the river's water. Second, in most of the Zambesi basin, located on the Central African plateau, the waters flow slowly through low plains and swamps providing few suitable sites for dams. Third, and most relevant for this study, was the pronounced seasonality of Zambesi flows and the annual flooding which the Portuguese government claimed had a serious impact on the riverine communities and their natural habitat, as well as for the European sugar plantations located near the mouth of the river. Indeed, flood control was one of the presumed advantages of building the dam.

But there is another side to this story. Waters from the flooding river typically inundated the banks of the Zambesi during the rainy season, beginning in December and ending in March. When the waters receded, they left a rich deposit of nutrients along the shoreline. In lowland areas this spillover extended, depending on the locale, over a several-kilometre stretch of land. Peasants throughout the valley considered these rich dark *makande* soils of the floodplains to be the most desirable agricultural sites in the region. Beatriz Maquina, an elderly women who had farmed her entire life, stressed that the "*makande* land located near the banks of the river always gave us good production. We cultivated a great deal of sorghum as well as some corn."[8] All the elders with whom we spoke distinguished *makande* from the more common sandy, rocky *ntchenga* soils which did not retain water and were difficult to farm.[9]

Given the low and irregular rainfall in the Zambesi Valley, access to the *makande* river-fed soils was critical to insure household food security. Much of Tete district and the Lower Zambesi

Valley has a semi-arid climate and savanna-like environment. The average annual rainfall is only 600 millimetres.[10] Droughts occur regularly, often with devastating consequences to the crops. Without *makande* lands peasant households faced the prospect of periodic crop failures on a regular basis and, even in the best years, little likelihood of producing a second annual crop. This vulnerability was true down river as well, where rainfall was more appreciable, but still erratic.

Peasant utilization of river-fed fields was also an integral part of a complex and highly adaptive indigenous agronomic system. Drawing on a rich repertoire of farming practices, born out of years of trial and error and detailed micro-ecological knowledge, local communities creatively adapted to the uneven soil quality, fluctuations in rainfall and challenges of flooding. Carlos Soda Churo, who was forced to relocate because of the dam, described farming practices prior to the impoundment, in some detail:

> Before Cahora Bassa each family had several fields. The number and size varied depending on strength of a person and the size of his family. The land near the river was very good. It was called *makande*. When the river rose and then receded in June, the area that had been covered with water was very good for farming. There we first planted maize. We cultivated beans in the same field as the maize. Beans needed something to rest on and the maize stalks served well. Nearby we cultivated a second small plot with sweat potatoes, tomatoes, cabbage and more beans. We harvested our gardens in September and October before the rains and flooding. By November we were working in our larger fields away from the river. On the *ntchenga* soils we planted sorghum, which does not require as much water. The mixed *ntchenga - makande* soils were better for maize, which needs more moisture than sorghum. Some people planted peanuts in their maize fields. We harvested these crops in June and July and then returned to our gardens.[11]

Churo's accounts underscore highlight three important features of the indigenous agronomic system. First, and foremost, the food production systems of local agriculturalists co-evolved with the seasonal cycle of the river's flooding patterns. Decisions regarding the spatial and temporal patterns of food production - including selection of the most appropriate crops and amounts planted, with reference to the season and different micro-ecological zones - were finely tuned to changes in the river's discharge rates as well as variations in soils and sunlight. Second, intercropping was an effective labour saving device, since several crops could be tended simultaneously. Cultivating peanuts in maize fields had the added advantage of restoring badly needed nutrients to depleted *ntchenga* soils. Finally, households spent most of the year engaged in agricultural production in order to minimize labour bottlenecks and to ensure an adequate supply of food.

To be sure, February and March were the "hungry months". It was a period in the agricultural cycle when typically peasants had less to eat, having consumed much of the previous years' grains which were the mainstays of the diet. But unlike other regions of Tete, where Africans did not have access to rain-fed fields, the situation of peoples living in the river valley was never precarious.[12]

The free flowing Zambesi was a source of sustenance in two other important respects. Before Cahora Bassa approximately 60 species of fish inhabited the river.[12] While the density of most species varied, elders recalled that the Zambesi provided an abundance. They relied on a variety of fishing techniques. Some used nets, made from sisal and cord, which they laid in the main channel of the river. Others paddled their canoes to rich fishing grounds, where they deposited poisons from local plants into the water.[11] Most fishermen used locally produced weirs (*mackonga*) which they placed at strategic points near the shoreline.

> We fished with *mackonga* in which we placed bits of *massa* [porridge]. The fish, attracted by the *massa*, would enter the *mackonga* and they would be trapped. The next morning we would return. The *mackonga* would be filled with fish, some of which we traded with our neighbours for sorghum, maize or even a chicken. People who had nothing to trade could buy a large fish for five *escudos* [about 17 cents].[13]

The river also attracted large herds of animals from the nearby forests. Impala, gazelle, elephants, buffalo and eland regularly watered on the banks of the Zambesi and adjacent wetlands, where they became easy prey for skilled hunters.

> Not everyone hunted, only those who were good with *gugudas* [muzzle-loading rifles]. Those hunters we called *Nkumbalume*. There was a lot of game, so they did not have to go far to find it. The hunters traded most of the meat for maize. Sometimes they sold a piece for sixpence. These hunters also had fields. They gave some of the game to the people who worked in their gardens.[11]

Meat was an important part of the local diet. As a relish accompanying the evening porridge, it provided an important source of protein. Peasants also consumed game in larger amounts at important social occasions and at rituals propitiating the ancestor spirits. All of this changed, however, with the construction of Cahora Bassa.

Building the Dam: The Plan and the Project

Since the seventeenth century, the majesty of the Cahora Bassa Gorge, located 400 miles from the mouth of the Zambesi River, had both awed and frustrated Portuguese colonial planners. But it was only in the 1950s, after the British had constructed a dam 100 miles up-river at Kariba, that Portuguese colonial planners began to contemplate a similar undertaking at Cahora Bassa Gorge.

In 1956 the Overseas Minister, Raul Ventura, dispatched a team of hydrologists to Cahora Bassa. The scientists surveyed the Zambesi region in May and June 1956 and issued a highly influential and optimistic report.

> It is clear that the utilization of those possibilities, accompanied by a well-defined policy of industrial development and mineral prospecting the mining could complete transform the economic prospects for the Province of Mozambique and, in consequence, for metropolitan Portugal. Rarely have conditions occurred which are so favorable for the economic development of a region.[14]

Within a year the Salazar regime had established a river basin authority to co-ordinate research, evaluate a wide array of feasibility studies, and lay out the blueprints for the development of the Zambesi basin. The Missão de Fomento e Povoamento de Zambeze (MFPZ) was modeled on the Tennessee Valley Authority. Five years later the MFPZ produced a voluminous fifty-six volume final report, which confirmed the previous assessment.

What is most striking in these documents is the magnitude of the project and how radically different it was from its British counterpart. Kariba was designed simply as a hydroelectric scheme to provide cheap energy for the copper mines in colonial Zambia, and for the European farming and industrial sectors in colonial Zimbabwe. By contrast, Portuguese planners conceived of Cahora Bassa as a multipurpose project designed to expand agricultural productivity, develop mining and promote forestry, reduce Mozambique's dependence on foreign imports, enhance the living conditions of the indigenous populations, and above all else, increase the size of the white settler community. They estimated that as many as 80,000 immigrants would settle in the region and that armed settler communities could provide the first line of defence against exiled African guerrillas seeking to overthrow the colonial regime.

Economic realities compelled colonial officials to scale back their ambitious plant. With little prospect of investment in the project from metropolitan investors and little evidence of settler interest in this malarial-infested region, a multiple purpose dam seemed less viable. Mozambique's inability to consume even ten per cent of the anticipated 2075-megawatt output from Cahora Bassa made the project even more problematic. Military considerations further complicated matters.

FRELIMO had vowed to sabotage Cahora Bassa. In 1968 they initiated a guerilla offensive in Tete, district home of the Cahora Bassa.

The economic uncertainty and security threats compelled proponents of the dam within the Portuguese state to lobby for an energy and military agreement with South Africa which would guarantee a market for Cahora Bassa's surplus power and incorporate Mozambique into South Africa's security zone. Based on projections that its power requirements would double between 1967 and 1980, the apartheid regime needed a secure supply of cheap energy and was anxious to blunt the "black onslaught.".

In 1969 Lisbon signed a $515 million agreement with ZAMCO – a South African-dominated consortium with partners in West Germany, France, Italy, and Portugal – to build the dam. This agreement reconfigured the Cahora Bassa project into a single purpose hydroelectric scheme financed by the sale of cheap electrical power to South Africa.[15]

While Portuguese authorities invested a great deal of time and energy on the financial and security dimensions of the dam, state planners paid scant attention to the impact of the hydroelectrical scheme on African peasants and their environment. Colonial authorities presumed that increased economic activity would have a trickle down effect on 'subsistence' African cultivators living in the Zambesi basin. Officials expressed confidence that the riverine communities would benefit from the introduction of new farming techniques new markets for their commodities and new job opportunities.[16] [17] Peasants displaced by the dam were to be regrouped into modern villages as part of a vaguely defined community development scheme.

State planners gave even less consideration to the ecological consequences. In 1973, the Missão de Ecologia Aplicada do Zambeze (MEAZ) commissioned a pre-impoundment survey of water quality, vegetation, soils and climate.[18] The following year a small team of researchers affiliated with the University of Lourenco Marques conducted a biophysical survey of the Lower Zambezi Valley.[19] Underfunded, poorly conceived, and often of shoddy quality, these investigations yielded little insights into the Zambesi's ecosystems.[19] One scientist closely associated with the project decried "the lack of ecological specialists with local knowledge who should be dealing with the interdependence and interrelationships of the whole Lower Zambesi as one integrated system. This danger is compounded as authority in both land-use planning and decision-making is vested in non-ecological experts."[20] Economic planners and civil engineers committed to completing the construction of the hydroelectric project without delay simply ignored the research from scientists that ran counter to their grand design.[19]

Elsewhere I have written about the actual construction of the dam.[21] Suffice to say that the colonial region built the hydroelectric project on the backs of 3000 African labourers who constituted the bulk of the work force.[22] Although nominally free, the Africans were recruited into a highly regimented and racialized labour regime. It was a regime in which: (1) Africans were assigned the most grueling and dangerous tasks; (2) work obligations were often secured through coercive extra-legal methods; (3) Africans were prohibited by custom and practice from holding supervisory positions, the living and working condition of black labourers was distinctly inferior to that of their European supervisors; (4) the living and working condition of black labourers was distinctly inferior to that of their European supervisors; and (5) Portuguese officials brutally stifled any dissent. Júlio Calecoetoa described the unrelenting pressure to work longer and harder.

> The Africans who broke the boulders were organized into groups of six men, each with its own *capitão* [overseer]. The *capitão* was African. The boss over all of them was Silva. ... The Africans worked very hard. The *capitães* were chosen because they could speak Portuguese. They told the workers what to do; they did not do the work themselves. From time to time the *capitão* would beat people he did not think were working hard enough. There were never any strikes. If people had struck, they knew they would be beaten with the *palmatoria*. When Africans were angry they work more slowly, but if the boss noticed that he would tell the *capitão* to whip them.[23]

Harsh conditions and the pressure to meet deadlines created a precarious work environment. To this day, memories of industrial accidents remain etched in the minds of many former workers. Listen to words of Pedro da Costa Xavier.

> At the intersection of two tunnels a large rock, larger than this house, fell, collapsing the tunnels and trapping the men and machines inside it. Company officials were unable to rescue the men and they ultimately had to seal off the tunnel. Many people died. It was more than twenty. There were smaller accidents as well. When the ropes, which were harnessing men working on the edge of the dam broke, people would fall to their deaths. Sometimes, when we were working in the tunnel, there would be rockslides and people would be killed. Others would suffer serious injuries. It was very dangerous work, not only for the Africans, but for the Europeans who worked with them.[24]

Despite such catastrophes, occasional labour unrest, and FRELIMO ambushes of traffic carrying essential equipment to Songo, work on the dam proceeded on schedule. By 1974 the widely acclaimed dam was virtually complete. Lost in the shuffle were the thousands of Mozambican peasants who had been forcibly relocated from the flood plains.

Dissolution of Livelihoods and Ecosystems: The Immediate Effects of Cahora Bassa

Colonial planners maintained that the long-term benefits of the dam would far outweigh any short-term disruptions in the lives of the riverine communities. They stressed that the social cost would be relatively inconsequential, given Tete's low population density.[25] This air of self-confidence was reflected in a 1968 meeting of the Governor of Tete district, the regional representative of the secret police (PIDE) and local administrators called to plan the eviction. The Governor emphasized that "the problems of relocation were purely technical".[26]

Building Cahora Bassa had immediate, multiple and far-reaching consequences for the displaced communities whose homelands and farms were flooded to create the massive lake behind the dam. Yet it was not simply being evicted from their homes and ancestral lands that proved so devastating. Unlike other powerless groups around the world displaced by hydroelectric schemes, the Zambesi peasants were herded into strategic hamlets (*aldeamentos*) which were an integral part of Portugal's broader counterinsurgency program designed to cut FRELIMO off from its rural base of support.[27]

Under the guise of protecting the peasantry, the Portuguese began to evict peasant communities in 1972, two years before the actual impoundment of the river. Local authorities, facing the pressure of an expanded war and construction deadlines, rarely even bothered to pay lip service to the notion that peasants should be persuaded to move voluntarily. Pezulani Mafalanjala, Mauricio Alemão and Bernardo Tappuleta Potoroia of Masecha remembered what happened the day they were told to move:

> They came and told us that the water was going to rise and that we would have to leave... Among us there were people who complained and did not want to move. They were very angry because they had fields and houses here and their whole life was here. But they had no choice.[28]

The amount of intimidation and violence accompanying the move seems to have varied considerably. In some cases, such as in Masecha, peasants were simply told that they had to move in three months time, before their lands were flooded. On the designated day Pezulani Mafalanjala remembered that "they sent trucks and took us away."[29] In other areas such as Chinyanda, the threat of force was more explicit. "The soldier came and brought us here".[30] To underscore the point that defiance would not be tolerated, colonial authorities burned the homes and possessions of those who refused to relocate.[15 31] Although colonial authorities initially claimed that only 25,000 Africans would be displaced, by the end of 1973 the number had jumped to over 42,000.[32 33]

For all the colonial rhetoric of development, conditions in the camps were rudimentary at best. A typical *aldeamento* contained between one thousand and fifteen hundred residents. They lived in mud and wattle huts laid out in a grid enclosed by a barbed wire fence. The original plans called for each *aldeamento* to include a school, health clinic, water pumps, grist mills, warehouse for food reserves, social hall and football field, costing more than $9 million.[34 35 36] Few of the "protected villages" had all, or even most, of these amenities.[27]

Peasants were literally penned in. Their only egress was through a military check point manned around the clock by local militia. Pezulani Mafalanjala described the sense of being held captives under constant surveillance:

> We started working in our *machambas* [fields] at six in the morning and we were required to leave the fields between twelve and twelve thirty. We had to be back in the *aldeamento* by one. We were always accompanied by the patrol. They guarded us to make sure that we did not have any contact with strangers in the bush. If you came back late you were interrogated about where you were and what you were doing. The militia then handed over the tardy individual to PIDE and they were charged with secretly providing food to FRELIMO. The accused were severely beaten and threatened before they were allowed to return home. Some even died in the interrogation.[37]

As Mafalanja's account suggests, the fields were often far from the strategic hamlets. To make matters worse, the designated lands which the government had cleared were rocky, hard to work and not very fertile. The stood in sharp contrast to the lands left behind. Jack Sobrinho, who was forced to relocate to an *aldeamento* in Estima summed up the general consensus: "The land at Chicoa Velha was good land. The land here was hard and full of rocks, so it produced nothing."[38] The arid conditions and absence of rain-fed lands dramatically reduced agricultural yields. So too did the colonial policies which limited each household to one small plot–typically less than a hectare in size. Peasants were forbidden from farming multiple fields in different ecological zones to take advantage of variations in soils, sunlight and moisture availability and to minimize risks.

Given restrictive government policies and the harsh environment, it is hardly surprising that the uprooted communities experienced increased food shortages and malnutrition. The elders were right. Without rain, good lands and sufficient time to work the fields, there could be no corn or sorghum.[39] There were also fewer opportunities to make up food deficits through hunting and fishing. Despite government plans to protect the herds which roamed in the river valley and adjacent forests, [40] large numbers of animals drowned when the Zambesi was impounded. Even in those areas where game survived, Portuguese military authorities prevented peasants from carrying rifles and severely restricted their movement.[41]

Food shortages were not the only social problem these rural communities experienced after their relocation. As in other dam projects throughout the world, sickness and death rates seem to have increased markedly especially among the very young and very old.[42] It is important to stress that the evidence before and after the impoundment is fragmentary and that health and sanitary conditions varied from one strategic hamlet to another. Moreover, the colonial regime did try to inoculate at-risk populations to prevent tuberculosis and yellow fever and provided medication to limit the debilitating effects of malaria.[43] Nevertheless, the data suggest that inadequate rural diets, combined with problems caused by poor sanitary conditions regularly exacerbated by heavy rains in January and February, left many rural communities reeling from cholera. In *aldeamentos* located near Lake Cahora Bassa water-borne parasitic illnesses such as schistosomiasiasis and malaria posed new health threats.[44] Bernardo Tapuleta Potoroia recalled that time with anguish:

> There was a great deal of hunger and many people also suffered from diseases during this period. There were serious problems with cholera, small pox, and malaria. Many people died. No one knew why or how this happened, just that many people were dying.[45]

The commonly held explanation for these misfortunes–that the flooding of sacred shrines and burial sites had alienated powerful royal ancestor spirits (*mhondoro*)–underscores the sense of cultural obliteration and vulnerability which the uprooted peasants experienced. Most *mhondoro*, through their earthly spirit mediums (*svikiro*), opposed the impoundment because sacred sites would be inundated and lost to posterity.

> When the people were forced to move, the *svikiro* left as well. The *svikiro* wanted to live elsewhere, but the government forced them to live in the *aldeamentos*. The *svikiro* became angry and warned of calamities. When the *svikiro* died, the *mhondoro* disappeared and never returned.[45]

Other *svikiro* refused to move and fled into the bush where they were never heard from again. Their absence was devastating. "There were no spirits to make it rain."[46]

In the final analysis, displacement adversely affected everyone, but it did not necessarily affect them equally. Chiefs, who received state subsidies and the choice tracts of land, suffered less than their subjects. Peasant women probably suffered more than their male counterparts. Most were forced to cope as best they could with the hardships of daily life inside the strategic hamlets. This meant caring for children, walking long distances to gather firewood and fetch water, assisting the sick and the elderly and performing a wide array of other household chores in addition to working on their household plots. By contrast, most men simply farmed. Others clandestinely fled to Zimbabwe or found employment at the dam site. However, difficult the living and working conditions of these male labourers, they were appreciably better than being penned up in the *aldeamentos*.[47]

Cahora Bassa's profoundly negative social effects were mirrored and exacerbated by its devastating impact on the hydrology and ecology of the riverine zone. From the outset, the building of Cahora Bassa fragmented and disrupted the riverine ecosystem. The dam cut off sediments and nutrients that were formerly transported to floodplains and the delta, it transformed seasonal flows from highly variable to relatively constant throughout the year, and it completely altered geomorphological patterns of erosion and deposition in the river channel. Combined, these changes transfigured the pre-dam ecological conditions of the Lower Zambesi. In doing so, the dam effectively destroyed the complex food production systems of the peasant communities living adjacent to the river whose agricultural and other livelihood strategies depended on the seasonal cycles of the undammed Zambesi. The rising waters of Lake Cahora Bassa obliterated whole riparian ecosystems and the agricultural lands, hunting grounds and traditional fisheries of those peasants displaced by the dam. From the Cahora Bassa gorge to the river's delta, the annual flooding which formerly deposited nutrients and sediments ideal for the production of a wide variety of food crops no longer inundated plains adjacent to the river. Communities up and down the riverine zone who depended on fish as a crucial component of their diet found greatly reduced variety and abundance of species.

While many of these changes became apparent only after several years, some of the detrimental effects of the dam's construction were felt almost immediately. Following completion of the construction phase, state officials–disregarding the suggestions of environmental scientists contracted to study the dam's likely ecological impacts–rushed to fill the reservoir as rapidly as possible. Their mandate was to begin generating electricity as originally scheduled, regardless of the consequences. Lake Cahora Bassa began to rise from the river bed on 5 December 1974 and reached nearly-full capacity approximately four months later at the end of March 1975. The extreme haste of the operation is clear when compared to the filling of Lake Kariba, which took nearly four years to reach full capacity.[48] In an extraordinarily short period of time the dam generated a lake where before there had been a river, destroying lands and animals critical to the local human ecology of riverine communities and creating a drastically different ecological setting.

The most immediate effect was the permanent inundation of 2,700 square kilometres of land which formed Lake Cahora Bassa. It was not simply the vast amount of land that was lost, but that what disappeared were the highly productive floodplains effectively used by peasant communities for centuries. Pezulani Mafalanjala, Maurício Alemão and Bernardo Tapuleta Potoroia recalled the high moisture retention capacity of the river-fed *mkande* soils which made it ideal for the culture of a variety of food stuffs in the Masecha region. "All the crops grown on the *makande* had a good supply of water and nutrients. In some lowland areas the river deposited sediments on banks for 2-3 kilometres from the river."[49] Other residents remembered stretches of the river where floodwaters would extend for seven-to-nine kilometres on either bank during the flood season.[50]

The inundated floodplain habitats constituted some of Mozambique's most diverse ecosystems. The lands near the river had supported numerous trees whose leaves would fall and act as natural fertilizer upon decomposition. Some idea of the diversity of tree species in the region is provided by accounts of people who, during times of drought, would forage for wild fruit.[51] These riparian ecosystems also supported substantial numbers and types of animal species, including *goma*, bush pig, buffalo, nyasa, gazelle, elephant and rhinoceros.[52] Despite the government's highly publicized plan (termed "Noah's Ark") to protect wildlife,[53] little was done. The effects were devastating. Bento Estima and Joseph Ndebuvchena remembered that

> After the flooding began many animals were stranded on Tanzwa and Manherere which are islands in the Zambesi. Some died on these islands because they could not get enough food. As the water kept coming higher, many animals were swept away if they couldn't swim to the other side of the river. [54]

Supia Sargent and Carlos Soda Churo of Estima evoked a similar scene, describing how "after the lake rose, animals fled to the islands and died of starvation. Others, who could swim, swam to the banks of the rising lake and ran away."[55]

In addition to the rapid and permanent inundation of ecologically important riverine lands, the decision to fill the Cahora Bassa reservoir at break-neck pace also had far-reaching consequences for human communities and ecological systems downstream. Despite the hydrologic fact that the portion of the river below Cahora Bassa was highly dependent on the main channel for continued flows, dam operators refused to allow compensatory releases through the dam during the filling of the reservoir. The flow rate of less than 60 cubic metres per day for over three months had catastrophic results below the dam.[56] The river was stopped in December precisely when the annual inundation of floodplains for agricultural production typically occurred. This was also a time when many fish species of the Lower Zambesi begin to spawn in adjacent floodplains. With closure of the dam and discharge reduced to ten per cent of its average, the fish were stranded as flood waters receded. Local farmers who depended on fish for supplemental protein harvested them in large numbers during this period, placing further pressure on fish populations. One scientist lamented, "Our work was ignored by the very people who requested the [ecological] survey in the first place."[57]

The downstream mayhem caused by completely shutting off the Zambesi's flow for the several months was quickly followed by cataclysm of a different kind. The filling of the reservoir occurred so quickly that by March 1975 water was overflowing the crest of the dam prior to its full completion.[19] In April, with construction of the dam and turbines almost complete, HCB engineers discovered a small defect in one of the turbines deep in the water of the almost full reservoir. Without any warning or consultation, they opened the turbines and sluice gates to full capacity and delivered an unnatural coursing of floodwaters downstream from the dam without any warning. Numerous small-scale farmers, at the time residing close to the river's edge to take advantage of the fertile soils, lost significant numbers of cattle and small poultry, and, in many cases, almost lost their lives. Upon repair of the fault, the dam operators again shut off the river's water causing a severe water shortage in the city of Tete and everywhere else downstream.[58]

By May and early June, the gates of Cahora Bassa were being opened and closed on a daily basis, timed to the power generation schedules of HCB engineers. The transformation of the river's annual cycle from a punctuated, highly seasonal flow regime to one characterized by constant flows throughout the year was complete. The Lower Zambesi had become a regulated river.

The Regulated River: Some Long-Term Effects of Impounding the Lower Zambesi

While the immediate effects of the dam at Cahora Bassa were devastating in several ways,[59] preliminary research suggests that the long-term effects may be even more consequential. The construction of Cahora Bassa adversely affected the economic security of hundreds of thousands of peasant households, irrevocably altered the biophysical relations of the Lower Zambesi from the reservoir to coastal regions. According to a recent United Nations report , "Cahora Bassa has the dubious distinction of being the least studied and possible least environmentally acceptable dam project in Africa."[60]

Prior to construction, the mosaic of ecological systems that comprised the Lower Zambesi (for example, river channel, floodplains, delta, estuary) was adapted to large fluctuations in river flows. Following the arrival of rains in December, the river would swell to several times its normal flow until March and then recede to fairly low rates during the dry season from May to October.[10] In the area near the Cahora Bassa gorge and in the floodplain zones stretching from the gorge to the delta, the rising waters triggered breeding among a substantial portion of fish species and these inundated zones provided critical habitat for fish reproduction and survival. Subsequently, the moisture and sediment left behind by receding floodwaters stimulated plant growth and hence supported diverse animal populations in the riverine ecosystems.

Peasant communities residing near the river, derived sustenance from these environmental resources through fishing, agriculture, gathering and hunting. The construction of Cahora Bassa sounded a death knell for this particular set of socio-ecological relations by regulating the river. Flow rates became much lower than normal during the former flood season and much higher than normal during the dry season. Moreover, the river was subject to erratic, unseasonal flooding as a result of dam operators' manipulation to generate hydroelectricity.[61] The consequences led one ecologist, who had periodically worked in the valley for the past three decades, to conclude that the Lower Zambesi "has been abused to a degree that has, fortunately, few parallels anywhere else in the world."[62]

Many of the long-term biophysical changes witnessed in Lake Cahora Bassa – for example, invasion of the reservoir by the Lake Tanganyika sardine, or *kapenta*, in the late 1970s[63]– were not predicted by project planners and consultants. Other effects associated with the reservoir include the growth of invasive weed species, development of a fishery involving greater types and numbers of fish adapted to lacustrine conditions, and nutrient enrichment of the lake via the accumulation of agricultural pesticides from upstream sources.[19] The precise impacts of these changes on the long-term integrity of the reservoir ecosystem are difficult to pinpoint–a situation made even more problematic by the extreme unpredictability of fluctuations in water levels due to unscheduled releases by dam operators.[64]

What is clear is that the management of the reservoir's water levels, has already produced adverse ecological and social effects. Most studies predicted drawdowns of between four and six metres whereas actual drawdowns have frequently exceeded 12-14 metres, a pattern that has not followed the course predicted in pre-impoundment studies.[65] The resultant large annual fluctuations in reservoir water level have impeded the growth of riparian vegetation resulting in lower fish productivity and have led to pools of stagnant water that provide ideal conditions for diseases.[66] Preliminary evidence suggests the likelihood of an increase in debilitating water-borne diseases, particularly schistosomiasis and malaria, typically associated with dam and reservoir.[67] Moreover, communities situated on the edge of the reservoir to take advantage of fishing opportunities are unable to effectively utilize the fertile margins of the lake for agriculture due to the unpredictability of reservoir level changes. Whole villages, built near the high water mark of the lake, suddenly find themselves over two kilometres from the shoreline after unannounced drawdowns.[68] Agricultural production in the area around the lake plummeted. Senteira Botão, an elder from Chipalapala put it starkly:

> In those times [before the dam] occasionally there was hunger but not like the hunger of today. There were shortages of maize, but then we had sorghum. If the rains were poor and we had a shortage of both, we eat sweet potatoes and other foods from our gardens. Hunger never threatened our community. No one ever died from hunger.[69]

Even the introduction of the *kapenta* sardine has proven to be a mixed blessing. While the fish is in great demand in neighbouring countries, South African and Zimbabwean fishermen, using large boats and nets, have come to dominate the reservoir's sardine fishery. They displaced local fishermen , many of whom have been forced to work on these boats for minimal wages in order to survive.[70]

Because it could reduce the storage capacity of Lake Cahora Bassa and damage the dam's turbines, sedimentation also poses a serious long-term threat.[71] Nevertheless, as of the early 1980s, project design consultants, construction firm representatives and the dam operators seemed oblivious to this problem. They had failed to collect even the most rudimentary data about actual sedimentation rates in the reservoir,[72] which might raise thorny questions about the project's actual lifetime.

With the construction of Cahora Bassa, the lifeblood of the floodplains, delta and estuary regions was placed in the hands of the dam operators.[73] The consequences have been profoundly negative. Flows that once reached rates of 28,000 cubic metres per second during the flood season and averaged 2,000-3,000 during the rest of the year were eliminated and replaced with flow rates of 900 cubic metres that varied little from month to month.[74] A recent assessment concluded that erratic and mistimed discharges have been "catastrophic" for downstream wetlands where vegetative growth and animal populations depended on annual flooding that brought nutrients and sediments.[60] By 1996, the geomorphology of the Lower Zambesi itself–formerly a wide river system with "open mosaics of marsh, pond, oxbows and shallow wetlands"–had been converted to a system with "choked wetlands, tree and bullrush encroachment along margins" and impoverished marshlands. The overall result is less diverse, less productive riverine ecosystems.[19] A disinterest in the downstream hydrological effects spurred by the dam also reflected a tacit disregard for the peasant and fishing communities for whom the floodplain system was a critical resource. According to the current director of the Zambesi Valley Authority,

> Pre-dam livestyles [sic.] of hundreds of thousands of local residents were dependent on annual flooding which sustained a diversified production system that incorporated flood recession agriculture, livestock management, fishing, gathering and hunting. Flooding was especially important for providing otherwise unavailable grazing during the latter months of the dry season and for maintaining the productivity of the riverine fishery.[75]

The impacts on riverine fish populations are especially pernicious, due to the apparent loss of biological diversity and to the drastic reduction of an important food source for riverine communities. While information concerning the dam's effects on other aspects of downstream floodplains is sketchy, the impacts on fish have almost certainly been devastating.

> In the Lower Zambesi valley, floodplain regions have, in the past, been a more important source of fish than the river itself, despite the fact that the floodplain is much smaller than might be expected from a river this size...the seasonal inundation of the Zambesi floodplain created large areas of warm shallow water which were rich in nutrients and provided ideal breeding grounds for fish.[76]

The breeding cycle of many fish species, triggered by early flooding, has in all likelihood been irretrievably interrupted.[14] The large numbers of fishing camps and drying racks that were clearly present in the 1970s no longer operate, indicating a general, if not disastrous, decline in the delta freshwater fisheries.[77] According to one anthropologist who has worked in the region, "Villages correctly attribute… a drop in the productivity of their fishery to dam construction."[60]

Even the predicted benefits of the dam in terms of downstream flood control did not materialize. Despite the apparent regulating effect of Cahora Bassa, a massive flood in 1978 resulted in over 40 lost lives, 200,000 homeless and destroyed more than 60,000 hectares of crops. The estimated damage was over $60 million. This devastation stands in stark contrast to the claims made in early documents of the Kariba and Cahora Bassa projects, touting the period of "total control" over discharges that the dams would usher in.[78]

Some of the most far-reaching and difficult-to-measure ecological impacts of large dams have occurred in the delta and estuary zones of impounded rivers where there has been a marked reduction of biological diversity, simplified landscapes and continued threats to ecologically and economically important biota. The delta region's vast populations of large mammals have been devastated by the effects of the dam. The Zambesi Delta, a massive zone of flooding and silt deposition, covers 18,000 square km, stretches almost 300 km along the coastline of Mozambique, and extends 100 km inland to the Zambesi's confluence with the Shire River. Earlier this century, the region supported an estimated population of nearly 130,000 water buffalo.[79] With reductions in seasonal flooding, the riverine area became easily accessible to commercial poachers as well as by RENAMO and government soldiers. As a result the buffalo population fell to less than 4,000 as of 1990 and other mammal species including zebra, hippopotamus, and waterbuck and several bird species have experienced a similar decline.[80] The destruction of an indigenous sub-species of zebra (*Equus burchelli selousi*), extinct in other parts of Africa, poses dire long-term consequences.

The lack of flooding in the region, creating a generally drier environment with less vegetation, had other devastating effects as well."[77] Grassland burning has intensified through the dry season. It is estimated that 90% of the lower Zambesi floodplain now burns every year. Moreover, the irregular flooding patterns have disrupted the breeding cycles of many delta species, including the endangered water cranes.[60]

The adverse effects of Cahora Bassa extends to the very mouth of the Zambezi. The sharp decline in silt transported down river has hampered the Zambezi's important delta-building function. As deposition of silt decreases with unknown implications for the coast's vulnerable communities of natural vegetation, the estuary is subject to greatly increased wind and sea erosion. This in turn almost certainly generates negative impacts on the estuarine fisheries that evolved under more stable conditions and depended on the annual flooding cycle that brought nutrients and sediments.20 Mangrove forests and shrimp fisheries, critical elements of the delta and estuarine system, have been particularly degraded at least in part if not wholly by the altered character of Zambesi flows. The full effects of the decrease in sediment transfer to the Zambesi delta on the coast's mangrove ecosystems are little known, although a recent survey showed that only a single channel of the Zambesi exhibited healthy mangroves. Throughout the rest of the coastal region there are large gaps in the mangrove forest with evidence of dried-out areas, dead mangroves and severe coastal erosion.[19]

What is clear is the devastating effects of the reduced wet season water flows of the Zambesi on Mozambique's strategic shrimp industry. The Sofala Bank, a broad and shallow shelf just outside the mouth of the Zambesi, is the site of a productive shrimp fishery that is one of Mozambique's most important sources of foreign currency. Since the early 1980s, catches of the coastal fisheries' two most important shrimp species (*Penaeus indicus* and *Metapenaeus monoceros*) have declined substantially due to both environmental factors and increasing fishing effort.[81, 82] There is a high degree of correlation between wet season river run off and the abundance of these economically important shrimp species. The dam at Cahora Bassa both reduces the amount of water discharged by the river and alters the seasonal pattern of run off. After the dam, the wet season runoff was reduced by about 40 per cent.[81, 82] Significant decreases in the amount of water released, particularly during the onset of the flood seasons when shrimp normally migrate towards the ocean, could drastically reduce the shrimp population by impeding an important stage in their life cycle.[83] For local fishermen as well as for the impoverished nation, hard-pressed for foreign currency, this loss of shrimp revenue is highly significant.

Conclusion

This paper has explored the deleterious socio-economic and environmental changes brought about by the Cahora Bassa dam, which are inextricably linked to broader questions of social and ecological sustainability and human security. It is part of an alternative history of Cahora Bassa, and argues that the historical memories and lived experiences of these riverine communities must figure prominently both in any scholarly analysis of the effect of Cahora Bassa and any new initiatives to remedy the situation.

References

[1] The author together with Arlindo Chilundo, as well as a number of other Mozambican scholars, is beginning to explore the complex and varied impacts of Cahora Bassa in a project funded by the John D. and Catherine T. MacArthur Foundation

[2] Adams, W. 1990. Green Development: Environment and Sustainability in the Third World. London.

[3] Chambers, R., and G. Conway. 1992. "Sustainable rural livelihoods: practical concepts for the 21st century." : Institute of Development Studies.

[4] Isaacman, A. 1997. "Historical amnesia, or the logic of capital accumulation: cotton production in colonial and post-colonial Mozambique." *Environment and Planning D: Society and Space* 15:757-790.

[5] Bolton, P. 1986. "Mozambique's Cahora Bassa Project: en environmental assessment." In *The Social and Environmental Impacts of Large Dams*, edited by E. Goldsmith and N. Hildyards. Cornwall.

[6] Vines, A. 1991. *RENAMO Terrorism in Africa.* London

[7] Throughout this paper we have cited some preliminary research on the effects of Cahora Bassa many of which were presented at a Ford Foundation-sponsored conference, 'Workshop Sobre O Uso Sustentavel da Barrágem de Cahora Bassa e do Vale do Rio Zambese' held in Songo from 29 Sep-2 Oct 1997, which was organized by the Arquivo do Patrimonio Cultural (ARPAC).

[8] Interview with Senteira Botão, Eliot Jumbo Muatisembero Sargento and Beatriz Maquina, Chipalapala, 26 May 1998.

[9] *Ibid*; Interview with Supia Sargent and Carlos Soda Churo, Estima, 22 May 1998; interview with Sene Simico, Mauzene Dique and Mzwengane Mafala-Njala, Nyatapiria, 27 May 1998; interview with Bento Estima and Joseph Ndebvuchena, Estima, 19 May 1998.

[10] Davies, B. 1986. "The Zambezi River system." Pp. 233 in *The Ecology of River Systems*, edited by B. Davies and K. Walker. Dordrecht.

[11] Interview with Supia Sargent and Carlos Soda Churo.

[12] Jackson, P., and K. Rogers. 1976. "Cabora Bassa Fish Populations Before and After the First Filling Stage." *Zoologica Africana* 11:377.

[13] Interview with John Paul and Khumbidzi Pastor, Estima, 21 May 1998.

[14] Bolton, P. 1983. "The Regulation of the Zambezi in Mozambique." Edinburgh: University of Edinburgh, pp. 445-446.

[15] Middlemas, K. 1975. *Cabora Bassa: Engineering and Politics in Southern Africa.* London, pp. 20-30.

[16] Bolton 1983, *op.cit.*, p. 158.

[17] Vidigal, M. 1970. "Cabora Bassa: Historia. Perspectivias. Justificacão. Aspectos Economico – Financeiros. Interesse Nacional do Empreendimento." *Electricidade* 14:7-20.

[18] Hall, A., I Maria, C. Valente, and B. Davies. 1977. "The Zambezi River in Moçambique: the physico-chemical status of the Middle and Lower Zambezi prior to closure of the Cabora Bassa Dam." *Freshwater Biology* 7:187-206.

[19] Davies, B. 1996. "Rehabilitation programme for Cahora Bassa and the Lower Zambezi." : International Crane Foundation/Ford Foundation.

[20] Tinley, K. 1975. "Morromeu: wrecked by the big dams." *African Wildlife* 29:24

[21] Isaacman, Allen, and Chris Sneddon. 1999. "Towards a Social and Environmental History of the Building of Cahora Bassa Dam." in *Conference on African Environments: Past and Present, July 1999*. Oxford.

[22] There were approximately 750 Europeans who held all the technical, administrative as well as most clerical positions.
[23] Interview with Júlio Calecoetoa, Songo, 18 May 1998.
[24] Interview with Pedro da Costa Xavier, 23 May 1998, Songo.
[25] The commonly used colonial figure of 4.5 people per square kilometre was actually quite misleading. The population in the area to be flooded by Cahora Bassa Lake was almost triple the official figure as compared to 1.5 inhabitants per square kilometre in other parts of the district not flooded (Bolton 1983, p. 358).
[26] Arquivo Histórico de Moçambique (AHM) Secção Especial (SE) 111, a.p. 10, no, 237, Governo do Distrito de Tete, Servicos Distritais de Administracão Civil, Actas das Sessões da Reunião dos Administradores e do Intendente com o Governo do Distrito, 27-28 December 1968.
[27] Jundanian, Brendan. 1974. "Counterinsurgency in Mozambique." *World Politics* 6:519-540.
[28] Interview with Pezulani Mafalanjala, Maurício Alemão and Bernardo Tapuleta Potoroia, Masecha, 25 May 1998.
[29] *Ibid.*
[30] Interview with Peter Size and Fedi Alfante, Chinyanda Nova, 25 May 1998.
[31] Interview with Padre Claudio Gremi, Songo, 20 May 1998.
[32] GPZ. 1974. "Relatorio de Actividade 1973." . Lisbon: Gabinete do Plano do Zambeze, p. 28.
[33] This figure includes Africans relocated down river at Caia. It is difficult to determine the actual number since there was a great deal of secrecy surrounding the forced villagization program. One author estimated that upwards of 200,000 peasants in Tete district were interned (Jundanian 1974).
[34] Bolton 1983, *op.cit.*, p. 363.
[35] GPZ. 1971. "Relatorio de Actividade 1970." . Lisbon: Gabinete do Plano do Zambeze, p. 46.
[36] GPZ. 1972. "Relatorio de Actividade 1971." . Lisbon: Gabinete do Plano do Zambeze, p. 20.
[37] Interview with Pezulani Mafalanjala, Maurício Alemão and Bernardo Tapuleta Potoroia.
[38] Interview with Jack Sobrinho and Wiseborn Benjamin, Estima, 20 May 1998.
[39] Interview with Pezulani Mafalanjala, Maurício Alemão and Tapuleta Potorioa
[40] For a discussion of this ill-fated plan see A.H.M., Governo Geral, Cota 864, 'Plano Base Para Salvamento e Transferencia da Fauna Brava da Albufeira de Cahora Bassa em Moçambique', K. L. Tinley, March 1973.
[41] nterview with John Paul and Khumbizi Pastor.
[42] McCully, Patrick. 1997. *Silenced Rivers: The Ecology and Politics of Large Dams*. London: Prometheus.
[43] GPZ 1974, *op.cit.*, pp. 59-63.
[44] Bolton 1986, *op.cit.*, pp. 161-162.
[45] Interview with Pezulani Malalanjala, Maurício Alemão and Bernardo Tapuleta Potoroia.
[46] Interview with Sene Simico, Mauzene Dique and Mzwengane Mafala-Njala.
[47] Interview with Vernácio Leone, Estima, 22 May 1998; Interview with John Paul and Khumbiidzi Pastor.
[48] Bond, W., N. Coe, P. Jackson, and K. Rogers. 1978. "The limnology of Cabora Bassa, Moçambique, during its first year." *Freshwater Biology* 8:445.
[49] Interview with Pezulani Mafalanjala, Maurício Alemão and Bernardo Tapuleta Potoroia.
[50] Joint interview with Supia Sargent and Carlos Soda Churo.
[51] Interview with Vernácio Leone, Estima, 19 May 1998.
[52] Joint interview with Jack Sobrinho and Wiseborn Benjamin.
[53] A.H.M., Governo Geral Cota 804, 'Base plan for rescue and translocation of wildlife from the Cahora Bassa Dam in Moçambique' K. Tinley, March 1973.
[54] Interview with Bento Estima and Joseph Ndebvuchena.
[55] Interview with Supia Sargent and Carlos Soda Churo.
[56] This decision was made in spite of recommendations from the environmental research team, particularly Davies, that the filling of the reservoir occur at a pace to ensure a minimum discharge rate from the dam of between 400 and 500 cubic metres *per second*. Davies and Dr. Luis Bareto, then Director of the Instituto d'Agronomia in Laurenco Marques, strongly requested (in Davies' words "almost demanded") several measures in addition to a minimum flow rate to head off what they believed would be the disastrous effects of the planned releases. These included: a two-and-one-half year minimum filling period; releases from the dam timed to match dry and wet season flows in the river; and no filling during the next summer flood (December-January). None of these suggestions were implemented (Davies 1996).

[57] Davies, B. 1975. "They pulled the plug out of the lower Zambezi." *African Wildlife* 29:26-28.
[58] Jackson, P. 1997. "Ecological studies on the Middle Zambezi prior to Kariba and Cahora Bassa and the need for surveys of the Lower Zambezi prior to the creation of further hydroelectric dams." in *Workshop Sobre O Uso Sustentavel da Barrágem de Cahora Bassa e do Vale do Rio Zambese, 29 Sept-2 Oct*. Songo.
[59] It is often difficult to distinguish the environmental and social effects caused by the dam from those caused by the civil war, and specifically RENAMO's campaign of rural terror and poaching.
[60] Beilfuss, Richard. 1999. "Can This River Be Saved: Rethinking Cahora Bassa Could Make A Difference for Dam Battered Zambezi." *World Rivers Review* 14:8-11.
[61] Jackson, P. 1986. "Fish of the Zambezi system." pp. 269-288 in *The Ecology of River Systems*, edited by B. Davies and K. Walker. Dordrecht, NL.
[62] Davies 1986, *op.cit.*, p. 258.
[63] Interview with Padre Claudio Gremi.
[64] Davies 1986, *op.cit.*, p. 256.
[65] Bolton 1983, *op.cit.*, pp. 219-220.
[66] Bolton 1983, *op.cit.*, pp. 389-391.
[67] Bolton 1983, *op.cit.*, p. 379.
[68] Bolton 1983, *op.cit.*, p. 163.
[69] Interview with Sentiera Botáo, Eliot Jumbo, Mautisembero Sargento and Beatriz Maquina.
[70] Interview with Padre Claudio Gremi; interview with Vernácio Leone; interview with Senteira Botão, Eliot Jumbo, Muatisembero Sargento and Beatriz Maquina.
[71] Bolton 1983, *op.cit.*, p. 402.
[72] Bolton 1983, *op.cit.*, pp. 312-313.
[73] We distinguish the delta (the typically broad, flat zone near the mouth of a river where there are high levels of deposition) from the estuary (the zone where the river's plume extends into the ocean and where fresh and salt waters mix).
[74] The Kariba Dam also had some effect on the flooding regime of the Lower Zambesi, but not nearly the drastic impact of Cahora Bassa (Davies 1986, pp. 235-242; Muhai 1997, p. 4).
[75] Muhai, A. 1997. "Cahora Bassa and Lower Zambezi Workshop." Pp. 4 in *Workshop Sobre O Uso Sustentavel da Barrágem de Cahora Bassa e do Vale do Rio Zambese, 29 Sept-2 Oct*. Songo.
[76] Bolton 1983, *op.cit.*, p. 395.
[77] Chande, B., and P. Dutton. 1997. "Impacts of hydrological changes in the Zambeze Delta to wildlife and their habitats with special attention to the large mammals." Pp. 4 in *Workshop Sobre O Uso Sustentavel da Barrágem de Cahora Bassa e do Vale do Rio Zambese, 29 Sept-2 Oct*. Songo.
[78] Bolton 1983, *op.cit.*, pp. 397-98.
[79] This figure is based on an observed 16,000 water buffalo remaining in the Morromeu Reserve in the mid-1970s (Tinley 1975).
[80] Beilfuss, R., D. Allan, and C. Bento. 1997. "Waterbird and wetland surveys in the Zambezi-Pungue floodplain system, Mozambique." Pp. 4 in *Workshop Sobre O Uso Sustentavel da Barrágem de Cahora Bassa e do Vale do Rio Zambese, 29 Sept-2 Oct*. Songo.
[81] Gammelsröd, T. 1992. "Improving shrimp production by Zambezi River regulation." *Ambio* 21:145-147.
[82] Hoguane, A. 1997. "Shrimp abundance and river runoff in Sofala Bank-the role of the Zambezi." Pp. 4 in *Workshop Sobre O Uso Sustentavel da Barrágem de Cahora Bassa e do Vale do Rio Zambese, 29 Sept-2 Oct*. Songo.
[83] Freshwater stimulates recruitment of shrimp to the shelf area and provides nutrients to the coastal waters. The shrimp spawn at sea, but require brackish water to serve as nursery areas. At an early stage in their life cycles, the immature shrimp move inshore, most likely during the dry season. As river runoff and/or rainfall in the area increases at the onset of the flood season, older juveniles that are unable to tolerate increasing amounts of freshwater migrate offshore where commercial fishing interests are able to take advantage of this seasonal increase in abundance (Gammelsröd 1992).

WATER AND HUNGER IN AFRICA

Aharon Zohar

The African continent is not homogenous in terms of water and land resources: while most of the countries which are located in the tropical climatic regions enjoy relative abundance of water and arable land, most of the countries located in the desert climatic region, south and north of the tropical climate, suffer from a severe shortage of these resources, which poses a threat to their existence. However, even those countries which are relatively rich in water and land resources suffer from severe constraints on agricultural development, stemming in large part from human constraints, some of which are no less serious than physical constraints.

Major Physical Constraints

Across the continent there is considerable fluctuation in the amount of precipitation: the duration and distribution of precipitation range from rain which falls throughout the entire year, to areas which are totally arid. Most of the agricultural crops are rainfed, dependent on the amounts of rainfall, its duration and frequency.

Major climatic constraints:

- variability and frequent unreliability of rainfall especially in arid and semi-arid areas: deficiency in rainfall in over 54% of the continent;
- unpredictable periods of drought: high to very high drought risks in 67% of the continent's area;
- high soil temperatures that adversely affect crops;
- cloudiness in humid areas, reducing photosynthetic efficiency;
- wind and storm damage to crops and erosion to soils;
- high temperatures and humidity, favourable to pests and diseases;
- problems of desertification, principally in the Sudan-Sahelion zone, where rainfall deviation from normal exceeds 30%.

There are no significant aquifers in Africa. About one-third of the continent comprises crystalline metamorphic rock with low water retention capacity. About one-fifth of the continent is calcareous rocks which have some water storage capacity but these regions are found mainly in the north and south edges of the continent.

The aquifers which can be most easily exploited are found in the alluvial fills of rivers as they are close to the surface, making it is easy to pump water from them. Unfortunately, these areas are small, with the alluvial layer in them thin. Below large tracts of the Sahara Desert there are fossil aquifers but extraction of water from them is so costly as to make it unfeasible as a realistic solution in Africa. In those places where this water is utilized for irrigation, the water is decreasing because of the non-renewable nature of these aquifers.

The Food and Agriculture Organization (FAO) studies on soil fertility, length of growing season, extent of frost hazard and water availability show that 38% of the land suitable for agricultural cultivation (about 244 million hectares) receives sufficient rain to facilitate the cultivation of agricultural crops. About 18% of the area fit for agricultural cultivation is suitable for irrigation, with half located in regions with annual precipitation levels ranging between 700 and 2,000mm, where there is a surplus of water which can be utilized for irrigation of smallholder plots. Notwithstanding the above, the use of water for irrigation across the continent is minimal and much of the irrigated area (about 5% of the 18% of the land fit for agricultural cultivation) is located along the Nile Valley in Egypt and Sudan.

In many fertile regions and in the river flood plains health hazards, primarily malaria and water-borne diseases, restrict the use of water.

The densely forested area in Africa constitutes about 10% of the continent area while sparsely forested area constitutes about 40% of the continent area. The rate of deforestation on the continent is estimated at about one half a per cent annually, though there are countries where it is higher than 5%.

Main vegetation problems:

- loss of vegetation cover;
- deterioration of rangelands;
- degradation of rainfed crop lands;
- waterlogging and salinization of irrigated lands.

Of all the physical factors influencing agricultural production in Africa, the land factors are apparently the most restrictive and also include the following problems:

- high degree of weathering, sandiness, deficiency in clay, high fragility and erodibility: land areas in Africa which were lost due to unchecked degradation exceed 85%;
- rapid rates of organic matter decomposition;
- high levels soil acidity and high tendencies of P-fixation;
- highly subject to multiple nutrient deficiencies under increasing intensity of use and shortened periods of fallow;
- proneness to leaching and high risk of erosion under prevailing tropical rainstorms;
- low inherent fertility: 81% of African soils has fertility limitations;
- serious salinity problems under irrigation, especially under poor water control or management;
- large areas exhibit physical and chemical constraints.

Major Human Constraints

Agricultural development hinges on human factors no less than on physical factors, that is, on human beings that are healthy, active, satisfied, educated, well-trained, and who receive a steady supply of food, water, sanitation, clothing and shelter. In addition to the need for educational advancement, there is also a need for a cadre of experts in the fields of agricultural research and training.

The illiteracy rate amongst the farmers in Africa exceeds 70%. Living conditions are difficult not only because of the poverty, hunger and malnourishment but also because of the prevalence of diseases which afflict the African population mercilessly, primarily malaria, AIDS and various intestinal disease. Mortality from these diseases has considerably lowered the average natural growth rate on the continent, which at the end of the 1980s stood at about 3.2% annually.

Among the other factors which strongly influence the production of food are:

- traditional cultivation methods;
- weakness of the support systems: agricultural research and training;
- physical (mainly transportation) and administration (marketing, credit, etc.) infrastructure;
- political instability and corruption.

Many countries in Africa are unable to impart to their citizens education on an appropriate level. Education-related problems include:

- inadequate funding and lack of good facilities;
- imbalance between theory and practice;
- more are trained in the arts and humanities than in science and agriculture;
- lack of vocational education;
- methods of teaching and curricula not adapted to current needs;

- lack of textbooks and visual aids.

Despite the significant growth in agricultural research expenditures and workers, no significant impact has been achieved in agricultural productivity because of:

- lack of adequate funding, facilities and equipment;
- low quality of staff;
- poor management and co-ordination of research.

How Can Food Production Be Increased?

As seen above, the water shortage is not the only factor responsible for the dire situation of agriculture in Africa. Furthermore, in many countries on the continent there is a water surplus, but this surplus is not utilized in most cases. The problem of potable water exists mainly in the north of the continent.

The range of significant factors detailed above have combined to create a situation where Africa stands helpless in the face of the future which promises only hunger, poverty and disease.

Increased food production is possible through adopting some or all of the following strategies:

- expanding the cultivated areas;
- increasing the yield per unit area;
- genetic improvements in crops and animals;
- use of appropriate technology;
- integrated pest, disease and weed control;
- improvements in post-harvesting;
- improvement in management and utilization of forest and grazing areas and aquatic resources.

Only a drastic change can steer the African continent away from its current course of dearth to a course which holds hope for the future of its inhabitants and its environment.

TO WHALE OR NOT TO WHALE: IS THAT THE QUESTION?

Brendon Gooneratne

"Save the Whale" Campaign (1970s)

In the 1970s the battle to "Save the Whale" became the biggest worldwide ecological issue to date. The dwindling numbers of great whales, brought to the brink of extinction by the ferocious whaling nations, created shock and dismay all over the world. Norway, the USA, Britain, Australia, Germany, Holland, Spain, Denmark in the 18th and 19th centuries, and the Norwegians, Icelanders, Japanese, Australians and Americans, together with other nations "new" to whaling (such as Peru, Korea and Argentina) had relentlessly hunted these mammals, which have inhabited the oceans for 50 million years until the Blue Whales, Humpbacks, Fin Whales and Sperm Whales had dropped to less than five per cent of their original numbers: in some species to less than one per cent.

The warning signs were very clear: unless commercial whaling was banned indefinitely, these whales would vanish from the oceans. Since we still do not really know the impact of their ecological significance involving plankton, other fish-eating mammals, and plankton-eating fish, nor do we possess accurate information regarding the oxygen production of the plankton and its consumption in this cycle, extinction became a frightening reality. A world without whales became an unthinkable idea to most people, and the campaign to save the whale took on immense momentum. This campaign finally achieved results with the moratorium on whaling being set in place by the International Whaling Commission (IWC) in the mid-1980s.

Project Jonah Australia was in the forefront of this battle, and was largely instrumental in forcing the Australian government of the day to set in place an inquiry into whaling. A former Chief Justice of Australia, Sir Sidney Frost, headed this inquiry which, after two years of intensive investigation, produced two large volumes of documentation supporting its conclusion that whaling ought to be banned in Australia (there was still some hunting of Humpback whales in the much-depleted oceans of Western Australia, even in 1978), this ban to be communicated internationally with the intention of halting any destruction of whale populations worldwide.

This was a landmark decision in the ecological history of the world, and set the standard for many more battles to come, if the conservation of the world's resources was to be achieved, and ecological disasters were to be avoided.

Mankind - the world's ordinary folk - is now taking part in the preservation of its own future.

There has been a highly orchestrated and expensively funded campaign by the whaling nations over the past 12 years to resume commercial whaling. The statements made are seemingly rational, but are in fact riddled with distortions of the truth. The statement that there was "very little serious policy discussion or scientific evidence involved when whaling was banned in the late 1970s" is, for example, quite incorrect. Evidence was collected and submitted at the Frost public inquiry by individuals and organizations from all over the world. The inquiry was conducted in many of Australia's capital cities, and evidence was gathered from overseas as well as from local sources.

The Frost Commission Report incorporated the scientific evidence submitted, and was published by the Australian Government in two volumes, with hundreds of pages. It is considered to be a landmark in the history of man's relationship with, and place in, the environment of our planet. In 1980 the Whale Protection Act was passed by the Australian parliament, and whaling was completely banned. Australian citizens found to be involved in any form of whaling (including the capture and/or slaughter of dolphins) are liable for prosecution, heavy fines, and confiscation of their boats.

Another statement suggests that there is "no danger at all to the species of whale that is proposed for use as food". The correct position is as follows. Over the last 300 years, and over the last century in particular, with the advent of technology (for example, helicopters, radar-fitted spotter-aircraft, factory ships, harpoon guns fitted with grenades, and so on) the history of commercial whaling has produced a tragic end-result. The horrendous slaughter of the Great Whales has put species after species on the endangered list. We believe that some, such as the Blue Whale, are on the verge of extinction. This has largely been done by American and European whalers in the eighteenth and nineteenth centuries. In the twentieth century, Japanese whalers have joined in the slaughter.

The Blue Whale is the biggest and most majestic animal to have ever existed on our planet. Up to 110 feet in length and 170 tonnes in weight, it is three times larger than the largest dinosaur that has ever lived. What a tragic indictment on the whaling nations, the moratorium having come too late, that such a situation has come to pass! The numbers of Blue Whales have fallen from an estimated 200,000 to less than 1,000. Fin Whale numbers have fallen from 500,000 or more to less than 15,000; some estimates put the numbers at 10,000 or less. Humpback Whales around the coast of Australia fell in number from 10,000 to 200 - and all this was done under the aegis of the International Whaling Commission, and under the IWC's supervision. Mercifully, the Humpbacks have slowly begun to recover, and are now helping Australia by becoming the focus of a profitable and *non-intrusive* whale watching industry. Whale watching is becoming popular in Japan, too, while in the United States it is already a thriving industry.

History shows that whaling has been ferocious, avaricious, irresponsible, and carried out with complete disregard for the ecological consequences to the oceans (whales play a key part in the phytoplankton/krill/oxygen cycle). No moral or ethical considerations have been brought into play in assessing this significant life form in our oceans from whom we still have so much to learn to our advantage. Is it any wonder that we, who are very concerned about the plight of our planet, are unwilling to even consider a lifting of the moratorium?

In fact, Project Jonah Australia has proposed that the moratorium established in 1986 should be extended for at least 50 years, like the recent Antarctic Treaty, which banned mineral and oil exploration in the Antarctic for 50 years. That was one instance in which nations previously committed to exploitation of this fragile continent realized the danger involved, and changed their stance. We trust that similar good sense can prevail with regard to whaling.

Another so-called justification of the lifting of the moratorium is that a resumption of commercial whaling would "allow the larger species adequate access to food". The facts are as follows: the food supply previously consumed by numbers of the Great Whales is *not* now being totally consumed by the rising population of the Minke whales. On the contrary, the supply is being largely utilized by the massively increasing numbers of Adelie Penguins and Crab-eater seals (now numbering in the millions). Nature has its own way of filling the vacuum created when man upsets the natural balance of millennia.

Additionally, the increase in the extent of the hole in the ozone layer is causing a reduction of phytoplankton, which is in any case going to seriously hinder any recovery of the Great Whales. Tens of thousands of dolphins and whales are accidentally caught in, and drowned by, the huge walls of death in the form of gillnets, some 10-15 kilometres long and 30 metres deep, that vacuum-clean the oceans for shoals of fish. Increasing public concern and local opposition, together with media coverage of confrontation by organizations such as Greenpeace, have helped enormously in backing public opinion to condemn this practice and enforce bans on such wasteful slaughter. This has been particularly so in the tuna fishing industry, because of the close association of schools of dolphins with shoals of tuna which shadow the dolphins.

It is correct to state that Alaskan Inuit (Eskimos) are today permitted to hunt the endangered Bowhead Whale in a trade-off involving the politics of whaling. But we are of the opinion that this is wrong. We are opposed to this quota, and believe strongly that the Bowhead Whale, a very endangered species, should be left alone.

It has been repeatedly asserted that the issue of whaling is one of "sentiment *versus* sanity". We believe that it is commercial sentiment which keeps the now limited whaling industry ticking over. It is a pity that the advocates of such views cannot attend sessions of the International Whaling Commission (IWC) as I have, in order to hear the emotional appeals made by delegations as they state and re-state the sentimental attachment to whale meat of coastal village whalers from whaling nations, or the "hunger" of school-children for whale meat.

Despite the moratorium coming into effect only 12 years ago (in reality Norway continued to whale in 1988/1989, and Japan has continued its charade of "scientific whaling") the whaling nations continue to clamour for a resumption of commercial whaling. It is remarkable that they attempt to justify their attitude, given the evidence on record that the decimation of the Great Whales has occurred due to whalers' disregard for the ecological consequences of their own greed and avarice.

Much is made of research into whales that have been killed: the justification of the "scientific whaling" programme. By its own admission, the Cetacean Research Institute has over 10,000 specimens of the reproductive organs of female whales awaiting research studies on the physiology and pathology involved. When I questioned the Institute's Director about this in 1990, and inquired why no real research had been done on these specimens, he told me that the Institute is short of both funds and staff. I asked him why, if this were the case, more whales are being killed for future research when there is so much material already available.

An additional matter for consideration is that biopsy specimens as small as 1 cm x 1 cm are providing scientists with a lot of information. Autopsy material from whales dying from both natural and unnatural causes could yield a great deal of information without any need for killing whales.

One of the most telling arguments against commercial whaling is that if Minke whales are to be killed on a quota system and their meat taken or sold, how is that meat to be distinguished from the flesh of any other Great Whale on the endangered list? Even the inspectors in the markets would not be able to distinguish one from the other. This would certainly revive pirate whaling - that other spectre of the "bad old days" when whaling outside the quota system of the IWC added enormously to the official IWC quotas of whales "legally" caught.

Another important point to be considered is that despite claims made to the contrary, *there is no known method of humanely killing a whale*. A creature as large as a whale, swimming freely in the deep oceans, cannot be humanely killed by any means short of the shattering of its body with large and powerful multiple grenades, leading to an agonizing and lingering death. This is still the method employed to kill the Great Whales in the limited quota system.

The evidence, in fact, is overwhelming against the resumption of commercial whaling, bar the greed and avarice of some of the world's nations. We strongly suggest that all whaling (including Aboriginal subsistence whaling) be stopped for at least 50 years to give these magnificent, gentle giants a chance to recover their numbers.

Modern humanity has hung its head in shame over the condoning and practice, prior to the latter half of the twentieth century, of many such cruelties to animals as well as to human beings. (In the case of the latter, we have only to look back on the history of human sacrifice, religious persecution, ethnic "cleansing", slavery, child labour, and the denials of human, political and property rights to women.) Are we to continue into the next century the practice of whaling, which is considered by many of the world's peoples to be barbaric, and conceded by most to be unnecessary to humankind's survival, since all the products of whaling can now be found in nature, or can be synthetically produced? Or are we going to grow up, developing responsible attitudes to the planet and compassionate feelings towards these helpless creatures who do us no harm, but instead add majesty, beauty and grace to our oceans, playing an important part in ocean ecology as they have done for 50 million years?

WORKING GROUP 5

INTERNATIONAL GOVERNANCE

A WORLD FREE FROM WAR

John Avery

Fear of creating a global tyranny was probably the main reason why the statesmen who drafted the United Nations Charter made the Charter too weak to be really effective. Even the disastrous experience of World War II was not enough to convince them that true government is needed at the global level. However, the Charter was signed before Hiroshima; and during the 50 years which have passed since then, modern weapons have become so destructive that a new world war would mean the collapse of civilization. We can no longer afford anarchy at the global level in an increasingly interdependent world. This is true not only because the institution of war is incompatible with modern weapons, but also because the positive functions of government are needed - for example in the fields of public health, development, ecology, population stabilization and food security.

If we compare the present United Nations with a true government, we can understand why the UN has not yet succeeded in eliminating war, although it has made many important contributions to peace. The present United Nations lacks the power to impose taxes; and thus it is perpetually in a state of financial crisis. The United States has added to this crisis by refusing to pay its dues. The motive behind this refusal seems to be a wish to control the UN by selectively giving financial support to those interventions of which the US approves. In order to remedy this situation, it has been proposed that the UN should be given the power to impose a very small tax on international currency transactions. The amount of money involved in these transactions is so immense that even a tiny fraction of a per cent in tax would be sufficient to solve the UN's financial problems. With more money available, the World Health Organization could work more effectively to combat infectious disease and to stabilize the global population; and the UN organizations promoting development and environmental protection could function more effectively. These positive functions of government would give the UN valuable prestige which would help it to solve political problems.

To work more effectively for peace, the United Nations also needs a legislature with the power to make laws which are binding on individuals, and courts with the power to arrest individuals who violate these laws even if they are heads of states. At present the General Assembly of the UN acts to enforce its resolutions by imposing sanctions on entire nations, although the burden of these sanctions often falls on the weakest and least guilty of the citizens, and although often the only effect of the sanctions is to unite a nation behind its guilty leaders. The Permanent World Court which has recently been established at the Hague is an important step in the right direction. One can hope that its jurisdiction will gradually be extended to cover crimes other than genocide and war crimes.

The governments of nations and federations are able to maintain internal order because each of them has a monopoly on heavy weapons within its own boundaries. Whenever large independent private armies exist within the borders of a nation or federation, the central government is unable to maintain peace, as has been demonstrated very frequently in recent history, for example in Lebanon and in Northern Ireland. The present United Nations is certainly extremely far from having a global monopoly on heavy weapons; and under these circumstances it is not surprising that the UN, in spite of its best efforts, has been unable to achieve a world free from war.

Instead of reforming and strengthening the United Nations, the rich western industrial nations have placed their reliance on NATO. This organization, which has recently celebrated its 50th anniversary, has an extremely high budget and a correspondingly large arsenal of modern weapons. The problem with NATO is that it does not represent all of the people of the world, but only part of them. The critics of NATO see it as being heavily dominated by the influence of the United States.

Thus, even if NATO were able to impose peace within its region of operation, those states which are not members of NATO would not see the peace as being just or democratic; and if NATO acted in Africa or the Middle East or Asia, its impartiality would be even less credible.

Thus, if we wish to build a stable and peaceful global society, strengthening and expanding NATO appears to be a false step because the organization lacks impartiality. By contrast, the Secretary General of the United Nations enjoys a high reputation for impartiality; and the General Assembly speaks for the whole of humanity. The task of reforming the United Nations and of building it into a just and humane global government will not be an easy one. We can find guidance and courage for this great task in the degree of peace and good government which has already been achieved locally in many parts of the world.

For example, Scandinavia today is characterized by honest democratic government, the absence of poverty, respect for human rights, a high educational level, equality between men and women, low birth and death rates, and a stable population combined with a high level of public health. These characteristics of today's Scandinavian society are linked with each other and with the absence of war in the region in a circle of mutual cause and effect. In addition, the close ties of trade, communication and cooperation between the Scandinavian countries have contributed to the disappearance of war in the region.

Not only Scandinavia, but also other large regions of the world where war has been eliminated locally can serve as models, as we work to achieve a war-free world. We can gain understanding from the fact that war has been eliminated locally within most nation-states and within a number of large federations. As examples of federations, we can think of Canada, the United States, the former Soviet Union, Brazil, Switzerland, the Federal Republic of Germany, and Australia. Some of these federations are so extensive in area, and include so many ethnic and linguistic groups, that it seems probable that the methods by which they have achieved internal peace can be globalized. China and India are examples of nations which are so large that together they make up a third of the population of the world, and they include people of many cultures and languages. The fact that these two enormous nations have achieved internal order gives us hope that the institution of war can be eliminated on a global scale.

We can also gain courage by thinking of the history of slavery. The institution of slavery was once so widespread that it was considered to be an inevitable part of human nature; but today slavery has been abolished in almost every part of the world. The example of the dedicated men and women who worked to abolish slavery can give us courage to approach the even greater and more urgent task which history has given to our generation - the abolition of war.

THE PRESENT PLANS TO RESTRUCTURE THE SECURITY COUNCIL OF THE UNITED NATIONS

Bengt Broms

Following the example of the Covenant of the League of Nations, the 1945 Conference of San Francisco adopted a similar stance when it was decided to include provisions for amendments to the Charter of the United Nations. It was also widely believed that the permanent members of the Security Council would not want to prevent amendments to the Charter which would be regarded as reasonable. There was, however, no indication of the future worsening of the relations between the Great Powers, a fact which was likely to complicate efforts to strengthen the activity of the United Nations in light of the provisions of Chapter XVII on amendments to the Charter.[1] In accordance with Article 108 amendments shall come into force when they have been adopted by a vote of two-thirds of the members of the General Assembly and ratified in accordance with their respective constitutional processes by two-thirds of all members, including all the permanent members of the Security Council. As a result of this wording, any permanent member may prevent the entering-into-force of any amendment to the Charter. Article 109 includes a provision to convene a General Conference of all members to review the Charter by a two-thirds vote of the members of the General Assembly and by a vote of any nine members of the Security Council. Any amendments recommended by a two-thirds vote of the General Conference shall take effect when two-thirds of members of the United Nations have ratified them including all the permanent members of the Security Council. Even in this case the permanent members have the possibility to prevent amendments adopted by the General Conference from entering-into-force. Therefore, it is easy to understand that the General Conference has never been held.

Nevertheless, the Charter has been amended when the General Assembly on 17 December 1963 adopted amendments to Articles 23, 27 and 61 which came into force on 31 August 1965. A further amendment to Article 109, which was technical and caused by an oversight in 1963, was adopted by the General Assembly on 20 December 1965 and it came into force on 12 June 1968. The amendment to Article 23 enlarges the membership of the Security Council from eleven to fifteen members. The amended Article 27 provides that decisions of the Security Council on procedural matters shall be made by an affirmative vote of nine members instead of formerly seven members and on all other matters by an affirmative vote of nine members instead of formerly seven members, including the concurring votes of the five permanent members of the Security Council.

The idea of a general review of the Charter was not entirely forgotten, however, and in 1969 Colombia asked that an item on the general revision of the Charter be added to the agenda of the General Assembly. This item was called “Need to consider suggestions regarding the review of the Charter of the United Nations”. When the item was taken up by the General Assembly it became quite clear that the viewpoints on the issue were divided. As a result of this, the General Assembly decided to ask the governments of member States to present their views to the Secretary-General of the United Nations.

It was not until 1974 that this item was debated at great length. During the debate the divisions of opinion became clear. On the whole, the non-aligned countries, including the majority of the Latin-American States, were strongly in favour of establishing a committee to study the issue. This opinion was shared by some members of the Western Group, notably Italy and Japan. China was of the same opinion. The other permanent members of the Security Council were, however, strongly opposed to the establishment of a committee. They argued that the setting up of a committee was premature, and such a decision could at least be held over till the next session of the General Assembly, if a decision were to be reached at all.

At the end of a long debate the General Assembly, however, adopted on 17 December 1974 Resolution 3349 (XXIX) including a decision to establish an *Ad Hoc* Committee on the Charter of the United Nations, consisting of 42 States to be appointed by the President of the General Assembly with due regard to the principle of equitable geographical distribution. The aims of the Committee were defined as follows:

(a) to discuss in detail the observations received from Governments;

(b) to consider any additional, specific proposals that Governments may make with a view to enhancing the ability of the United Nations to achieve its purposes;

(c) to consider also other suggestions for the more effective functioning of the United Nations that may not require amendments to the Charter;

(d) to enumerate the proposals which have aroused particular interest in the *Ad Hoc* Committee.

When the *Ad Hoc* Committee met from 28 July to 22 August 1975, the basic attitudes that had been expressed by the member States during the debates in the Sixth Committee were quite closely reflected. Whereas some members regarded revision as liable to lead the organization into a state of chaos, there were other members who concluded the exact opposite. They said that it was in the interest of the United Nations to develop the Charter to meet with the requirements of the present time. What, then, were the arguments used by the opposing sides?

Taking first the arguments in favour of amending the Charter, the most popular argument was that thirty years had passed since the Conference of San Francisco and that the world had undergone significant changes since then. It was also pointed out that the new members admitted to the United Nations had not had any share in the drafting of the Charter.

There were, however, members who drew from these facts the contrary conclusion, saying that the Charter had proved to be sufficiently elastic and it had been on the basis of the present Charter that so many new States had been born.

Another important line of argumentation of the proponents of a Charter review underlined that the United Nations was nearing a point of stagnation and it was therefore necessary to resolutely amend the Charter. Foremost of the examples cited was the composition of the Security Council together with the principle of veto which were regarded as the prime causes for the inefficiency of the Security Council. Indeed, proposals related to the composition and voting proved to be two of the most delicate problems. France, the Soviet Union, the United Kingdom and the United States supported by various other members were very sensitive about the provisions of the Charter on the Security Council. In their opinion neither the composition of this organ nor the right of veto constituted the primary causes of the inability of the United Nations to function well. These States concluded that attempts to revise the Charter "would merely result in 'confrontation' between large countries without giving the small ones any more special privileges than they already have".[2]

The realization of the factual situation led the proponents of a revision of the Charter to underline that they did not want to change the position of the existing permanent members of the Security Council. They were rather proposing that some additional permanent seats ought to be created in order to obtain a better balance in the composition of the Security Council. Accordingly, there was a proposal to create two new permanent seats to be distributed between the countries representing the Afro-Asian and Latin-American groups of States.

This proposal was coupled with the idea that the right of veto be established on a regional basis. Accordingly, Nigeria proposed that one veto should remain to be exercised by Western Europe, Eastern Europe and Asia, respectively, and the privilege of veto be extended to the regional groups of Africa and Latin America. Apparently the intention of Nigeria was that France and the United Kingdom were to be replaced by two new permanent members. A variation of this proposal was put forward by the delegate of Sierra Leone, who suggested the same redistribution of seats as Nigeria but with an important addition. He said that, should it not be possible to deprive the United Kingdom and France of their permanent seats, then the possibility of adding two more permanent seats should be considered. The African and Latin-American States were to determine who should have the veto power at any one time. These two geographical groups of States were expected to apply some kind of rotation within their respective groups. The holder of the rotating permanent seat was, of course, to be entitled to the right of veto.

Some proponents of the Charter review were more moderate and proposed another alternative. They suggested that, while the composition of the Security Council could remain the same as now and while the permanent members would continue to have the right of veto, the use of this right could be limited either by means of an amendment to the Charter or, should this not be acceptable, by means of a gentlemen's agreement. The permanent members were to promise not to make use of the right of veto in certain matters like, for instance, the admission of new members to the organization.

Among the permanent members of the Security Council, France, the Soviet Union, the United Kingdom and the United States were generally unwilling to consider any proposals which would have meant amendments to the Charter. They stressed that the right of veto had been always a prerequisite for their participation in the organization adding that the right of veto had many times saved the United Nations from making grave political mistakes. Therefore, the right of veto was not to be abolished. Certain other members adopted a similar position.

Although there were several other proposals to amend the Charter, those concerning the Security Council were the most important in determining the general attitude of many members of the *Ad Hoc* Committee. This trend has continued in the debates of the succeeding Committee established by the General Assembly. By Resolution 3499 (XXX) of 15 December 1975 a new Special Committee on the Charter of the United Nations and on the Strengthening of the Role of the Organization was created. This Special Committee with a membership of 47 States had a double mandate to make proposals, to amend the Charter and, without amending the Charter, to make recommendations to strengthen the United Nations.

At the first session the debate showed that, insofar as the Security Council was concerned, the previous disagreements had not disappeared. It should be noted also that several delegates concentrated rather on proposals to strengthen the organization than on those related to the Security Council. For example, Romania was of the opinion that the membership of the Security Council should be enlarged. But Romania did not propose any new permanent seats. Furthermore, Romania considered that it was essential to retain the right of veto, although undue use should not be made of the right of veto simply in order to further the interests of a few countries at the expense of the general interests of peace.

Among the permanent members of the Security Council the delegate of China was the only one to stress the need to review the Charter. In the opinion of China, this was necessary with a view to changing the unjustifiable situation of superpower manipulation of the United Nations and implementing the principle of equality of all States.[3] Nevertheless, on behalf of the United States a warning was given against a review of the Charter. Such a review was seen as unproductive and as one which might maximize areas of disagreement and harden positions, thus hindering the flexible development of the United Nations.[4] In a similar fashion, the delegates of France and the Soviet Union preferred to stress the importance of the strengthening of the role of the organization at the expense of the review procedure.

Although there have always been many members of the Special Committee underlining the importance of amending those provisions of the Charter which govern the Security Council, there has prevailed an understanding that in order to avoid an open clash it is better to wait and hope that a consensus would one day exist.

The Present Situation

It was not until 3 December 1993 that the General Assembly adopted Resolution 48/26 to establish a Working Group which would be open-ended to discuss all problems related to the Security Council. Although the President of the General Assembly will act as the Chairman, the Working Group has selected two Vice-Chairmen to direct the work.

From the beginning of the activity of the Working Group it has been evident that the basic differences of opinion which have been described above still persist. While a majority of the delegations participating in the negotiations supports the idea of increasing the number of members of the Security

Council and certain changes of procedure, there is nowadays a group of delegations, led by Italy and Pakistan, which has decided to oppose all such reform proposals. The permanent members of the Security Council have, however, by and by adopted a stance which favours limited changes. Under these circumstances it has not been possible to reach consensus. Nevertheless, today the various views are easy to identify.

To take up first the problem of increasing the number of members of the Security Council, a specific proposal by the President of the Fifty-first session of the General Assembly, Razali Ismail, may be regarded as a detailed basic proposal. According to this proposal the membership of the Security Council should be raised to twenty-four. There should be three new permanent members representing Africa, Asia and Latin-America while two seats would be allocated to industrialized States. These two States were meant to be Germany and Japan, although the original proposal did not include their names. The remaining four new non-permanent seats would be reserved for the other four geographical groups of States leaving out the Western Group. According to the proposal the new permanent members would not possess the right of veto. Furthermore, ten years after the adoption of the proposal a conference would convene to consider the impact of these amendments. At a later stage the proposal was amended so as to give a collective veto to two new permanent members of the Security Council. A principle of rotation was also to be applied to the new permanent members. It turned out, however, that the group did not reach a consensus on its proposals. The task of the Working Group has continued until now and it is again possible to see the main differences of opinion. They concentrate still on problems related to the composition and voting procedures of the Security Council. In fact, one may conclude that the differences are nowadays smaller than before.

To take first the present permanent members, it appears that they still object to any curtailment of their own rights by means of amendments to the Charter. Nevertheless, for the time being, the composition of the Security Council could be enlarged to something between twenty and twenty-four members. While Russia and the United States have so far defended the lower figure, France and the United Kingdom could accept the higher figure. China is especially interested in enlarging the representation of developing States in the Security Council.

The Organization of African Unity (OAU) has continually insisted on an amendment allowing Africa to have two permanent seats in the Security Council. These seats would then rotate according to a specific system dividing Africa into five zones.

The Non-Aligned Movement (NAM) has proposed different solutions to enlarging the composition of the Security Council. In any case, the NAM would like to increase the total membership by eleven new members to twenty-six. Unless there is a consensus among the members of the General Assembly as to the number of new permanent seats, all new seats should be non-permanent. This view has been shared also by Canada, which would not want to have any new permanent members in the Security Council.

The Nordic States have supported an amendment to the effect that there should be five new permanent members in addition to three non-permanent members. Thus, the total number of members would be twenty-three. The proposal is somewhat flexible insofar as the total number of seats is concerned.

The Latin American States have different views, as some of them, like Mexico, do not accept the idea of increasing the number of permanent members. Some other Latin American States, including Argentina, have also been mainly interested in increasing the number of non-permanent seats.

To turn to the voting methods of the Security Council there are two basic proposals. The present permanent members would prefer the old system of using the right of veto without any limitations. There is, however, a large group of States representing the opinion that, as long as there is no final amendment of the Charter limiting the use of the right of veto, the present permanent members should agree not to use this right except in cases of vital importance. By now there is still no general agreement between the permanent members of the Security Council and the other members of the organization as to the cases to which the voluntary renunciation of the right of veto would apply.

Conclusions

While I have above discussed the work of the *Ad Hoc* Committee and the Special Committee solely in view of the amendments proposed to renew the composition and procedure of the Security Council, it should be realized that the ongoing work has resulted in many important changes in the activity of the United Nations which have not required formal amendments to the Charter. This is also proof of the elasticity of the Charter insofar as there are several provisions which are simply no longer applied and their formal continued existence has not done any harm to the activities of the organization.

What has made the proposed amendments to the Charter relating to the Security Council most important is the fact that once the present differences of opinion have been solved there will open up a possibility to review many other provisions of the Charter to strengthen the activity of the organization. Only a few weeks ago, Secretary-General Kofi Annan announced that in his opinion the time is ripe to solve the problems of amending the Charter to improve the activity of the Security Council.

Looking at the most recent stands of different members of the United Nations one finds satisfaction in the fact that the existing proposals for Charter review are not as radical as many delegations feared, at least when the matter was first taken up by the General Assembly. Similarly, the proponents of Charter amendments have by now noticed that even those permanent members of the Security Council that have expressed their fear about amending the Charter have become interested in the development of the United Nations in order to cope with increasing criticism. A common denominator has been found in the interest of the United Nations itself to fulfil its tasks as efficiently as possible.

Thinking of the present possible consensus it is here submitted that it should be possible for the different groups to agree on an increased membership of the Security Council to find a generally acceptable compromise somewhere between twenty and twenty-four members. This concerns also the future division between permanent and other members. There seems to be a wide understanding that at least Germany and Japan should become new permanent members representing the industrialized States while the African, Asian and Latin-American geographical Groups should become represented among permanent members and with an additional member among the non-permanent members. Also there are valid arguments to apply a system of rotation among those new permanent members which represent specific regional groups. In any case, all rotating members would have during their terms the same rights as the other permanent members.

I would like to add a word of caution against the proposal to limit the right of veto by means of an amendment to the Charter. Such a change and even insistence on it may jeopardize the expectations of a consensus. Instead, there may exist a possibility to find a compromise in agreeing in principle on which items the permanent members should, as a rule, abstain from resorting to their right of veto. Another important possibility to improve the activity of the Security Council would be to insert into the Charter a formal amendment to the effect than an abstention by a permanent member from voting in the Security Council is not equal to the use of the right of veto. To adopt this formal amendment would only confirm an existing practice which has, however, also brought some criticism.

Should the hope expressed by the Secretary-General be fulfilled, a most difficult series of political disputes within the membership of the United Nations would have come to an end. With the new challenges the organization is faced with, for instance, in its most important task of maintaining international peace and security, a solid performance would become considerably easier to attain.

References

[1] See Bengt Borms, The United Nations, Helsinki 1990, pp. 848-869.

[2] See Doc. A/10033, 94 and 100-101.

[3] Doc. A/AC. 182/SR.12, 27. On the reply by the delegate of the Soviet Union, see *Ibid.*, 14.

[4] UN Doc. A/AC. 182/SR. 8, 6.

REFORMING THE UNITED NATIONS

Joseph Rotblat

What Churchill said about democracy ("the worst form of government except for all ...other forms") applies equally, at the global level, to the United Nations. The UN suffers from many shortcomings, and its performance has often been dismal, but it is the only institution we have with the mandate to maintain peace in the world, and nothing should be allowed to undermine its authority and make it even more inadequate.

Saying this does not, however, mean that we should not expose its weaknesses and criticize its performance, as long as this is done in the positive spirit of seeking ways to enable the UN to function properly. In this note, I start with a criticism of several aspects of the UN organization, and then go on to outline some measures that seem to me to be necessary to make the UN a more effective institution.

Weaknesses of the UN Charter

I believe that eventually some form of World Government will have to be established, if the scourge of war is to be eliminated completely. The United Nations is the closest approximation to a world governance system that can be realistically expected at the present time. But for it to fulfil this role will require amendments to its Charter.

The chief reasons for reforming the UN derive from the fact that the world today is radically different from the world in 1945, when the Charter was adopted. Examples are: the ideological issues that divided the world at that time have all but gone; the structure of the membership of the Security Council no longer reflects (if it ever did) the political and economic forces in the world; the number of member states has hugely increased and there is continuous pressure to increase the membership still further; and there is a strong and growing interdependence between nations (to a large extent as a consequence of scientific and technological innovations).

Sovereignty

A criticism of the UN Charter of a fundamental nature – evident even in 1945 – is the basing of the organization on the principle of the sovereignty of nations. Thus, Article 2.1 of the Charter states: "The organization is based on the principle of the sovereign equality of all its Members." And Article 2.7 goes on: "Nothing contained in the present Charter shall authorize the United Nations to intervene in matters which are essentially within the domestic jurisdiction of any state...."

The principle of sovereignty of nations seems to me to be almost in contradiction to the main objective of the UN: preventing war. Sovereignty is the foundation of the nation-state whose main function is to ensure the security of its citizens against threats from other states: this is taken to mean possessing the ability to wage war. Anatol Rapoport (in a chapter of the Pugwash Monograph *World Citizenship: Allegiance to Humanity*) put it as follows:

> It is regarded as axiomatic that the autonomy of a state is secured by its sovereignty, and sovereignty by its war potential. Thus, a war establishment is universally held to be an indispensable institution in

> practically every state . . . the survival prospects of the institution of war, on the one hand, and of the nation-state, as we know it, on the other, are closely related.

I believe that so long as the institution of war is recognized as an integral part of the nation-state, occurrence of war is almost unavoidable. The adherence to the Roman dictum "If you want peace, prepare for war" is largely responsible for the fact that war has been endemic throughout history.

In any case, the concept of the absolute sovereignty of the nation-state is completely anachronistic in view of the ever-growing interdependence between all nations, and the acceptance by the great majority of nations of international treaties, which limit national rights on many important issues.

Intervention

The prohibition of any intervention in matters of domestic jurisdiction is intolerable in a caring world community. It makes it practically impossible for the UN to take action to prevent suppression of minorities and the abuse of the rights of the individual. This iniquitous system leads to absurd situations, such as that not long ago when the UN went to war to restore the sovereignty of a feudal state, but did not lift a finger when the same aggressor horribly murdered many of its own population, belonging to a minority ethnic group.

The recent celebration of the 50th anniversary of the Universal Declaration of Human Rights has made us all conscious of the importance of this issue. *Surely, human rights must take precedence over the rights of nation-states.*

Self-defence

Another dangerous concept in this nuclear age is that of self-defence. Article 51 of the UN Charter states: "Nothing in the present Charter shall impair the inherent right of individual or collective self-defence if an armed attack occurs against a Member of the United Nations." (It goes on to stipulate "until the Security Council has taken measures necessary to maintain international peace and Security," but with the veto principle in operation this becomes meaningless in practice.) We all know how easy it is to create a pretext of self-defence to justify waging an aggressive war. Just how dangerous the application of the self-defence principle can be became evident in the recent utterance of the International Court of Justice on the legality of the use of nuclear weapons. About half of the members of the Court were unable to decide whether the use of nuclear weapons would be unlawful in an extreme case of self-defence, thereby placing the interest of the nation-state above that of mankind as a whole.

Veto in the Security Council

The principle of the veto by the permanent members of the Security Council makes the Council practically impotent when it comes to dealing promptly with disputes between nations. This has been demonstrated time and time again in the past, most recently over the Kosovo crisis.

Structure of Peace-keeping Operations

The present system of initiating and executing peace-keeping and peace-enforcement operations renders the UN quite ineffective and consequently encourages the retention of military alliances, such as NATO, which arrogate to themselves some of the vital functions of the UN.

Desirable Ways of Reforming the United Nations

Sovereignty and Related Issues

Ultimately, the most important element of state sovereignty – the right and capability to wage war – will have to be eliminated. The only coercive power in the world should be vested in the United Nations in the form of an international police force. *This means no national military forces.* The achievement of this objective will, however, take a long time. In the meantime some changes should be made in the UN Charter to further this objective.

(a) Delete the word "sovereign" in Article 2.1

(b) Repeal Article 2.7 and replace it with something like: "Notwithstanding the right of member states to regulate their domestic affairs, the Security Council shall be empowered to take the necessary action to prevent suppression of minority groups and the abuse of human rights.

(c) Repeal Article 51 and put in its place a text which should make it difficult to use the guise of self-defence in order to perform an act of aggression. In any case, it should *prohibit the first use of weapons of mass destruction in any action of self-defence*.

Security Council

The revision of the composition of the Security Council has been a subject of discussion for many years, and some changes have been made since 1945. But the situation is still unsatisfactory. In my opinion, the main necessary change is in the composition of the membership with the permanent seats. As already observed, the present distribution (three out of the five members are European states; all five are nuclear powers) does not reflect the balance of political and economic power in the world, nor is it equitable.

If the number of permanent seats were to be increased to seven, a possible equitable distribution might be: two from Europe (East and West); two from America (North and South); two from Asia (North and South) and one from Africa.

Right of Veto

It is intolerable that a single nation – no matter how powerful – should be able to frustrate the will of all other nations. I feel strongly that the veto principle should be abolished and replaced by that of consensus, in the meaning of a large majority. In the case of the Security Council, a proposed resolution should not be rejected unless it is opposed by at least two of the permanent members.

An alternative way of dealing with the veto problem would be a rule that if a proposed resolution is consistently defeated by one member (in, say, three sessions), and thus the Security Council fails to exercise its primary responsibility, the issue be passed to the General Assembly where it could be enacted by a two-thirds majority.

Consideration should also be given to the involvement of the International Court of Justice (but acting at a sensible speed), by granting it the right to make binding decisions rather than give advisory opinions.

UN Peace-keeping Forces

With the ultimate aim of phasing out national defence forces and replacing them by an international police force, a start should be made by setting up a permanent UN military force. Designated as the UN Police, it should consist of personnel recruited from different countries, but all being paid by the UN, wearing UN uniforms, and serving under the direct command of UN officers.

At the start, this force will be relatively small and its task would be to maintain the various ongoing UN peace-keeping units. Situations requiring larger forces will have to be handled by the current method of using contingents from national armies. But the principle should be established immediately that *no military intervention be permitted by any nation or a military alliance of nations, except under a specific resolution of the Security Council.*

A Secure Financial Basis for the UN

The setting up of a permanent UN Police force would entail an additional huge expenditure. This could be secured by increasing the dues paid by member states, but past experience with collecting the dues (often subject to the whims of domestic political wrangling) gives little hope of achieving this.

In any case, if the UN is to be able to fulfil the functions for which it was set up, it should have a secure source of income, sufficient to maintain not only the military force but also the whole of its organizational apparatus.

The simplest and most equitable way is to make everybody pay for the cost of keeping the peace. This could be achieved by a small addition to the income tax or the VAT (value added tax) imposed in many countries on various transactions. However, any increase in taxation is unpopular and will be strongly resisted.

A better solution that has been suggested is a very small (in relative terms) tax on international transactions of money (the Tobin Tax). It has the additional advantage of being collected automatically.

Another way of securing an independent source of income (which would also have an important symbolic value) would be by prospecting nature's riches that have not yet been appropriated by individual nations. A possible field of exploration is the mineral and biological resources of the deep oceans, which should be declared the property of the peoples of the world, and administered by the United Nations. The necessary initial outlay of capital could be met by a one-off levy on member states; they would be asked to accept it on the understanding that they would not have to pay any further dues to the U.N.

INTERNATIONAL GOVERNANCE

Goran von Bonsdorff

Great decisive questions are of a global nature. They can be solved only by united efforts of the whole of mankind. The only institution capable of representing the whole of mankind is the United Nations. The United Nations can make the appropriate decisions. Therefore, strengthening the authority of the UN is very important in the present situation.

Efforts for this purpose have been made already for many years. They have produced certain partial results, such as, for example, the enlargement of the Security Council. Many other endeavours, such as changing the use of the veto in the Security Council, or the establishment of a permanent, multinational UN standby force have, however, not resulted in lasting reforms.

An important reason for this failure may be that the UN is an organization of sovereign states. The members stand up for their national interests, which frequently conflict with each other and therefore make decisions more difficult. An effective method for reducing those conflicts could be the formulation of common interests stronger than the special national interests.

Historical Development Processes

By studying historical development processes we can identify and characterize common central interests. We start from the fact that the existence of people on the Earth depends on climate and natural resources. We can observe changes in the climate and how they influence the life of all human beings. Even the natural resources change, and some of them are diminishing with serious consequences for the future. Animals and plants are also diminishing. These phenomena are global; they cover the whole earth.

Methods for utilizing natural resources are called technology. From a historical point of view, the distance between different groups of people has given rise to different technologies and with them gradually different civilizations. With the emergence of industrialization about 300 years ago, prerequisites were created for a global technology. This has been rapidly developing in our time and is assuming new forms.

The global technology has again created prerequisites for a global economic system. We can say that this system began developing seriously about 150 years ago. The economy, like the technology, continuously finds new forms.

In connection with technology and economy, a global political system has also been developing adapted to the situation of humanity. This system began to develop about 100 years ago. Like technology and economy it is advancing both on the global and the regional level.

We can thus speak of a progressive process of change in the direction of a comprehensive global system which can be analysed from the following points of view: a) general prerequisites such as climate and natural resources; b) technological stage and development; c) economic and social systems; and d) political forms of organization. All these components of the system are dependent on each other.

Maintenance of Common Values

We must realize that a process of integration is going on and that people are little by little integrating into the world society. During the ongoing process new values are created, whereas old values can get a new meaning, weaken, or perhaps wholly disappear. Special measures are necessary in order to protect important values running the risk of disappearing.

One of these values is democracy, the possibility for the individual citizen to influence political decisions. The classic expression for representative democracy is a popularly elected assembly which can make binding decisions. From a global point of view we should need a world parliament. It should represent the peoples as a complement to the UN which represents the sovereign states, in other words, the governments.

The idea of a world parliament has been thoroughly discussed and many proposals have been elaborated. Personally, I think that the parliament could consist of approximately 12,000 members, or one for every 500,000 inhabitants on the earth. The role of the parliament would be to deal with the great questions concerning the future existence of mankind and in this respect represent the voice of the ordinary citizens.

Is this something quite unrealistic? The answer may be that we already have an assembly representing international democracy on a regional level. That is the parliament of the European Union. Its existence implies a step towards more democracy in a part of the earth which is comparatively highly developed with respect to technology, economy and political structure. Perhaps Europe could be an example of the realization of international democracy for other regions in the world.

In a situation where regional parliaments existed not only in Europe but also in Africa, America and Asia, the realization of a world parliament would be much easier than now.

The EU Parliament and the UN

The EU Parliament is no direct substitute for a popularly elected authority in the framework of the UN, but it should be a component in the UN system and cooperate closely with the UN. We should assume that regional institutions also in the future would be an integral part of the global political structure.

We could consequently regard the EU parliament as an important step in the direction of a world parliament in the framework of the UN system. The citizens participating in the elections of the parliament could be regarded as an ideological vanguard of mankind in the efforts to create a world order meeting the needs of our historical age. The most actively engaged among those persons could continue working to build a new world political system.

With respect to other regional parliaments I think that the best chances for establishing them should be in Africa and America. In Asia it could be more difficult because of the existence of enormous states such as China and India. Also Russia is in this respect a problem which we should discuss with the political scientists in that country.

Finally, I want to emphasize that the comparatively low participation in the EU elections shows that the integration is proceeding slowly, but this is not especially surprising. It is nothing catastrophic or even disquieting. It is more of a warning signal for universities, high schools, citizens institutes and societies that they should give more attention to the EU and its role. In this way an understanding about the global community would become increasingly incorporated into the consciousness of the citizenry.

My message is, consequently, that the EU should not isolate itself or try to create a new European nationalism, but much more than up to now serve as a positive factor of integration in the global world community.

THE RESORT TO MILITARY FORCE, INTERNATIONAL LAW, AND THE JUST WAR TRADITION

Micheline Egge Grung

Two Normative Standards

By normative standard I mean directives for how states should conduct their relations with other states. There is no doubt that such directives exist, but it is not always certain whether the directives actually *legally* bind states. Nor is it certain whether they *effectively* bind them, because there is a lack of sanctioning power in the international system.

Another question is whether the directives should *bind* states. The idea that each state is bound by international norms is inconsistent with sovereignty understood as *raison d'état,* an approach which says that states should primarily pursue their national interest, more or less regardless of the effects this has on other states. On this account of international relations, values and conceptions of right and wrong have no bearing.

I believe, however, that it is possible and necessary to talk about right and wrong actions in the international system. A decision to use military force is perhaps the most important decision that governments may take. Therefore, it is important to reflect on the norms which ought to guide political decision-makers.

International law is a result of a harmonizing process leading to formal consensus between states. The result is an explicit agreement (international convention). However, international law is also reflected in the longstanding practice of states, so-called customary international law. Contemporary international law is a detailed and comprehensive system of rules regulating state behaviour. Norms regarding the use of military force are only one part of this complex system.

Now a few words on the just war tradition. The origin of this tradition is generally traced back to St. Augustine (354–430 AD) in the early Middle Ages. The question of whether it is morally right to go to war has been a perennial concern for thinkers in theology, philosophy, history, law and political science. The result of the centuries-long reflection is not a single, elaborated theory, but a collection of different normative theories regarding the use of military force.

The just war tradition consists of two pillars: *ius ad bellum* and *ius in bello.*

- *Ius ad bellum* are principles for the decision to go to war; that is, circumstances that legitimize the resort to military force.
- *Ius in bello* are principles for just warfare, a moral evaluation of the means employed in fighting a war.

Both pillars are reflected in contemporary international law: the *ius ad bellum* is the chief concern of the UN Charter, while the *ius in bello* is the main concern of the Hague and Geneva Conventions stipulating the protection of civilians, for example.

According to the just war tradition, several criteria must be fulfilled for a war to be just (see below).

The sharp distinction I have just drawn is obviously somewhat forced. Contemporary international law about the resort to and the use of military force has in fact the just war tradition as its basis. With the emergence of the state system in the 17th century, the just war tradition was "translated" into that period's international law, among others by Hugo Grotius.

International Law and the Resort to Military Force

The UN Charter of 1945 consists of several rules that specifically deal with the right and restriction of the use of military force. These rules pertain to the UN's member states, nearly 200 countries. However, it is generally held that the rules are also binding on non-UN members because they have been accepted as a general rule of international law (*jus cogens*).

Article 2 (4) states the Charter's main rule concerning the resort to force: no member state is permitted to use or threaten to use force against other states. There are, however, two exceptions to the main rule.

- According to Article 51, all members states have the right to protect themselves individually or collectively in case of an act of aggression directed from another state.
- According to Article 42, the Security Council may undertake coercive measures when international peace and security is threatened. The procedural rules in the Security Council say that all of the five permanent members, in addition to ten of the fifteen rotating members, must agree.

Since Article 42 itself does not define which situations "qualify" as a threat to international peace and security, it has been up to the Security Council to interpret the rule. The Council's practice seems to show that human rights have become an accepted and legitimate reason for interventions (for example, Northern Iraq in 1991, Haiti in 1991/92, and Somalia in 1994), although the Council has always referred to "international peace and security" in its mandates. There is no longer any doubt that human rights are a matter of international concern. On the other side, for action actually to take place on this basis it seems as if great powers generally have to have some strategic interest in the outcome. There have been many "forgotten" humanitarian catastrophes.

However, according to the UN Charter, the Security Council must approve intervention. In spite of the lack of a mandate from the Security Council, NATO chose to intervene in Kosovo, Serbia and Montenegro in 1999. Leading members of NATO appealed to moral (not legal) principles as the chief motivation for NATO's forcible intervention. According to the alliance, it would have been morally wrong not to intervene, although the Security Council had not given it a mandate.

In the case of Kosovo, there were good reasons for Russia and China to be sceptical about intervening. It remains the case, however, that circumstances could arise when a lack of consensus arises from weak or invalid reasons. Imagine, for example, a veto resulting from an unstable internal political situation in Russia, and not from "a reasoned scepticism". In such situations, should the international community remain inactive in the face of a humanitarian catastrophe?

One could argue that in such situations the letter of the UN Charter must be set aside. This is what NATO did. There are other ways of determining whether intervention is legitimate or not. One approach is the just war tradition.

The Just War Tradition and the Resort to Military Force

According to the just war tradition, several criteria must be fulfilled for a decision to go to war to be just:

1. ***Just cause***. According to Augustine's and Aquinas' frameworks there are three just causes for going to war: to restore order, to punish evil, and to take back something that has been illegally taken.
2. ***Right intention***. A just cause is not enough. The decision-maker must also have the right intention for waging war; their wish or motive must be that the war is fought for good purposes or causes. For example, a war must not be fought for revenge.
3. ***Legitimate authority***. The decision-maker must have competence to take the decision to go to war. Private individuals do not have the competence. One would probably regard the UN and sovereign states as legitimate authorities. As for sub-national groups the case is unclear.
4. ***Proportionality***. The war's costs in terms of victims and material destruction must not exceed its benefits.

5. ***Last resort***. One must have undertaken all other possible means before actually carrying out acts of war.

The just war tradition and the rules in international law about the use of force set two different standards. Thus, an "illegal" act according to international law may be legitimate according to the just war tradition, and vice versa.

I believe there are important reasons why the just war tradition could supplement international law.

If a state or a number of states set aside the requirements in the UN Charter, the just war tradition may serve as a moral perspective for the decision-makers to take their bearings from.

The just war tradition may also illuminate important aspects that might be neglected in international law, such as restoring *justice*. The UN Charter's main stated aim is to maintain international *order*.

However, I tend to believe that international law is presently a more effective normative standard than the just war tradition. Since states are the creators of international law, it is easier to hold them responsible for unlawful behaviour. In addition, the rules in international law are more explicit. Using the criteria in the just war tradition in decision-making requires considerably more deliberation.

UNITED NATIONS: MANAGER OF INTERNATIONAL PEACE AND SECURITY OR A SILENT SPECTATOR?

Jasjit Singh

The primary purpose of the organization of the United Nations, as laid down in Article I of its Charter, is "to maintain international peace and security", and to that end, "to take effective collective measures for the prevention and removal of threats to peace, and the suppression of acts of aggression or other breaches of peace, and to bring about by peaceful means, and in conformity with the principles of justice and international law, adjustment or settlement of international disputes or situations which might lead to a breach of peace." The actual record of the UN, as might have been expected from a less than perfect organization, and an even less equitable international order in which the erstwhile great powers sought to maintain autonomy in pursuit of their own national and strategic interests, was bound to be a mixed one. The UN represented the power structure of the international political architecture towards the end of the Second World War. This structure, and the realities of the post-war power equations, led to the emergence of the bipolar framework where two of the great powers among the victors of the war assumed the status of "superpowers" soon after the UN was established. UN responses to issues and problems affecting global security, therefore, were conditioned by the interests and goals of the superpowers and other great powers.

Peaceful Resolution of Disputes and Conflicts

The UN, with the experiences of the League of Nations to guide the thinking for a post-war dispensation of management of international peace and security, was structured on modifications to the collective security framework by relying more on preventive diplomacy to address pacific settlement of disputes under Chapter VI. The Security Council was given powers designed to encourage pacific settlement of disputes likely to endanger the international peace and security, but is not specifically required to do so. Resort to force, in pursuit of international peace and security through collective security goals of the United Nations, was to be regulated by an executive component in the shape of the Security Council operating under provisions of Chapter VII. The elaborate provisions of Chapters VI and VII of the UN Charter, however, have produced mixed results during the past fifty years.

The most notable positive contribution of the UN, which often gets ignored, has been to further peaceful change in what could have been a far more turbulent process in the period of decolonization.

> Bearing in mind the scope of changes in the world since 1945 - the process of decolonization, the decline of old powers and the emergence or reemergence of new ones, shifts in the sources and nature of economic power, the rapid development of military technology, the collapse of regimes in eastern Europe and the Soviet Union in 1989-91 - it is striking that many developments that could easily have occasioned major wars did not do so.[1]

India's example of non-violence as the root to independence and as a tool for socio-political change has been the most significant contribution to the predominantly peaceful process of decolonization. India acted as a bridge between the conflict of ideologies and balance of power, although the superpowers tended to seek the use of this bridge for their own national interests. India's policies (especially non-alignment, so grossly misunderstood by the West in general and the USA in particular), and efforts at and through the UN to seek greater accommodation, cooperation, and peaceful resolution of disputes, made a major contribution to this peaceful change.

It can be argued that the direct impact of the UN on decolonization was small. But the fact is that the Charter and UN General Assembly provided strong support to the withdrawal of colonial powers. Although the influence of the UN on this process (of decolonization) was only small and the overwhelming majority of the 800 million people still under colonial domination after World War II gained independence without any major participation on the part of the UN, the Organization actively supported the struggle for self-determination from the late 1940s onward when the non-aligned nations, with India at the forefront, provided strong support to the process. With the passing of General Assembly Resolution 1514 (XV) of 14 December 1960 and the installation of the Committee of 24 in the following year, constant pressure started to build up on the remaining colonial powers. The self-determination of the peoples of Rhodesia, Namibia, and the Portuguese colonies, as well as those of other territories, was at least accelerated by numerous UN activities.

Not all the cases of decolonization were peaceful, or the UN effective. The most agonizing breakdown of peaceful decolonization was the two-decade long Vietnam War (1954-1973). But even here, the UN played an important role although none of the parties involved, except the USA, were members of the UN.[2] Prime Minister Jawaharlal Nehru, recognizing this handicap, had proposed on the eve of the Geneva Conference on India-China (April-July 1954), a six-point proposal for the peaceful settlement of the India-China conflict, including the suggestion that the decisions emerging out of the conference should be reported to the UN. India had also played a seminal role in the decolonization of Indonesia, especially through persuasion and support through the international community, including at the UN.

The UN started with 50 members in 1945. Twenty-five years later, UN membership had risen to 127. The growth in the number of sovereign states members of the UN altered many of the assumptions of the Dumbarton Oaks framework which provided the basis of the UN Charter. Some of the structural problems of the UN came to the fore. Western dominance in the early years, both in the Security Council and the General Assembly, accentuated the Cold War and, as a consequence, led to the paralysis of the Security Council. This led the United States to seek the famous "Uniting for Peace" resolution of the General Assembly to enable military response to North Korea's invasion of the South in 1950. The precedent was to prove embarrassing for the West when Yugoslavia sought a similar approach during the Suez crisis in 1956. Both incidents highlighted the paralysis of the Security Council because of veto when the interests of the five permanent members were involved, and the legal weakness of the authority of the General Assembly. These factors have, from the very beginning, continued to limit the effectiveness of the UN in maintaining international peace and security.

Preventive diplomacy to achieve peaceful settlement of disputes as laid down in Chapter VI has rarely been successful. The UN failed miserably to find an objective and just solution to India's complaint of Pakistani invasion of the state of Jammu and Kashmir, which had acceded to India in accordance with the legal and constitutional provisions of transfer of power by the British in 1947. The Security Council has a quasi-judicial role in settling disputes that endanger peace and security. Although a Commission for India and Pakistan was set up to look into the case, it sought to find *ad hoc* solutions more in accordance with great power interests than the merits of the case. Even more important was the UN failure to ensure compliance by Pakistan of the resolution governing cease-fire wherein Pakistan had agreed to withdraw its forces (regular and irregular) from the territories of the state.[3]

The record of the UN in other cases has not been any more impressive. By one assessment, 191 disputes that involved fighting were referred to the UN between 1945-1990.[4] Only 47 of these (representing 24.6 per cent) were successfully resolved. The success rate during the first four decade was even lower, at 23 per cent. Overall, the success rate kept declining from over 30 per cent in the early 1950s to come down to the very low figures of around 15 per cent during the decade starting around 1972. It is generally believed that the UN has become more effective after the Cold War started to wind down. The number of disputes referred to the UN and the success rate in resolving the referred disputes rose notably after 1985. But this appears to be more due to the significant increase in the number of less intense/insignificant disputes referred to the UN since 1985. As compared to 36 per cent of the referred disputes being insignificant during the forty years before 1985, the percentage of such disputes jumped to 65 for the five year period after that.[5] During this period, the United States was the target of complaints no less than nine times, followed by Israel (five times) and South Africa (four times). The United States also used the veto more frequently than any other permanent member of the UN Security Council to ward off complaints against itself and to protect Israel.

Partly because of the low success rate in the resolution of disputes by the UN, and substantively because states have sought to retain their autonomy in decision-making, the number of disputes and armed conflicts which have *not been referred* to the UN has been substantial. This is particularly so in respect of intra-state conflicts, with or without external intervention. The UN emphasized the principles of sovereignty and non-interference in internal affairs of states. The combination of these two factors, and major reservations regarding the ability of the UN to take an objective position shorn of strategic and geo-political interests of external actors (especially the great powers), has proved to be a major dis-incentive to states seeking pacific settlement of disputes through the UN. India's experience with its attempts to seek justice from the UN went a long way in shaping India's attitudes in this regard. The other territorial disputes, and resolution of the Indo-Pakistan wars of 1965 and 1971, essentially remained outside the UN framework, even where the cease-fire was instituted pursuant to the Security Council resolutions. By 1987, the two superpowers had begun to cooperate in the resolution of conflicts and the UN resolution (No. 598) calling for a cease-fire in the long Iraq-Iran war was symbolic of the change from Cold War dynamics. But the UN essentially remained irrelevant to the large number of conflicts affecting international peace and security. Between 1985-1990, more than 43 intra-state armed conflicts (with external intervention in the majority of cases) had remained un-referred to the UN.

Collective Security?

The collective security scheme contained in Chapter VII was never implemented after 1945. The obvious reason was the inability of the five permanent members of the Security Council to reach agreement across what had rapidly become the Cold War divide. Article 43 provisions to place military forces of states at the disposal of the UN have never been agreed upon. The failure to do so was not due to any technical problems, but "primarily the result of a political impasse."[6] In the post-Cold War period, when greater cooperation among major powers has been witnessed, there has been little response to the Secretary-General's proposal for establishing a standing volunteer force or pre-assignment of national forces to the UN (although Article 45 specifically proposes that). The Military Staff Committee authorized under Article 47 never got off the ground. "During the Cold War, the permanent members could never agree, and since then the United States has been reluctant to empower this committee, lest the UN gain such legitimacy and effectiveness that it would constrain US freedom to act unilaterally."[7] The ideological conflict is often cited as the reason for the failure of collective security mechanisms. But the more dominant factor was the jealous guarding of sovereignty and their own national interests by the great powers, beside the unwillingness of the superpowers to place their military contingents outside their national command and control. This continues to be a significant problem in the organizing of UN peacekeeping operations. National security arrangements, therefore, either remained autonomous, especially of the neutral and non-aligned group of countries, or were linked into military alliances, including larger bloc systems like the NATO and WTO.

The collective security concept, especially if it is meant to be applied universally to all members of the organization, is fundamentally untenable in an international system composed of sovereign states since, to be effective and credible, it must be applied impartially. Ideal collective security arrangements would also entail automatic and binding commitments to respond to aggression with force. The issues involved in establishing a collective security system under the UN, especially after the failure of the arrangements under the League of Nations, had been discussed during the deliberations that led to the setting up of the UN.[8] Contrary to what some scholars believe, the UN is not really a collective security system.[9] And given the asymmetries of power in the international system, it would be naive to expect impartiality especially where great power interests and goals are involved. Direct aggression and war between the great powers (until recently, only the five permanent members of the Security Council), during the Cold War was avoided by them because of the risk of a nuclear holocaust. But wars and armed conflicts continued during the Cold War, essentially in the territories of the developing world. The great powers exacerbated many of them with direct and indirect interventions.[10] With ever more modern weapon systems, including nuclear weapons, the superpowers indulged frequently in "coercive diplomacy" in pursuit of their interests.[11]

Figure 1: **Number of Vetoes in the Security Council, 1945-1994**

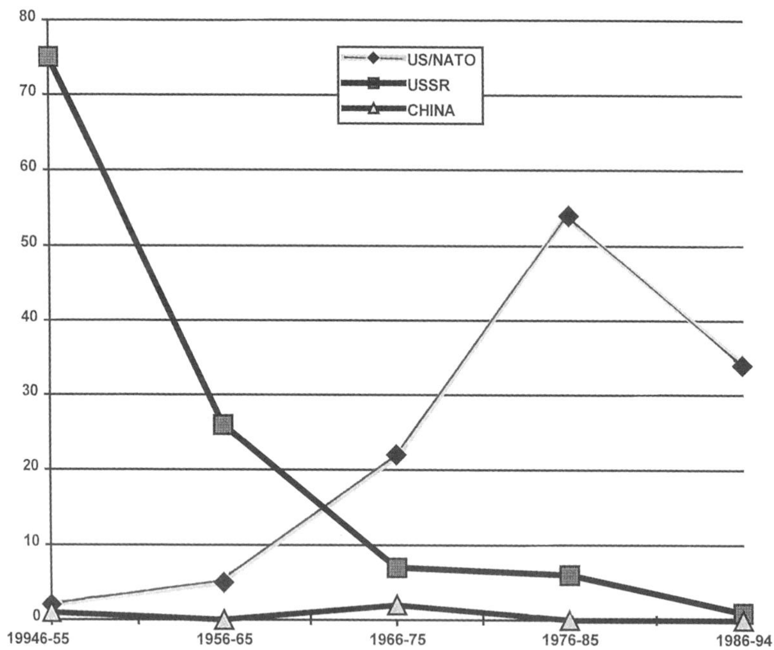

The structural problems that the UN faced in trying to implement a system of collective security can best be illustrated by the experience of the functioning of the UN Security Council which carries the primary responsibility to ensure international peace and security as the executive organ of the United Nations. The Security Council played an important role in easing or containing numerous crises, and it provides a high level for diplomatic contact and negotiations. The Charter provides for the decisions to be taken on the basis of unanimity among the five permanent members holding veto powers. This had been built into the Charter consciously finally through the compromise reached at Yalta.[12] It was inevitable that the exercise of veto powers by any of the permanent members, or the constant possibility of this happening, would dominate the functioning of the Security Council. Between 1945 and 1993, a total of 195 resolutions were vetoed, many by more than one permanent member. Figure 1 provides the correlation of vetoes by the US/NATO allies, USSR/Russia, and China. It may be seen that throughout the history of the UN, a very high number of vetoes were cast either by the Soviet Union (especially during the first two decades), or by the United States and its allies (during the remaining years). China stands out as the only permanent member which has been extremely economical with its vetoes.

Conventional wisdom claims that the UN was paralysed during the Cold War era, especially after the mid-1970s. This empirical evidence indicates that functioning of the Security Council was affected by vetoes by the Soviet Union during the early years, when the General Assembly and the Security Council was dominated by the West. But from mid-1950s, the Western powers started to use veto extensively. While the USSR/Russian vetoes account for a little less than half the total number, the vetoes by the US, UK, and France combined account for more than half. In terms of time horizon, 110 out of the total (115 up to 1994) Soviet vetoes came in the first twenty years. The highest number of US vetoes (34) in any decade came during 1976-1985. This also contributed significantly to the increasing US distrust of the UN. Interestingly, the number of vetoes cast came to its lowest figures in mid-1960s. This was also the period which was being described as the "beginning of the end of Cold War".

UN and Intra-state Conflicts

The birth of the UN was followed immediately by that of the Cold War which stymied UN ability to deal with inter-state conflicts. But its record of managing intra-state conflicts has been mostly disappointing. The structural limitations, principles of sovereignty and non-intervention in domestic affairs were major reasons. But more important was the fact that the five permanent members of the Security Council, especially the two superpowers, intervened in many of them directly or indirectly. As noted above, two-thirds of the conflicts between 1945-1982 were intra-state, and external intervention was a significant factor in over two-third of these.[13] Superpower involvement and interventions prolonged and vitiated many of these civil wars and intra-state conflicts of which the long wars in Vietnam, Afghanistan, and the Horn of Africa were typical examples. Only in some cases did the UN play a positive role, as in the Congo in 1960-61.

The UN was not really designed to address the problem of civil wars. To that extent, even where civil wars tended to expand and involve the interests of other great powers, the UN appears to have generally played a passive role. The provisions of the Charter itself inhibit the UN in playing an active role in civil wars and intra-state conflicts. The UN is a state-centred organization where the Charter establishes strong taboos against intervention in domestic affairs and, and hence, "civil wars". The frequency of UN interventions in the name of "peacekeeping" has increased markedly since the end of Cold War. But this appears to be more a case of misplaced focus of the UN, unable to deal with inter-state conflicts (which is its primary role) attempting to expand its role into an even more complex field of human conflict.

Civil wars, and more specifically, the internal breakdown of states, represent grave human tragedies. But their implications for "international peace and security" are not always clearly established except where there is an external involvement. It is specifically in such case that the UN has tended to shy away. Civil wars and militant violence (including sustained terrorism) of different types were going on in Afghanistan, Bosnia, Russia (Chechnya), Northern Ireland, India (Jammu and Kashmir), Pakistan (Sindh/Karachi), Sri Lanka, Yemen, Tajikistan, Angola, Rwanda, Somalia and other hot spots in Africa and a host of other places in 1994. But few, if any, were likely to escalate or spill over so as to endanger international peace and security. After the initial enthusiasm for interventions in civil wars and "humanitarian peace keeping", the United States has recognized this underlying reality as symbolized in President Clinton's address to the UN General Assembly on 27 September 1993:

> In recent weeks, in the Security Council, our nation has begun asking harder questions about proposals for new peacekeeping missions: is there a real *threat to international peace*? ..... The United Nations simply cannot become engaged in every one of the world's conflicts. If the American people are to say yes to UN peacekeeping, the United Nations must know when to say no." (emphasis added).

This is equally applicable to all members of the United Nations. The answer, therefore, lies in the UN paying greater attention to prevention rather than post fact actions. The UN, for example, will need to consciously discourage ethno-nationalism, and be less permissive in providing legitimacy to separatism and secessionism.

The modification of the earlier peacekeeping role into what has come to be called "peace-enforcement" role has many problems. But what is often ignored, and where the UN has not initiated any action (except in the case of the problem of landmines), is in relation to the means with which these "civil wars", armed militancies, insurgencies, and other forms of armed violence including terrorism are fought - that is, *small arms and light weapons*.[14] The landmines issue emerged on the international consciousness finally as a result of the attention focused by non-governmental, especially human rights, organizations.[15] But it is difficult to escape the conclusion that the attention was also made possible because the socialist countries had been primarily responsible for the spread of landmines. On the other hand, few governments, especially of the Western world, with the USA leading, are innocent of the proliferation of small but highly lethal weaponry across the globe which today threatens international peace and security, seriously challenges the stability of states in many cases, and is spreading like cancer into civil society. The democratic states are particularly vulnerable to the consequences of spread of small arms into their societies. The problem is that many states and sub-state groups have sought to use armed violence to alter the socio-political content of the state system in accordance with their own ideological (often based on ethno-nationalism or religious radicalism) or national goals and objectives. The UN has recognized the destabilizing influence of ethno-nationalism. But it has not come to grips with the problem of the emerging world in respect of political exploitation of religion. Even more significant has been its failure to address the serious threats to international peace and security caused by the global spread of small arms and light weapons, although the dramatic increase in the frequency and role of UN peacekeeping missions has been caused by conflicts where ready availability of lethal weapons is a significant factor. Unfortunately, although this problem assumed acute proportions by the early 1980s, and the superpowers left a great deal of such weaponry as a debris of the Cold War they were prosecuting for four decades, the Secretary-General's *Agenda for Peace* (1992) does not even mention the issue except in a passing reference in the follow on report of January 1995.[16] It is difficult to visualize effective measures for maintaining international peace and security today without a concerted attempt at dealing with the problem of proliferation of small arms and light weapons.

The UN in the 21st Century

The UN represents an evolutionary process in the efforts of the international community to seek ways and means of establishing international peace and security through resolution and harmonization of the numerous conflicting demands of the international political architecture founded on the principle of the sovereign nation-state system. The record so far has been mixed, and the actual functioning deviated a great deal from the goals and objectives set by itself in the beginning. The UN will need to review its Charter and functioning if it is to meet the challenges to international peace and security in the 21st century. Among a large number of measures needed for restructuring the United Nations, two fundamental needs stand out.

Firstly, the conceptual base for addressing the issues of international peace and security has to be constructed so as to harmonize the contradictions of the sovereign state system with national security needs. The attempts to find a solution in collective security, as has been sought so far, is unlikely to provide real solutions. What is needed is a paradigm shift from competitive security to cooperative security. The former is generated and reinforced by the concept of national sovereignty, a principle enshrined in the foundations of the UN. The UN cannot change the international system of sovereign states; but it also remains a static organization structured to meet past threats in accordance with the geopolitical framework of the past. There is a view that sovereignty is a dying concept and this process must be speeded up. The example of West Europe is the easiest to cite. If this was the trend, peace and security in former Yugoslavia should have been easily found. Theoretically, of course, dilution of sovereignty should reduce the demand for national security at the expense of security of another state. But, in reality, it will have to be diluted to an extent of making the existing system of states irrelevant before governments will be ready to forego their national security interests. The United Nations, by definition is a state-centred organization. Its role in maintaining international peace and security will have to remain within this framework.

But what is feasible is to move inter-state security into a cooperative paradigm. This would require cooperation on matters of peace and security not only with friendly states, but more so with unfriendly and adversary states at progressively higher levels. The broad principles for such cooperation exist in the UN Charter. There have other attempts to define the parameters and basis for cooperative security.[17] The fact that a level of cooperation among the two superpowers had existed even at the peak of the Cold War confirms that it is feasible to move along this direction, and that this would establish stability (absence of war) and peace on a more durable basis than the model of collective and competitive security that has been sought so far.

Secondly, it has been apparent for many years that changes in the Security Council will be necessary if the UN is to meet the demands of international peace and security in future. The central role of the Security Council will have to continue. But its present composition reflects the geopolitical order obtaining at the close of World War II. The UN, on the other hand, must be able to respond adequately to the challenges of peace and security not only in the 1990s, but in the early decades of the 21st century at the very minimum. This demands a Security Council that reflects the geopolitical realities of the coming decades, and the General Assembly's resolutions to be treated with greater respect to ensure that democratic (and normative) deficits of the UN are removed.[18]

The international political architecture had been evolving into a polycentric framework even before the Cold War ended. It can be hypothesized that this was an important factor in the demise of the Cold War confrontation. The international order is likely to move towards multi-polarity over a period of time. But this is also likely to bring about polarization in the international community. This is not likely to be helpful in the management of international peace and security even if provides a level of stability. The end of Cold War has etched out the contours of the polycentric geopolitical order more clearly. There is a need, therefore, for the Security Council to transform into a structure that is more representative. It is obvious that countries like Japan, Germany and India have to be included among the permanent members of the Security Council at an early date.[19] However, this cannot be the total list, and the restructuring of the Council should be based on objective criteria. The question of veto will have to be decided on the basis of equity among the permanent members, and the use of veto should be restricted only to enforcement provisions where military force is sought to be applied against the member state. There is also a need for greater public accountability of the UN, especially the Security Council and the requisite checks and balances in an institution where "No organ exists that can call the Council to account under the rule of law to protect the interests of members, or to protect other persons with no voting power, or to protect the long-term interests of law itself."[20]

Diffusion in the relative power among significant powers was in evidence over the previous decade; and this was also manifested in the growing tri-lateralism of US-Japan-West Europe (with Germany central to it). The disintegration and collapse of the USSR expedited the re-arrangement and realignment in the international political architecture towards a polycentric system. The end of the Cold War marks the growing importance of centres of power (like Japan, Germany, and India) who have not built up their national strategies around military power and its projection as an instrument of policy. Going by conventional wisdom about attributes of national power, the half-a-dozen leading powers in the coming years would be the USA, Japan, the European Union, India, China, and Russia. The current five permanent members of the UN Security Council also happen to be the five recognized nuclear weapon states. If non-nuclear weapon states like Japan and Germany are admitted as permanent members of the UN Security Council, a profound shift away from military power as the central factor in the power and security calculus would take place, especially since the UN offers new opportunities to improve international peace and security in future.

In the final analysis it is clear that a comprehensive approach to reforms in the UN system is needed. Prime Minister Ingvar Carlsson of Sweden has cautioned that "There is a risk, though, that changes would be made piecemeal in a way which would leave the United Nations much the same as it is now, and which would not address some of the urgent security demands of these closing years of the century."[21] The organization will have to remain state-centred and managed by governments for a long time to come. The concept of "strengthening" the UN has been proposed to overcome some of the existing deficiencies.[22]The key to reforms does not lie in merely tinkering with it under the rubric of "strengthening" it. While some aspects require strengthening, there is a need for some basic changes. For example, the UN must set for itself a goal for abolishing war and armed conflict. As it is, war as an instrument of state policy is losing its 19th century rationale and cannot be sustained into the 21st century. But what is needed is to move beyond that, to ensure that norms against use and threat of use of force, by states or non-state actors, are strengthened to ensure durable and comprehensive international peace and security.

References

[1] Adam Roberts & Benedict Kingsbury, *Presiding Over a Divided World: Changing UN Roles, 1945-1993,* Lynne Rienner, Boulder, 1995, p. 59.

[2] For a detailed account of the conflict, role of the UN, and that of India, see MS Rajan, note 1, pp. 72-107.

[3] Resolution of the UN Commission for India and Pakistan, August 18, 1948, under which Pakistan's vacation of the occupied territories was accepted (even by Pakistan) as a pre-condition for a plebiscite to decide the final accession of the state to India or Pakistan. Pakistan continues to hold these occupied territories 47 years later.

[4] Ernst Haas, "Collective conflict Management: Evidence for a New World Order?", in Thomas Weiss (ed.), *Collective Security in a Changing World*, Lynne Rienner, Boulder, 1993, p.64.

[5] Ibid, p. 67.

[6] Joel Larus, ed., *From Collective Security to Preventive Diplomacy*, John Wiley & Sons, New York, 1965, p. 224.

[7] Robert C. Johansen, "Building World Security: The Need for Strengthened International Institutions", in Michael T. Clare and Daniel C. Thomas, eds., *World Security: Challenge for a New Century*, St. Martin's Press, New York, 1994, p.392.

[8] Roland N. Stromberg, "The Idea of Collective security", in Joel Larus, ed., *From Collective Security to Preventive Diplomacy*, John Wiley & Sons, New York, 1965, pp.273-303.

[9] See John J. Mearsheimer, "The False Promise of International Institutions", *International Security*, Vol. 19, No. 3, Winter 1994/95, pp. 5-49.

[10] According to Istvan Kende, direct foreign military intervention took place in over 73 per cent of the total 148 wars computed by him during 1945-1982 ("Wars of ten Years, 1967-1976", *Journal of Peace Research,* Vol. XV, No.3, 1978; and "New Features of Armed Conflicts and Armament in Developing Countries", *Development and Peace*, Spring 1983).

[11] According to Berry Blechman, Stephan Kaplan et al, (*Force Without War: US Armed Forces as a Political Instrument,* Brookings Institution, Washington DC, 1978), and Philip Zelikow, (Force Without War 1975-82", *The Journal of Strategic Studies*, Vol.7, No.1, March 1984, pp.29-54), the US employed its military forces on 259 occasions for coercive purposes to achieve political goals during 1946-1982. The USSR similarly used its military power for coercive diplomacy on 167 occasions between 1946-1979. (See Stephan Kaplan, *Mailed Fist, Velvet Glove: Soviet Armed Forces as a Political Instrument*, Brookings Institution, Washington DC, 1979).

[12] Bruno Simma, ed., *The Charter of the United nations: A Commentary*, Oxford University Press, 1994, pp. 9-10.

[13] Istvan Kende, *op. cit.*

[14] See Prashant Dikshit, "Proliferation of Small Arms and Minor Weapons", *Asian Strategic Review 1992-93*, IDSA, New Delhi, 1993, pp.205-231.

[15] *Landmines: A Deadly Legacy*, The Arms Project, Human Rights Watch, New York, 1993.

[16] Boutros Boutros-Ghali, *Agenda for Peace* 1995 (second edition) United Nations, New York, 1995.

[17] See the Panchsheel agreement between India and China, June 27, 1954; the Simla Agreement between India and Pakistan, July 2, 1972; India-China agreement on maintaining peace and tranquillity on the borders December 18, 1988, and mutual and equal security agreement of September 1993. For one outline of cooperative peace and security, see *Towards a New Asia*, Report of the Commission for a New Asia, 1994, pp. 11-19.

[18] J. D. Sethi recommended some changes in the statute to rectify this. See J. D. Sethi, "A Broken Totality", in U. S. Bajpai, *Forty years of the United Nations*, Lancer International, New Delhi, 1987, p. 38.

[19] Coral Ball, "The fall and Rise of the United Nations", in Coral Ball ed.,*The United Nations and Crisis Management: Six Studies,* Strategic and Defence Studies Centre, Canberra, 1994, p. 16.

[20] Jose E. Alvarez, "The Once and Future Security Council", *The Washington Quarterly*, 18:2, Spring 1995, p. 18.

[21] Ingvar Carlsson, "Roles for the UN in International security after the Cold War", Security Dialogue, Vol. 26(1), pp. 7-18.

[22] Anthony Parsons, *From Cold War to Hot Peace*, Michael Joseph, London, 1995, pp.244-260.

UNITED NATIONS PEACEKEEPING OPERATIONS: THE CHALLENGE OF CHANGE

Jasjit Singh

The UN Secretary General had remarked very appropriately that "Peacekeeping can rightly be called the *invention* of the United Nations"[1] (emphasis added). It is an invention because there is no real legal, constitutional, or structural basis for such operations, and the UN Charter has no provisions for it. This is particularly so in respect of peace-enforcement within states. In many respects, peacekeeping missions have increased rapidly, partly because the Security Council has now provided legitimacy to intervention in other states. The comparison between what happened in Grenada (1983) and Panama (1989) with Haiti now are typical examples.[2] It can be argued that a traditional role for peacekeeping had grown over a period of time as a compromise approach and *ad hoc* solution to specific issues affecting international peace and security. But the expansion of the peacekeeping operations in recent years, in many cases, is directly contradictory to the very basis of the UN Charter itself creating a fundamental problem of legitimacy. This is especially relevant in case of UN interventions in intra-state conflicts. As John Gerard Ruggie says: "This growing misuse of peacekeeping does more than strain the United Nations materially and institutionally. It has brought the world body to the point of outright strategic failure - indeed, in Bosnia the line has already been crossed."[3] If the United Nations peacekeeping operations are to play any meaningful role in future, some fundamental issues related to the qualitative issues will require serious consideration.

Armed Conflicts

The role of the United Nations had assumed a higher profile and importance with the decline and end of the Cold War. This is confirmed by relevant statistics as seen in Table 1. This has its relationship with quantitative and qualitative changes in peacekeeping operations by the United Nations. Conventional wisdom would lead us to believe that there has been an upsurge in the number of armed conflicts. This perception is largely due to the fact that there has been an upsurge of armed conflicts in Europe which had enjoyed comparative peace from intra-state conflicts since the Second World War. But empirical evidence indicates that the number had plateaued and started to reduce in the late 1990s. While many of the changes in peacekeeping commitments, especially in terms of their scope and direction of expansion of the role, had been conditioned by the uncertainties of the transition from the Cold War period to what was erroneously believed to be a "unipolar" international order, it should be possible to start addressing the issues now with a degree of objectivity so that solutions for the future can be crafted consistent with principle and practice of the Charter of the UN. This will also point to the direction and substance of reforms necessary in the UN as it completes half a century of its life.

As regards the trend in armed conflicts, it needs to be noted at the outset that an accurate assessment is not always possible except in inter-state wars, and even there, the causes and casualties mostly remain clouded. However, the Department of Peace and Conflict Research at Uppsala University in Sweden has been carrying out a systematic collation of data related to armed conflicts

While it shows many deficiencies (for example, it does not include the armed conflict in Sindh/Karachi, in Pakistan, where over 2,000 people died in 1994 due to armed violence and the government forces were involved), it perhaps represents the best attempt in the field. According to their study, armed conflicts appear to be declining in intensity and occurrence. (Armed conflict here has been defined as a contested incompatibility which concerns government and/or territory, where the use of armed force between two parties results in (i), a *minor conflict* with at least 25 battle-related deaths during the year but less than 1,000 deaths overall; or (ii), an *intermediate armed conflict* which results in at least 25 battle-related deaths per year with over 1,000 cumulative total number of deaths; and (iii), a *war* is defined as an armed conflict with over 1,000 battle-related deaths per year.) Data covering 290 armed conflicts from 1989 to 1994 indicates that the total number of armed conflicts, which stood at 47 in 1989, came down to 42 in 1994.[4] Even more significant has been the shift in the density of conflicts down towards lower levels. Wars, which had maintained a certain consistency at around 19-20 wars in each of the years from 1989 to 1992, dropped dramatically to a mere eight by 1994: in Azerbaijan, Georgia (Abkhazia), Turkey, Yemen, Afghanistan, Myanmar, Algeria, and Angola. All these are intra-state wars, mostly following their involvement in the Cold War conflict. There is no UN peacekeeping mission in these countries except in Afghanistan and Georgia, where they are present more in an observer role than a peacekeeping or enforcement mission. The UN's role in Afghanistan, in fact, has been negative, especially in 1992, when it triggered off a more intensified stage of civil war and made solutions more difficult. The reduction in the number of wars has resulted in intermediate conflicts becoming the largest category, with the minor ones continuing.[5]

Table 1
Some Statistics on United Nations Activities
Related to Peace and Security: 1988 to 1994

	As on 31 January 1988	As on 31 January 1992	As on 16 December 1994
Security Council resolutions adopted in the preceding 12 months	15	53	78
Disputes and conflicts in which the UN was actively involved in preventive diplomacy or peacmaking in the preceding 12 months	11	13	28
Peacekeeping operations deployed:			
Total	5	11	17
Classical	5	7	9
Multifunctional	-	4	8
Military personnel deployed	9,570	11,495	73,393
Civilian police deployed	35	155	2,130
International civilian personnel deployed	1,516	2,206	2,260
Countries contributing military and police personnel	26	56	76
United Nations budget for peacekeeping operations (on an annual basis) - $Usm	230.4	1,689.6	3,610
Countries in which the UN had undertaken electoral activities in the preceding 12 months	-	6	21
Sanctions regimes imposed by the Security Council	1	2	7

Source: Boutros Boutros-Ghali, *Agenda for Peace 1995*, second edition, UN, New York, 1995, p.8.

The category of minor and intermediate armed conflicts during 1994 was composed essentially of intra-state conflicts and civil war situations. These numbered 36, with five conflicts going on in Europe: in Bosnia-Herzegovina (three distinct conflicts), Western Bosnia, Russia-Chechnya; 15 in Asia: in Egypt, Iraq, Israel, Cambodia, India (six in Kashmir, Assam, Punjab,

Manipur, Nagaland, and Naxalites/PWG/MCC groups), Myanmar, Papua New Guinea, Philippines (two), Sri Lanka, and Tajikistan; 11 in Africa: in Angola, Chad, Djibouti, Liberia, Mali, Niger, Rwanda, Sierra Leone, Somalia, Sudan, and Uganda; and four in the Americas: in Colombia, Guatemala, Mexico, Peru. UN peacekeeping operations were going on in six of these countries related to these conflicts, besides operations covering some of these countries *not related to these conflicts*, but in the context of other inter-state wars. The conclusion becomes inescapable that the UN peacekeeping was covering 17 per cent of the armed conflicts (minor, intermediate, and wars). The reason for this, no doubt, is the very nature of the United Nations, and the nature of these armed conflicts, which are mostly internal armed conflicts, although an external factor has been operative in almost all cases in exacerbating the conflicts and making their political resolution more difficult.

Peacekeeping Operations

There had been a sudden upsurge in the demand and commitment to UN peacekeeping operations in the world in the early 1990s (Fig. 1). However, it is also clear that these operations and commitments have reached a plateau. Countries are increasingly reassessing the political, economic, and human costs of peacekeeping. These are being measured against perceived direct national interests. Nearly 650,000 people have served in 35 UN peacekeeping operations over the past five decades to 1995. A total of 1,145 have died in these operations, including 85 of India's finest youth. The UN has spent over $11 billion on these operations. At present, nearly 70,000 personnel from 68 countries are serving in UN peacekeeping operations across four continents, including Europe. As of December 1994, there were 5,209 Indian men and women serving with six UN peacekeeping missions. The tempo of peacekeeping had died down by the early 1960s. The decade of the 60s saw a total of 6 new operations, which came down to 3 in the 1970s - all in the Middle East.

There were no new peacekeeping operations during 18 years after June 1974 until 1988! Since then, a total of 22 operations had been launched by January 1995. The Cold War was certainly a major factor in this process since the role of the UN itself was downgraded and the superpowers were resorting to proxy wars and other means to pursue their own perceived interests, where the larger international cooperative approach was subordinated to these interests.

Collective security had taken a back seat, while collective defence of the two conflicting blocs assumed primacy. wars and conflicts, however, had continued. In fact, nearly 250 wars have been recorded since 1945. But they were dominated by great power interventions ranging from Vietnam, Afghanistan, to Somalia (in the 1970s and 80s) and Lebanon, with little or no role for UN peacekeeping, except when the superpowers agreed to it. Whatever peacekeeping operations took place during the quarter century after 1964 were under bilateral agreements, for example, like those undertaken by India in Sri Lanka (in 1971 and 1987-90), and the Maldives in 1988 at the request of legitimate host governments.

UN Peacekeeping and Observing

UN peacekeeping operations may be broadly classified under three categories. The *first* is the traditional one, where the UN force acts as a firebreak, mostly between states, and essentially to monitor conflict termination arrangements under truce and cease-fire agreements. They constitute what has come to be termed as the classical operations and, in essence, cover peace observing and peacekeeping. These inevitably are constructed on the consent of the parties to the conflict, and hence possess greater legitimacy and pose lesser challenges. In some cases, threats to international peace and security developed when consent by one party was withdrawn, as happened in 1967 in the Middle East. This itself held out enough lessons with respect to the risks and complexities of UN peacekeeping without consent of parties concerned. The majority (21 out of 35) of UN peacekeeping operations fall under this category. At the end of 1994, 17 UN peacekeeping operations were under way. Of these, nine constituted the traditional or classical peacekeeping operations. Many resulted from the withdrawal of superpowers from conflicts in the developing world. This in itself provides the optimism that the need for such operations will reduce in the coming years, unless polarization of the international order picks up momentum. The remaining eight were multifunctional falling under the second and third categories noted below. These are more complex than the traditional ones.

Figure 1
Number of Peacekeeping Missions- 1945-1994

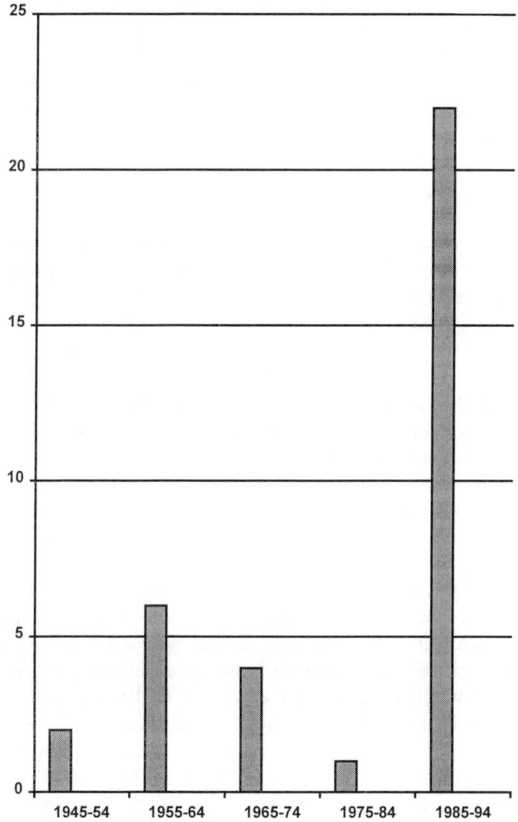

Traditional peacekeeping had started with small observer missions starting in 1948 in the Balkans, Indonesia, South Asia, and the Middle East, in areas outside the East-West confrontation in Central Europe. These were also consequent to agreements to halt armed conflicts that had followed the process of withdrawal of colonial powers. The task of the UN was to stabilize the cease-fires to which parties had consented. These activities were premised on Chapter VI, and were part of the contribution to creating a more peaceful environment in the disputes between states. The first mission to be labelled "peacekeeping" explicitly was the establishment of the UN Emergency Force (UNEF I) in the Sinai to oversee the separation and withdrawal of forces and to create a buffer zone between Egypt and Israel following the 1956 Suez crisis.

The total number of traditional peacekeeping operations by the UN have increased during recent years from 13 new operations commenced till 1988, to a total of 22 operations initiated since then (to end-1994). The overwhelming majority of peacekeeping operations undertaken to 1988 were of the traditional type. Such operations have also tended to be of comparatively long duration. Nine out of the 13 operations were undertaken in the Middle East region, mostly related to Arab-Israeli wars. There were 17 ongoing operations at the end of 1994, out of which 6 had been initiated before 1988 and represented traditional peacekeeping operations, including peace observing. The real change in the nature of peacekeeping is reflected in the fact of (i) the quantitative jump in the number of operations as noted earlier, but, more importantly, in (ii) the qualitative change which is indicated by the fact that compared to 85 per cent of the pre-1988 operations being traditional, since then only 2 out of the 22 operations after 1988 (that is, a mere nine per cent) were of a nature that could be considered traditional. These two related to monitoring cease-fire agreements in Iraq-Kuwait, and Tajikistan-Afghanistan frontiers. But even here, the problem is that peacekeepers are deployed within one country rather than between countries making the mission more difficult and less effective.

Peacekeeping and Imploding States

The *second* category concerns actions by the international community under UN auspices essentially to deal with the circumstances arising out of a partially or fully "imploded state", where the normal institutions of a state are near collapse or have collapsed and a state of anarchy exists. In most cases this has resulted from prolonged Cold War tensions/conflict, especially in regions and territories where state-ness was fragile to start with (as in most of former colonial territories) followed or accompanied by civil wars, and possibly also including protracted natural disasters, like famine. UN peacekeeping operations to deal with such situations have normally been referred to as peace-building operations. The UN has often sought to provide humanitarian assistance and seek reconstruction of the state organs. These are often turning into situations which require interjection of force to ensure the safety of humanitarian operations. Consent, by definition, would not be available in the case of imploded states. Somalia and Rwanda stand out as typical examples.

Progressive erosion of the institutions in a country, with or without external interference, often leads to the failed state. Such situations pose a serious challenge to the international community. The basic reality is that the state system still provides the best organizing principle for peoples. Notwithstanding the frequent occurrence of intra-state conflicts and civil wars, the state is hardly obsolete. But the problem has been: how to deal with a failing and imploded state? The attempts so far have concentrated on humanitarian assistance and an expansion of the peacekeeping role. These are situations which have posed the greatest challenges, especially where the local authority has substantively disappeared, is contested, or has totally collapsed. UN intervention has often started with humanitarian assistance which requires protection. Somalia is a classic example. While consent and cooperation is a basic pre-requisite for successful peacekeeping operations, in such situations, and unlike even civil wars, there may be no identifiable parties to any internal conflict. More often than not, the number of parties claiming rights are numerous with debatable authority.

The UN has sought to deal with such difficult situations through peacekeeping under Security Council authority. But there is a need to consider whether there any other options for an organization that had been set up to deal with inter-state conflicts and not the implosion of states. Some experts have suggested a modified system which draws upon the UN trusteeship arrangements, since the issue no longer remains that of use of force or peacekeeping, but more of adequate governance.[6] Peacekeeping operations as an adjunct to temporary trusteeship arrangements may offer greater scope for success than the present system, which assumes a functional local authority but in reality finds it non-existent, or difficult to cooperate with.

Peacekeeping and Civil Wars

The *third* category are those that aim to deal with intra-state conflicts and civil wars to essentially seek their de-militarization so that peace and security can be restored. These have often been described as peace-enforcement operations. These are the most difficult contingencies to handle by UN peacekeeping operations. Many of these have resulted in the UN getting involved in what are essentially governmental functions. Some of these have resulted in perceived success (like Cambodia), while others have resulted in serious complications and perceived failures.

Fundamentally, the UN has no role in internal conflicts and civil wars, and hence the level of legitimacy remains very low or questionable. But more often than not, such operations, linked to humanitarian operations or independent of them, are required to de-militarize existing armed conflicts within states (no doubt, frequently with an external factor strongly influencing the internal conflict), where ideology, religion, ethnicity and a host of other motivations drive the conflict. Cambodia and Somalia are typical examples. Requirement for peacekeeping under this category has often arisen out of armed conflicts born out of local factors and external involvement, like support of secessionist armed militancy. UN intervention in Congo in 1961, which was a point of departure from the traditional peacekeeping operations, was undertaken to uphold the institution and integrity of the state. The intervention in former Yugoslavia, on the other hand, came about to manage the disintegration of a state member of the UN the inviolability of whose borders had been guaranteed under the Helsinki process and the CSCE agreements. This type of peacekeeping - which requires interventionary force which will be faced with the prospect of getting involved in armed conflict directly - is the most complex, and carries the greatest risks, politically and physically, especially since such operations may often have to be undertaken without prior, full, and continuing consent of the parties to the conflict. The challenges that invite such interventions impose greater responsibilities on the international community to ensure that such operations are reduced to the minimum through prevention strategies, and when they become inevitable, that they are carried out with greater efficiency and wider international support.

References

[1] Boutros Boutros-Ghali, *Agenda for Peace 1995*, second edition, United Nations, New York, 1995, p.57.
[2] See Anthony Parson who argues in favour of the changed role *in From Cold War to Hot Peace: UN Interventions 1947-1994,* Michael Joseph, London, 1995, p.267.
[3] John Gerard Ruggie, "The United Nations: Stuck in a Fog Between Peacekeeping and Enforcement", in William H. Lewis, *Peacekeeping: The Way Ahead*, McNair Paper 25, National Defense University Press, Washington DC, 1995, p.1.
[4] Peter Wallensteen & Margareta Sollenberg, "After the Cold War: Emerging Patterns of Armed Conflict 1989-94", in *Journal of Peace Research*, vol.32, no.3, 1995, pp.345-360.
[5] *Ibid*, p.345.
[6] Thomas G. Weiss, "The United Nations at Fifty: Recent Lessons", *Current History*, May, 1995, p.227.

UN REFORM AND CIVIL DISOBEDIENCE

Ivo Slaus

In spite of many failures and various shortcomings during the last 50 years, the UN organizations - General Assembly, Security Council, UNDP, WHO, UNEP, UNESCO, UNICEF, FAO, and so on - have many acomplishments to their credit and the world is - because of them - certainly a better place. There is no doubt that the UN system is much more successful than the League of Nations ever was. (The foundation of the League of Nations is due to President Wilson. The idea goes back to Jeremy Bentham (1748-1832).) Therefore, why even attempt to improve the system, particularly since in a process of modification many good aspects could be ruined or at least hindered?

Nevertheless, there are compelling reasons why it is imperative to improve the UN system and here I will list those that seem to me to be the most crucial.

1. Many political problems and dangers facing the contemporary world,[1] notably totalitarian regimes, weapons of mass destruction, terrorism and violence cannot be adequately coped with by the UN system as it is now.

2. The UN system is based on the concept of the nation-state interests rooted in the sovereignty of nation states, on Richelieu's idea of *raison d'etat*, where states are no longer restrained by a pretence of a moral code, on the balance of power and on so-called realpolitik.[2] However, the damages of such politics to the very interest of the nation state are enormous. The very idea of Lord Palmerstone that nations do not have permanent friends but only permanent interests is inadequate. Ethics, specifically global ethics are essential in national, international and global politics and economics.[3]

3. Nation states are not the only, in many cases not even the dominant, political force in the contemporary world. Many transnational companies, even some individuals, and many non-governmental organizations (NGOs) are presently politically more important than many nation states. The role that some nation states play, as well as the role of some transnationals, of some NGOs and of some individuals, is commendable. The actions and philosophy of some of them are threatening peace, stability and human rights. Nevertheless, since they are actors in the international and global arena, it is desirable that they should be appropriately included in the UN structure.

4. The engines of globalization and fast change are science and technology. A. Toffler argues,[4] and I agree with him, that the political power has shifted and it is now largely determined by science and technology. A salient feature of science, and even technology is that it is international. Therefore, the system based exclusively on nation states is inadequate. Proposals to include academies as well as scientific and scholarly associations in the UN system (and within the nation-state political structure) have already been made and certainly have merit.[5]

5. S. Huntington argues that the contemporary world is faced with a clash of civilizations,[6] and for him the civilizations are determined by religions. Though I disagree with his argument, it is reasonable to involve religions in the international, global system, and certainly recent efforts on global ethics go in this direction.[7]

6. There is a gross disproportion between the number of nation states (about 200) and the number of different cultures and ethnic groups (about 5,000-10,000). Sustainable development requires maintaining and strengthening both biological and cultural diversities. If one would argue that cultural identity is best preserved and developed within and by a nation state, one would face a variety of problems: self-determination, a surge of new nation states, a problem of borders, and eventually,

how to cope with the UN representation of thousands of nation states, some of them very small and some extremely powerful.

7. There is always a danger of the abuse of power, and, therefore, various restrictions, moderations and restraints have been developed throughout history: constitutions and laws, division of power, limitations of power by unassailable human rights and by participation in power of those subjected to it.[3] These restrictions are much better developed (though also abused) within nation states than on the international arena. International courts are one example, but it is obvious that their effectiveness so far is rather limited.

8. It became fashionable to talk about and write books on the theme the "end of history", "end of science" and "end of politics".[8] This leads to political inertia. Not only does this keep the status quo, but it enhances all those dangers and threats we listed above: dictatorship, terrorism and violence.

I argue that the salient characteristics of the contemporary world ask for novel democratic and pluralistic forms of institutions and activities to complement and enrich the structure based on representation of nation states. (Even international courts are based on the structure of nation states). An example which demonstrates the inadequacies of the present structure is the fact that the last part of the twentieth century has been plagued by dictators who grabbed power in a rather small country and then transformed such a country into a threat to world stability. The reaction of the world community, particularly of the UN, has always been much too slow and barely adequate. Most actions have been only partial and all have resulted in enormous collateral damage, for example, in the case of economic sanctions, and much more when a military action has been required. Both of these attempts at solving the problems - military action and economic sanctions - are based on the argument that it is the responsibility of the people to change their government, and, therefore, the people are co-responsible for the actions of their governments. This is only partly true, particularly since the international political structure in the initial phase reinforces these dictators. A novel approach calls for simultaneous action of both citizens of the country in question and the international community to remove such dictators from political power.

The notion that individuals have inherent rights and, therefore, freedom to choose whether to obey the state or not - is a rather recent phenomenon. Here we are interested only in one specific form of disobedience - civil disobedience (CD) - and we will focus on its role in facing present political problems. Is CD a useful political tool? Can it help to reduce any of the above problems and dangers? I will not attempt to discuss various definitions of CD. It suffices to state that CD is a public defiance of a law or policy enforced by established governmental authorities; that the action of CD is premeditated; understood by the actors to be illegal; and carried out in a civil way, that is to say, by way of carefully chosen and limited means.[9] (Uncivil disobedience is any act of disobedience that does not recognize such limits, including extreme cases such as revolutionary terrorism, but not state terrorism or antirevolutionary terror in defence of the state regardless of whether the ties between the regime and the terrorists are open.) CD is a public, nonviolent, conscientious yet political act contrary to law, usually done with the aim of bringing about a change of law or of the policies of the government.[10] One of the frequent criticisms made of CD is that it cannot be generalized. For, if any person claims the right to disobey the law, than everybody can claim the same right: it would be terrible if everyone disobeyed the law. (However, it is rightly said that it would be also terrible if everybody always obeyed without questioning - that would be the death of freedom[11] and, we should add, the death of creativity.) This criticism is totally misplaced, since civil disobedience is not advocating disobedience to all laws, and it is not argued that one act of disobedience calls for and/or justifies total disobedience. CD is not breaking all laws, but it is directed to a specific law. The means and ends of CD must be public. One may find analogies with the way science works where some laws are challenged, proven to be incomplete, and replaced by more accurate and more general laws, and sometimes even totally different laws. Often, in the process, the entire worldview can and does change. None of this hurts science. On the contrary, it assures the progress of science.

Though we have quoted the accepted definitions of CD, a brief historical survey would help to distinguish CD from other forms of dissent, for example, revolution, resistance, violence without hate, nonresistance, conscientious objection, nonviolent action and pacifism. These examples might also help us to appreciate whether and to what extent CD can enrich international and global politics.

The essential argument in support of CD was expressed already by Socrates[12] who claimed that the state cannot force anyone to act unjustly, and we must be prepared to suffer evil, but not do evil. In spite of efforts of all totalitarian systems to the contrary, Socrates's view is one of the pillars of our ethics (for example, "unjust laws are acts of violence, rather than laws... and such laws do not bind us" - St. Thomas Aquinas, and "Whenever legislators endeavour to take away the property of people, they put themselves into a state of war with the people, who are thereupon absolved of any further obedience."- John Locke). During the last 150 years, both theory and practice of CD manifested significant new approaches. There were sit-ins at restaurants, churches and libraries in the American South, labour strikes as conceived by Eugene Debs and used in the 1930s (workers refused to leave their workbenches), free speech movements, various forms of protests against the Vietnam war, various teach-ins. The following five seem to me to be most important: First, H.D. Thoreau's essay "On the Duty of Civil Disobedience", based on his lecture given in 1848.[13] His opening sentence "That government is best which governs least" echoes the ancient Chinese saying that the best rulers are those that we even do not know they exist. Thoreau continues:

> Law never made men a whit more just, and by means of their respect for it, even the well disposed are made daily the agents of injustice. A natural result of an undue respect for law is that you may see a file of soldiers marching to the wars, against their wills, against their common sense and conscience...Now, what are they? Men at all? or small movable forts at the service of some unscrupulous man in power?

Second, the movement for racial equality led by Martin Luther King. King formulates the criteria[14] by which one can determine that a law is unjust and that therefore, it is our duty to disobey it. Third, the resistance to nuclear weapons, which developed into the Pugwash movement.[15] Fourth, *aggiornamento* initiated by Pope John XXIII, who emphasized that laws violating human rights are not only immoral, but also without force of law. Fifth, the teachings and activities of Mahatma Gandhi.

The first mass CD movement was the November 1913 march to Transvaal in South Africa led by Mahatma Gandhi. Gandhi argues that every citizen is responsible for every act of his government and this is the reason why it is the duty of the citizen to challenge his state. A just and democratic state deserves active loyalty, but the citizen always retains the right to disobey particular laws that he regards unjust and repressive. It is then even his duty to disobey, since a citizen cannot relinquish even a portion of his moral responsibility in the name of social contract, and/or sovereignty. Therefore, Gandhi integrates the demands for loyalty to the state with the demands for challenge, including disloyalty to government. It is impossible to be loyal unless a person is ready to challenge his own government decisions and actions. Here are two aspects that Gandhi emphasized: one is the commitment to a political action, and the other is his uncompromising commitment to nonviolence - *ahimsa* and *satyagraha* (*santyagraha* can be translated as love-force, truth-force). A *santyagrahi* must never forget the distinction between evil and evil-doer. A *santyagrahi* will always try to overcome evil by good, anger by love, untruth by truth.[16]

What we have presented here does not represent a complete history of recent CD. Nevertheless, several important conclusions can be drawn: CD is deeply rooted in democratic tradition rather than in the revolutionary or anarchic tradition, which both can generate totalitarian regimes. The practice of CD is fairly developed and it has been reinforced by Gandhi, himself a lawyer. One can distinguish three prongs: one is to mobilize and to recruit the support of spectators; the second is to educate; and the third is to reform those in power. Is CD a human right? Certainly not. Bayard Rustin argues that it is the duty to be a civil disobedient with the objective of revealing inconsistencies in the society and of correcting them.[17] Though this statement appears at first as an attractive one, it can either become a truism, or it can degenerate into enforcement, and CD is contradictory to compulsory requirements.

How can CD contribute to reducing the main political dangers? Here is my list.

1. One of the greatest menaces to freedom is a politically inert people. The political inertia often results from the feeling that nothing can be done. This is generated by the cumbersome and slow process of most (except revolutionary and coup d'etat) political actions. CD opens a niche of political activities which is timely, it does not require existing political organization to back it up, and it is

focused on a precise and often single issue. These characteristics make CD a tool of the modern time of information and communication technologies.

2. Every act of CD is an attempt at civic education: from the acts of CD a correct lesson has to be drawn by all involved: civil disobedience, spectators and those who are in command. Through CD we are developing our value system (though there are basic ethical values, some aspects of our value system are not constant but evolve over time). Two features of CD are of particular importance for the development of the value system. One, CD is not a revolutionary, sudden and complete change, but rather a set of incremental changes, which allows for a built-in correction mechanism. Second, a person involved in CD is willing to suffer himself a legal punishment of his illegal conduct and this helps to guarantee his motives, which can hardly be selfish.

3. Incremental contributions of CD: it can create and establish a just law, and it can force the implementation of the existing law which is ignored.

4. CD develops and ensures civil liberty possibly better than anything else can do. It builds and strengthens a civil society by engaging citizens in concrete actions. CD is complementary and not in contradiction to existing democratic institutions and processes: elections, activities of the independent branches of the government: legislative, juridical, and executive, as well as to all other pillars of the civil society: free press, educational, cultural and scholarly institutions.

5. CD is a constant watchdog on any possible abuse of political power. Therefore, it significantly reduces the danger that a regime becomes totalitarian, and eliminates at least one possible reason for the use of weapons of mass destruction. In the society which includes CD at least some justifications for terrorism and violence are eliminated.

Democratic and pluralistic, international and global politics have to be based on checks and balances involving all actors. Presently, it is almost exclusively nation states, that is to say, their representatives in all various regional and international bodies. Even the mechanism of CD is now restricted solely within a specific nation state. It is essential that all actors are as politically independent as possible. Nowadays, many academies, even some NGOs are controlled, influenced or blackmailed by their governments. Civil disobedience can also be manipulated by the governments. So-called "people happenings" in ex-Yugoslavia have been manipulated by those who wanted (and succeeded in grabbing) power. CD can be also manipulated by trasnationals and NGOs, or any other group. It is necessary to invent ways of incorporating CD in an international and global system. CD is fully realized as an activity of a group,[18] and it is necessary to construct mechanisms that will prevent the abuse of these groups, for example, using a CD group as a means for realization of objectives of either nation states, NGOs, transnationals or any other. CD is an expression of the free will of the human being, it underscores the relationship between us and laws, even between us and ethics. The laws and the states, all social institutions, are for the human being, to serve a person and to enable a person to fully express her/his free will. The human being is the beginning, the centre and aim of all laws. In many cases we all agree that laws are to protect the individual and as an illustration: in Rio de Janeiro the City authorities decided not to fine drivers who run through the red light from 10 pm till 5 am, since - while waiting at the red light - many have been robbed and even killed.[19]

It may be argued that I am actually discussing civil involvement, rather than CD. Certainly, I am strongly arguing for involvement of citizens. However, I want to stress that the involvement has to be critical. Therefore, though it may not be easy to invent legal mechanisms for including CD in an international and global political system, it is certainly necessary that the spirit of the thoughts and actions of Socrates, Thoreau, Martin Luther King, Bertrand Russell, John XXIII and Mahatma Gandhi become one of the basic pillars of our political mechanisms. The need to be critical, to challenge the authorities - and to do it in a civil way -is very well described - in my view - in the following two writings which I will use here instead of any conclusion:

What is Law? Martin Buber argues that the Old Testament should be read as a continuous dialogue between the people and God. It is through such dialogues that Jews became the chosen people. Jewish laws are not dogmas, they are questions to be pursued. Though the laws are expressed as grammatically imperative sentences, if the subjects are free persons, who can object and disobey, then the law is at least to some extent a question. Governments derive their just powers from the consent of the governed. The word consent contains within it the meanings of both assent and dissent. When this is forgotten, the law loses its power to persuade reasonably, and it is no longer law. The power to persuade correlates with the power to question, and so the law is a dialogue. A good law not only is a question, but it keeps the big question of justice open.[20]

Paulo Coelho wrote in his novel The Fifth Mountain:

> That night somebody entered Jacob's tent and wrestled with him until daybreak. Jacob accepted the combat even knowning that his adversary was the Lord. And in the morning the Lord said "Let me go", and Jacob replied "I will not let you go except when you bless me." And God blessed Jacob. Sometimes it is necessary to struggle with God....
>
> Elijah fled from doubts, from defeat, from moments of indecision. But the Lord was generous and has led him back to the abyss of unavoidable, to show him that the man has to choose - and not accept - his fate. When Elijah asked the Angel why must he choose, the Angel replied: "Because a man must choose. Therein lies his strength: the power of his decision. He who makes no choice is dead in the eyes of the Lord.[21]

References

[1] Ivo Slaus, contribution to the conference of the Swedish Pugwash on the occassion of the 90th birthday of Sir Joseph Rotblat, 1998; also in Proc. Intl. Conf. on Sustainable Development, Bled, Slovenia, 6 June 1999.

[2] H. Kissinger, *Diplomacy*, NY 1994.

[3] H. Kueng, *A Global Ethics for Global Politics and Economics*, NY, Oxford, Oxford University Press, 1998.

[4] A. Toffler, *Power Shift*, 1994.

[5] International conference on Peace, Human Rights and the Responsibility of Intellectuals, Opatija, 1994, specifically contributions by Ivan Supek and by Iorri Duursma.

[6] S. Huntington, The Clash of Civilization, *Foreign Affairs* 72 (1993).

[7] Explorations in Global Ethics, edited by S.B. Twiss and B. Grelle, Westwide Press, 1998, and references therein.

[8] *The End of Politics*, edited by A. Schedler, MacMillan Press Limited, London, 1997.

[9] *International Encyclopaedia of Social Sciences*, Vol 2, p. 473 (1968); J.G. Murphy, *Civil Disobedience and Violence*, Wadsworth Publ.Co, Belmont, 1971; *Dissent and the State*, edited by C.E.S. Frank, Oxford University Press, Toronto, 1989; R.D. Dixit, *Civil Disobedience*, GDK Publications, 3623, Chawri Bazar, Delhi, 1980; J.F. Childress, *Civil Disobedience and Political Obligations*, New Haven and London, Yale University Press, 1971.

[10] John Rawls, *A Theory of Justice*, Harvard University Press, Cambridge, Mass, 1971.

[11] Burton Zwiebach, *Civility and Disobedience*, Cambridge University Press, Cambridge, 1975.

[12] Plato, Crito and Apology, in J.G. Murphy, *Civil Disobedience and Violence*, Wadsworth Publ.Co, Belmont, Ca, 1971.

[13] H.D. Thoreau, "On the Duty of Civil Disobedience" in J.G. Murphy, *Civil Disobedience and Violence*, 1971.

[14] Martin Luther King, "The Criteria of an Unjust Law," in C. Crawford, *Civil Disobedience*, Th. Y. Crowell Co, N.Y., 1973.

[15] Bertrand Russell, "Civil Disobedience and the Threat of Nuclear Warfare," in H.A. Bedau, *Civil Disobedience*, Pegasus, NY, 1969.

[16] M. H. Gandhi, "Non-violence," in J.G. Murphy, *Civil Disobedience and Violence*, 1971.

[17] H.A. Freeman et al, Civil Disobedience, an Occassional paper, published by The Center for the Study of Democratic Institutions, 1966,

[18] H. Arendt, Civil Disobedience in Crises of the Republic,, A. Harvest HBJ Book, 1972.

[19] *Los Angeles Times*, Saturday, July 24, 1999, A2.

[20] S. Buchanan, in Freeman *et al., op.cit.*

[21] Paulo Coelho, *The Fifth Mountain*, translated by C.E. Landers, Harper Flamingo, 1998, Harper Collins Publ., NY.

SOME ASPECTS OF THE GOVERNANCE QUESTION IN SOUTH AMERICA

Mariano Egaña and Walter Scheuer

In a recent short and somewhat apocalyptic article by Oswaldo Rivero bearing the title "Non-governable chaotic entities"[1], a few South American countries are mentioned. Colombia is one of them because the author sees the country as entering a stage under the rule of warlords and drug-dealers, as being deeply troubled by very active conflicting armed groups, as having increased the mean annual income of its population during the approximately last thirty years by only 0.2 per cent, and also because of its relatively high urban population growth-rate. Peru is mentioned because of reasons similar to the last two given for Colombia, whereas Bolivia appears due only to its 4.1 per cent yearly urban population growth-rate (1990-96). Rivero assumes that these factors contribute to turning a country into an ungovernable chaotic entity.

In what follows, we shall take a brief, partial and unsystematic look at some present governance related aspects of the ten South American countries with Spanish or Portuguese roots, devoting some more detail to those closer to being ungoverned (such as Colombia) or having a relatively uncertain future in some area (such as Venezuela, on the political side). It will be seen that, due to the great interest of the USA in the matter, the Colombian drug production and related money laundering questions permeate much of the scenery.

We have limited ourselves to the South American region, since the broader Latin American one – also including Central America, the Caribbean Region and Mexico – differs, from the geopolitical point of view, from the South American one. Indeed, from the strategic perception of the USA, the dominant regional (actually, world) power, South America and the rest of Latin America are two geopolitical spaces with different significance. The former is a *region of strategic interest* for which decisions concerning continental security are adopted taking into account political and economical interests as well. Instead, Central America, the Caribbean conglomerate and Mexico form a *region of strategic security*, a space where the USA establish the security parameters unilaterally, as shown by their direct interventions in Panama and Grenada and their indirect ones in, for instance, El Salvador and Nicaragua.

A Few Features of the Present Situation in South American Countries

Colombia (37.1 inh.; 1,139 km^2; 1,738 US$; 53 pos. - 0.85 val.; 90 hom.) [*Inhabitants* (in millions, 1997); *surface area* (in thousands); *GDP* per capita (1997); *Human Development Index* (position and value, 1995); *Intentional homicides per 100,000 inhabitants* (1989-96, according to country)][2]

Over 40 per cent of its territory is out of governmental control. Two insurgent groups, the Revolutionary Armed Forces of Colombia (FARC) and the National Liberation Army (ELN), concentrate the military opposition against the Colombian regular armed forces, as well as against strong paramilitary groups that massacre peasants they suspect of cooperating with the guerillas. Analysts estimate the insurgent forces amount to over 20,000, equipped with last generation conventional weapons, acquired with funds of unspecified origin, although probably coming from the growing drug business (an alternative the USA government officially takes for granted). The rising daily death toll in the confrontations varies according to the source, but is nearing ten. This situation has negatively affected legal economic activity, mainly tourism; conversely, cultivation of coca has increased about 30 per cent between 1997-98.

The FARC, as a clear sign of their political growth and searching for recognition as regular belligerents instead of guerilla status, send "diplomatic representatives" to the countries of the region and communicate with the head of the New York Stock Exchange, who has recently visited the conflict zone. On the other hand, an example[3] of the current institutional crisis reflected in the loss of power and responsibility by the state, even in the military field itself, is illustrated by what follows. The Colombian Ministry of Defence, through an agreement signed with two foreign oil companies, has set up two *ad hoc* battalions for the protection of their plants (730 men in all). In return, the companies provide important general equipment, arms and salaries for those troops. The corresponding institutional figure resembles that of state-dependent mercenaries meant to protect specific private property and profit, so as to prevent the loss of foreign investment, a situation close to that of an emerging privatization of the Colombian official armed forces. The Constitutional Jury declared the procedure unconstitutional.

The presence of the USA in Colombia and its concern about the development of the internal situation are growing daily, and perhaps so are the chances of its direct participation in the conflict. However, such participation would depend on an active involvement of the countries of the region. At the end of August 1999, the head of the USA Drug Enforcement Agency (Barry McCaffrey) visited Bolivia, Peru, Brazil and Argentina, officially only to press them to adopt harder positions in reference to growing money laundering, but also to explore their political willingness to cooperate in an eventual external military intervention, even though the USA government repeatedly denied such a possibility. To conciliate McCaffrey's exploration with the above-mentioned visit of the head of the New York Stock Exchange, one may have to keep in mind the clear distinction between government and the private sector prevailing in the USA.

Venezuela (22.8 inh.; 912.1 km^2; 3,267 U$; 46 pos. - 0.86 val.; 12 hom. (1989))

The election to the presidency of former lieutenant colonel Hugo Chávez (known as having staged a failed coup against civilian rule in 1992) meant the end of a democratic institutional system that governed during 40 years through two repeatedly alternating political parties: COPEI and Acción Democrática. After only six months in office (and keeping a popular support of 88 per cent) the party of Chávez – Nuevo Polo Patriótico - won 121 out of 128 seats in the powerful National Constitutional Assembly, whose aim is to reform the 1961 constitution. The undeniable support he was given by all levels of society can be understood as a reaction against the high levels of corruption that ravaged the country and were mutually tolerated by both parties.

Chávez's ideological position is ambiguous. He says he holds with the US conception of Simón Bolivar and advocates the building of a "humanitarian" capitalist model of free market and respect for public rights, surveyed by the regulating presence of the new state.

Not having turned into productive inversions the large benefits arising in the 1970s from the exports by the state-owned oil industry, nowadays 80 per cent of the Venezuelan population still has unfulfilled basic needs. This tells of a troubled economy, now in recession since 1998. On the other hand, Venezuela is the main oil supplier of the USA; moreover, about 50 per cent of its exports go towards the US market, and from there comes about the same percentage of its imports. Because of these figures, Venezuela holds a certain special interest for the USA.

A border of 2,200 km with Colombia implies the risk of an overflow of the Colombian conflict.

Ecuador (11.9 inh.; 283.6 km^2; 1,392 US$; 73 pos. - 0.77 val.; 10 hom.)

For many months the country has been in serious economic problems and troubled by social protests coming from a variety of sectors, which has led the high ranks of the armed forces to address a severe warning to the political class (this seems to confirm the long-held belief that those forces are the real power behind the constitutional government). Thus, even if the government of President Mahuad has overcome several political crises during the last months, institutional continuity could come to be endangered. The border with Colombia is about 600 km long.

Peru (24.4 inh.; 1,285 km^2; 2,209 US$; 86 pos. - 0.73 val.; ? hom.)

President Fujimori's intentions of holding on to power threaten the institutional system. The country is directly affected by the conflict in Colombia, separated from Peru by a 1,600 km long border. This is shown in the fact that at present the Peruvian army is training about 300 Colombian officers in anti-guerilla warfare, taking advantage of the experience acquired during their successful struggle to defeat the local guerilla. On the other hand, in contrast to what happened in Colombia, Peruvian coca cultivation is said to have diminished over 25 per cent between 1997-98, an improvement which seems in danger of being reversed by the more recent emigration to Peru of Colombian drug-dealers escaping from their country because of the violence there.

Brazil (163.4 inh.; 8,512 km^2; 3,239 US$; 62 pos. - 0.81 val.; 20 hom. (1989))

Due to rising recession and following a maxi-devaluation in 1999, active social protests are increasing, headed by small and large farmers, industrial workers, public servants, and the so-called Movement of the Landless (over 4 millon peasants excluded from the economic system). Recession, increasing the already great public insecurity, and the action of small private armed groups that oppose popular demonstrations and demands, are some of the factors politically destabilizing President Cardozo's government.

As Brazil shares 1,600 km of its Amazonian border with Colombia, the Colombian conflict triggered the decision of the Brazilian government to transfer troops to the border region and to make it clear that, in case of FARC penetration towards Brazil, there would be a military response.

Bolivia (7.8 inh.; 1,099 km^2; 942 US$; 116 pos. - 0.59 val.; ? hom.)

The bad social-economic situation – its GDP per capita as well as its HDI are the lowest of the region – probably contributed to the decision of the great industrialized nations to include Bolivia, as the only South American country, into the initiative addressed to alleviate the international debt of the highly-indebted poor countries.

In spite of its weak economic situation, the country seems to have reasonable stability after the intense repression of the once strong trade unions at the end of the 1980s. The government – with foreign financial support – led a campaign which succeeded in diminishing the coca cultivated area by 15-20 per cent, between 1997-98.

Paraguay (5.1 inh.; 406.8 km^2; 1,489 US$; 91 pos. - 0.71 val.; ? hom.)

Paraguay is the largest marijuana producer in the world, which implies a context of probable future conflicts. But already the recent murder of Vice-president Argaña and the following street riots that caused the fall of the government as well as the escape of Lino Oviedo, long-time strongman of Paraguay, evidenced the great weakness of the current highly corrupted democratic system. Today, under political asylum in Argentina, Oviedo keeps enough operative possibilities to intensify this weakness.

Besides, the large illegal arms and drugs trade is growing in the country, in a mafia context of intense smuggling of all kinds of goods, which, in the near absence of regular economy, has flourished for many years (especially in the region where the borders of Paraguay, Brazil and Argentina meet).

Chile (14.6 inh.; 756.6 km^2; 3,957 U\$; 31 pos. - 0.89 val.; 3 hom. (1989))

The Chilean economy is the most stable and planned of the region, though it has led to a highly unfair income distribution. The current recession, the worst of the last 14 years, is a consequence of the Asian economic crisis, since Chile has based its development on large exports into that region; following recession, unemployment rose close to a 10 per cent rate. This, in turn, led to increasing industrial workers, teachers, and students mass demonstrations.

Another sign of social discord comes from the fact that the Chilean society is deeply polarized in their vision of the 1973-90 military regime. The international judicial conflict about ex-dictator Augusto Pinochet is now intensifying this internal contradiction.

Argentina (35.7 inh.; 2,780 km^2; 6,512 U\$; 36 pos. - 0.89 val.; 4 hom.)

After ten years of President Menem's administration and only three months before the taking over of a new government, the country is deep in recession and severely affected by the structural crisis of many bankrupt provincial states. These conditions contribute to the rise of the already large unemployment rate (15 per cent). Daily insecurity is growing, as well as administrative corruption. To the protest demonstrations of low-income sectors, are increasingly added those of small industry owners and of the agribusness sector, small and large, protesting against the absence of credit, and against the high interest rates when credit is eventually granted. The expectations of the citizenship are not high about the chances of the next administration to solve economic and security problems in the short term, but there is confidence regarding a significant decrease in corruption.

Following the unforgettable Mothers of Plaza de Mayo pattern of civil action, started during the 1976-83 death-laden military dictatorship, at the beginning of the present decade people began to organize peaceful weekly street demonstrations in connection with criminal, social or environmental injustice. These repeatedly achieved success, in some cases after going on for months or even years in connection with one and the same question. Although they were often unheared, nevertheless they have shown that within an inefficient and/or corrupt political, security and juridical system, sometimes it is possible to change injustice into justice, which certainly is a positive aspect of civil disobedience.

Uruguay (3.2 inh.; 177.4 km^2; 3,437 U\$; 38 pos. - 0.89 val.; 4 hom.)

The country is also suffering from the regional economic recession, which increasingly provokes the people's reaction in demonstrations, recently mainly of public servants and the agribusiness sector.

Mercosur (without associated countries: 207.4 inh.; 11,876 km^2; 1,489 to 6,512 U\$; 36 to 91 pos. - 0.89 to 0.71 val.; 4 to 20 hom.)

In 1991, the Asunción Treaty created the at present dominant economic regional space in South America. It comprises 68 per cent of the surface of the ten countries we are taking into account and 64 per cent of their population. Full members Argentina, Brazil, Paraguay and Uruguay, joined by Chile and Bolivia as associates, have strengthened this custom union of geopolitical importance, and in only eight years trade increased four times.

But the economic and production asymmetries of its main partners (Brazil and Argentina), added to the frequent international financial crises since 1998, in particular the already-mentioned Brazilian devaluation, have recently sparked difficult-to-solve conflicts that require changes in the original treaty. Central among them are the effective prohibition of non-agreed-upon macro-economic measures and the setting up of a so far non-existent arbitration process for handling minor disagreements. However, at present, the local industrial and agribusiness sectors press their governments for the adoption of protectionist measures.

Moreover, this regional market has been negatively affected both by Menem's untimely and unconsulted proposal of "dollarizing" the country's economy and Argentina's equally untimely and unconsulted application to become a member of NATO. An eventual future increase in the number of permanent members of the UN Security Council will probably become a source of additional friction in the Mercosur context.

Some Broader Aspects of the Regional Situation

At present, in the South American countries mentioned above, there are common processes which – although in a different degree according to the country – in some cases affect the present and, in others, the future governance of their societies. In our opinion, when talking about "governance in South America" at this time, it implies considering first the social factors (often a consequence of strongly inequitable economic situations) and then the institutional ones. Among these factors, we would like to stress the following:
(a) Growing social violence and civil insecurity. (Of low incidence in Chile and Uruguay)
(b) Increase of the gap between socially included and excluded sectors. (All countries)
(c) Economic recession and thus high unemployment. (All countries)
(d) Crisis of the institutional democratic system (Current and intense in Colombia; current and high in Venezuela, Peru, and Paraguay)
(e) People's generalized perception that most of the institutional sector neither represents them, nor works to improve their quality of life. (All countries)
(f) People's generalized perception that regional security is endangered by a military escalation following a possible overflow of the Colombian conflict. (All countries)

As item (d) shows, in our opinion, at present the crisis of the democratic system is particularly intense, although in different degrees, in "only" four countries. No clear correspondence relates this situation to their GDP or HDI values, although only Ecuador and Bolivia have worse GDP values than any of these four countries, and only Bolivia has a worse HDI.

Closing Remarks

South American countries are undergoing a historical stage characterized by social and political instability. In spite of their efforts to consolidate genuine democratic systems in the region in the course of about the last 15 years, factors of a diverse nature hinder their integral development. We mention the following, in the selection of which we could not avoid the influence of our own Argentine experience:
(a) Weakness of democratic institutions not yet consolidated after decades of authoritarian rightwing governments, almost always supported by the USA and other foreign powers.
(b) The state in crisis as a consequence, among other factors, of the opening of national economies, privatization of public services [4] and removal of restrictions for laying off personnel.
(c) Discredited, bureaucratic, and inefficient political parties.

(d) Judicial systems willing, even in their highest levels, to please the government in office, frequently offering protection in connection with corrupt business.
(e) Absence of long-term state policies, reached at by consensus with representative sectors of society.
(f) High levels of administrative corruption, public and private, including high tax evasion by the more affluent social sector.
(g) General lack of control entities responsible for establishing reference parameters, in particular for banks and the privatized services.
(h) Scarce development of a civic culture that, through the accumulation of "social capital", may promote the consolidation of the Civil Society (the Third Sector).
(i) Increase in new threats to security: drugs trade, massive illegal immigration, conflicts of an environmental nature.
(j) Impossibility to honour the payment of the interests of foreign debts, which leads to permanent negotiations with the international financial organizations.

The juxtaposition of (a) and (b) is not hazardous. The crisis mentioned in (b) is particularly severe in the region, in comparison with the one going on in the "first world", and this is partially due to the fact that the world process of what Ricardo Petrella [5] calls the "Dispossession of the State", takes place in South America under the conditions described in (a).

References

[1] Oswaldo Rivero; *Le Monde Diplomatique, South Cone Edition, N°2, p. 5*; Editorial Tres Puntos S.A.; Buenos Aires (August 1999).

[2] According to the GDP per capita values, all 10 countries are included in the intermediate income group; according to the Human Development Index (HDI), 6 of them are included in the group of high development, 4 in the group of medium development and none in the low development group (groups as defined by UN). The highest/lowest 1995 HDI values are 0.96 (Canada)/0.19 (Sierra Leona). Data taken from recent publications by UN and the Inter American Development Bank.

[3] Taken from an article by Martín Lozada in the Patagonian newspaper *Río Negro* (21 August 1999, p. 13).

[4] Privatizations which often – although not always – improved the services, but always meant higher costs for the user.

[5] Ricardo Petrella; *Le Monde Diplomatique, South Cone Edition, N°2, p. 1*; Editorial Tres Puntos S.A.; Buenos Aires (August 1999).

INFORMATION TECHNOLOGIES AND INTERNATIONAL GOVERNANCE: POSSIBILITIES AND PROBLEMS

Anna Moden

With the increased influence of intergovernmental organizations (IGOs) and multinational corporations, where the decision-makers are not directly, or not at all, elected by the people, individual citizens may seem to have been removed from the centres of power. At the same time, we see a rising demand for transparency on the national and international levels. When voters cannot directly vote a mistrusted politician or party out of office, we have to rely more on another form of public opinion. This is where information technologies can play an important role in empowering individual citizens and groups and give them a voice in decision-making in which they formally do not have a direct part. The Internet and other information technologies have provided a new way for individuals and groups to engage in political action and take a more active part in some aspects of international governance. This papers is an attempt to bring up some of the possibilities that these new technologies offer and also some of the problems that need to be considered.

Possibilities

Information technology offers the public a tool to exercise increased influence over issues such as environmental degradation, arms control and human rights that require international governance. The Internet, probably the first example of information technology that comes to mind, facilitates the distribution of information and a rapidly growing number of private persons now have access to documents from governments and international organizations at their home computers.[1] There is increased pressure on governments, IGOs and corporations to make this information available to the public. They seem to be fairly responsive, both to improve public relations and for the purpose of increasing public participation.

The Internet is also a very efficient and inexpensive networking and opinion-building tool. The global effort to ban landmines successfully used information technologies to garner support for their mission and to set agendas for international cooperation and national policies. A powerful coalition of 600 NGOs, labour organizations and businesses, including the AFL-CIO, the Sierra Club and the Western Governors' Association, managed to bring the negotiations on the Multilateral Agreement on Investment (MAI) to a halt, with the help of a global Internet campaign.[2] Whether this was a desirable outcome of the use of the Internet is under debate. With the growing demand for transparency, governments and IGOs are forced to take public opinion into account. In the words of Ann Florini of the Carnegie Endowment for International Peace: "The information revolution provides for translational collective action: networking and transparency or regulation by revelation."[3]

Patti Whaley, the deputy general secretary of Amnesty International, spoke about the role of communications technologies in the Kosovo conflict at Student Pugwash USA's 20th anniversary conference.[4] She told the story of Radio B92, which shows how important the Internet can be as a tool to build international public support. Radio B92, the leading independent broadcaster in Yugoslavia, was banned by the Yugoslav government this spring when NATO air strikes began. To back Radio B92, a group in Amsterdam almost immediately launched a "cyber support campaign" and a website, www.helpB92.xs4all.nl, that received 15 million hits in seven days. (Following the end of the NATO bombing, Radio B92 returned to the air with a new website, www.freeB92.net [5])

Whaley also mentioned her organization's plans to use the Global Positioning System (GPS) and the Internet to provide "virtual accompaniment" for human rights workers at risk. The human rights worker in question would carry a GPS receiver, and with the help of radio signals transmitted by one of the 24 GPS satellites in orbit the exact position of the person can be determined.[6] This information could then be used by groups on the Internet who would take turns "accompanying" the human rights worker and be ready to immediately alert Amnesty International if there were reasons to believe that he or she was at risk. Amnesty would then mobilize its international network to lobby on behalf of the human rights worker.

Imagery generated by remote sensing satellites offers another way for non-governmental actors to obtain information that can be used to advance their agendas and push for policy changes. Imagery that has the capability of showing objects on the ground that are as small as one metre in diameter is now commercially available and can be used to detect refugee movements, discover mass graves, and track environmental degradation. Until recently, these kinds of images were only accessible to the military.[7]

The emergence of the technologies described above might serve to hinder people in positions of power from abusing it. The mere risk that someone may be watching what you are doing, or that it is easier to reveal information about abuses, might discourage atrocities or environmental destruction, among other things.

Questions and Problems

The question of access to these information technologies is central. Who can use these technologies to pressure governments and international organizations to pay attention, and what issues do they promote?

A large majority of the world's population does not yet have the possibility to benefit from these technologies. There are efforts being made to narrow this gap between developed and developing countries, but obviously much remains to be done. An interesting example is a new project led by the UNDP and Cisco Systems, a leading US corporation providing computer network hardware, to use the Internet to fight poverty and assist in development and democratization in the developing world. "Their objective is to marry Cisco's technology with the UNDP's global presence to bridge the immense gap between what the UN organization calls the 'knows and the know-nots.'"[8] The project will create the world's largest Internet site, www.NetAid.org, and the partners plan to provide local Internet access at UNDP centres all over the world to connect non-governmental organizations, private groups, and businesses to promote development.

Censorship by repressive government also hinders access to these potentially powerful technologies. We have seen, for example, how the Chinese government blocks access to the Internet and monitors e-traffic when it suspects that they are being used by dissidents.

In addition to the positive uses of satellite imagery, it can also be used to plan terrorist attacks and for industrial and other types of espionage. The government of Israel, for example, has expressed concerns about safety and the US Senate has voted in favour of restricting the availability of imagery of Israel.[9] The implications of the availability of this type of imagery for sovereignty needs to be examined. What happens to the concept of sovereignty when international groups and corporations can obtain information that until recently only security agencies of national governments had access to?

The increased speed of information exchange, propelled by the global media and the growth of the use of the Internet, where the public can follow developments almost in real time, forces fast decision-making with higher risks of making mistakes. The quality of the information available is also important. P J Simmons warns that the information revolution allows for the rise of "global idiots." Today, anyone with a computer and a modem can launch a global campaign and have a disproportionately large influence on public debate.[10]

With so much information out there, it is also often difficult to determine what is legitimate and what is not. However, the same technologies used to spread false or misleading information can be used to reveal attempts to manipulate opinion. An example is the website launched by supporters of the government of Tunisia, www.amnesty-tunisia.org, praising the country's human rights record, using the well-respected human rights organization's name to, one has to assume, build credibility. Amnesty International, which has a different view of the human rights situation in Tunisia, countered by setting up a new website, www.amnesty.org/tunisia, that showed the screen of the Tunisian propaganda site and that of the Amnesty International site at the same time.[11] The site of the Tunisian government supporters eventually dropped the use of the word "amnesty" in its name and changed their site to www.rights-tunisia.org On the home page of the new site is an article named "Efforts by Tunisian government in the cause of human rights."

Conclusion

Because the influence of information technologies will only increase in the future, and will have a growing impact on international governance, it is important to examine both the possibilities and problems connected with these technologies. More transparency in decision-making and more public participation, aided by information technologies, will lead to more democratic international governance. However, to achieve a tolerably well-functioning system for global governance, we also have to realize the limits of the kind of public participation in decision-making that information technologies make possible.

References

[1] See, for example, the government of South Africa's website www.gov.za or the European Union site www.europa.eu.int

[2] Stephen J. Kobrin, "The MAI and the Clash of Globalization," *Foreign Policy*, Fall 1998.

[3] Ann M. Florini, speech at the Initiative for Social Action and Renewal in Eurasia, 28 July 1998, www.isar.org/carnegie

[4] Patti Whaley, speech at Student Pugwash USA's conference "Science and Social Responsibility in the New Millennium," 29 June 1999.

[5] Regardless of Frontiers – Protecting the Human Rights to Freedom of Expression on the Global Internet. Report from the Global Internet Library Campaign, www.gilc.org/speech/report

[6] Information on how GPS works from www.aero.org/publications/GPSPRIMER/WhatisGPS.html

[7] Yahya A. Dehqanzada and Ann M Florini, "No More Secrets? Policy Implications of Commercial Remote Sensing Satellites," Carnegie Endowment for International Peace, Working Paper Number 1, July 1999, www.ceip.org/programs/governance/RemoteSensingConf/NoMoreSecrets2.htm

[8] Thomas W. Lippman, "An Unlikely Net Alliance," *The Washington Post*, 12 August 1999.

[9] Federation of American Scientists, Public Eye Program, www.fas.org/irp/congress/1996_cr/h960730r.htm

[10] P J Simmons, "Learning to Live with NGOs," *Foreign Policy*, Fall 1998.

[11] www.amnestyusa.org/news/1999/53000199.html

GOOD GOVERNANCE IN AN AFRICAN PERSPECTIVE: AN INQUIRY INTO ACTORS AND ACTIONS

Bas De Gaay Fortman

> *As to our problems, we mention war, diseases, malnutrition, etc. In regard to child rights, in Africa we lack education. We need your support in order to fight against poverty and end wars in Africa.*
>
> Yaguine Koita and Fodé Tourikara[1]

Reviewing publications from the late 1960s and early 1970s, the period just after most of sub-Saharan Africa achieved independence, what strikes me is first of all our belief in the new politics as a basis for development. Not without reason colonialism had been regarded as the basic constraint to improvement of people's living conditions and to political integration.[2] The independent African state not merely signified political liberty, it would also constitute the primary instrument towards the uplifting of its own people.[3] A remaining constraint was, of course, white minority power in much of Southern Africa but liberation was certain. "One day", my colleague Richard Sklar exclaimed, "African nations will negotiate from strength."[4]

Sklar wrote this in a book entitled *Politics in Zambia* which appeared in 1974: after the transition to "the Second Republic" with its "one-party system of democracy". In that publication we do find concern with the quality of Zambian politics but, remarkably, not with the character of the state and the quality of *government* in post-independence Zambia. Yes, we had read René Dumont's *L'Afrique noire est mal partie*[5] with its attack on the African "elites" who had just continued their colonial predecessors' attempts at surplus extraction from the rural masses. We did review the barter terms of trade between people in rural and urban areas, showing that peasants had to produce twice as much maize to afford a shirt or a pair of trousers than before independence, and we did criticize Zambia's excessive spending on a huge copper-decorated Conference Hall, constructed especially for the non-aligned summit of 1970. But we did not look at the exercise of political authority, nor did we examine the quality of government as such. My own book on Zambia's economic reforms entitled *After Mulungushi: The Economics of Zambian Humanism*[6], describes and explains, among other things, why the Zambian government had to take a majority interest in a number of "key enterprises" but nowhere does it examine the capacities of the state as an intended instrument of development. In the non-competitive markets of a former colony, private enterprise was seen as a source of exploitation and the state as the protector of its people; it was as simple as that. The statist approach to development that was generally adopted was based on a general belief in the illegitimacy of private profit-making, shared by intellectuals and policymakers alike[7], combined with a failure to look behind the outward appearance of the state. In Charles Elliott's *Constraints of the economic development of Zambia* (1971)[8], to mention just one more publication to which I personally contributed, there is nothing on government and the capacities of the state. Faced by constraints in the international economic order and in the national economic structure, the primary challenge in our eyes was to find the right economic policies.

Ten years later, publications began to appear on matters such as *Corruption and Inefficiency in Zambia.*[9] A rapidly growing attention was paid to the terms of the political order and the ways in which political authority is exercised. A new terminology emerged with words such as "the failed state", "the collapsed state", "a rent-seeking society", "the facade of the African state", "the role of kleptocratic elites", "pocketalism", "the patrimonial state", "the predatory national state". My former colleague Sklar, too, now went beyond the abstraction of "African Nations", pointing at the control of dominant institutions in society by "ostensibly privileged power holders and their immediate families".[10] "African exceptionalism", as it has been aptly termed[11], had come to an end. What this implies is that African political discourse inherent in, for example, *African socialism*, *Zambian humanism* and *Negritude*, was no longer regarded as an acceptable basis for models of government that would make Africa exempt from democratic principles such as accountability, transparency, and substitutability of persons holding office.

Illustrative of the shift in focus is the way in which industrialization through import substitution is looked at as a strategy for development. While in the sixties and seventies the emphasis was primarily on diversification in economic "mono-cultures" and the need for protection through import restriction in the "infant industry" stage, the literature of the eighties and nineties focuses on "the competition for licenses and the governmental positions that dispense them."[12] Indeed, two decades later the language of development in Africa appears to have changed completely. Good government and "governance" had become "almost an obsession" in the development debate.[13]

The background of this new focus on *governance* lies in dysfunctional market economies in connection with failed economic policies. In the mid-eighties the failure of *development engineering* based on an instrumentalist view of the state and its law- and policy-making capacities had been clearly noted. The emphasis shifted from state-led development to a revived belief in the market. In 1986 a Conference was held in Nairobi on "the Enabling Environment for Private Sector Contributions to Development".[14] This was held to imply primarily a shrinking public sector so as to create space for the market economy. The general failure of the post-independence para-statals was acknowledged with ensuing public sector reforms including outright privatization. In order to acquire desperately needed foreign exchange African governments had to accept "structural adjustment programmes" that encompassed such public sector reforms as well as a severe cutting in subsidies, including those on basic food stuffs. These were meant to dismantle the old system of government controls. This consisted of not only a huge para-statal sector but also high protective barriers, state-operated credit schemes, price and incomes control, exchange control (with the accompanying black markets in foreign currency), and a system of state-set producer prices for agricultural commodities.

However, "an enabling environment" was also interpreted as a government that would provide the necessary supporting services. In its 1989 report on *Sub-Saharan Africa: From Crisis to Sustainable Growth* the World Bank specified the role of the state in *an enabling environment* as follows:

> The State has an indispensable role in creating a favourable economic environment. This should, in fact, be its primary concern. It is of the utmost importance for the State to establish a predictable and honest administration of the regulatory framework, to assure law and order, and to foster a stable, objective, and transparent judicial system. In addition it should provide reliable and efficient infrastructure and social and information services, whether private or state-owned.[15]

In its summit on *Social Development* in Copenhagen (1995) the United Nations re-emphasized the need for creating enabling environments for development. Recently, the United Nations Development Programme (UNDP) has linked the term to human rights. In this connection, it implies that rather than waiting for concrete violations to be addressed, actors in the field of human rights should enhance an environment within which people are able to exercise their rights under the law. In terms of policy, UNDP has identified some "entry points" for its enabling environments approach. These include promotion of democratization and the institutions of good governance.[16]

The conventional meaning of *environment* is the circle of people around us. In the context of good governance it refers to the institutional setting within which power is exercised. Particularly relevant in this connection are the economy, the polity, society and culture. For institutions such as the World Bank and the International Monetary Fund it is the *economy* that comes first. Good governance is primarily *effective* governance in the sense of an exercise of political authority conducive to the execution of economic policies held to be essential for a smooth functioning of the market economy.

Actually, in its policy implications the notion of good governance came into operation when it appeared that economic objectives could not be attained through economic reforms alone. In its policy documents the World Bank later expanded its view of good governance to include democratic principles, and a strong civil society while even showing an open eye to the necessity of rooting African government in Africa's own cultural traditions, beliefs and the structure of its component societies.[17]

Naturally, there is nothing against a government able to execute its own policies including the collection of taxes, nor against transparency, accountability and responsive governance in the sense of government responding to the needs and demands of civil society.[18] On the contrary, from the African peoples' own perspective there is every reason to applaud the end of African exceptionalism. Yet, there are two interrelated aspects to the governance debate that demand an urgent shift in attention on the part of all actors and their actions. The first one concerns context, the second content.

The *context* of good governance is, as we have seen, the creation of enabling environments. "Enabling whom?", one is inclined to ask. The answer lies in the background to the whole debate: *Sub-Saharan Africa: From Crisis to Sustainable Growth,* as it is expressed in the title of the 1989 World Bank report that launched the notion. The actors whose environment should be "enabling" are all those involved in the modern economy tuned to growth of Gross Domestic Product per capita. Government must provide "rules to make markets work, and ensure property rights", it is stated in the 1992 World Bank report on *Governance and Development.*

Looking now at the contemporary African setting, what we see is a huge majority of people excluded from this environment. More than 60 per cent live in situations of abject poverty, lacking any access whatsoever to the modern sector of their economies.[19] They are either subsistence farmers or unemployed workers in the shanty towns of the cities. Even in South Africa 70 per cent of the inhabitants of the townships are jobless. But, it might be asked, would not these benefit from an improvement of the environment of the modern market economy? In response to a similar question Engelhard has recently pointed to three important considerations.[20] Firstly, there are no indications that conventional economic growth in the African context reduces poverty. In Mali, for example, a six per cent GDP growth rate since 1995 has been accompanied by a 28 per cent increase of the number of people living below the poverty line. Secondly, a focus on the modern export-oriented sector will affect only very few actors. In fact, none of the sixteen African success stories in income-generation mentioned by the French Development Agency is in the realm of the world market. Thirdly, even if a focus on the modern sector resulted in a five per cent growth rate, it would still take a quarter of a century for the peoples of Sub-Saharan Africa to attain an average income per head at the level of Tunisia today.

Actually, the environment of those living in the outlying rural areas and the squatter compounds of the cities is far from "enabling". Millions of Africans live in *adverse environments*: misallocation of resources, extreme socio-economic inequality and exclusion, violence in and among communities, corruption, cultures of domination and submissiveness. This is the usual setting of abject poverty, famine and other gross and systematic violations of basic human dignity. In such conditions good governance is a term so far removed from the realities of the day that it has lost all meaning. Access to clean water, education, health services, food of sufficiently nutritious value – these are the pertinent issues.

A primary adverse or "disabling" factor (as opposed to the qualification "enabling" in "enabling environments") is, indeed, a malfunctioning economy. This has, however, two aspects: economic ineffectiveness in the sense of misallocation of resources and economic injustice meaning exclusion of people from access to resources and entitlements to basic goods and services. The second question, then, is "What is the *content* (meaning) of development as a goal that should be enabled by an environment of good governance?" A review of conventional development-oriented approaches to poverty and destitution[21], based purely on increasing productivity as they are, reveals that responses merely focusing on market freedom tend to be insufficient at the least and sometimes even counterproductive. What is needed, then, is a view on development based on economic justice. This requires a proper analysis of existing *entitlement systems* - defined here as regularized arrangements in which people's participation in the economy is rooted - and their functioning. What structures are behind the little the poor can get from day to day and what blocks them in efforts to improve their livelihood? How to explain the apparent entitlement failures behind abject poverty? Clearly, if resource constraints are to be tackled through attempts to increase productivity these will have to be supplemented by entitlement-oriented approaches. For one thing, such entitlement analyses are likely to point to the necessity of supporting credit schemes for the hundreds of thousands of micro-enterprises in Africa, rather than merely focusing on the operations of a modern banking system.

In this connection, it should be noted that the African masses are primarily dependent on the functioning of rural subsistence economies and urban informal economies. The role of law here is, first of all, *customary law* in an agricultural setting and *living law* in the context of an urban informal economy. This implies a focus on actual rather than just formal control of power albeit with an open eye to legitimacy in processes of executing power. In countries in which there is no meaningful labour law, for example, workers may still exert a significant influence upon working conditions.[22] Living law is *not* state-law. Noteworthy in connection with the still growing attention to failed and collapsed states is a certain improvement in the living conditions in Somalia in a situation of almost complete ungovernment.[23]

Another important distinction in regard to the content of development is between *primary* and *subsidiary* entitlement systems. Subsidiary titles to claim goods and services materialize only after failure to acquire on the basis of rights that guarantee immediate access. An example is social welfare to support people who fail to get earnings from labour; another illustration is food aid in case of famine. Naturally, people tend to prefer primary entitlement in the sense of having access to resources and rights to goods and services on the basis of their integration into the community rather than subsidiary entitlement as compensation for their marginalization. Illustrative is a shift in government thinking about poverty in South Africa: "Previously the emphasis was on redistribution of incomes... closing the gap. Now the focus is on more jobs".[24] Such shifts in approach reflect a growing attention to people as subjects of *rights* rather than charity.

In order to be meaningful in a development context, good governance, then, is to be related to the living conditions of the poor masses. Consequently, this notion should be lifted out of a narrow Weberian bureaucracy perspective and connected with people's daily struggles for livelihood. Illustrative in this connection is my own experience with one of the weekly "Donors on governance, democracy and human rights" meetings that are held in the office of the deputy permanent representative of UNDP in Lilongwe (Malawi) each Monday morning. On the day I attended this session a survey was presented which the Carter Centre for Democracy had carried out in Mwanza district, a truly outlying rural area. Focus group discussions had revealed that these people did not have any nostalgia for the authoritarian rule of dictator Banda, although their mistrust in politicians was revealed by the request that any development assistance should reach them through their chief and headmen rather than government ministers or MPs. Asked if they had anything to say on their own behalf they expressed a major worry about the tremendous rise in the cost of living since Banda had been removed from office and replaced by a democratically elected president. Apparently, the meeting saw no way to react to this confrontation with "the economic prerequisite of democracy" (Dan Usher). Upon hearing the report from Mwanza district, it simply proceeded to the next item on its agenda: the introduction of television in that country, something which Banda had always prohibited. Obviously, there is very little connection between the agenda of the donors on good governance and the daily anxieties of the poor masses. Of course, corrupt government does affect the lives of the poor but their first concern remains the purchasing power of the little money they can obtain by selling fish or a few bags of maize. Indeed, GG (good governance) cannot be separated from PPP (the purchasing power of the poor). If good governance is to mean anything to the poor, it should have a meaning in terms of economic, social and cultural rights.[25]

Finally, I should like to come back to our neglect of the issues of good governance in the pioneering period of the sixties and seventies. It is true that many mistakes were made as a result of our blindness to the actual use of political authority and the terms of the political order. But the present obsession with good governance conceals the persistence of constraints to a development beneficial to the masses that we then did notice in those pioneering days: a world economy in which Africa suffers from unequal terms of trade, a dual local economy with continuous marginalization of the subsistence sector and a painful daily manifestation of economic injustice inherited from colonialism. Certainly, these preoccupations have not gone out of date. Good governance is one thing; it is not the one big thing. Trying to explain the seemingly permanent African crisis, for example, the quality of African government is certainly not the only phenomenon that strikes the eye. Other major explanations lie in the false start of Western colonialism[26], exogeneous factors such as adverse developments in world commodity prices[27], and the disastrous "development cooperation" policies of foreign donors.

In the setting of Pugwash our primary concern is with peace and security. Indeed, human security also presupposes a realization of civil and political human rights. (Civil rights, for one thing, imply "personal security", a notion signifying that one cannot be lifted from one's bed in the middle of the night with no-one hearing anything any more.) However, as the United Nations Development Programme has shown in its *Human Development Report 1994*, it also refers to people's daily struggles for livelihood. Good governance was the slogan following the implementation of structural adjustment programmes that promised Africans "short term pain for long term gain". The removal of basic food subsidies and the way in which market pricing has affected access to health services and to education did indeed bring pain. So far the gain has rarely manifested itself. In conclusion, then, it is not primarily good governance but a full realization of people's economic, social and cultural rights - food, clothing, housing, clean water, health services, education - that should be seen as a major challenge confronting the 21st century.

References

1 Letter of 21 July, originally supposed to have been written in Conakry before these two African boys hid themselves in the landing equipment of an airplane to Europe in which they subsequently froze to death. Whether constructed or authentic, these words do represent the desperate human conditions in much of Africa today.

2 See, for example, Bas de Gaay Fortman and Paschal Mihyo, A False Start - Law and Development in the Context of a Colonial Legacy, in *Law and Politics in Africa, Asia and Latin America*, Vol. 26, Nr 2, 1993, pp 136 - 163.

3 "Uplifting" is the literal meaning of *Chitukuko*, the term used for development in Chinyanja, a language widely spoken in Southern Africa.

4 Richard L. Sklar, Zambia's response to the Rhodesian unilateral declaration of independence, in William Tordoff (ed.), *Politics in Zambia*, Manchester: Manchester University Press, 1974, p. 362.

5 René Dumont, *L'Afrique noire est mal partie*, Paris: Editions Seuil, 1961, translated in English as *False Start in Africa* (1968).

6 Bastiaan de Gaay Fortman, *After Mulungushi. The Economics of Zambian Humanism*, Nairobi: East African Publishing House, 1969.

7 See also Michael F. Lofchie, The new Political Economy of Africa, in D.E. Apter and C.G. Rosberg, Political Development and the New Realism in Sub-Saharan Africa, Charlottesville: University Press of Virginia, 1994, p. 178.

8 Charles Elliott, *Constraints of the economic development of Zambia*, Oxford: OUP, 1971.

9 Klaas Woldring, Corruption and Inefficiency in Zambia: A Survey of Recent Inquiries and Their Results, in *Africa Today*, Vol. 30: 3, 1983.

10 Richard Sklar, Social Class and Political Action in Africa: the Bourgeoisie and the Proletariat, in Apter and Rosberg (eds.), , *op.cit.*, p. 120.

11 See David Apter and Carl Rosberg, Changing African Perspectives, in Apter and Rosberg, *op. cit.*, p. 46.

12 Lofchie, *op. cit.*, 174.

13 David Williams and Tom Young, Governance, the World Bank and Liberal Theory, in *Political Studies*, Vol. XLII, p. 84.

14 See Sklar in Apter and Rosberg (eds.), *op. cit.*, p. 123.

15 World Bank, *Sub-Saharan Africa: From Crisis to Sustainable Growth*, Washington DC, 1989, p 55.

16 James Gustave Speth, Poverty: A Denial of Human Rights, in *Journal of International Affairs*, Fall 1998, Vol. 52, No 1, 288-291. See also United Nations Development Programme, *Integrating human rights with sustainable human development. A UNDP policy document*, New York: UNDP, January 1998

17 See a publication by its Senior Policy Adviser in the Africa Region Technical Department, Pierre Landell Mills, *Governance, Civil Society and Empowerment in Sub-Saharan Africa*, paper for the Annual Conference of the

Advancement of Socio-Economics 1992, quoted in David Williams and Tom Young, *op. cit.*, p. 95.

[18] The term *civil society* refers to "that segment of society that interacts with the state, influences the state, and yet is distinct from the state" (Naomi Chazan, 'Africa's Democratic Challenge', *World Policy Journal*, 1990, p. 281). It implies organizations that take an interest in the affairs of governance beyond their own group interests. There is an obvious interaction between civil society and good governance as mentioned above. As Judd has put it: "Civil society needs government to be open, responsive, and accountable. And an active civil society, acting not only as a check and balance on government but also informing political debate is essential and indispensable to politically sustainable development" (F. Judd, 'Fair's Fair: Democracy, Social Justice, Sustainable Development and International Synergy', *Development; Journal of the Society for International Development*, 1992 No 3, p. 7).

[19] This figure as well as following facts on the situation of people living today in Sub-Saharan Africa is taken from a stimulating paper in *Le Monde Diplomatique* by Philippe Engelhard: "Vers un développement à l'africaine" (September 1999, p. 21.

[20] See Engelhard, *Ibid.*

[21] See Berma Klein Goldewijk and Bas de Gaay Fortman, *Where Needs Meet Rights. Economic, Social and Cultural Rights in a New Perspective*, Geneva: World Council of Churches, 1999, Chapter 3.

[22] A historical example of *living law* as opposed to formal law is the situation in 18th century Barbados where it was impossible for slave-owners to sell husband and wife separately. This "prohibition" was not based on any formal indictment let alone that in case of infringement those whose well-being was at stake here, could have taken any contentious action. (Slaves were not legal subjects but merely objects of ownership.) The practice of selling spouses separate from each other had simply become impossible, based on social relations in which those who formally could not have any rights informally had achieved legitimation of certain power. Behind this "living law" was, of course, the fear of slave resurrections, or, in sociological terms *potential collective action*. Thus, it was the manifestation of individually powerless people as a *collectivity* that had apparently changed social relations.

[23] See *The Economist*, 28 August 1999.

[24] R. Mogotlane, *Developing South Africa in Peace: Migration in Perspective,* Paper for Pugwash Meeting no. 218, Lahti, 1996, p. 3.

[25] See Berma Klein Goldewijk and Bas de Gaay Fortman, *Where Needs Meet Rights*, op. cit.

[26] See De Gaay Fortman and Mihyo, *op. cit.*

[27] Zambia, for example, suffered from a fall in copper prices which give a tonne of copper bars today around one quarter of the exchange value in the first ten years after independence. Notably, copper is its major earner of foreign exchange, counting for a considerable proportion of both exports and government income.

COMMUNITY PARTICIPATION A *SINE QUA NON* FOR AN EFFECTIVE ANTI-CORRUPTION STRATEGY: YOUTH PARTNERSHIP FOCUS

Nceba Gomomo

Integrated mechanisms are required to fight corruption. These should include both reactive and preventative machinery to be effective: the former represented by the legislative and institutional framework. And the latter represented by civic and moral education programmes to expose the citizen to the nature, cause and the impact of corruption on democracy. Youth[1] must be identified as the first point of entry for these policy reforms as they form an integral part of future governance. History proves that the youth sector could be an effective role-player in mobilizing communities around social issues due to the fact that, in general, the sector is accessible and better educated. This paper's point of departure is that all other traditional means, often reactive, of fighting corruption, hinge upon the moral/ethics education system a country implements to ensure compliance rather than enforcement. The World Bank President, James Wolfensohn, not only calls for diagnostic surveys on the provision of public services and business practices, and the training of judges, parliamentarians, auditors and journalists on how to identify and combat corruption but emphasizes the training of local communities to enable countries to fight corruption in a practical way[2]. The involvement of local communities provides the youth sector with a responsibility to be visible as active role-players in the "developmental discourse" to eradicate corruption. To make policy recommendations around the involvement of the youth sector it is imperative that the following aspects of the corruption debate are examined.

Firstly, the paper examines the theoretical aspect of corruption, namely, the legal and social conceptualization of corruption. Secondly, the paper discusses the concept of participatory development. Thirdly, causes and consequences of corruption are briefly discussed. Fourthly, the effectiveness of traditional means to combat corruption with a specific reference to the South African policy framework as its stands today is discussed. Fifthly a survey of the attitudes to corruption of youth in Hong Kong and South African involvement during the period of anti-apartheid mobilization provides a basis for my argument that the youth sector stands at a threshold for a sustainable anti-corruption strategy in the next millennium. A revisit to the South African National Crime Prevention Strategy (NCPS), as point of reference to the relevance of civic education, is also discussed in some detail. Consequently, the paper endeavours to provide some policy recommendations for an effective and sustainable strategy which links education and development as necessary paths to bring about better governance and effective participation in democratic processes to fight corruption.

Theoretical Aspect of Corruption

Legal and Social Conceptualization of Corruption

Among the numerous definitions of corruption the World Bank defines it as "the abuse of entrusted power for personal gain" or for the benefit of a group to which one owes allegiance. This interpretation presupposes the notion that corruption is occurring whenever there is abuse of entrusted power but leaves out individual corruption in the private sector for personal gain.

Legally, corruption is often defined to constitute an act of bribery, that is a general term covering the misuse of authority as result of considerations of personal gain, which need not be monetary. The South African Corruption Act[3] defines corruption as follows: "active corruption consist in unlawfully and intentionally giving or offering a benefit of whatever nature which is not legally due to someone else upon whom some power has been conferred or who has been charged with some duty with the intention to influence the latter to commit or omit to do some act in future in relation to such power or to reward the latter for some act or omission in the past in relation to such power".

A person commits passive corruption if he is somebody upon whom some power has been conferred, or who has been charged with some duty and he unlawfully and intentionally receives a benefit of whatever nature which is not legally due, with the intention that he should commit or omit to do some act in future in relation to such power, be rewarded for some act, or omission by him in the past in relation to such power.

On the other hand, sociologists like John Makum Mbaku try to give a more practical and contextual interpretation. He defines corruption in Africa as a practical problem involving the "outright theft, embezzlement of funds or other appropriation of state property, nepotism and the granting of favours to personal acquaintances, and the abuse of public authority and position to exact payment and privileges[4]. Typical of many definitions, this description of corruption denotes a view that corruption is a public sector phenomenon.

By and large, there is a consensus that we require a more workable definition, which would be able to drive and inform concrete and effective national and international measures. In other words, there is general acceptance that corruption is found in all spheres of society, that is, the public sector, private sector, and civil society.

Another dimension to conceptualizing corruption at the socio-economic level is how the corrupt activities affect development and growth, in particular its impact on redistribution of resources.[5] Furthermore, it is imperative to consider the correlation between corruption and security to be able to give a holistic analysis of the nature of corruption. We have seen a paradigm shift take place away from the traditional model of security that sees security only from a military dimension of security.[6] Security, as we conceptualize it today, encompasses conditions in which individual citizens can live in freedom, peace and safety; participate fully in the process of governance; enjoy the protection of fundamental rights; have access to resources and basic necessities of life; and inhabit an environment which is not detrimental to their health and well being. Corruption needs to be understood within this context, that is, as a general impediment to security and development[7].

As such there is no doubt that an anti-corruption strategy forms part of a broader developmental discourse for sustainable development. Therefore, participatory development would enhance the chances for better human security.

Causes and Consequences of Corruption

Much research has been conducted on the causes and consequences of corruption. Corruption is generally known to be caused by three fundamental factors. Firstly, lack of effective judicial and administrative institutions, devoted to providing clear and transparent mechanisms for private and public sector interaction, creates a fertile ground for corruption. Secondly, lack of direct participatory democracy is a major problem in today's political systems. And thirdly, a lack of commitment to upholding principles of good governance or good corporate governance in the public and private sector is another challenge in devising methods of combating corruption. The last two factors emphasize the absence of individual and group integrity to uphold moral standards that will ensure societal compliance with anti-corruption practices.

The causes of corruption have a direct consequence for the kind of societal values we accept to live under, unless there is a change to perceptions that corruption is a necessary tool for getting things done in developing countries. Even though there seem to be arguments from the economic literature that suggest corruption does not necessarily hinder economic growth,[8]there is increasing evidence that economic corruption is both unfair and inefficient. Politically, corruption leads to loss of faith in the government and non-compliance with laws and regulations. Eventually, it undermines democratic values and political equality and threatens human security.

Corruption is not only found in developing countries or emerging economies, although these economies may be more susceptible to corruption. International experiences indicate that the policy a state adopts towards fighting corruption will determine whether it takes a risk of marginalization in a world of rapid economic integration or creates opportunity for its people to prosper. The fact that there is a divergence of views on whether or not corruption hinders growth and development, means that countries adopt different policies to fight corruption. However, the evidence is towards movement to a more harmonized approach to bring about uniformity in this regard. This is evident from the general societal reaction, which is increasingly becoming uniform in dealing with corruption.

Traditional Means To Combat Corruption

There are numerous types of strategies[9] to minimize corruption. These include, *inter alia*, societal, legal, market, and political strategies.

- Societal strategies place emphasis on the determination of a common standard of morality against which corrupt behaviour can be measured. Vigilance by members of society and education to make it relatively costly for individuals to engage in corruption is also emphasized.
- Legal strategies emphasize enactment and creation of legislative and institutional mechanisms to combat corruption. Law is seen as a means to constrain corrupt behaviour by appointing commissions of inquiry or special prosecutors can be chosen to investigate individuals and groups suspected of corruption. And, where evidence points to corruption, the judiciary system can adjudicate over the case and punish the guilty according to national laws[10]. National laws and institutions have been created in many countries (including South Africa) to proscribe corruption.
- South Africa has enacted legislation and policies, which are necessary to control corruption/commercial crime. In the last four years we have see documents like Public service codes of conduct enacted to foster a corruption-free public service and a Prevention of Organized Crime Act and the National Prosecuting Authority Act enacted into law. And various institutions have been created to combat corruption such as the Heath Special Investigating Unit, National Director of Public Prosecution, Public Protector, Auditor General's offices, Public Service Commission and Anti-corruption Unit-SAPs.
- Furthermore, various gatherings have been convened nationally and internationally in the past 12 months to tackle the corruption malady. On the other side, civil society has vigorously challenged the sustainability of government initiatives, especially they are more reactive in nature and there is nothing practical they have done to involve and educate the public on this crime.
- Market-related strategies assume that there is a relation between the structure of the market and the incidence of corruption. The prescribed remedy is less government regulation and greater reliance on the markets for the allocation of resources. Such an approach, however, appears to emphasize the manipulation of outcomes within existing rules instead of proper reform of the rules. This presupposes that the market does not contribute to the incidence of corruption but the problem is with the legal system that regulates the market. Since rules define market outcomes, greater reliance on markets for the allocation of resources without reforming existing rules will have little or no outcome on the transformation of society. Bayley (1966) argues that unless there

is effective reform of the socio-political rules within which the markets function, incentives for opportunism will remain and corruption will continue unabated[11].

- Political strategies for corruption clean-up emphasize the decentralization of the public sector. This flows from a reasoning that concentration of power in the hands of a few politicians and bureaucrats results in corruption and this can be monitored and minimized by a system that encourages greater public participation. And under this approach (which is the crux of this paper) an effective anti-corruption strategy emphasizes a human rights path of sustainable human development.

However, all the last three strategies are dependent on a societal and political willingness and readiness to combat corruption. In the last decade international efforts have largely emphasized creating a legislative and institutional environment to combat corruption rather than a holistic approach that considers reorientation of communities as the first step to a sustainable strategy. One distinctive flaw of the present state policies is the lack of clear involvement of the youth sector in this regard.

Youth is the hope for our future. Their disengagement (as evidenced by lack of civic education focused to fight corruption) on issues of grave importance for stability and prosperity of their countries should be discouraged. In Africa (including South Africa) the Youth has been in the forefront when forces of liberation fought against imperialist and colonialist tendencies, that include concentration of resources to a group of people. My thesis is that corruption equates to anti-development. And youth cannot be allowed or encouraged to negate their role of enhancing and defending democracy by ensuring that corruption-preventing measures are in place. Educating a society on causes of corruption and its impact on development would give a state the required advantage over criminals by isolating them from a society, as their activity could be seen as anti-development. Others have moved a motion that corruption must be declared a *crime against humanity.*[12]

The fact that the youth sector is the most educated and accessible sector (in particularly in Africa) would make the dissemination of information easy and well targeted. It is within this context that I will explore why the youth, as an important stakeholder on issues of development, reconstruction and growth, cannot stand aside on issues of corruption. Firstly, my observation on involvement of youth will consider the Hong Kong model of an anti-corruption youth drive and, secondly, look at the South African political struggle as a basis for community mobilization to ensure that corruption is seen as a crime against humanity and a practical problem for development.

Participatory Development

It is accepted that empowerment and democratization can only take place within a conducive institutional environment comprising properly functioning parliaments, legal and judicial systems and electoral processes[13]. Parliaments made up of freely and fairly elected members representing different parties are crucial to popular participation and official accountability. Effective legal and judicial systems protect the rule of law and human rights. Open electoral processes that build public confidence and trust are important for establishing political legitimacy.

Moreover, governments everywhere need to make themselves more responsive to global economic changes and to the political demands of citizens. Political decentralization gives more political clout in decision-making to citizens or their elected representatives, and is usually associated with representative government, citizen participation and democratization. Decentralization is frequently associated with democratization because it allows citizens, or their representatives to formulate and implement policies more effectively than if all decisions were made by a small political elite or only by national political authorities.

Greater community participation has been associated with development. Democratization cannot be disassociated from development as it is a single process[14]. In its original sense, community participation was about improving the sustainability of individual projects by mobilizing the support of indigenous, local organizations and traditional technologies in developing countries.

At the beginning of the 1980s, the importance of organization and improving both the sustainability of development itself and people's development capabilities started to become apparent from experience with structural adjustment support and mounting concern for poverty and environment issues. There is increased recognition of the necessity not only of community participation in individual projects but also of widespread citizen participation in national development activities in recipient countries. This broad participation involves the alleviation of poverty through economic, social and political action, the cultivation of the private sector and the cultivation of a society where people are able to dissent or lodge objections against their government. It has also become increasingly understood that the government should play a major role in building the organizational and physical infrastructure that makes this participation possible. This period of governance transformation brought about a dialectical shift in the very concept of participation.

The following contributing factors lie behind this conceptual shift:

- visible moves towards multiparty democracy in Africa;
- calls for reforms from the burgeoning middle classes in Korea, Taiwan and Thailand;
- the impact of structural adjustment policies on developing countries;
- a shift to economic liberalization and transitions to market economies after the fall of socialist regimes in the Soviet Union and Eastern Europe.

In its original sense, democracy means equal participation by members of society in the processes of making and carrying out societal decisions. In contemporary democracies, however, it principally means guaranteed participation in the process of electing representatives. A problematic or controversial aspect of participation has emerged in recent years as result of a spreading awareness that party politics is waning and that parliaments have lost their original function of representation. Today, new forms of people's participation in politics have evolved. People do not only take part in politics via elections. With the rise of pressure and interest groups, they also take part in politics on a daily basis through specific activities, for example, anti-pollution movements and consumer campaigns.

The goal of economic and social development in developing countries is to set in motion a process of self-reliant and sustainable development through which social justice will be realized. Self-reliant development is defined as a process of building the indigenous mechanisms of society that will enable developing nations ultimately to achieve growth without aid. "Sustainable" development means continuing a stable growth pattern in such a way as economic development is in harmony with the environment. The realization of social justice means equalizing and ensuring opportunities for people to participate in order to rectify disparities between regions, income levels and gender.

The focus on development should not only be on increasing the material production but also on fostering and improving the social capabilities of people involved in development. For this to be done, people involved in development should take an active part in the process of planning and implementing development activities as well as enjoying their benefits. This is what participation means. Participation in every sphere of politics, economy and society is important as both the goal and means of development.

This is important for countries in transition to democracy and which regard empowerment of the people to participate in the process of governance as the centre of its transformation policy. For example, South Africa accepts that people-driven processes will bring a kind of democratic system that considers the most vulnerable sector of society, the poor, as active participants in decision-making. By denoting the people-driven principle as a cornerstone of policies to further democratization and development is to accept citizens must become full participants, especially where people are no longer viewing the state to be the enemy but as a vehicle for change. The United Nations Development Programme (UNDP) has argued that in making the case for a different approach to the developmental agenda in which eradication of corruption is an integral part: "Sustainable human development seeks to expand choices for all people... while protecting the natural systems on which all life depends. Moving away from a narrow, economy-centred approach to development, sustainable human development places people at the core, and views humans as both a means and end to development. Thus sustainable human development aims to eliminate poverty, promote human dignity and rights and provide equitable opportunities for all through good governance."[15]

Therefore, what becomes the issue is the role the people should play in a newly democratic state when it comes to policy formulation and decision-making. To achieve true people participation the democratic state has a responsibility to encourage popular mobilization to ensure that people themselves, and in their own interest, become conscious activists for development and social transformation. In the developmental discourse a state is expected to increase its resources by empowering the public who will be able to possess an in-depth understanding of the evolution of the newly-established state and its social predicaments, such as corruption. Participatory development through the youth sector will ensure that the populace comprehends the nature of corruption in terms of its practical causes and consequences to the social fibre. I will briefly examine the causes and consequences of corruption to the type of knowledge youth should have to participate effectively in its eradication.

The Hong Kong Model for Youth Participation

Hong Kong has gone beyond just a legislative mechanism but realized that an effective anti-corruption strategy rests on the involvement of the youth and general citizens. The Hong Kong Independent Commission Against Corruption in its dual mission to curb corruption and to educate the public identifies the youth sector to be the future of any effective anti-corruption strategy. This flows from an understanding that public awareness about social uprightness dependent on the value systems inculcated in the youth sector as the representatives of future governance. In Xhosa culture there is a saying *"Umthi ungotwa uselulala"* meaning that you need to educate children about certain values at a very early stage of their development to be able to have an impact on their social behaviour. Thus the Hong Kong Federation of Youth (HKFYG) commissioned a survey on young people's views on corruption and bribery to be able to anticipate and the understand the nature of the problem. The survey was conducted in October 1993 and a total of 520 young people, aged from 15 to 29, were interviewed. The standard error of this survey was less than 2.2 per cent. The findings of the survey were the following.

Most respondents indicated that they would not solve problems by offering bribes whereas up to 10 per cent would do so. Their opinions reflect the commonly-accepted notion that giving bribes is morally wrong.

On the issue of corruption nearly 45 per cent of the respondents pinpointed private companies as the most seriously corrupt sector, followed by government departments. And most respondents agreed that corruption makes Hong Kong an unfair society.

With regard to punishment, the respondents were divided between a heavier penalty and uniform sentence for both the private and public sector. Half of them, particularly the younger and less educated, felt that corrupt civil servants should be punished more heavily whilst another 46 per cent, mostly older and more educated, agreed with uniform treatment. These results revealed two important issues: that there are higher expectations of integrity for civil servants, while on the other side there is clear indication that society will not tolerate corruption from private and public sectors[16].

A survey conducted by HKFYG in 1997 on the same issues indicate that most respondents still believe that they would not solve a problem by offering a bribe, however, a relatively significant number of respondents would offer a bribe or said it would depend on the situation. From this there is an indication that moral standards are declining among the youth[17]. This might be influenced by the increasing unemployment rate of young people and the impact of unemployment problems on a young person's family. According to the survey conducted by HKFYG in 1998 on the "impact of unemployment on young people" about 15.5 per cent of young people were found to be unemployed[18].

My general observations about young peoples' attitudes in Hong Kong is they have very strong value system due to acceptance of their traditional practice, acceptance of civic education and the government policies in this regard. For example, according to the survey conducted by HKFYG in 1997 [19]on youth value indicators, most young people had a fairly positive view of the government which they regard as trustworthy and not corrupt. And most of them consider more civic education to be a worthy remedy to the situation of decaying moral standards. Another very important observation is the impact of young people's economic situation on their moral or value system. Thus it is imperative that an anti-corruption strategy should be dynamic, to be able to address issues of levelling the economic playing field in a society, especially with the young people. The argument that economies of inequality bring about more poverty to the poor and instability within a society justifies the above interrelation of an anti-corruption drive with creation of viable economic opportunities for young people.

South African Youth Model in Community Mobilization

The involvement of young people in Hong Kong is an acceptance that the capacity to implement an effective anti-corruption policy can only be guaranteed if there is institutional re-alignment of the existing institutions, supported by community partnerships that are reflective of the demography of a society. However, it should be noted that the institutional re-alignment and partnership created will lack a developmental meaning, if elites and policy-makers define the anti-corruption agenda. These partnerships will be platforms where dialogue should be encouraged on issues of public policy (including corruption). In this way these community partnerships will contribute to widening the constituency for reform, increasing ownership, and strengthening the commitment to combat corruption.

A people-driven approach assumes that participation will bring about better understanding of policy issues and thereby increase public participation in policy formulation. Public participation will not automatically bring about transfer of skills to deal with corruption but a concerted effort to bring about true participation will determine the outcome.

Furthermore, the people-driven approach requires a continuous dialogue with various stakeholders and this is a shift from a top down prescriptive approach to a more flexible process.

A wealth of experience about community- or people-driven approaches exists from the South African struggle, which was always a struggle against an abusive state, which was in place before 1994. If it is accepted that corruption has been part and parcel of the Apartheid State which has generally concentrated power in the hands of the few and lacked transparency, then we can begin to understand how crucial it is to involve today's youth in the developmental agenda which includes promotion of accountability in our society. Youth involvement could draw a line of demarcation between a strategy which is superficial and designed only to bolster the image of the government and the one which is meant to build capacity to monitor and control corruption in the public sector, the private sector and the civil society.

The state, through its structures, such as the South Africa National Youth Commission and education institutions, should empower the youth to be able to mobilize the public behind *anti-corruption strategy*. Civil society and the private sector should participate in the government initiatives as their involvement may ensure that the end product of the activities to promote youth partnership are effective in combating corruption. These partnerships as envisaged by a youth workshop convened by the Institute for Security Studies and Transparency International-South Africa (TI-SA) on 18 March 1999 for South African youth organizations, will provide an environment where a youth sector will creatively, resolutely and consistently expose and fight corruption. Therefore, governments should encourage civic education as the ultimate weapon for a real political participation in a democratic system.

Civic Education as an Empowering Tool

Civic education in a democracy is preparation for sustaining self-government[20]. Democratic self-government means citizen participation based on informed, critical reflection. The goal of civic education is competent and responsible participation in local, state, and national political life. Such participation requires the acquisition of a body of knowledge, the development of intellectual and participatory skills, the development of certain dispositions or traits of character, a reasoned commitment to the fundamental values and principles embodied in their constitution and also to understand the social constraints they are faced with that includes corruption. In other words, an effective civic education programme would enable citizens to make wise choices in full awareness of alternatives and provides the kind of experience and understanding that foster the development of a reasoned commitment to those values and principles that enable a free society to exist. It is in this context, I argue, that youth are well situated to galvanize and mobilize the general public to participate in and create better governance. Civic education will deepen democracy in a society as fragmented as ours and further sideline anti-development forces. Civic education is also known for effective legal socialization programmes. In this regard, legal socialization is a form of socialization in which values, attitudes and behaviours with respect to the law are developed. The assumption behind legal socialization is that:

- those who know the law will be less likely to break it out of ignorance;
- knowledge will lead to greater awareness of consequences for lawbreaking; and
- greater knowledge will increase moral support for the law.

For this to be successful school curricula should incorporate courses covering law, corruption, democracy and human rights and Africanization (in case of Africa).

Revisiting the South African National Crime Prevention Strategy

There are certain important elements that underpin the National Crime Prevention Strategy (NCPS) which is the cornerstone of the government's strategy to fight crime (including corruption) in the country. The NCPS has the following objectives which resemble many other anti-crime strategies found in other countries.

- Establishment of a comprehensive policy framework which will enable government to address crime in a co-ordinated and focused manner which draws on the resources of all government agencies, as well as civil society.
- Promotion of a shared understanding and common vision of how we, as a nation, are going to tackle crime.
- Development of a set of national programmes which serve to kick start and focus the efforts of various government departments in delivering quality aimed at solving the problems leading to high crime levels.
- Maximization of civil society's participation in mobilizing and sustaining crime prevention initiatives.
- Creation of a dedicated and integrated crime prevention capacity which can conduct ongoing research and evaluation of departmental and public campaigns as well as facilitating effective crime prevention programmes.

The South African National Crime Prevention Strategy has a two pronged approach: (1) *Crime Control* - measures which deploy most of the resources towards responding after crime has already been committed; (2) *Crime Prevention* - which aim at preventing crime from occurring in the first place. The major focus (albeit only on paper) of the policy is to redirect the resources in the country to a system which regards public participation as one pillar for implementation. The question then is what makes it difficult for governments to deliver on the provisions of the policy.

In my view, government efforts to respond to international and national pressures tend to create and enact *ad hoc* mechanisms which prove to be academic for the long-term purposes of anti-corruption strategy. Reactive mechanisms are as important as preventative measures but do not ensure sustainability which is one fundamental principle required for a developmental policy.

Beside the legislative framework needed to fight corruption, the Lima Declaration (which is the landmark document agreed by the global community to articulate a broad strategy to fight corruption, at the international and national levels, in all sectors and with the participation and cooperation of all walks of life) provides each country with a task to develop an effective strategy in which public participation stands as a *sine qua non*. In response to the Lima Declaration countries have recently embarked on various National Anti-Corruption Initiatives without a visible contribution from the youth sector (South Africa). For example, South Africa has embarked on the following initiatives.

- In October 1998, a national moral summit was held to deal with issue of deteriorating ethics and values and to set out a programme of action to enhance the moral fibre in South Africa.
- The government launched the Public Sector Anti-Corruption Summit in November 1998 which will be strengthened by a National Anti-Corruption Summit.
- On the other side, business and government are involved in efforts to deal with commercial crime. Business Against Crime has proposed a programme of action to transform criminal justice system to address this problem through a specialized approach.
- TI-SA also has introduced the concept of an Integrity Pact in South Africa.
- The 9th International Anti-Corruption Conference will be held in South Africa, Durban in October 1999.
- South Africa might be the first African country to join the fight against corporate bribery, waged with the Convention on Combating Bribery of Foreign Officials in International Business Transactions.

Even though the South African government appears to be gearing its state machinery to fight corruption it left out an important role-player - the youth sector. Corruption is not simply the product of an inadequate or inappropriate regulatory environment but also of a climate that has evolved as a result of our specific historical circumstances. Therefore, enacting legislation and policies should not be seen as an end on its own. It is logical that a strategy, which does not intend to encourage public participation through the youth sector, will remain an academic exercise with the intention to have a trickling down effect. Therefore, no country could afford to reduce corruption to an academic issue. Corruption in both developed and developing countries is a practical and growing problem, especially that the economic costs of the continued diversion of public funds to personal use can be serious, undermining a country's stability, much as Mobutu had done in Zaire.

(1) Corruption undermines the legitimacy and integrity of newly-democratized systems and impedes the progress of countries in transition.
(2) Commercial crime has been identified as being a single factor which leads to closure of many big companies, emerging companies and organs of civil society. For example, the National Small Business Council's closure is indicative of the impact corruption could have on developmental policies.
(3) It has been indicated that corruption is one of the main impediments to multinational companies and individuals investing in Africa.
(4) The impact of corruption on poverty.

For any country to be able to reduce the above predicaments in the next century it will not be able to utilize state machinery as a panacea, but people-driven strategies are seen to be the way for future development. International organizations recognize this view and have adopted the participatory development approach in their projects to enhance and accelerate change. This requires a serious effort to include the public on policy reform through civic education to encourage broader participation and inclusivity. The contemporary experience on development projects intended for investment indicates that there is diversion of funds to the hands of the few. To prevent this, government should not negate its duty to educate society with regard to state processes that intend to bring about a better life. States will achieve this if they ensure that there is a collaborative effort by all stakeholders. Without education there can be no collaboration, rather a deliberate act of marginalization of the populace or civil society, in particular the youth sector.

Marginalization of Youth

The question I would like to address here is whether there is any room for marginalization in a democratic society. It is my firm belief that a society that regards participatory democracy as the core value of its political system would have no regard for any kind of systematic marginalization. However, it has been argued that political participatory rights do not necessarily lead to effective participatory democracy. The answer to the above question therefore lies in the dichotomy between liberal and illiberal democracies[21]. Liberal democracies are those democracies that limit ends and means of government and respect liberal rights and that also hold free elections. On the other side, illiberal democracies are those regimes based on more or less free elections that do not practice limited government, nor consistently respect fundamental human rights. To a large extent elections do not necessarily create a democratic regime, nor are liberal government necessarily democratic. By observing the role of the ordinary citizen in policy formulation, decision-making and monitoring of accountability within a society it could be assessed whether a state is truly democratic or simply represents a state with a constitution that cannot be enforced by the judiciary. If society is not empowered to participate from an informed position this may be interpreted as systematic marginalization by the state. Another issue is the degree of state limitation of fundamental rights in the constitution. If citizens can freely associate themselves with a particular cause, or where freedom of speech is not so limited as to negate the content of the very right. Arguments on systematic marginalization of society cannot hold water in the contemporary constitutionalism. A constitution supported by the education system is a tool that enables citizens to mobilize themselves around issues. On the other hand, if government fails to provide a basic education that encourages civic education this may be tantamount to systematic marginalization of public participation.

The youth sector in a country where basic education, freedom of association and freedom of speech are seen as fundamental rights cannot justify their claim of the systematic marginalization of the youth. Thus I argue that, where these rights are guaranteed, there is thin line between self-marginalization and a state of ignorance. The youth cannot stand aside and wait for government to engage them; they must take control of their future by fighting corruption in all its manifestations.

Conclusion

The world today has undergone tremendous economic, political, cultural and social changes that impose different kinds of interstate relationships and state-citizen relations. In the last decade, we have seen how the information age has undermined state sovereignty and how money laundering, corruption, commercial crime could be committed by a person from another state against another state. It is for this reason that it is imperative that emerging markets should strengthen their legislative and institution capacity to combat crime, but the state should not be trapped into believing the latter means of fighting corruption are a panacea. I have argued herein that corruption cannot only be understood as a legal concept; rather it has also an economic, political and moralistic dimension. Therefore, to be able to fight corruption effectively one needs firstly to ensure that all these spheres of human development are met. I have indicated in terms of the Hong Kong experience that moral/civic education may stand at the threshold of sustainable strategy if it is coupled with a clear economic/industrial strategy to uplift the living standard of a society, for example,. in case of youth unemployment rate which has a direct impact on the level of moral standards amongst the Hong Kong Youth Sector.

Furthermore, I have indicated how the government policy that fosters a participatory development can also ensure better governance. In addition, I have used the South African struggle to indicate that mobilization of the youth sector could bring about public awareness in society in general about the nature, causes and consequences of corruption. This in-depth knowledge could lead to community partnerships and state-community partnership against corruption.

Thus I argue that corruption as a legal and social problem remains an issue to be addressed through reactive and preventative measures and collaborative efforts of society. Imaginative infrastructures could only come into existence if the members of the community have an in-depth understanding of the objective of anti-corruption strategies and how these strategies translate to public efficiency and a poverty alleviation drive. I think the youth is that missing link between quick fix measures and sustainable strategies to combat corruption for better governance. Through the youth we can develop people-driven strategies because they are the most educated, accessible and easy to organize sector of society. For example, in the 1970s in South Africa it was this sector which challenged our community on use of alcohol and this proved at that time to be successful in refocusing the adult sector to fight apartheid. And it was the youth which took the liberation struggle to every corner of our neighbourhood. It is this that prompted the ISS and TI-SA to convene a meeting to find ways to bring the youth together on strategies against the cancer of corruption in South Africa. And we hope that this endeavour would have an impact not only on South African youth policies but also will have an impact on the Continent as a whole.

Knowledge about corruption should not be treated as a secondary element of an anti-corruption strategy; rather it should be a prerequisite for any country to engage its citizens to fight corruption. This implies that the task ahead of us could be achieved if an education/public awareness campaign in this regard is at the centre of anti-corruption strategy. In-depth understanding of corruption could be achieved through civic education, which is the key to effective public participation. Through the youth sector involvement, as I have indicated, better moral standards would be realized. Students must learn what organized crime is, why it exists, and what can be done about it. Most importantly, they must play an active role in informing the society about its role in combating counter-revolutionary forces that are constantly engaged in corruption. This requires school-based programmes and existing civic education needs to be expanded and focused on concerns about corruption and its impact. And youth partnership and coalitions around this need to be created to effectively address the corruption malady.

References

[1] The term "youth " refers to definition as envisaged by the South African National Youth Commission youth policy. Youth is referred to as young people (males and females aged from 14 to 35 years. It further recognizes that there may be some people who fall outside this age range but who may experience similar circumstances to other young people. Thus, its definition indicates the primary target group, without excluding those who may share similar circumstances.

[2] Wolfensohn, James (1998), "A back-to-anticorruption strategy", USIA Electronic Journal, Vol. 3, No 5, November 1998

[3] Act 94 of 1992

[4] Mbaku JM , " Bureaucratic Corruption in Africa: The Futility of Cleanups, The Cato journal Vol.16 No.1

[5] Corruption hinders proper redistribution of resources and concentrate to the elite class of our society.

[6] McNamara Robert, (1968), "The Essence of Security" defined security as follows: Security means development. Security is not military hardware, though it may include it; security is not military force though it

may involve it, security is not traditional military activity, though it may encompass it. Security is development; without development there can be no security.

[7] Southern Africa Political & Economic Monthly , April 1995 Vol.8, No.7. McNamara Robert, (1968)," The Essence of Security,

[8] http://www. Inform.umd.edu/iris/tlktch21.html " There is some debate whether corruption is a serious problem or even undesirable. Africa countries are commonly cited as examples of the high cost which corruption inflicts on the economic growth. However, some South East Asian economies are perhaps among the most corrupt in the world, and yet many of them have achieve spectacular economic growth. Some authors have argued that some corruption might be desirable and enhance economic growth. If corruption was a serious impediment to economic growth, they say it would be difficult to explain the extraordinary success of South East Asian economies.

[9] Gillespie, K, and Okruhlik, G (1991) " The Political Dimensions of Corruption Cleanups: A framework for analysis." Comparative Politics 24(1): 77-95

[10] Padhy, KS. (1986) "Corruption in Politics". Delhi, India: BR Publishing Corp.

[11] Bayley, DH (1966) " The effects of Corruption in a Developing Nation." The Western Political Science Quarterly 19(4): 719-32.

[12] Crime against humanity, is a sanction of international law against those who commits certain act forbidden by charter of United Nation. These Crime are regarded as those acts which are against the conscience and dignity of mankind which are contrary to the principle of the Charter".

[13] Participatory Development and Good Governance Report of the Aid Study Committee, 1995, Japan International Co-operation Agency, Since the 1980's there has been growing awareness that an essential precondition for sustainable development is a strong determination to achieve development on the part of developing country governments and the beneficiaries themselves. Equally important is that they share the responsibility and take the initiative to bring about their own country's development.

[14] If elections do not translate into a equitable wealth distribution then the hypothesis that democratization leads to development would be nugatory from a pragmatic point of view, Unreported paper on "Right to vote

[15] Governance is about the collective, democratic management of people's lives. Governance has to ensure that process of mandating, developing and monitoring of implementation of policies is inclusive and participatory." The ANC discussion document for the Commission on Governance"

[16] The Hong Kong Federation of Youth Groups, "Young People's Views on Corruption", Youth Poll Series No.10 1993

[17] The Hong Kong Federation of Youth Groups, Young's people Outlook on life" Youth Poll Series No.41, 3.1997.

[18] The Hong Kong Federation of Youth Groups, "The Impact of Unemployment on Young People" Youth Poll Series No. 54. 5 1998

[19] The Hong Kong Federation , Tuning in to Youth "The setting Up of Hong Kong Youth Indicators" Youth Study Series No. 12

[20] Linda Darling-Hammond: Achieving Democratic Education, 1997-98, North Central Regional Educational Laboratory. If we educate for democracy she notes, we give all people the tools to think independently and reach their own conclusions. She adds firmly, It's an empowering kind of education.

[21] Fareed Zakaria, 1998. "The rise of illiberal democracy", mentioned that the process of genuine liberalization and democratization is gradual and long-term, in which election is only one step. Without appropriate preparation it might even be a false step". Thus South Africa should ensure that electorate really influence the democratic process than reinventing the wheel and legitimize a government which is not in touch with people.

ABBREVIATIONS AND ACRONYMS

ABM	anti-ballistic missile
ACDA	(US) Arms Control and Disarmament Agency
AIDS	immune-acquired deficiency syndrome
ANC	African National Congress
APEC	Asia Pacific Economic Cooperation
APL	anti-personnel landmine
ARF	Asian Regional Forum
ASEAN	Association of South-East Asian Nations
BMD(O)	Ballistic Missile Defense (Organization)
BTWC	Biological and Toxin Weapons Convention
BW	biological weapon
BWC	Biological Weapons Convention
CD	Conference on Disarmament
CDM	clean development mechanism
CIS	Commonwealth of Independent States
CRS	Congressional Research Service
CSCE	Conference on Security and Cooperation in Europe
CTBT	Comprehensive Test Ban Treaty
CW	chemical weapon
CWC	Chemical Weapons Convention
DARPA	Defense Advance Research Projects Agency
DOE	(US) Department of Energy
DRC	Democratic Republic of Congo
ECOWAS	Economic Community of West African States
ESDI	European Security and Defence Identity
EU	European Union
FAO	Food and Agriculture Organization
FDI	Foreign Direct Investment
FMCT	Fissile Material Cut-Off Treaty
FRELIMO	Front for the Liberation of Mozambique
GCP	gross criminal product
GDP	gross domestic product
GIC	Geneva International Centre for Humanitarian Demining
GIS	geographic information systems
GPS	global positioning system
HEU	highly-enriched uranium
IAEA	International Atomic Energy Agency
ICBM	intercontinental ballistic missile
ICC	International Criminal Court
ICJ	International Court of Justice
IGO	intergovernmental organization
IMF	International Monetary Fund
ISDSC	Inter-State Defence and Security Committee
IWC	International Whaling Commission
KLA	Kosovo Liberation Army
LEU	low enriched uranium
MCP	Malawi Congress Party
MYP	Malawi Young Pioneers
NAM	Non-Aligned Movement
NATO	North Atlantic Treaty Organization

NGO	non-governmental organization
NMD	National Missile Defense
NNWS	non-nuclear weapon states
NPT	(Nuclear) Non-Proliferation Treaty
NTM	national technical means
NWFZ	nuclear-weapon-free zone
NWS	nuclear weapon states
OAS	Organization of American States
OAU	Organization of African Unity
ODA	Overseas Development Assistance
OECD	Organization for Economic Cooperation and Development
OPEC	Organization of Petroleum Exporting Countries
OSCE	Organization for Security and Cooperation in Europe
PNE	peaceful nuclear explosion
SADC	South Africa Development Community
SAL	structural adjustment loan
SARCO	South African Regional Policy Commissioners Co-ordinating Organization
SCP	Sustainable Cities Programme
START	Strategic Arms Reduction Talks
THAAD	theatre high altitude area defence
TMD	theatre missile defence
TRC	Truth and Reconciliation Commission
UNDP	United Nations Development Programme
UNEP	United Nations Environment Programme
UNESCO	United Nations Educational, Cultural and Scientific Organization
UNFCCC	United Nations Framework Convention on Climate Change
UNITA	*Uniao Nacional para a Independencia Total de Angola*
USEC	United States Enrichment Corporation
WHO	World Health Organization
WMD	weapons of mass destruction
WTO	Warsaw Treaty Organization

AUTHOR INDEX

Albrecht, Ulrich 224
Atiyah, Michael 58, 137
Avery, John 405
Bonsdorff, Göran von 416
Boutwell, Jeffrey 101
Broms, Bengt 407
Calogero, Francesco 187
Cammack, Diana 246
Chachiua, Martinho 106
Cilliers, Jakkie 82
Dupré, Annamarie 253
Egana, Mariano 441
Estrella Tampieri, Hugo 313
Falconer, Tannia 322
Ferreccio, Catterina 378
Fortman, Bas de Gaay 450
Ginzburg, Alexander 363
Ginzburg, Veronica 363
Gleick, Peter 329
Goldblat, Jozef 197
Gomomo, Nceba 456
Gooneratne, Brendon 400
Grung, Micheline Egge 418
Guinnessy, Paul 317
Haden, Eva-Jane 371
Hébert, Jean-Paul 215
Hopkins, Lynne 276
Hsieh, Shu Yuan 180
Isaacman, Alan 382
Kamenju, Jacob 109, 262
Kang, Young Hoon 183
Kaplan, Martin 59, 276
Kartha, Tara 218
Khattab, Moushira Mahmoud 77
Knapp, Vladimir 285
Koster, Karel 151
Liu Gongliang 159
Liu Min 185
Lock, Peter 112, 201
Luby, Stefan 315
Mechelynck, André 339
Mendis, D.L.O. 319
Miettinen, Jorma 238
Miller, Morris 293
Milne, Tom 163
Moden, Anna 447
Montero Cabrera, Maria Elena 325
Muller, Marie 53
Mulli, Lucy 266
Mwakasungura, Undule 269
Ngubane, B.S. 94
Nikitin, Alexander 121
Otim, Peter 346
Pinguelli Rosa, Luiz 373
Powell, Derek 131
Rathjens, George 61, 128
Rohr, Carsten 350
Rotblat, Joseph 66, 412
Scheuer, Walter 441
Selebi, Jackie 102
Sijde, Bart van der 145, 151
Singh, Jasjit 125, 171, 232, 421, 430
Slaus, Ivo 436
Solomon, Hussein 272
Stumpf, Waldo 67
Tenaglia, Giancarlo 187
Turton, Anthony 353
Zohar, Aharon 397
Zuma, Jacob 91